Spectroscopic Techniques for Polymer Characterization

Spectroscopic Techniques for Polymer Characterization

Methods, Instrumentation, Applications

Edited by Yukihiro Ozaki and Harumi Sato

WILEY-VCH

Editors

Professor Yukihiro Ozaki
Kwansei Gakuin University
School of Biological and
Environmental Sciences
2-1 Gakuen, Sanda
669-1337 Sanda, Hyogo
Japan

Professor Harumi Sato
Kobe University
Graduate School of Human
Development and Environment
3-11 Tsurukabuto, Nada-ku
657-8501 Hyogo, Kobe
Japan

Library of Congress Card No.: applied for

British Library Cataloguing-in-Publication Data
A catalogue record for this book is available from the British Library.

Bibliographic information published by the Deutsche Nationalbibliothek
The Deutsche Nationalbibliothek lists this publication in the Deutsche Nationalbibliografie; detailed bibliographic data are available on the Internet at <http://dnb.d-nb.de>.

Print ISBN: 978-3-527-34833-6
ePDF ISBN: 978-3-527-83030-5
ePub ISBN: 978-3-527-83032-9
oBook ISBN: 978-3-527-83031-2

Typesetting Straive, Chennai, India
Printing and Binding CPI Group (UK) Ltd, Croydon, CR0 4YY

C109035_131021

Contents

List of Contributors *xiii*
Preface *xvii*
General Introduction *xix*

Part I Recent Progress on Spectroscopic Techniques *1*

1 Polymer Spectroscopy – Spectroscopy from the Far-Ultraviolet to Far-Infrared/Terahertz and Raman Spectroscopy *3*
Yukihiro Ozaki and Harumi Sato
1.1 Introduction to Polymer Spectroscopy *3*
1.1.1 Outline of Polymer Spectroscopy *3*
1.1.2 Brief History of Polymer Spectroscopy *5*
1.2 Overview of Molecular Spectroscopy from the Far-Ultraviolet to Far-Infrared/Terahertz and Raman Spectroscopy in Polymer Research *6*
1.2.1 IR and Raman Spectroscopy Analyses *6*
1.2.2 FIR/Terahertz and Low-Frequency Raman Spectroscopy *8*
1.2.3 Near-Infrared (NIR) Spectroscopy *8*
1.2.4 SERS and TERS Spectroscopy *9*
1.2.5 FUV Spectroscopy *9*
1.3 Specific Examples of Molecular Spectroscopy Studies of Polymers *10*
1.3.1 Infrared, Raman, and NIR Spectroscopic Evidence for the Coexistence of Hydrogen Bond Types in Poly(Acrylic Acid) *10*
1.3.2 Low-Frequency Vibrational Modes of Nylon-6 Studied by Using IR and Raman Spectroscopies and Density Functional Theory Calculations *16*
1.3.3 NIR Spectra of Linear Low-Density Polyethylene and Their Chemometrics Analysis *21*
1.3.4 Study of the Crystallization Behavior of Asymmetric PLLA/PDLA Blend by IR and Raman Spectroscopy and Raman Imaging *23*
1.3.5 3D SERS Imaging Using Chemically Synthesized Highly Symmetric Nanoporous Silver Microparticles *28*

1.3.6 Tip-Enhanced Raman Scattering Spectroscopy Study of Local Interactions at the Interface of Styrene–Butadiene Rubber/Multiwalled Carbon Nanotube Nanocomposites *34*
1.4 Perspectives for Polymer Spectroscopy *39*
References *41*

2 FTIR Spectroscopy and Spectroscopic Imaging for the Analysis of Polymers and Multicomponent Polymer Systems *45*
Huiqiang Lu, Andrew V. Ewing, and Sergei G. Kazarian
2.1 Investigation of Polymers Using FTIR Spectroscopy and Spectroscopic Imaging *45*
2.1.1 Investigation of Miscibility in Polymer Blends *46*
2.1.2 Investigation of Intermolecular Interactions *47*
2.1.2.1 Investigation of Partially Miscible PMMA–PEG Blends Using Two-Dimensional Disrelation Mapping *48*
2.1.3 Investigation of Crystallization in Polymers *51*
2.1.3.1 Investigation of Solvent-Induced Crystallization in Polymers *51*
2.1.3.2 Investigation of the Crystallization Process of PHB, PLLA, and Their Blends *53*
2.2 Investigation of Polymers Subjected to High-Pressure or Supercritical CO_2 Using FTIR Spectroscopy and FTIR Spectroscopic Imaging *55*
2.2.1 Morphology of Polymeric Materials Under High-Pressure or Supercritical CO_2 *56*
2.2.2 Investigation of Interaction in Polymers Under High-Pressure or Supercritical CO_2 *59*
2.2.2.1 Investigation of the Effect of High-Pressure CO_2 on the H-Bonding in PEG–PVP Blends *60*
2.2.2.2 Investigation of the Mechanism of Interaction Between CO_2 and Polymers Through the Thermodynamic Parameters Produced from In Situ ATR–FTIR Spectroscopy *61*
2.2.3 Investigation of Crystallization in Polymers Under High-Pressure or Supercritical CO_2 *61*
2.2.4 The Investigation of Structural Changes and Crystallization Kinetics of Polymers Exposed to High-Pressure CO_2 Through In Situ High-Pressure FTIR and FT-Raman Spectroscopy *64*
2.2.5 Investigation of Swelling and CO_2 Sorption into the Polymers Under High-Pressure or Supercritical CO_2 *65*
2.3 Conclusion *67*
References *68*

3 Interfaces in Polymer Nanocomposites Characterized by Spectroscopic Techniques *75*
Liliane Bokobza
3.1 Introduction *75*
3.2 Types of Interactions at the Interface *76*

3.3 Characterization of the Interfaces *80*
3.3.1 Fluorescence Spectroscopy *82*
3.3.2 Solid-State NMR Spectroscopy *85*
3.3.3 Vibrational Spectroscopy *88*
3.3.3.1 Infrared Spectroscopy *89*
3.3.3.2 Raman Spectroscopy *91*
3.4 Conclusions *95*
References *96*

4 Far-Infrared/Terahertz and Low-Frequency Raman Spectroscopies in Polymers *107*
Harumi Sato
4.1 Introduction *107*
4.2 Intermolecular Hydrogen Bonds in the Low-Frequency Region of PHB by QCCs *108*
4.3 Several Types of Intermolecular Hydrogen Bonds in PCL *109*
4.4 Stress-Induced Crystal Transition of Polybutylene Succinate (PBS) *113*
4.5 The Differences in Intermolecular Hydrogen Bonding Between PET and PBT *115*
4.6 THz Imaging of Polymer Film *117*
4.7 Conclusions *120*
References *120*

5 Near-Infrared Spectroscopy and Imaging of Polymers *125*
Daitaro Ishikawa, Yuta Hikima, and Yukihiro Ozaki
5.1 Introduction to NIR Spectroscopy *125*
5.1.1 Principles of NIR Spectroscopy *125*
5.1.2 Characteristics and Advantages of NIR Spectroscopy *126*
5.1.3 Analysis of NIR Spectra *126*
5.2 Applications to Polymer Science and Engineering of NIR Spectroscopy *128*
5.2.1 Polarized NIR Spectroscopy Studies of Molecular Orientation of Polymers *128*
5.2.2 Isothermal Crystallization Kinetics of Poly(3-hydroxybutyrate) *134*
5.2.3 Crystallization of Poly(3-hydroxybutyrate-*co*-3-hydroxyhexanoate) During Melt Extrusion Promoted by Residual Crystals *140*
5.2.3.1 Outline of Online NIR Analysis and Online NIR Monitoring of the Residual Crystal Amount at the Extruder Outlet Nozzle *140*
5.2.3.2 Amount of Residual Crystals at the Extruder Outlet *141*
5.2.3.3 Crystallization of Extruded Strands *145*
5.2.3.4 Analysis of Extruded Strand Crystallization Using the Avrami Equation *146*
5.3 NIR Imaging for Polymer Sciences *148*
5.3.1 Introduction *148*
5.3.2 Theory of NIR Imaging *148*

5.3.2.1 Acquisition of Hypercube *148*
5.3.2.2 Data Transfer and Mapping *149*
5.3.2.3 Feature of NIR Imaging Devices *150*
5.3.3 Applications of NIR Imaging *151*
5.3.3.1 Monitoring of Crystal Evolution Combined with Chemometrics *151*
5.3.3.2 Quality Evaluation Potential for Wide Area *153*
5.3.3.3 Diffusion Process Monitoring *153*
5.3.3.4 Degradable Process Monitoring of Biodegradable Polymer *154*
5.3.3.5 Rapid Evaluation of the Water Content in PLA Pellets *156*
5.3.3.6 Nondestructive Detection of Degraded Polylactic Acid Moldings *157*
References *160*

6 Far Ultraviolet Spectroscopy for Polymers *165*
Yusuke Morisawa and Nami Ueno
6.1 Introduction *165*
6.2 Measurement of ATR–FUV Spectra of Polymer *166*
6.3 ATR–FUV Spectra of Nylons *167*
6.4 ATR–FUV Spectra of Poly(3-hydroxybutyrate) (PHB) and Its Graphene Nanocomposites *172*
6.5 ATR–FUV Study of Poly(ethylene glycol) (PEG) and Its Complex with Lithium Ion (Li^+) *176*
6.6 Summary *181*
References *181*

7 Synchrotron-Based UV Resonance Raman Spectroscopy for Polymer Characterization *183*
Barbara Rossi, Mariagrazia Tortora, Sara Catalini, Alessandro Gessini, and Claudio Masciovecchio
7.1 Basic Principles of Raman and UV Resonance Raman Spectroscopy *183*
7.1.1 Molecular Vibrations and Raman Effect *183*
7.1.2 Resonance Raman (RR) Scattering *191*
7.1.3 Fundamental Applications of UV Resonance Raman Spectroscopy *193*
7.2 Synchrotron-Based UV Resonance Raman: Basic Principles and Instrumentation *193*
7.2.1 Synchrotron-Based UVRR Setup on IUVS@Elettra *194*
7.3 SR-UVRR Characterization of Biopolymers *197*
7.4 UV Resonance Raman Studies on Polymeric Hydrogels *203*
7.4.1 Water Confinement in Polysaccharide Hydrogels *204*
7.4.2 Phase Transition in Thermo-Sensitive Polysaccharide Hydrogels *208*
7.4.3 Water and Polymer Dynamics in pH-Responsive Polysaccharide Hydrogels *212*
7.5 Conclusions *215*
Acknowledgment *217*
References *217*

8 **Sum Frequency Generation Spectroscopy for Understanding the Polymer Dynamics at Buried Interfaces** *227*
Daisuke Kawaguchi and Keiji Tanaka
8.1 Introduction *227*
8.2 Principle *228*
8.3 Examples *230*
8.3.1 Nonsolvent Interface *230*
8.3.1.1 Polystyrene *230*
8.3.2 Solid Interface *238*
8.3.2.1 Polystyrene *238*
8.3.2.2 Polyisoprene *240*
8.3.2.3 Poly(styrene-*co*-butadiene) Rubber [89] *244*
8.4 Conclusions *250*
Acknowledgements *251*
References *251*

9 **Application of Two-Dimensional Correlation Spectroscopy (2D-COS) in Polymer Studies** *259*
Yeonju Park, Isao Noda, and Young Mee Jung
9.1 Introduction *259*
9.2 Theory *260*
9.2.1 Background *260*
9.2.2 Properties of 2D-COS *260*
9.3 Applications of 2D-COS in Polymer Studies *261*
9.3.1 Applications of Conventional 2D-COS *261*
9.3.1.1 Biodegradable Polymers *261*
9.3.1.2 Thermo-Responsive Polymers *262*
9.3.2 2D Hetero-Spectral Correlation Analysis *267*
9.3.3 Two-Dimensional (2D) Gradient-Mapping Method *269*
9.3.4 Chemometric Techniques Combined with 2D-COS *270*
9.3.5 Smooth Factor Analysis *272*
9.3.6 Projection 2D-COS *275*
9.3.7 2D-COS for Hyperspectral Imaging *278*
9.4 Conclusions *284*
References *284*

10 **Molecular Dynamics in Polymer Science** *297*
Mateusz Z. Brela, Marek Boczar, and Marek J. Wójcik
10.1 Introduction *297*
10.2 Historical and Theoretical Background *299*
10.3 Applications *302*
10.3.1 Vibrational Spectra of Hydrogen-Bonded Polymers *303*
10.3.2 Studies of Interactions Between Polymers and Water *304*
10.3.3 Mechanical Properties of Polymers *306*
10.3.4 Interphase Interactions *307*

10.4 Summary and Perspectives *309*
Acknowledgment *311*
References *311*

11 Spectroscopic Analysis of Structural Transformations Associated with Poly(lactic acid) *317*
Shaw L. Hsu and Xiaozhen Yang
11.1 Introduction *317*
11.2 Spectroscopic Tools *319*
11.2.1 Vibrational Features of PLA Crystals *321*
11.2.2 Analysis of Disordered PLA Chains *323*
11.2.3 Description of Anisotropic PLA – Polarized Spectra *327*
11.3 Simulation Studies for Both Ordered and Disordered Structures *329*
11.4 Analysis of Conformational Changes in PLA During Deformation *334*
11.5 Aging Behavior in PLA *338*
11.6 Conclusion *340*
Acknowledgment *340*
References *340*

Part II Topical Polymers Studied by Spectroscopy *345*

12 Probing Molecular Events in Self-Healable Polymers *347*
Qianhui Liu, Lei Li, and Marek W. Urban
12.1 Introduction *347*
12.2 Microphase Separation *349*
12.3 Entropically Driven Self-Healing *353*
12.3.1 Free Radical and Cationic Recombination *355*
12.3.2 Van der Waals Interactions *360*
12.3.3 Chemical Sensing of Damage–Repair Cycle *361*
Acknowledgments *365*
References *365*

13 Recent Application of Vibrational Spectroscopy to Conjugated Conducting Polymers *367*
Yukio Furukawa
13.1 Introduction *367*
13.2 Carriers *369*
13.3 Optical Absorption Spectra Upon Chemical Doping *371*
13.3.1 P3HT *371*
13.3.2 Poly(2,5-bis(3-hexadecylthiophene-2-yl)thieno[3,2-*b*]thiophene) (PBTTT-C16) *372*
13.4 Raman Spectra of Positive Polarons and Bipolarons Generated Upon Chemical Doping *374*

13.4.1 P3HT *374*
13.4.2 PBTTT-C16 *375*
13.5 Carriers and Electrical Properties Based on ILGTs *377*
13.5.1 ILGTs *377*
13.5.2 Raman Spectra of ILGTs Fabricated with P3HT *378*
13.5.3 Raman Spectra of ILGTs Fabricated with PBTTT-C16 *380*
13.6 Carrier Mobilities *383*
13.7 Raman Images in the Channel Region *383*
13.8 Carrier Dynamics in Bulk Heterojunction Films *386*
13.8.1 Photoexcitation Dynamics on Femto- and Picosecond Time Scales *386*
13.8.2 Microsecond Recombination Dynamics of Long-Lived Carriers *387*
13.9 Conclusions *388*
References *388*

14 Vibrational Spectroscopy for Fluoropolymers and Oligomers *393*
Takeshi Hasegawa
14.1 Perfluoroalkyl-Containing Compounds *393*
14.1.1 Molecular Conformation on Phase Diagram *393*
14.1.2 Molecular Vibration of an R_f Group *396*
14.1.3 The SDA theory *400*
14.2 Spectroscopy for R_f Compounds *402*
14.2.1 ROA analysis of R_f Compounds *402*
14.2.2 Surface Modes of Phonon and Polariton *405*
14.2.3 Summary and Perspective *408*
References *409*

15 Probing Structures of Conductive Polymers with Vibrational Spectroscopy *413*
Jianming Zhang and Yuan Yuan
15.1 Introduction *413*
15.2 Application of Vibrational Spectroscopy *413*
15.2.1 Chain Packing/Aggregate Mode Identification *413*
15.2.2 Conformation-Sensitive Bands Identification *414*
15.2.3 Doping-Sensitive Bands Identification *415*
15.2.4 Thermally Induced Phase Transitions *417*
15.2.5 Structural Dynamics *418*
15.2.6 Chemical Composition/Morphology Analysis in Conductive-Polymer-Based Blends *420*
15.2.7 Surface/Interface Molecular Orientation *423*
15.2.8 Structure and Dynamics of Charge Carriers *425*
15.2.9 Electric-Field-Induced Structural Changes *429*
15.3 Conclusion *431*
References *431*

16 **Weak Hydrogen Bonding in Biodegradable Polymers** *435*
Harumi Sato
16.1 Introduction *435*
16.2 Weak Hydrogen Bonding in Poly(3-hydroxybutyrate) *436*
16.3 Comparison Between Weak and Strong Hydrogen Bonds *438*
16.4 Difference in the Side Chain Length; PHB and PHV *439*
16.5 Polyhydroxyalkanoate Copolymers *442*
16.6 Crystallization Process of PHB *443*
16.7 Other Kinds of CH···O Hydrogen Bonding *443*
16.8 Conclusions *447*
References *449*

Index *453*

List of Contributors

Marek Boczar
Jagiellonian University
Faculty of Chemistry
Gronostajowa 2, Krakow 30-387
Poland

Liliane Bokobza
Former Professor at ESPCI
92200 Neuilly-Sur-Seine
France

Mateusz Z. Brela
Jagiellonian University
Faculty of Chemistry
Gronostajowa 2, Krakow 30-387
Poland

Sara Catalini
European Laboratory for Non-Linear
Spectroscopy, LENS
50019 Sesto
Fiorentino, Firenze
Italy

Andrew V. Ewing
Imperial College London, South
Kensington Campus
Department of Chemical Engineering
London SW7 2AZ
UK

Yukio Furukawa
Waseda University
Graduate School of Advanced Science
and Engineering, Department of
Chemistry and Biochemistry
Shinjuku-ku, Tokyo 169-8555
Japan

Alessandro Gessini
Elettra Sincrotrone Trieste
Trieste 34149
Italy

Takeshi Hasegawa
Kyoto University
Institute for Chemical Research
Uji, Kyoto 611-0011
Japan

Yuta Hikima
Kyoto University
Graduate School of Engineering
Nishikyo-ku, Kyoto 615-8510
Japan

Shaw L. Hsu
University of Massachusetts (Amherst)
Department of Polymer Science
and Engineering
Amherst, MA 01003
USA

Daitaro Ishikawa
Fukushima University
Faculty of Food and
Agricultural Sciences
1 Kanayagawa, Fukushima 960-1296
Japan

Young Mee Jung
Kangwon National University
Chemistry, Institute for Molecular
Science and Fusion Technology
Chuncheon 24341
Korea

Daisuke Kawaguchi
Kyushu University
Department of Applied Chemistry
744 Motooka, Nishi-ku, Fukuoka
819-0395
Japan

Sergei G. Kazarian
Imperial College London, South
Kensington Campus
Department of Chemical Engineering
London SW7 2AZ
UK

Lei Li
Clemson University
Department of Materials Science and
Engineering
Clemson 29634
USA

Qianhui Liu
Clemson University
Department of Materials Science
and Engineering
Clemson 29634
USA

Huiqiang Lu
Imperial College London, South
Kensington Campus
Department of Chemical Engineering
London SW7 2AZ
UK

Claudio Masciovecchio
Elettra Sincrotrone Trieste
Basovizza, Trieste 34149
Italy

Yusuke Morisawa
Kindai University
School of Science and Engineering
Department of Chemistry
Higashi-Osaka 577-8502
Japan

Isao Noda
University of Delaware
Department of Materials Science and
Engineering
Newark, DE 19716
USA

Yukihiro Ozaki
Kwansei Gakuin University
School of Biological and Environmental
Sciences, 2-1 Gakuen
Sanda, Hyogo 669-1337
Japan

Yeonju Park
Kangwon National University
Kangwon Radiation Convergence
Research Support Center
Chuncheon 24341
Korea

Barbara Rossi
Elettra Sincrotrone Trieste
Basovizza, Trieste 34149
Italy

Harumi Sato
Kobe University
Graduate School of Human Development and Environment
Nada, Kobe 657-8501
Japan

Keiji Tanaka
Kyushu University
Department of Applied Chemistry
Faculty of Engineering
744 Motooka, Nishi-ku, Fukuoka 819-0395
Japan

Mariagrazia Tortora
Elettra Sincrotrone Trieste
S.S. 114 km 163.5, Basovizza, Trieste 34149
Italy

Nami Ueno
University of Innsbruck
Institute of Analytical Chemistry and Radiochemistry
Innrain 80-82, 6020 Innsbruck, Austria

Marek W. Urban
Clemson University
Department of Materials Science and Engineering
Clemson 29634
USA

Marek J. Wójcik
Jagiellonian University
Faculty of Chemistry
Gronostajowa 2, Krakow 30-387
Poland

Xiaozhen Yang
Institute of Chemistry
Chinese Academy of Sciences
Beijing 100080
China

Yuan Yuan
Qingdao University of Science & Technology, Key Laboratory of Rubber-Plastics
Ministry of Education/Shandong Provincial Key Laboratory of Rubber-plastics
School of Polymer Science and Engineering
53 Zhengzhou Road, Qingdao 266042
China

Jianming Zhang
Qingdao University of Science & Technology, Key Laboratory of Rubber-Plastics
Ministry of Education/Shandong Provincial Key Laboratory of Rubber-plastics
School of Polymer Science and Engineering
53 Zhengzhou Road, Qingdao 266042
China

Preface

It was our great honor and pleasure to publish this book **on the special occasion of the 101st anniversary of the polymer hypothesis of Herman Staudinger**. Polymer spectroscopy has a history of 80 years or so. Professor Shaw Ling Hsu of the University of Massachusetts kindly provided a general introduction to polymer spectroscopy for this book.

Recently, polymer spectroscopy has demonstrated remarkable progress in many aspects. For example, a couple of new spectroscopies joined the family of polymer spectroscopy, including far-ultraviolet (FUV) spectroscopy, tip-enhanced Raman scattering (TERS), and terahertz (THz) spectroscopy. Nowadays, spectroscopic techniques of polymers range from FUV to far-infrared/THz and Raman spectroscopy. IR, Raman, near-infrared (NIR), and FIR/THz spectroscopy have all made significant progress. You can find the progress of these spectroscopies detailed in this book. Of particular interest is the development of imaging techniques in IR, Raman, NIR, and THz spectroscopies. Three-dimensional imaging also emerged. Spectral analysis and data treatment methods have also advanced significantly. It is noted that the quantum chemistry approach has been introduced not only to polymer vibrational spectroscopy but also polymer electronic spectroscopy.

Currently, several important textbooks on spectroscopic techniques for polymers are available; they are all important books, but some of them are not new. We thought there was strong demand for a state-of-the-art textbook on polymer spectroscopy. Therefore, we have decided to prepare a modern book on polymer spectroscopy for a wide variety of readers. We have aimed at writing a book, which will find a place in history. This book consists of two major parts: Recent Progress on Spectroscopic Techniques and Topical Polymers Studied by Spectroscopy. The first part starts with an overview of polymer spectroscopy and then introduces seven kinds of modern spectroscopic techniques for polymer characterization. Two spectral analysis methods, two-dimensional correlation spectroscopy and molecular dynamics are also reviewed. In the second part, we have six reviews on polymers studied by spectroscopic techniques, which currently receive keen interest: biodegradable polymers, self-healable polymers, conducting polymers, and fluoropolymers. We attempted to prepare a book that is well balanced between basic science and applications. One of the goals of this book is to make a strong bridge between spectroscopists, polymer scientists, and engineers in academia and industry.

We succeeded in inviting active front runners in modern spectroscopic techniques for polymers from many countries. This book is useful for scientists, engineers, and graduate students in numerous areas of science and engineering. One can use this book as a text, for example, at a graduate school seminar.

We hope that many readers can learn much about spectroscopic techniques for polymers from this book, and that this book can inspire readers to utilize various kinds of spectroscopies for a variety of novel investigations.

Last but not least, we would like to thank Ms. Quraishi Sakeena of Wiley VCH for her continuous efforts in publishing this exciting book.

July 2021

Yukihiro Ozaki
Harumi Sato

General Introduction

On the special occasion of the 101th anniversary of the Polymer Theory by Hermann Staudinger.

2020 marks one hundred years of recognition of polymers as a branch of the materials discipline. During this period, polymers have moved from an amorphous concept first enunciated by Hermann Staudinger to one of the most important platforms for technology development. Coinciding with the development of polymer science and engineering are marked advancements in our understanding of the physics of spectroscopic transitions and incredible developments of associated instrumentations. This book shows how a segment of these spectroscopic advances have supported the advancement of polymer science as a whole. It intends to highlight the contributions of vibrational spectroscopy to the advances of polymer development.

There is little doubt that vibrational spectroscopy has developed into one of the most crucial characterization tools, complementing other techniques. Spectroscopy started with William Herschel discovery of infrared in 1800, long before any concept of quantum mechanics was established, and even before the recognition of molecular vibrational transitions in the early 1900s. It is difficult to imagine the incredible developments of infrared spectroscopy since the review first published by R. Bowling Barnes and Lyman G. Bonner in 1936. Even when I went to University of Michigan to study polymer spectroscopy with Professor Sam Krimm in the late 1960s, I remember looking incredulously at the large brackets outside the Randall Physics Laboratory that were used to hold the lens guiding sunlight into the basement as excitation for infrared spectroscopy. Nowadays almost all laboratories have a very reliable and accurate infrared spectrometer to analyze chemical compositions and polymeric structures. We now also treat the incredible Fourier transform technique as being an inherent part of any infrared instrument. The speed with which we gather data has also improved tremendously, allowing for routine kinetics experiments with temporal resolution in milliseconds or faster. There are numerous attachments, which make polymer surface analysis rather mundane in nature. It can be said if there is a need, infrared spectroscopy can fulfill it.

Raman spectroscopy has a shorter history, since the Raman effect was not reported until 1928 when Chandrasekhara Venkata Raman did, also using focused sunlight. The effect of having scattered radiation different from the excitation frequency was first predicted by theoretician Adolf Smekal in 1923. After a lull of several decades, there was a tremendous surge of interest in this sector of spectroscopy due to the development in 1964 of a laser that could provide intense and polarized radiation for excitation. Only then could the Raman effect be observed expediently with the use of multiplier tubes. Furthermore, the observation and developments of various Raman-associated phenomena, such as surface-enhanced spectroscopy, coherent anti-Stokes spectroscopy (named for Sir George Gabriel Stokes, a nineteenth-century British physicist), and the Fourier transform multiplex advantage, have made Raman spectroscopy a common technique in many laboratories, along with infrared spectroscopy. The weak Raman scattering associated with water makes structural analysis of biological samples quite routine in nature.

It is not an exaggeration to state that vibrational spectroscopy, including infrared absorption and Raman scattering, is now one of the most widely used characterization techniques in polymer studies. This book comprises studies that describe various applications that have benefitted significantly from the use of vibrational spectroscopy. These spectroscopic studies focus on examples such as molecular interpretation of adhesive properties, measurement of crystalline domain size, analysis of phase-separated structures and their formation kinetics, aging/phase transformation of polymers, surface studies, orientation behavior of polymers, and examples in which imaging properties have proven to be crucial. In the last one hundred years, advances in synthesis have provided numerous fascinating new structures that need to be characterized. Polymers confined to interface or interphase that dominate physical properties need to be differentiated from the ones in the three-dimensional state. The conformation and orientation of deformed polymers in processed goods need to be elucidated. The aging behavior and changes in physical properties of polymers have often been defined using macroscopic techniques. Vibrational spectroscopy is a versatile and appropriate tool for the analysis of polymers undergoing changes as a function of time, temperature, and environment. Chemical and physical systems undergoing change display spectral signatures from the radio frequency to the X-ray region of the electromagnetic spectrum. This has led to the development of time-resolved tetrahertz (THz) spectroscopy, which has added to the measurements of transient phenomena in the subpicosecond range. These developments are described further in later chapters.

The astronomical increase in computational power just in the last fifty years has changed the landscape of vibrational spectroscopy significantly. In the earlier days, the use of finite groups to analyze equilibrium molecular structures not only clarified the optical activities of Raman and infrared transitions, but also provided a means to calculate normal coordinates. This formed a basis for analyzing the active vibrations. Wilson and his coworkers provided the rigorous, elegant treatment of the mathematics involved in detailed vibrational analyses of polyatomic molecules that is still being practiced today with much higher speed. Because of the enormous computational capabilities commonly available nowadays, those earlier studies solving the diagonalization of secular determinants have evolved into sophisticated calculations capable of analyzing not only the optical activities but also the quantum mechanical calculations necessary to determine the frequency and intensities of optical transitions. Some of those advances in molecular dynamics and density-functional theory calculations will be described in this book. The lower-lying transitions can now also be used to analyze the thermal properties of polymers. These computational developments have also enhanced the capabilities of two-dimensional correlation spectroscopy, allowing for a better understanding of the intermolecular interactions.

The detailed analysis of the magnitude and specificity of intermolecular interactions in polymeric systems is extremely important. Though these interactions are weak in nature, they are important because all molecules in the condensed state possess such pairwise interactions. Their changes as a function of temperature and time are essential in determining changes in the physical properties of macroscopic polymers. Few explicit spectroscopic features are explicitly associated with the condensed state. However, when these features are found, vibrational spectroscopy provides the molecular interpretation of these intermolecular interactions, whether they are hydrogen bonding, electroscopic, dipolar, or van der Waals interactions. It is these interactions that determine miscibility behavior in blends, adhesive behavior associated with interface, engineering modulus or impact strength, and other properties. On the other extreme of frequency space associated with low-lying intermolecular vibrations are the high-frequency, more isolated near-infrared spectroscopy (NIR) bands. These NIR vibrations have proven useful to follow chemical reactions, permeation behavior of moisture into polymers, and chemical composition analysis.

An area of great interest in polymer studies is the application of vibrational spectroscopy, particularly Raman, to the analysis of disordered chains, which include polymers in solution and melt, as well as amorphous polymers in the condensed state. As virtually all commercial polymers are carbon-based, the change in polarizability along the backbone is large. Raman scattering is, therefore, particularly suitable for the characterization of polymer backbone conformation. Disordered chains lack long-range order but may contain short ordered sequences because of significant differences in the relative energy of different rotational isomeric states along the polymer chain. For disordered structures, the vibrational spectra observed may be complex due to changes in both band frequency and shape. In these cases, the vibrational mode can be quite complicated because the contribution of various internal coordinates to the vibration (the character of the vibration) may change significantly, unlike that of an ordered crystalline chain. In addition, these disordered chains may adopt a specific conformational distribution depending on geometric constraints such as surfaces, clathrates, or interfaces. Amorphous chains thus require a completely different treatment. Changing the frequency and character of several vibrations provides tremendous insight into the structure of polymers in solution or melt, or the disordered regions of semicrystalline polymers. Rather than analyzing one specific chain conformation for an ordered chain, the analysis of vibrational spectra arising from disordered chains requires the vibrational spectrum of a conformational distribution. An extremely large number of possible conformers exist, each with a unique spectrum often only slightly different from one another. However, the isotropic Raman spectrum for the disordered state can be simulated as a composite of contributions from the ensemble of chains generated by a Monte Carlo procedure that assigns both a conformation and a total probability for each chain. These calculations can only be carried out using the modern computational techniques.

In many cases, it is important to assess the spatial distribution of various chemical species or morphological structures in polymers. Based on their spectroscopic features, infrared and Raman techniques are useful for mapping the distribution of structural variations within polymer samples. This information is quite different from the exact atomic placements available from techniques such as scanning electron microscopy. However, in many applications, such as in polymer blends and composites, vibrational spectroscopy provides valuable information regarding the spatial distribution of individual components that simply cannot be obtained using other techniques. Since the mapping capability of spectroscopic techniques depends on specific signals from molecular entities, even the distribution of functional groups can be differentiated within each sample, in contrast to techniques like microscopy. In addition to the differentiation of chemical species, it is also possible to assess other differences, such as segmental orientation or different degrees of crystallinity in various polymer samples. Imaging capability is needed in many polymer applications where multilayered thin films are necessary. Mapping the structure of individual layers can be accomplished using a confocal technique. Different segmental orientations as a function of cooling profile (surface versus bulk) can be differentiated using vibrational spectroscopy. Thus, the fracture of composites can be analyzed.

It should be noted that the spatial resolution of traditional spectroscopic techniques depends on the wavelength of the probing radiation. Raman excitation generally uses lasers in the visible range (400–800 nm). In contrast, infrared generally uses radiation in the micron range. Based on the Rayleigh criterion, the spatial resolution achievable using Raman will be significantly higher than the resolution achievable using infrared. In this book, we will review the various applications in which traditional imaging techniques have played a valuable role. The confocal capability of a spectroscopic technique will be discussed using only Raman scattering as an example. Although quite different in capability, infrared spectroscopy has been used extensively to differentiate between surface functionalities and bulk structure using reflectance techniques such as attenuated total reflectance (ATR) or more specialized external reflectance spectroscopy. In these cases, the spatial resolution is a fraction of that achievable by Raman. In some specialized cases, Raman technique can provide exceptional resolutions because the signal-to-noise ratio can be large due to a resonance effect. In these cases, clearly identifiable Raman signals (backbone stretching vibrations) can be used to analyze the perfection of graphene used in various processes. Specialized Raman techniques have been developed for surface analysis. Surface-enhanced Raman spectroscopy (SERS) has proven useful for analyzing monolayers of adsorbed molecules onto metallic surfaces. The study of adsorbed molecules on metal surfaces using Raman spectroscopy, at one time an almost impossible task, has rapidly developed into an area of interest in recent years. SERS has developed a limited but totally different audience and has proven useful for the analytical characterization of various polymers, with applications ranging from food packaging to analysis of water quality. The development of hybrid techniques involving sharp probing tips, as in atomic force microscopy (ATM), overcame the limited spatial resolution usually associated with traditional spectroscopic imaging techniques. These new developments can be used to measure distribution of chemical features down to a scale of less than a hundred nanometers. Vibrational spectroscopy is now used broadly in almost all polymer science and engineering laboratories, due to its ability to characterize morphological features on all scales, from the smallest to ones in the hundreds of nanometers.

January 2021 *Shaw L. Hsu*

Part I

Recent Progress on Spectroscopic Techniques

1

Polymer Spectroscopy – Spectroscopy from the Far-Ultraviolet to Far-Infrared/Terahertz and Raman Spectroscopy

Yukihiro Ozaki[1,2] *and Harumi Sato*[3]

[1] *Kwansei Gakuin University, School of Biological and Environmental Sciences, 2-1 Gakuen, Sanda, Hyogo, 669-1337, Japan*
[2] *Toyota Physical and Chemical Research Institute, Nagakute, Aichi, 480-1192, Japan*
[3] *Kobe University, Graduate School of Human Development and Environment, Higashi-Nada, Kobe, 659-8501, Japan*

1.1 Introduction to Polymer Spectroscopy

Polymer spectroscopy has played a very important role in the investigation of the structure, physical and chemical properties, and reactions of polymers in the last half century [1–9]. As an analytical technique, polymer spectroscopy was born just before World War II and gradually became more commonplace in the 1950s. Throughout the 1950s and 1960s, polymer spectroscopy developed significantly in parallel with the development of infrared (IR) spectroscopy, although Raman spectroscopy was also used in that period [1–4]. The brief history of polymer spectroscopy is described in Preface and this chapter later.

The purpose of this chapter is to provide an overview of polymer spectroscopy. This chapter consists of an outline of polymer spectroscopy, a brief history of polymer spectroscopy, an overview of molecular spectroscopy for polymer research, and a review of examples of studies based on polymer spectroscopy. In the last part of this chapter, we describe the perspectives for polymer spectroscopy.

1.1.1 Outline of Polymer Spectroscopy

Polymer spectroscopy is largely based on optical spectroscopy, which involves spectroscopy in the ultraviolet (UV), visible (Vis), and IR regions. The UV region ranges from 10 to 380 nm and may be divided into four regions: vacuum ultraviolet (VUV, 10–120 nm), far-ultraviolet (FUV, 120–200 nm), deep ultraviolet (DUV, 200–300 nm), and UV (300–380 nm) [10]. Spectroscopy in the UV region is important because it provides the electronic spectra of molecules. However, compared with IR spectroscopy, UV spectroscopy has rarely been used for polymer research. Further, recently, among the UV regions, FUV spectroscopy has been applied to polymers to investigate the electronic and molecular structure and intermolecular

Spectroscopic Techniques for Polymer Characterization: Methods, Instrumentation, Applications, First Edition. Edited by Yukihiro Ozaki and Harumi Sato.

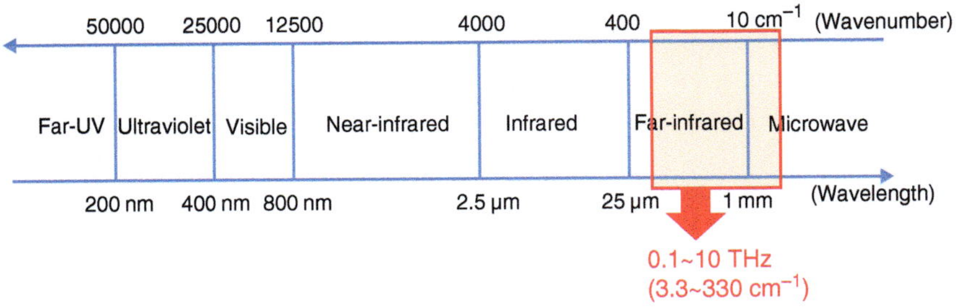

Figure 1.1 The region of electromagnetic wave from 200 nm to 1 mm.

interactions of polymers [11–14]. In Chapter 6, FUV studies of polymers are introduced. In addition, UV resonance Raman spectroscopy has also been used to investigate polymer structure and functions and will be reviewed in Chapter 7.

The IR region (800 nm to 1 mm, 12 500 to 10 cm^{-1}) is so wide in terms of energy that it is divided into three regions: near-infrared (NIR, 800–2500 nm) [15–17], IR (or mid-infrared; MIR, 4000 to 400 cm^{-1}) [18–20], and far-infrared (FIR, 400 to 10 cm^{-1}) [21, 22], as shown in Figure 1.1. Spectroscopic techniques in these regions have developed independently over the years, although the developments of NIR and FIR spectroscopies generally remained far behind those of IR (MIR) spectroscopy. However, in the last three decades, remarkable developments have been made in NIR spectroscopy [15–17]. Recently, advances in FIR spectroscopy have also been made. For example, terahertz spectroscopy was initiated at the end of the 1990s because of developments in new light sources and detectors in the FIR region [23–25].

IR, NIR, and FIR/terahertz spectroscopies are basically vibrational spectroscopic techniques [26]. IR spectroscopy is concerned mainly with fundamental vibrational modes [18–20], NIR spectroscopy is the spectroscopy of overtones and combinations of fundamentals [15–17], and FIR/terahertz spectroscopy treats low-frequency vibrational modes such as skeletal vibrations, torsional vibrations, and lattice vibrations [23–25]. Crucially, IR spectroscopy also involves the overtones and their combinations, NIR spectroscopy is concerned with electronic spectroscopy, and FIR/terahertz spectroscopy involves rotational spectroscopy. Thus, there is a clear border between IR and NIR spectroscopy because NIR spectroscopy is not related to the fundamentals; in contrast, the border between IR and FIR is not always clear. However, IR, NIR, and FIR spectroscopies are not three sisters; rather, IR is the mother of NIR spectroscopy because the overtones and combinations originate from the fundamentals.

Visible spectroscopy is typically only used as an ancillary tool in polymer studies, for example, to study the electronic structure of conductive polymers (Chapter 13). As an optical spectroscopic technique, fluorescence spectroscopy is also used for polymer research but is not considered in this book. In addition to the above optical spectroscopies, Raman spectroscopy is very important in polymer studies [27–30]. This is one of the key spectroscopic techniques in polymer research. In this book, normal Raman spectroscopy, resonance Raman spectroscopy, UV-resonance Raman spectroscopy, surface-enhanced Raman scattering (SERS) [31–33], tip-enhanced Raman scattering (TERS) [31–33], and Raman imaging [34] are introduced as important tools for polymer science. In addition, there is no doubt that nuclear magnetic resonance (NMR) spectroscopy is very important in polymer spectroscopy, but it is beyond the scope of this account.

1.1.2 Brief History of Polymer Spectroscopy

In the second part of this chapter, we provide an overview of the history of polymer spectroscopy in relation to the 100th anniversary of polymer theory in 2020. Just before World War II, studies of polymers vibrational spectroscopy started [18–20]. For example, Kirkwood [35] and Whitcomb et al. [36] performed normal vibrational calculations on polyethylene. After World War II, developments in IR spectroscopy progressed significantly because of advances in its light sources, spectrometers, and detectors. Therefore, even in the 1940s, many polymer scientists started using IR spectroscopy for polymer studies. For example, Shimanouchi et al. [37] reported normal vibrational calculations of polyethylene. Many pioneering papers on IR studies of polymers were published in the 1950s. Shimanouchi [38], Krimm [39], Eliot [40], and Hummel [41] conducted systematic studies on the vibrational spectra of polymers. Of course, there were many other scientists who advanced polymer spectroscopy in the 1950s. In particular, several important books concerning the IR spectra of polymers were published in this period [5]. Eliot [40] wrote a very important review of polymer spectroscopy in 1960, and, in the 1960s, many research groups were involved in IR studies on the characterization of polymers [4].

In the 1960s, Schachtschneider and Snyder [42] conducted normal vibrational calculations on polyethylene; Tasumi and Shimanouchi[43] studied the vibrational spectroscopy of polyethylene using the modified Urey–Bradley force field; and Miyazawa and coworkers [44] reported IR studies of polyoxymethylene, polyethylene glycol, polypropylene, and polyethylene and also introduced neutron scattering in these studies [45]. Tadokoro and coworkers [46] investigated the IR spectra of polyoxycyclobutane, polyamides, and other polymers in the 1960s. They also published several papers on FIR studies of polymers, such as polyamides and polyesters [47]. Krimm and Bank [48] also reported FIR studies on polyethylene.

In contrast, Raman spectroscopy studies of polymers started with those of long-chain *n*-alkanes. For example, Mizushima and Shimanouchi [49] reported the Raman spectrum of *n*-paraffin in 1949, and Hendra [50], Koenig [51], and others were involved in Raman studies of polymers in the 1960s. Since the introduction of lasers to Raman spectroscopy at the end of the 1960s, Raman studies of polymers have shown enormous progress [4, 5]. For example, the longitudinal modes (LAM) of polymers were investigated using Raman spectroscopy by Mizushima and Shimanouchi [49] and later by Schaufele and Shimanouchi [52] for a series of *n*-alkanes. Studies on LAM have been conducted in several directions, including the investigation of the effects of interactions at the lamella interface on the LAM frequencies. Stroble and Eckel [53] and Hsu and Krimm [54] made prominent contributions to these studies.

In the latter half of the 1950s and 1960s, several interesting NIR studies of polymers were reported [4]. For example, in 1956, Miller and Willis [55] used an NIR method for the analysis of butadiene–styrene copolymers based on the evaluation of the aromatic and aliphatic CH combination bands at 4580 and 4250 cm^{-1}, respectively. Takeuchi et al. [56a] used NIR spectroscopy to predict the acrylonitrile content in styrene–acrylonitrile copolymers. Tosi [56b] found a correlation between the frequency of the first overtone of the CH_2 stretching mode at 5800 cm^{-1} and the copolymer composition and interpreted the results in terms of the methylene sequence lengths of the ethylene–propene copolymers prepared using different catalysts. Many researchers have employed NIR spectroscopy for the investigation of polymeric structures and analytical purposes [4].

In this report, we do not describe the history of vibrational spectroscopy of polymers since the 1970s because many research groups have published excellent studies. However, it is important to mention two renaissances of vibrational spectroscopy that occurred in and after the 1970s. One is the introduction of lasers to Raman spectroscopy [28–30]. This made Raman experiments much easier and enabled various kinds of Raman spectroscopy, such as resonance Raman spectroscopy, SERS, Raman microscopy, and time-resolved Raman spectroscopy. Another is the development of Fourier transform (FT)IR in the 1970s and 1980s [18, 19]. This made IR measurements much easier and more precise. Moreover, various kinds of IR spectroscopy were born because of the development of FTIR spectroscopy; for example, micro-IR, time-resolved IR, and photoacoustic IR. These two renascences were truly great events that opened the doors for current vibrational spectroscopy.

1.2 Overview of Molecular Spectroscopy from the Far-Ultraviolet to Far-Infrared/Terahertz and Raman Spectroscopy in Polymer Research

In this chapter, because the characteristics and applications of IR and Raman spectroscopy in polymer research are reported in many chapters of this book and have been discussed in many other studies, we only provide a brief overview [4–9]. We describe the characteristics and applications of FIR/terahertz/low-frequency Raman spectroscopy, NIR, FUV, SERS, and TERS in polymer studies. Although applications of FIR/terahertz/low-frequency Raman spectroscopy, NIR, and FUV are introduced in Chapters 4–6, respectively, a description of them is presented in this chapter.

1.2.1 IR and Raman Spectroscopy Analyses

IR [18–20] and Raman spectroscopy [28–30] are key vibrational spectroscopies. They are mainly concerned with fundamental vibrational modes, although some bands arising from overtones and combinations of fundamentals also appear in both spectra. IR and Raman spectroscopy are often complementary, and the differences between IR and Raman spectroscopy lie in their selection rules. One of the two selection rules in IR spectroscopy is that a vibrational mode that experiences a change in the dipole moment during a molecular vibration is IR active, whereas that of Raman spectroscopy is that a vibrational mode with a change in molecular polarizability during a vibrational mode is Raman active. In general, asymmetric vibrations yield strong IR absorptions, whereas symmetric vibrations give rise to intense Raman bands. In IR spectra, bands arising from the stretching modes of polar functional groups such as C = O, OH, and C = N and bands arising from local vibrations, such as the CH_2 scissoring mode, are intense, whereas bands derived from a whole molecule or a large part of the molecule are weak. In Raman spectra, bands arising from the stretching modes of heavy atoms, such as S–S, C–S, and C–Cl stretching, are strong. In addition, bands originating from the stretching modes of multiple bonds, such as those arising from C = C, N = N, and C≡N stretching modes, are strong, and in-phase vibrations derived from a whole molecule or a large part of the molecule produce intense bands in Raman spectra. The ring-breathing mode of the benzene ring and the accordion vibrations of long-chain molecules are good examples of these.

IR spectroscopy uses IR absorption, as well as IR reflection (e.g. attenuated total reflectance [ATR], external reflection, reflection–absorption [RA], diffuse reflectance [DR], emission, and photoacoustic spectroscopy) [18–20]. Among them, ATR and RA spectroscopies are particularly useful for polymer research. An IR spectrometer can also be connected to optical microscopy (micro-IR, Chapter 2) and atomic force microscopy-infrared (AFM-IR) equipment, as well as other instruments. The polarized IR spectra of stretched polymers are useful for investigating the molecular and crystal structures of polymers, particularly their orientation [4, 5, 7, 8]. IR spectroscopy is also useful for studying the crystallinity, and so-called crystallization-sensitive bands can be used as indicators of crystallinity [4, 5, 7, 8]. The symmetry properties of polymers can be used to analyze their vibrational spectra.

Raman spectroscopy is a nondestructive analytical tool that enables the in situ analysis of a sample in its original form, and it is possible to use fiber optics for the Raman measurements. There are various kinds of Raman spectroscopy [28–30]; among them, normal Raman, Raman microscopy, Raman imaging, SERS, and TERS are often used for polymer studies. Resonance Raman spectroscopy has been used to study conductive polymers (Chapter 13). UV resonance Raman spectroscopy has recently been employed to study polymers (Chapter 7), and polarized Raman spectroscopy is useful for molecular orientation studies. In this chapter, we will demonstrate examples of normal Raman spectroscopy, Raman imaging, and TERS.

In polymer research, IR and Raman spectroscopies yield information about the chemical nature and physical properties [4–8]. Concerning the chemical nature, each spectroscopic method is useful, for example, for the identification of polymers and the investigation of structural units and the type and degree of branching. It is also possible to identify additives and impurities. As for the latter, both spectroscopic techniques give rise to information about stereoregularity, conformational order, polymer orientation, hydrogen bonding, intermolecular interactions, lamellar thickness, and state of order such as crystalline and amorphous (AMP) phases. Polymer reactions can also be monitored using both IR and Raman spectroscopies.

1.2.2 FIR/Terahertz and Low-Frequency Raman Spectroscopy

The low-frequency region is of great interest in the studies of the state of order of polymers because low-frequency modes often reflect the structure of the backbone and the intermolecular interactions of the polymers [4, 5]. In the low-frequency region, the following types of vibrations are observed.

1. Stretching vibrations of heavy atoms, e.g. I–I stretching mode.
2. Bending vibrations of the polymer skeletons.
3. LAM.
4. Torsional vibrations.
5. Lattice vibrations (intermolecular vibrations between adjacent chains within a crystallographic unit cell).

In recent years, low-frequency vibrational spectroscopy has become a subject of keen interest because of the marked progress in terahertz time-domain spectroscopy (THz–TDS) [23–25] and low-frequency Raman spectroscopy using volume Bragg notch filters. Currently, high-quality low-frequency spectra (FIR, THz–TDS, Raman) of various materials can be obtained, allowing the investigation of high-order conformations, crystal polymorphism, intermolecular interactions including hydrogen bonds, and the dynamics of various systems. As targets of these techniques, polymers are particularly interesting because their low-frequency vibrational modes can sharply reflect the interchain interactions and higher-order polymer conformations [57] (Chapter 4). In this chapter, we will introduce an FIR, low-frequency Raman, and quantum chemical calculations of nylon-6 in detail [57d]. Moreover, Chapter 4 reports the low-frequency vibrational spectroscopy of polymers.

1.2.3 Near-Infrared (NIR) Spectroscopy

The applications of NIR spectroscopy to polymer studies began with quantitative analysis of polymers, such as the determination of the hydroxyl number, water content, and residual carbon–carbon double bonds [4]. Since chemometrics and optical light fibers were introduced to the analysis of the NIR region, the usefulness of NIR spectroscopy in polymer research and analysis has dramatically increased (Chapter 5 and this chapter). Currently, NIR spectroscopy is widely used as a non-destructive and in situ analytical and structural probe of polymers [4, 15, 16]. One of the most active applications of NIR spectroscopy to polymers is online analysis and quality control using NIR light-fiber optic spectroscopy [58, 59], although there are many other applications, some of which are described in Chapter 5.

Although the practical applications of NIR spectroscopy in the polymer industry are extensive, an understanding of the principles behind their analysis remains insufficient. Recently, efforts to unravel complicated NIR spectra more clearly and to interpret the chemometrics models involved have taken several directions [15]. One is the systematic study of the NIR spectra of various polymers from the perspective of basic studies such as those of hydrogen bonding [60]. Another direction is the theoretical study of the assignment of bands arising from overtones and combinations in the NIR spectra of molecules using density functional theory (DFT) calculations [61].

1.2.4 SERS and TERS Spectroscopy

SERS is sensitive to single molecules and is molecule-specific and, thus, has been used in materials science, biomedical science, surface science, environmental analysis, and beyond [31–33]. The enormous enhancement of SERS signals arises mainly from two mechanisms: electromagnetic (EM) and charge-transfer (CT) effects. EM enhancement stems from localized surface plasmon resonance on a metal substrate, whereas CT enhancement is a resonance-like process in which the CT resonance depends on the energy difference between the Fermi level of a metal substrate and the molecular orbitals of an adsorbate.

Recent rapid progress in TERS has resulted from developments in nanotechnology [33], which has enabled spectral measurements with a spatial resolution below 10 nm and control at the single-molecule level. Currently, in nanotechnology, measurements are frequently based on scanning tunneling microscopy (STM), scanning electron microscopy (SEM), and transmission electron microscopy (TEM), but these methods cannot provide information about molecular structure. On the other hand, TERS has very high spatial resolution and, at the same time, allows one to explore molecular structure. TERS is the product of the organic integration of scanning probe microscopy (SPM) techniques and a plasmon-enhanced effect.

SERS suffers from a limited spatial resolution of hundreds of nanometers owing to the diffraction of light. Therefore, the information obtained from SERS is the average of large amounts of different surface sites. On the other hand, TERS is a newly developed Raman technique with spatial resolution beyond the diffraction limit of light; it can provide site-specific information and can be performed on any material surface regardless of morphology. TERS can be applied in various environments, such as ambient air, ultrahigh vacuum, solutions and liquids, and electrochemical environments. An example of a TERS study is reviewed in this chapter.

1.2.5 FUV Spectroscopy

FUV spectroscopy is concerned with electronic transitions, as in the case of DUV and UV spectroscopy, but FUV spectroscopy is clearly different from the other UV spectroscopies [10–12]. First, in contrast to the DUV and UV regions, which are free from oxygen absorption, the FUV region requires either a vacuum evaporation system or a nitrogen-gas-purging system. In the DUV and UV regions, absorption bands originating from various types of lower-energy electronic transitions, such as π–π^* and n–π^* transitions, can be observed, but higher-energy n–σ^* transitions and Rydberg transitions do not appear, whereas in the FUV region, in addition to the bands arising from π–π^* transitions, those from n–σ^* and σ-Rydberg transitions are observed [10–12]. Compared with that of DUV spectroscopy, the development of FUV spectroscopy of the condensed phase has made very slow progress. The main reason for this is the very high absorptivity of molecules in the FUV region [10–12]. To overcome this difficulty, we introduced the ATR technique to the FUV region [10–12]. The ATR technique allows the measurement of spectra similar to those of transmittance spectra with a very short optical path length. The development of ATR–FUV spectrometers has enabled the measurement of the spectra of liquid and solid samples in the nearly full FUV region without problems such as peak saturation. ATR–FUV spectroscopy has been recently applied to polymer studies to investigate the structure, electronic transitions, hydrogen bonds, and intermolecular interactions of polymers [13, 14]. Chapter 6 reports some examples of FUV studies on polymers.

1.3 Specific Examples of Molecular Spectroscopy Studies of Polymers

In this chapter, six examples of molecular spectroscopic studies of polymers are reviewed. They include the applications of IR, Raman, NIR, FIR, low-frequency Raman, TERS, Raman, and SERS imaging studies.

1.3.1 Infrared, Raman, and NIR Spectroscopic Evidence for the Coexistence of Hydrogen Bond Types in Poly(Acrylic Acid)

Dong et al. [62] used IR, Raman, and NIR spectroscopies to reveal the coexistence of various types of hydrogen bonds in poly(acrylic acid) (PAA) (Figure 1.2). The IR, Raman, and NIR spectra of a cast film of PAA were measured between 40 and 140 °C to investigate the hydrogen bond structures and their dissociation. The different hydrogen bond structures persisted even when the temperature rose well above the glass transition temperature. The Raman spectra confirmed the presence of different types of hydrogen bonds. It was found from the temperature-dependent intensity changes in the band arising from the first overtone of the OH stretching mode of PAA that the COOH groups dissociate significantly at high temperatures. Dong et al. [62] revealed the coexistence of various possible hydrogen bond forms analogous to those in the small aliphatic acids in PAA (Figure 1.2). The oligomeric chains of the COOH groups in PAA may explain the previously proposed cooperative hydrogen bonds in PAA or polymer blends containing PAA.

PAA and its copolymers are very useful as thickening agents for latexes and adhesives, as well as ingredients in pharmaceutical, cosmetic, coating, and agricultural chemical formulations. As part of the investigation of the interesting properties of PAA, for example, its special viscosity behavior in solutions, many researchers have extensively studied the conformational changes and hydrogen bonds between the macromolecular chains [63, 64]. The hydrogen bonds in PAA play a major role in determining its properties. Using IR, Raman, and NIR spectroscopy, it was found that although PAA is usually assumed to be strongly self-associated, it is a complex hydrogen-bonded mixture with a considerable number of dissociated species [62].

Figure 1.3 shows the temperature-dependent IR spectra of PAA in a cast film measured from 40 to 140 °C [63, 64]. As shown in Figure 1.3, the IR spectrum of PAA is characterized by a broad OH stretching band (3600 to 3000 cm^{-1}), overtones and combinations of bands at 1413 and 1248 cm^{-1} enhanced by Fermi resonance with the broad OH stretching band, C = O stretching bands (1750 to 1700 cm^{-1}), and relatively broad and conformationally sensitive backbone bands (1500 to 800 cm^{-1}) [62]. Based on the comparison of the spectra in Figure 1.3 with previously reported

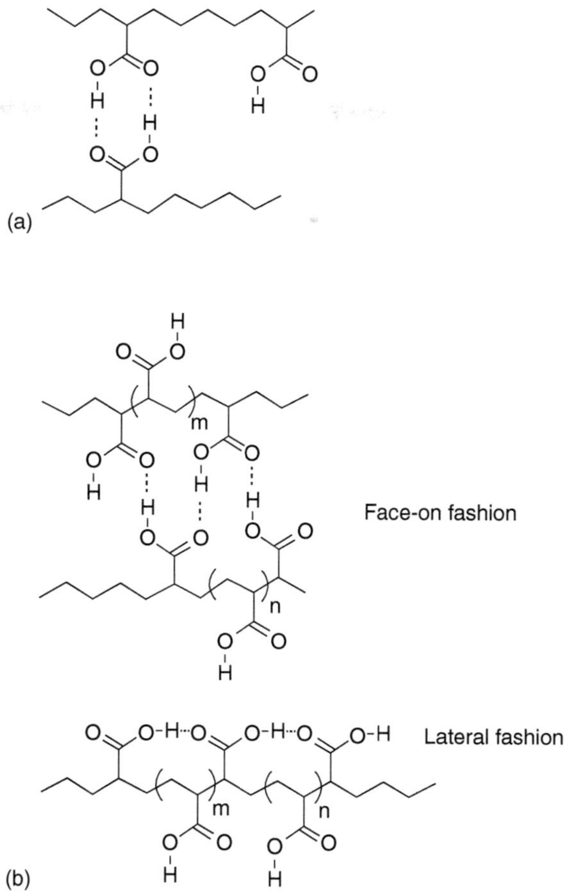

Figure 1.2 Four possible hydrogen-bonding arrangements of the carboxylic acid groups of PAA. (a) Cyclic dimer and free monomer of COOH groups. (b) Inner and terminal COOH in linear oligomeric forms. Source: Dong et al. [62].

IR spectra of PAA, it was found that the spectra in Figure 1.3 are a result of nearly atactic PAA.

Figure 1.4a,b shows an enlargement of the C = O stretching band region measured at 40 and 140 °C, respectively [62]. Figure 1.4c,d displays the second derivatives of the spectra in Figure 1.4a,b, showing three bands for both spectra collected at 40 and 140 °C. Dong et al. [62] determined the number of band components contributing to the band envelope in the 1800 to 1600 cm^{-1} region to be four using principal component analysis (PCA) [62].

Figure 1.4a,b also shows the results of curve fitting. These results suggest the presence of four bands in this region. For all the spectra obtained from 40 to 140 °C, four bands gave the best fit for the CO envelope, whereas a three-band fit always left some discrepancy. The resolved spectra shown in Figure 1.4a,b yielded four component bands at 1686, 1705, 1723, and 1742 cm^{-1} for the sample at 40 °C and those at 1699, 1706, 1725, and 1743 cm^{-1} for the sample at 140 °C, respectively. The bands at 1705 and 1742 cm^{-1} were assigned to the C = O stretching modes of the cyclic hydrogen-bonded COOH group in the dimeric form and free (non-hydrogen-bonded) COOH group, respectively (Figure 1.2a) [62]. The position

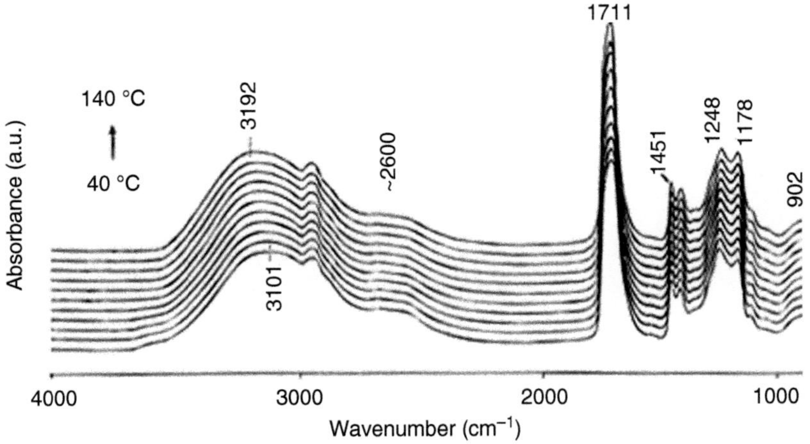

Figure 1.3 FT-IR spectra of PAA measured from 40 to 140 °C (10 °C intervals at rate of 1 °C min^{-1}). Source: Dong et al. [62].

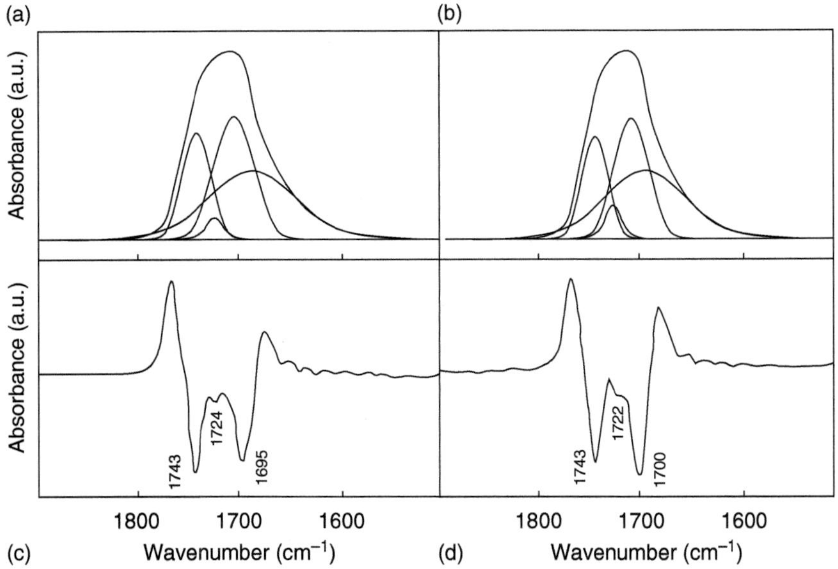

Figure 1.4 FT-IR spectra of PAA in the C = O stretching region measured at (a) 40 and (b) 140 °C; (c) second derivative of spectrum (a); and (d) second derivative of spectrum (b). Curve-fitting results are also shown in the spectra (a) and (b). Source: Dong et al. [62].

of the free C = O band near 1742 cm^{-1} changes little with temperature and is very close to those previously found for free COOH groups in various systems. The frequency of the C = O asymmetric stretching band of the cyclic dimer changes little with temperature because of its fixed configuration.

Of interest is the appearance of a C = O band near 1685 cm^{-1}. This band shifts to the higher-energy side with increase in temperature. The band at 1685 cm^{-1} suggests the presence of an additional hydrogen bond. Several research groups have investigated the structure of self-associated acetic acid using various techniques

[65, 66]. Based on their studies, the C = O band near 1686 cm^{-1} can be ascribed to the inner hydrogen-bonded groups of the oligomeric COOH groups, whereas the residual band near 1725 cm^{-1} was assigned to the terminal group in the oligomeric COOH chains (Figure 1.2b). Of note is that the C = O band widths of the inner hydrogen-bonded groups are much broader than those of other species, reflecting a broad distribution of oligomer chains with variable lengths. Moreover, Dong et al. [62] considered that the less restricted motion of the linear dimeric/oligomeric COOH compared to that of the cyclic dimer results in a slightly lower frequency observed for linear dimer/oligomers. In addition, it is very likely that the C = O groups in the oligomeric chains couple with two adjacent C = O groups in contrast to the C = O group in the cyclic dimer, in which only two C = O groups couple with each other. This may also account for the appearance of the C = O stretching band of the oligomers on the lower wavenumber side than that of the cyclic dimer species [62].

Tanaka et al. [66] compared the dissociation of PAA with that of its monomer analogs in aqueous solutions and found that the dissociation of COOH groups into free species is less favorable for PAA than that for glutaric acid (model compound for the dimer of the PAA repeating unit) and 1,3,5-pentanetricarboxylic acid (model compound for the trimer of the PAA repeating unit). They suggested that some sort of cooperative or nonadditive hydrogen bond is formed in PAA solutions, which makes the dissociation more unfavorable than that in its monomer analogs. Dong et al. [62] proposed that the oligomerization of COOH groups in two- or three-dimensional arrays of the group linkage may account for such cooperative hydrogen bonds.

The Raman spectrum of PAA is shown in Figure 1.5 [62]. In the enlarged Raman spectrum shown in Figure 1.6a [62], an asymmetric broad band arising from C = O stretching modes was observed around 1690 cm^{-1}. The second derivative of the spectrum in Figure 1.6b shows at least three bands in the C = O stretching region, whereas the curve fitting gives four bands at 1742, 1715, 1680, and 1660 cm^{-1}. The

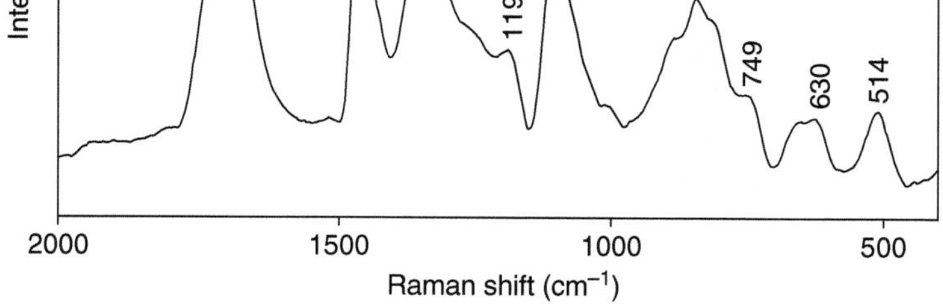

Figure 1.5 An FT-Raman spectrum of PAA measured at 64 °C. Source: Dong et al. [62].

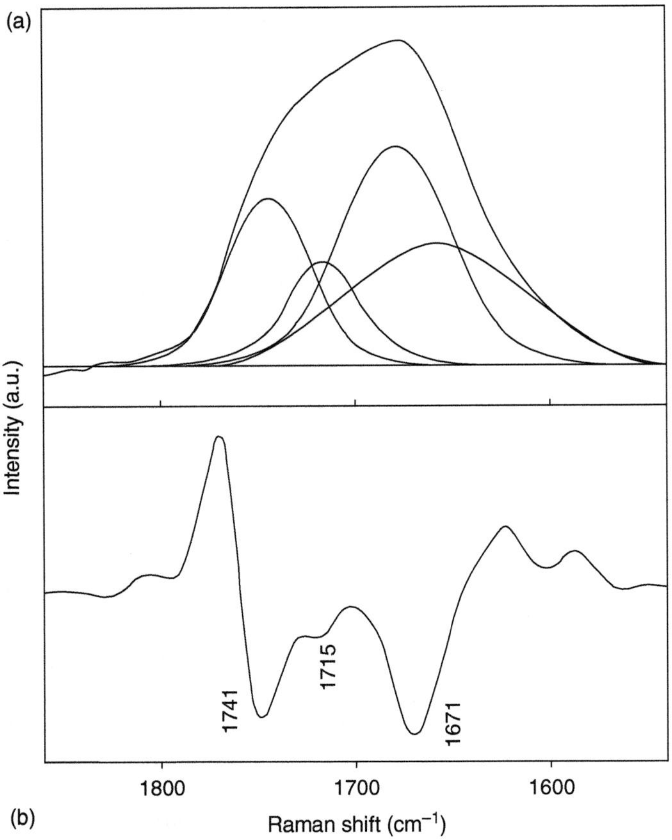

Figure 1.6 Enlarged Raman spectrum of PAA between 1850 and 1550 cm^{-1} showing the components contributing to the C = O stretching band: (a) a curve-fitting result and (b) second derivative plot. Source: Dong et al. [62].

1680 cm^{-1} band was assigned to an in-phase C = O stretching vibration of the cyclic dimer form, whereas the other bands at 1742, 1715, and 1660 cm^{-1} were ascribed to the C = O stretching modes of free COOH and terminal and inner hydrogen-bonded COOH in the oligomeric forms, respectively [65, 66]. It should be noted that the frequency of the dimer species in the Raman spectrum does not coincide with the counterpart in the IR spectrum. However, this observation is reasonable because the Raman band originates from a normal mode involving in-phase $(CO)_2$ stretching, whereas the IR band results from the out-of-phase $(CO)_2$ stretching in the dimer.

Coleman et al. [67] investigated various kinds of hydrogen-bonding species in polymers and adopted Flory's lattice filling methodology to describe two sets of chains, one covalent and the other hydrogen bonded, both of which share the same segments. Interestingly, the association of the repeating units is independent of the covalent chain length and can be treated as a "small" molecule (one containing a single functional group) of equal molar volume. Dong et al. [62] postulated that the hydrogen bond structure in the oligomeric forms takes a face-on arrangement [65, 66] for acetic acid, propionic acid, and *n*-butyric acid or a lateral arrangement [68, 69] for fatty acids adsorbed on a solid surface (Figure 1.2b). It is rather difficult to distinguish these two types of arrangements by IR and Raman measurements, and, thus, it was difficult to conclude which arrangement of COOH chains is favored in PAA. It should be noted that the formation of oligomeric COOH groups does not require hydrogen bonding between the adjacent repeating units in the same hydrocarbon chain because both oligomers and dimers can be formed between COOH groups located on different parts of the chains, as shown in Figure 1.2b.

The species formed by monomeric COOH groups are not well resolved in the IR OH stretching and C = O stretching regions in IR and Raman spectra. Thus, NIR spectroscopy has been used to clarify these species. Figure 1.7 shows the FT-NIR transmission spectra of PAA measured over a temperature range of 40–140 °C [62]. The broad band at around 6820 cm^{-1} is due to the first overtone of the OH stretching mode of the free carboxylic group. The long tail on the lower wavenumber side of this band can be attributed to the terminal free carboxylic groups in the oligomeric forms with different chain lengths. The intensity of the tail increases with temperature, which can be easily verified from the difference spectra of this region. In the NIR region, bands due to hydrogen-bonded OH groups are extremely weak, whereas those arising from the free OH group are relatively strong. The large displacement of the relatively light proton in the free OH group allows it to be in the "free" state relative to the hydrogen-bonded OH groups, resulting in a greater breakdown of the harmonic oscillator approximation and, hence, the selection rules that, in principle, do not permit the appearance of overtones and combinations of bonded species.

As can be seen from the enlargement of the 7000 to 6300 cm^{-1} region (the inset in Figure 1.7), the intensity of the 6820 cm^{-1} band gradually increases with temperature. Broad bands in the 4600 to 4500 and 4060 to 4030 cm^{-1} regions, both associated with COOH combination vibrations, also increase with temperature, whereas the intensities of other bands irrelevant to the COOH group vibrations are temperature independent [62]. In the inset of Figure 1.7, the band area of the free COOH species increases significantly above 100 °C. It is quite reasonable that above the glass transition temperature (T_g) of PAA, the segmental motion accelerates the considerable gain of the free COOH groups as the macromolecular chains disentangle at the expense of the hydrogen-bonded forms. In the IR region, neither the OH stretching band arising from free COOH (near 3500 cm^{-1}) nor the intensity change of the free COOH group (the 1742 cm^{-1} band) is clearly detectable. However, this is not a problem in the NIR region.

In this study, IR, Raman, and NIR spectroscopy were utilized to explore the hydrogen bonding in PAA [62]. In addition to the presence of monomer and cyclic dimer forms of COOH, additional spectral features were interpreted in terms of the presence of oligomeric species. The coexistence of various hydrogen-bonded species of the COOH groups in PAA may provide new insights into the thickening mechanism of the polymer.

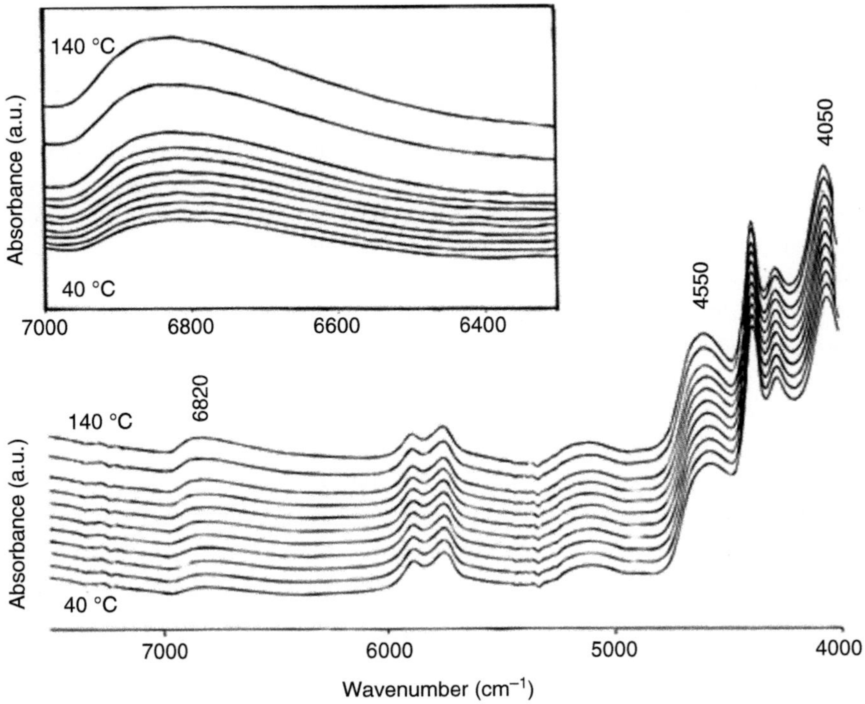

Figure 1.7 FT-NIR spectra of PAA measured from 40 to 140 °C (in 10 °C intervals at a rate of 1 °C min^{-1}). Inset: enlarged NIR spectra showing the first overtone of an OH stretching mode. Source: Dong et al. [62].

1.3.2 Low-Frequency Vibrational Modes of Nylon-6 Studied by Using IR and Raman Spectroscopies and Density Functional Theory Calculations

Low-frequency vibrational spectroscopy has recently been used extensively to investigate the structures and physical properties of polymers [57d]. In Chapter 4, one can find several good examples of this subject; here, we present one example. Figure 1.8 shows the FIR spectra in the region of 350 to 50 cm^{-1} of (a) nylon-6 in the α-form measured in the temperature range of 29–194 °C and (b) their second derivative spectra at 29 and 194 °C [57]. The FIR spectrum of nylon-6 at 29 °C shows characteristic crystalline bands at 294, 222, and 111 cm^{-1} (Figure 1.8a). In the region of 125 to 80 cm^{-1}, as the temperature increased, the maximum peak position shifted to 98 from 111 cm^{-1} with a decrease in intensity at 111 cm^{-1}. The second-derivative spectra indicate that the position of the band at 111 cm^{-1} changes little with temperature.

Figure 1.9a–c depicts the temperature-dependent changes in the FIR intensity of nylon-6 at (a) 294, (b) 222, and (c) 111 cm^{-1} from 22 to 210 °C [57d]. The intensity of the FIR band at 111 cm^{-1} did not change until approximately 60 °C, which corresponds to the glass transition of nylon-6 at approximately 54 °C. Above about 60 °C, the intensity decreased linearly and became constant at temperatures above approximately 170 °C. This temperature is close to the Brill transition of nylon-6 at 160 °C.

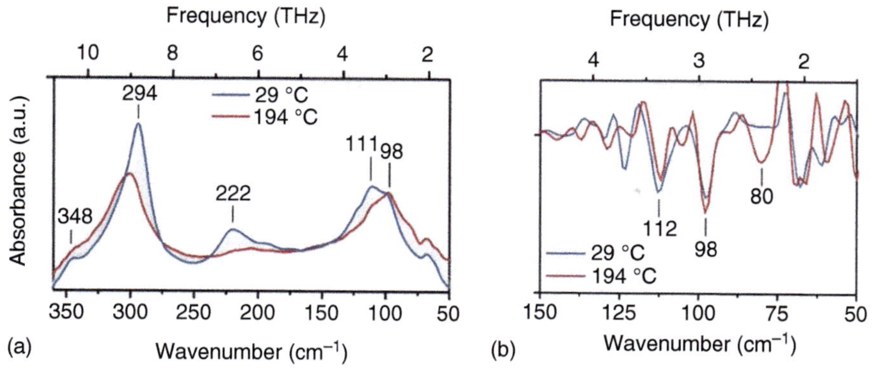

Figure 1.8 (a) FIR spectra of nylon-6 in the α-form measured from 29 to 194 °C and (b) their second derivatives at 29 to 194 °C. Source: Yamamoto et al. [57d].

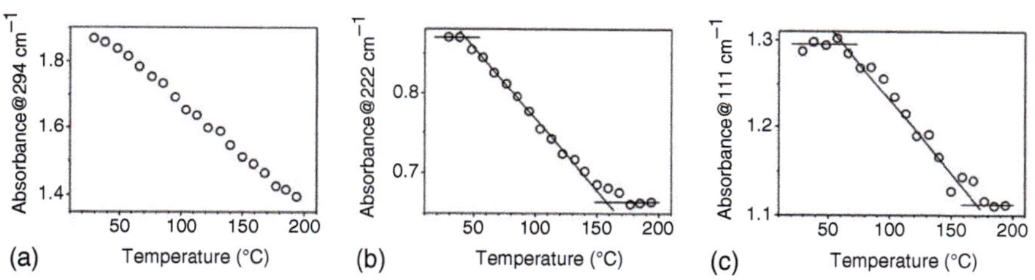

Figure 1.9 Temperature dependence of the FIR intensity of nylon-6 at (a) 294, (b) 222, and (c) 111 cm^{-1} from 22 to 210 °C. Lines are eye guides. Source: Yamamoto et al. [57d].

It is known that after the glass transition, the lattice lengths of the *a*- and *c*-axes of the α-form of nylon-6 gradually approach each other and finally reach unity after the Brill transition. This trend is similar to that of the FIR intensity at 111 cm^{-1}. Hence, the intensity at 111 cm^{-1} may be used as an indicator of the lattice length of nylon-6 in the α-form.

The intensity of the weak bands in the 240 to 190 cm^{-1} region decreased with temperature in a similar manner to that of the band at 111 cm^{-1}. The temperature dependence of the intensity at 222 cm^{-1} (Figure 1.9b) yields transition temperatures of c. 50 °C (glass) and 160 °C (Brill). The intense band at 294 cm^{-1} became weaker and broader with increasing temperature; however, the peak top intensity only gradually decreased and did not show any clear transition (Figure 1.9a).

Figure 1.10 shows (a) the Raman spectra of nylon-6 in the α-form in the 180 to 70 cm^{-1} region at 22, 90, 150, and 210 °C and (b) their second derivatives at 22–150 °C [57d]. In the Raman spectrum obtained at 22 °C, a broad and strong Raman peak was observed at 104 cm^{-1}. On increasing the temperature to 150 °C, the maximum peak position shifted to 100 cm^{-1}. The second-derivative spectra (Figure 1.10b) reveal that this shift is induced by a relative change in the intensities of the two peaks.

Figure 1.11 presents (a) the polarized Raman and (b) FIR spectra of nylon-6 in the α-form at room temperature with polarization perpendicular (red lines) and parallel (black lines) to the stretched direction of the nylon film [57d]. Both the strong

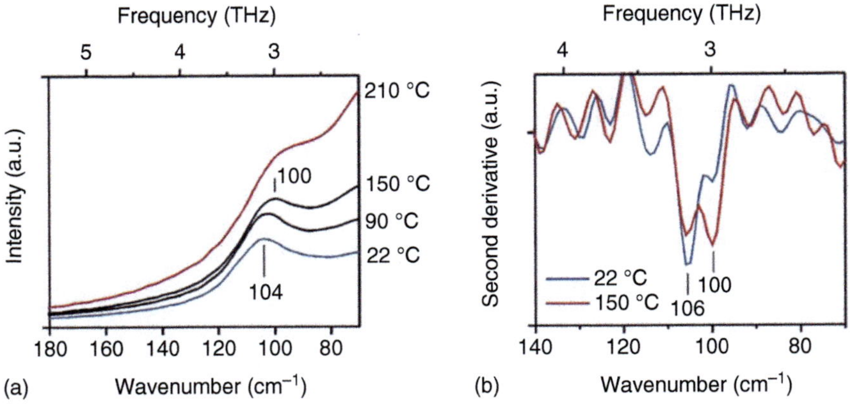

Figure 1.10 (a) Raman spectra of nylon-6 in the α-form collected at 22, 90, 150, and 210 °C and (b) their second derivatives at 22–210 °C. Source: Yamamoto et al. [57d].

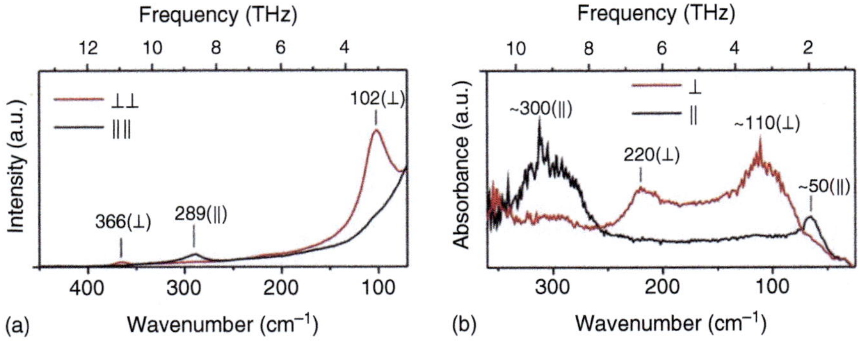

Figure 1.11 (a) Polarized Raman and (b) FIR spectra of nylon-6 in the α-form at room temperature with polarization perpendicular (red lines) and parallel (black lines) to the stretched direction of the nylon film. Source: Yamamoto et al. [57d].

Raman band at approximately 102 cm^{-1} and FIR band at approximately 110 cm^{-1} are polarized perpendicular to the chain axis, suggesting correspondence of these two bands. Comparisons of the experimental (top) and calculated (bottom) (a) Raman and (b) FIR spectra of crystalline nylon-6 in α-form are shown in Figure 1.12 [57d]. The calculations were carried out using the fragment methodology at the CAM-B3LYP-GD3BJ/6-311++G** level of theory. On the basis of the peak positions and polarization directions, the calculated bands at 365, 273, and approximately 210 cm^{-1} were ascribed to experimental Raman bands at 365, 289, and 213 cm^{-1}, respectively. In the lower wavenumber region, perpendicularly polarized strong Raman bands were calculated at 127, 108, and 89 cm^{-1}. It is very likely that the calculated strongest band at 108 cm^{-1} corresponds to the experimental band at 104 cm^{-1}.

The agreement of the FIR spectra was also satisfactory in terms of the band positions, spectral shapes, and polarization directions. The FIR bands at 294 and 222 cm^{-1} were calculated to be 274 and 216 cm^{-1}, respectively, consistent with

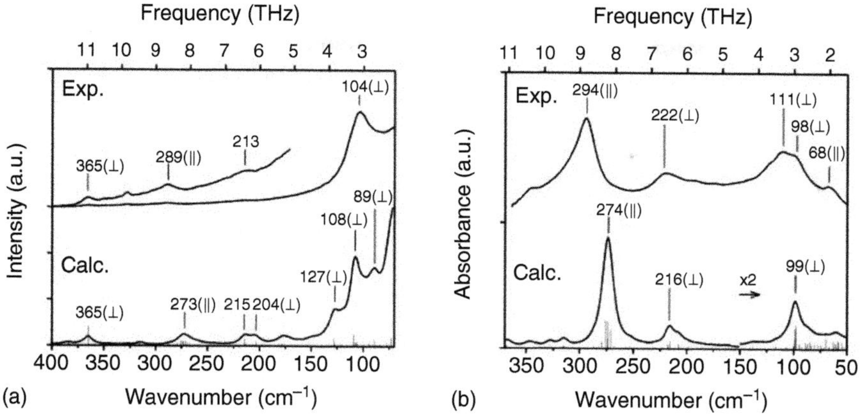

Figure 1.12 Comparisons of the experimental (top) and calculated (bottom) (a) Raman and (b) FIR spectra of crystalline nylon-6 in the α-form. The calculations were carried out using the fragment methodology at the CAM-B3LYP-GD3BJ/6-311++G** level of theory. Polarization directions are indicated as parallel (||) or perpendicular (⊥) to the chain axis of nylon-6. The calculated FIR intensity below 150 cm^{-1} is doubled for clarity. Source: Yamamoto et al. [57d].

the polarization directions. In the lower-wavenumber region of the calculated spectrum, one can find an FIR band at 99 cm^{-1} showing perpendicular polarization and a weak shoulder at approximately 80 cm^{-1}. These calculated bands were ascribed to the experimental bands at approximately 111 and 98 cm^{-1} based on the polarization and intensities.

Figure 1.13 illustrates the atomic motion corresponding to the FIR active modes of nylon-6 at (a) 247, (b) 216, and (c) 99 cm^{-1} viewed from the origin of the a-axis [57d]. The two chains along the a-axis are overlapped in this figure, and the bold letters in (c) indicate atoms in the front chains. The calculated atomic motions corresponding to the FIR band at 99 cm^{-1} (Figure 1.13c) are a mixture of the torsional motion of the methylene groups and the transverse mode of amide groups, in which the NH and O atoms oscillate out of the amide plane. This assignment is partly in agreement with the previous one based on the normal-mode analysis [70, 71], which is the torsional motion of the methylene chain. However, Yamamoto et al. [57d] found that amide groups also contributed to this mode. As discussed later, the amide groups govern the intensity of the FIR band. Our own calculations indicate that the translational motion of the amide groups is not involved in this mode, although its involvement has been reported [72]. As discussed above (Figure 1.9c), the temperature dependence of the FIR intensity at 111 cm^{-1} yielded two transition points corresponding to glass and Brill transitions. The torsional nature of this band can explain why the intensity is sensitive to changes in the lattice length. The calculated Raman band at 108 cm^{-1} has similar atomic motions to the FIR band at 99 cm^{-1}: the torsion of methylene groups and the transverse motion of the amide groups.

Figure 1.14 shows the simulated (a) Raman and (b) FIR spectra of nylon-6 in the α-form with atomic tensors of amide groups zeroed out (red) and exact (black) [57d]. Notably, except for the Raman band at 270 cm^{-1}, the intensities of all the other

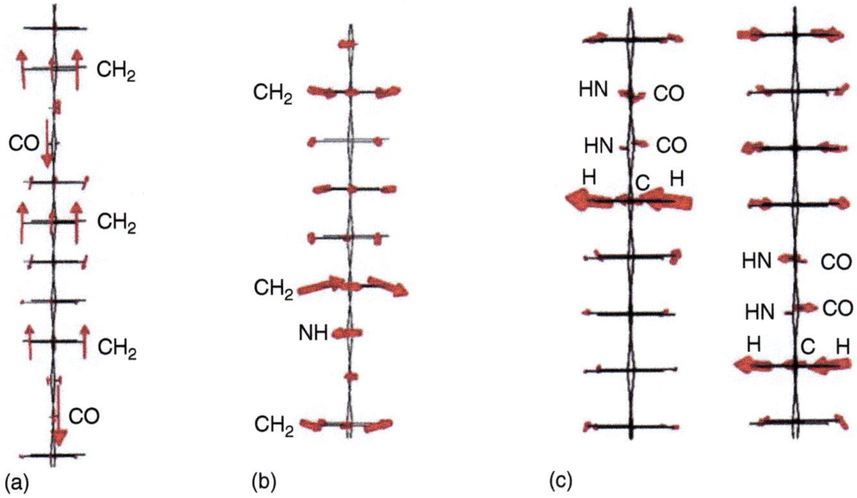

Figure 1.13 Atomic motion of the FIR active modes of nylon-6 in the α-form calculated at (a) 247, (b) 216, and (c) 99 cm^{-1} viewed from the origin of the *a*-axis. Two chains along the *a*-axis are overlapped in this figure, and the bold letter in (c) indicates atoms in the front chains. Source: Yamamoto et al. [57d].

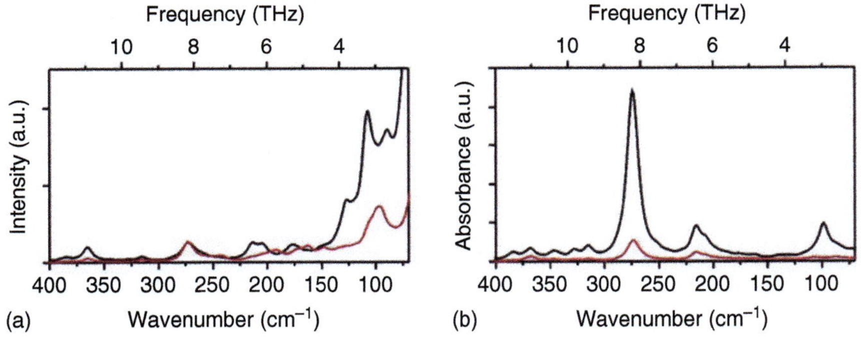

Figure 1.14 Calculated (a) Raman and (b) FIR spectra of nylon-6 in the α-form with atomic tensors of amide groups zeroed out (red) and exact (black). Source: Yamamoto et al. [57d].

major Raman and FIR bands are decreased markedly by zeroing out. These results indicate the dominant contribution of the amide groups to the low-frequency intensities. In particular, the FIR bands at approximately 100 cm^{-1} were nearly perfectly erased, suggesting that the FIR band is highly sensitive to the chemical environment of the amides. Interestingly, the atomic motions corresponding to the FIR band at 99 cm^{-1} were dominated by the methylene group, but their absorption intensity was governed by the amide groups. These calculation results indicate that the experimentally observed decrease in absorption at 111 cm^{-1} with increasing temperature is a result of changes in the chemical environment of the amide groups during the change in crystal lattice length. In contrast, the corresponding Raman intensity at about 108 cm^{-1} is derived not only from the amide groups but also from the methylene part.

1.3.3 NIR Spectra of Linear Low-Density Polyethylene and Their Chemometrics Analysis

NIR spectroscopy has been used not only for basic studies of polymers, such as hydrogen bonding, intermolecular interactions, and physical properties, but also for practical applications [15–17]. Several examples of both can be found in Chapter 5. However, here, we present one particularly good example of an application: a combined chemometrics and spectral pretreatment method yielding band assignments based on loading plots of PCA and partial least squares (PLS) analysis [73]. NIR DR spectra were measured for pellets of 16 types of linear low-density polyethylene (LLDPE) with short branches and polyethylene (PE) to enable band assignment and build calibration models that predict their density [73]. The density of the LLDPE samples was in the range of 0.911–0.925 g cm^{-3}. The original NIR spectra data set, their second derivatives, and the spectra after multiplicative scatter correction (MSC) treatment were used to develop the models. The correlation coefficients were calculated to be 0.961, 0.965, and 0.970 for the original NIR spectra, their second derivatives, and those with the MSC treatment, respectively, and the standard error of prediction (SEP) was found to be 0.001 g cm^{-3} for all cases. It was found from the regression coefficient (RC) plots for the calibration models that bands at 1192, 1376, and 1698 nm originating from the overtone and combination modes of the CH_3 groups contribute significantly to the density prediction [73].

Figure 1.15 compares the NIR DR spectrum of one of the LLDPE samples with that of PE [73]. As shown in Figure 1.15, the NIR spectrum of LLDPE with short branches is significantly different from that of PE without a branch. For example, PE showed intense bands at 1728 and 1764 cm^{-1}. The band assignments were not straightforward; however, based on previous NIR studies on alkyl compounds and polymers, Shimoyama et al. [73] assigned bands in the 1140–1250 nm region to the second overtones of CH_2 stretching modes or their combination modes: specifically, those in the 1350–1450 nm region to the combinations of 2 × C–H stretch (str) + C–H deformation (def) (CH_2 or CH_3), and those in the 1700–1800 nm region to the first overtones of CH_2 stretching modes or their combination modes. They assigned bands at 1146, 1186, 1374, 1644, and 1698 nm to CH_3 group vibrations, as will be discussed later. PE shows two characteristic features in the 1720–1770 nm region, originating from the CH_2 vibrational modes. It should be noted that the bands arising from the CH_2 groups are relatively stronger in the spectrum of PE, whereas those arising from the CH_3 groups are more prominent in the spectrum of LLDPE.

Figure 1.16a,b shows the NIR spectra of the 16 types of LLDPE and PE before and after MSC treatment, respectively. Notably, the baseline varies from one sample to another for the pellet samples, and the MSC treatment is useful for baseline correction [73]. The second derivatives of the NIR spectra shown in Figure 1.16a are presented in Figure 1.17. The second derivative spectra identified weak bands at 1146, 1186, 1300, 1374, 1634, 1696, and 1730 nm, which are not always clearly observed in the original spectra. The second derivative spectra clearly differentiate the spectra of LLDPE from that of PE in the 1160–1220, 1360–1400, and 1630–1780 nm regions.

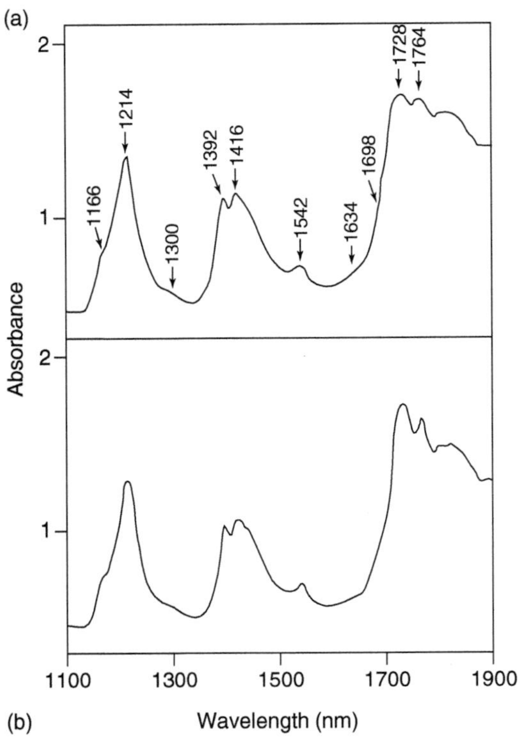

Figure 1.15 (a) A NIR DR spectrum of LLDPE sample and (b) that of PE. Source: Shimoyama et al. [73].

Figure 1.18 displays a score plot of PCA factors 1 and 2 for 51 MSC-pretreated NIR spectra in the 1100–1900 nm region of the 16 types of LLDPE and PE (NIR spectra were measured three times for each polymer) [73]. The NIR spectrum of PE was clearly discriminated from the rest using two factors. The polycarbonate (PC) weight-loading plot of factor 1 for the score plot shown in Figure 1.18 is presented in Figure 1.19 [73]. The loading has three upward peaks at 1192, 1380, and 1698 nm and three downward peaks at 1542, 1728, and 1764 nm. Shimoyama et al. [73] ascribed the upward peaks to the overtone or combination modes of the CH_3 groups and the downward peaks to those of the CH_2 groups. Notably, the bands at 1192, 1380, and 1698 nm are very weak or almost missing in the raw spectra in Figures 1.15 and 1.16, but they play key roles in the discrimination of PE and the LLDPEs. Thus, it is very likely that factor 1 reflects the existing ratio of the number of CH_3 and CH_2 groups.

Shimoyama et al. [73] developed a PLS regression calibration model to predict the density of LLDPE using their MSC-treated NIR spectra. It was found from the plot of the residual validation variance of the PLS regression that four factors were sufficient for prediction. They attempted to build principal component regression (PCR) and PLS regression calibration models based on the raw spectra, their second derivatives, and the spectra after MSC treatment. The PLS regression analysis yielded significantly better results than the PCR analysis, and, among the PLS results, the MSC treatment gave somewhat better results than the second derivative.

A loading plot of the RCs for the model obtained from the NIR spectra after MSC is depicted in Figure 1.20 [73]. The RC plot is characterized by three downward peaks

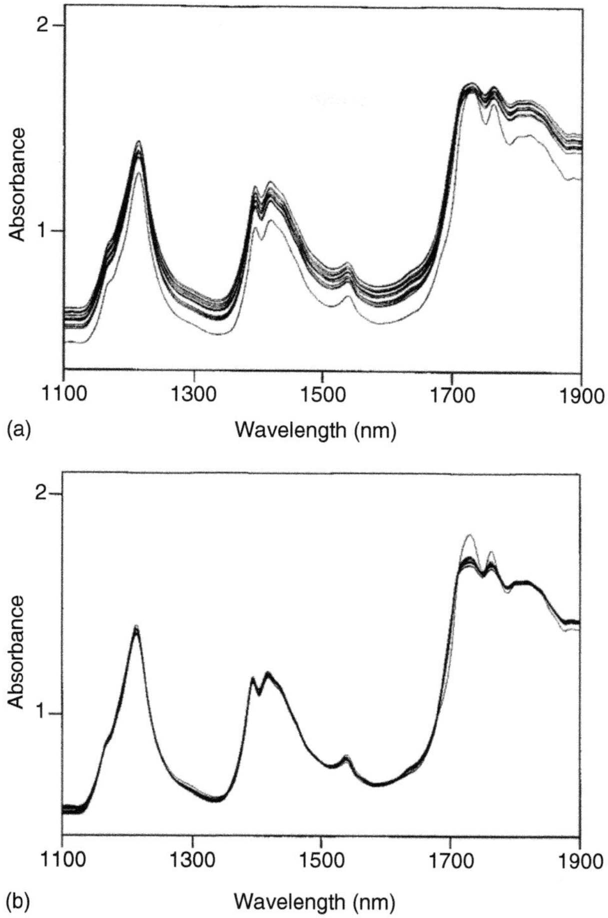

Figure 1.16 NIR spectra of the 16 kinds of LLDPE and PE investigated (a) before and (b) after MSC treatment. Source: Shimoyama et al. [73].

at 1192, 1376, and 1698 nm arising from the CH_3 groups. These bands are not clear in the raw spectra but are prominent in the RC plot. An upward peak at 1542 nm corresponds to the band derived from the CH_2 groups. As shown in Figure 1.20, the RC plot is related to the relative intensities of the CH_3 and CH_2 bands. The density of LLDPE is controlled by the number and types of branches attached to the main chain of the PE. The lower the density, the larger the number of CH_3 groups. Therefore, it seems reasonable that NIR spectroscopy can predict the density of LLDPE in terms of the number of CH_3 groups.

1.3.4 Study of the Crystallization Behavior of Asymmetric PLLA/PDLA Blend by IR and Raman Spectroscopy and Raman Imaging

In this study, the isothermal crystallization of an asymmetric poly(L-lactide)/poly(D-lactide) (PLLA/PDLA) blend (4/1) at around 100 °C was investigated by IR,

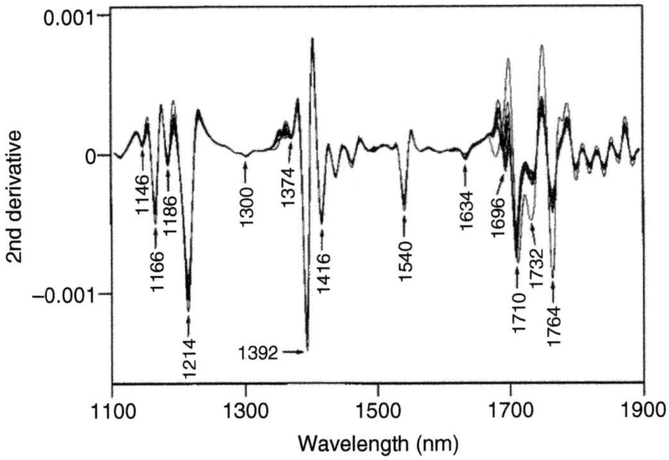

Figure 1.17 Second derivative of the NIR spectra shown in Figure 1.16a. Source: Shimoyama et al. [73].

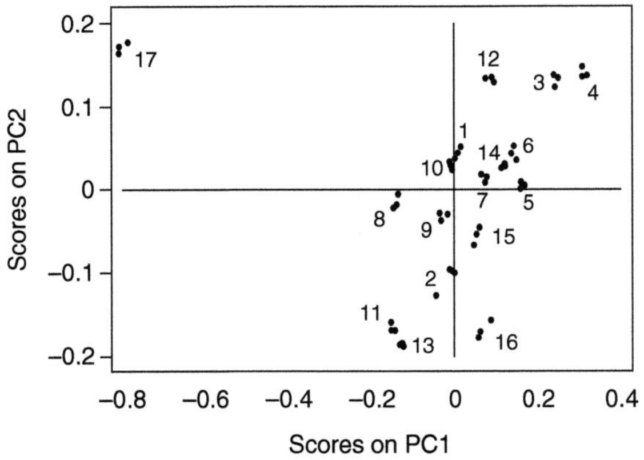

Figure 1.18 Score plot of PCA factors 1 and 2 for the 16 kinds of LLDPE and PE based on 51 NIR spectra recorded in the 1100–1900 nm region (NIR spectral data were pretreated by MSC). The numerals adjacent to each point indicate the sample number. Source: Shimoyama et al. [73].

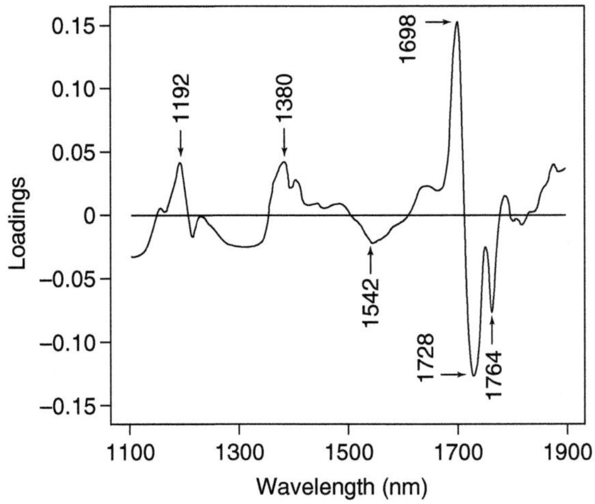

Figure 1.19 A PC weight-loadings plot of factor 1 for the score plot shown in Figure 1.18. Source: Shimoyama et al. [73].

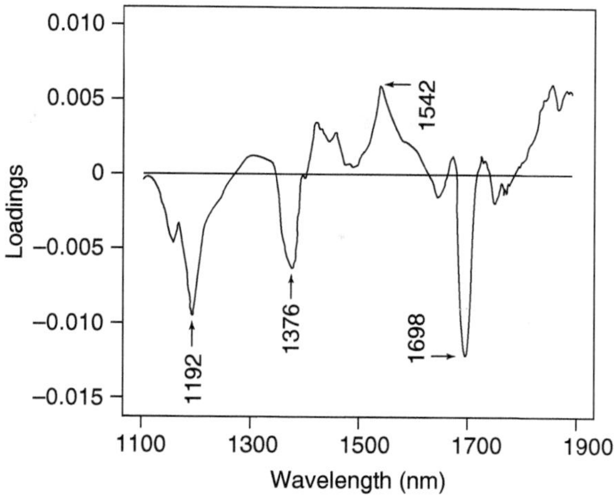

Figure 1.20 Loadings plot of regression coefficients for the PLS calibration model from the MSC treated NIR spectra. Source: Shimoyama et al. [73].

Raman, and Raman imaging [74]. The IR results indicate that the crystallization of the homocrystal (HC) of the blend was delayed owing to the formation of a large amount of the stereocomplex (SC) in the isothermal process. A unique spherulite morphology was observed through polarized optical microscopy (POM), and the distributions of SC and HC in and out of the large spherulites were revealed by Raman imaging measurements. These results demonstrate that the SC crystals are homogeneously dispersed in both the AMP and spherulite regions. In contrast, the HC crystal was located mainly in the large spherulite area. A possible isothermal crystallization process for an asymmetric PLLA/PDLA blend was proposed.

Figure 1.21 shows the time-dependent variations in the IR spectra in the 970 to 890 cm^{-1} region of asymmetric PLLA/PDLA 4/1 blend collected during the isothermal crystallization process at 100 °C [74]. Formation of both SC and HC poly(lactic acid) (PLA) was found in the C–C stretching region. It was reported that the band at 908 cm^{-1} is due to the 3/1 helix conformation in PLA SC crystal lattice, whereas the band at 921 cm^{-1} derives from the 10/3 helix conformation, which exists in the PLA α'/α HC [74]. As shown in Figure 1.22, the band at 908 cm^{-1} appears before that at 921 cm^{-1}, indicating that the SC appears first at the isothermal temperature of 100 °C [74]. Thus, clearly, their crystallization rates and completing times are asynchronous.

Figure 1.23a–c displays POM images of the asymmetric blend sample of PLLA/PDLA (4 : 1) isothermally crystallized at 100 °C for different times [74]. As shown in Figure 1.23a–c, the growth of spherulites starts from the inside to the outside and may consist of many small crystallite domains. Figure 1.23d shows the round shape of a large spherulite observed from the melt, and Figure 1.23e shows the result of the crystallization of a neat PLLA HC at the same temperature (100 °C). As shown in Figure 1.23d, a light circle corresponding to a spherulite was observed with a radius of approximately 50 μm. However, a black Maltese cross was not observed, in contrast to the HC of neat PLLA under the same crystallization conditions (Figure 1.23d). To provide new insights into the composition, Zhang et al. conducted Raman imaging experiments.

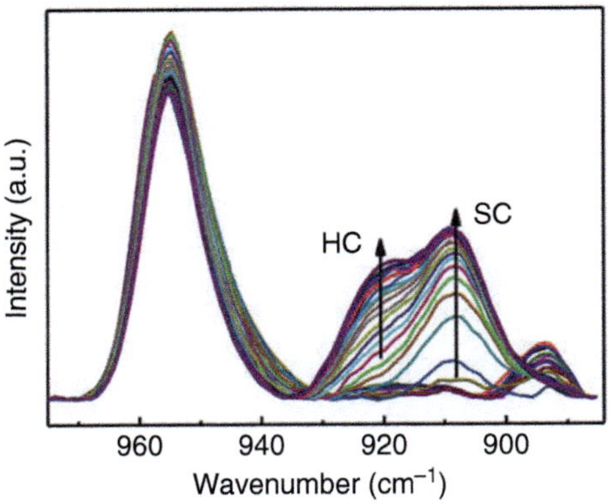

Figure 1.21 Time-dependent IR spectra of asymmetric PLLA/PDLA 4/1 blend, which were measured in the isothermal crystallization process at 100 °C. Source: Hu et al. [74].

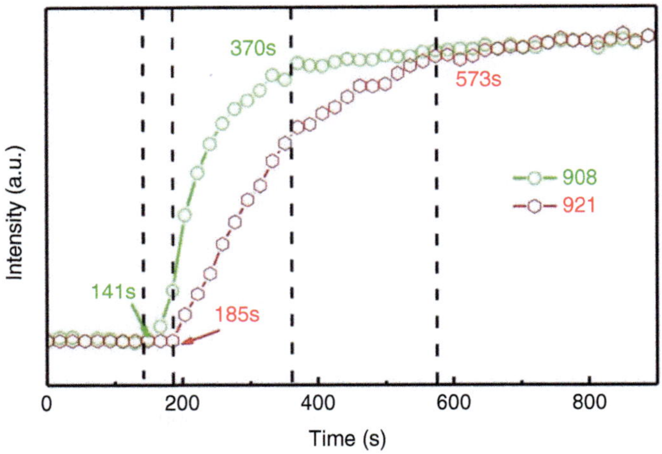

Figure 1.22 Time-dependent variations of IR band intensity of SC at 908 cm^{-1} and that of HC at 921 cm^{-1} at an isothermal temperature of 100 °C. Source: Hu et al. [74].

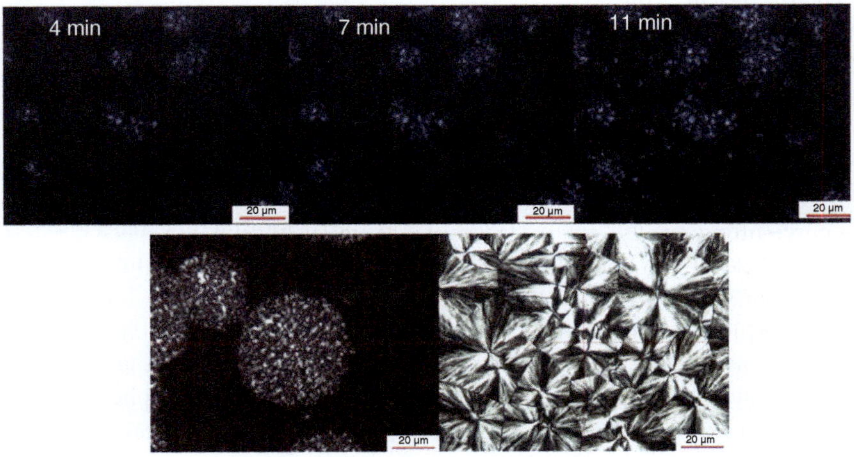

Figure 1.23 POM images of the asymmetric blend sample of PLLA/PDLA (4 : 1) collected at a melt-isothermal temperature of 100 °C for (a) 4, (b) 7, and (c) 11 mm; (d) whole spherulite morphology and (e) PLLA spherulite morphology obtained at an isothermal temperature of 100 °C. Source: Hu et al. [74].

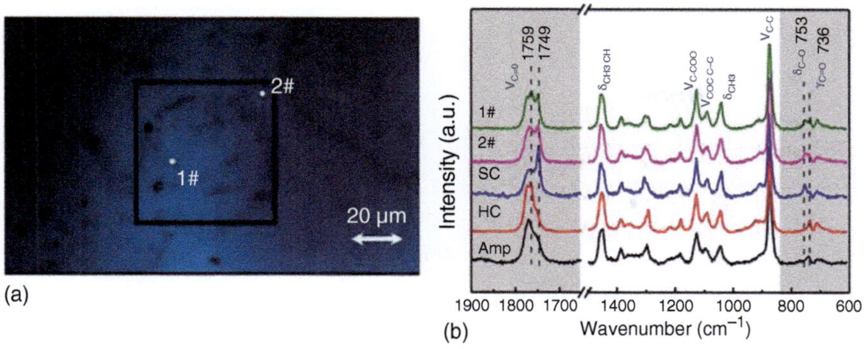

Figure 1.24 (a) Optical image of the blend sample after isothermal crystallization at 100 °C for 15 minutes (black framed region is the imaging region). (b) Raman spectra measured at different positions and those of PLA different crystal forms; #1: point located within the spherulite region, #2: point out of the spherulite, SC: stereocomplex 1 : 1 PLLA/PDLA blend isothermal crystallization at 200 °C, HC: Homocrystal PLLA isothermal at 100 °C, Amp: amorphous PLLA quenched sample from melting. Source: Hu et al. [74].

Single-point Raman measurements focusing on the large spherulite region were carried out first. Figure 1.24a shows an optical image of the blend sample after isothermal crystallization at 100 °C for 15 minutes (the black framed region is the imaging region) [74]. Figure 1.24b shows the Raman spectra collected at different positions marked in #1 and #2 in Figure 1.24a. The Raman spectra of SC, HC, and AMP PLA measured at different states are also shown in the same figure. Examining the whole Raman spectra, the five spectra in Figure 1.24b are very similar to each other; however, there are significant differences in the $\nu C=O$ (1800 to 1700 cm^{-1}) and $\gamma C=O/\delta C—O$ (800 to 700 cm^{-1}) regions. In the former region, the spectrum of the #1 point is similar to that of #2, and they are distinctly different from the spectrum of SC. HC and Amp yielded similar spectra in this region. In the region from 800 to 700 cm^{-1}, the relative intensities of the bands at 733 and 755 cm^{-1} could be used to distinguish SC from HC.

Zhang and coworkers [74] developed Raman images using both the 1800 to 1700 and 800 to 650 cm^{-1} regions. Here, we explain the results for the images based on the 800 to 650 cm^{-1} region. Figure 1.25a shows the Raman spectra and their curve-fitting results in the 800 to 650 cm^{-1} region measured at points #1 and #2, and Figure 1.25b presents Raman spectra in the same region of PLA SC, HC, and AMP with curve-fitting results [74]. As shown in Figure 1.25b, there are specific bands for HC and SC at 736 and 754 cm^{-1}, respectively. Images constructed using the band intensities at 754 and 736 cm^{-1} are displayed in Figure 1.25c,d, respectively. Figure 1.25c illustrates the intensity distribution of the band at 754 cm^{-1} based on the curve-fitting result shown in Figure 1.25a. As shown in Figure 1.25d, the spherulite region had the highest intensity for the band at 736 cm^{-1}. Combined with the discussion above, Zhang and coworkers [74] concluded that the SC disperses everywhere in the sample, whereas HC mainly exists within the large spherulite region.

In summary, the IR results reveal that the formation of the SC occurred prior to that of the HC during the isothermal process. POM measurements indicated that

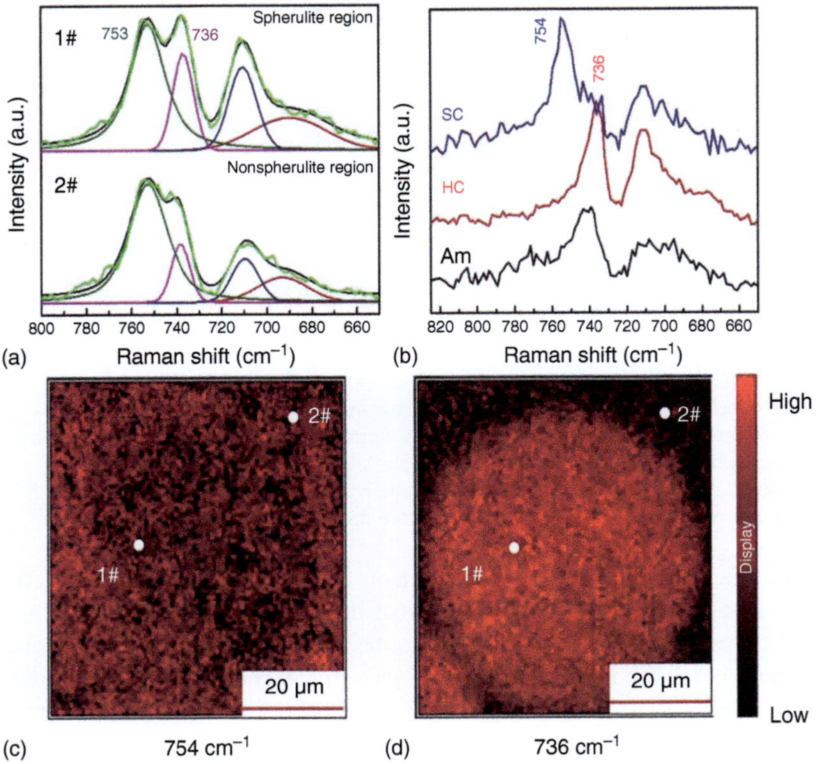

Figure 1.25 (a) Curve-fitting results in the 800 to 600 cm^{-1} region at single points (positions #1 and #2) and (b) Raman spectra of SC, HC, and Amp of PLA; (c, d) Raman images developed using the band intensities at (c) 754 and (d) 736 cm^{-1}. Source: Hu et al. [74].

large spherulites appeared after the formation of small crystallites. However, the crystallization of HC does not correspond to the already-formed SC crystals. This observation is unlike the normal case in which crystals exclude other components from their own domains. Based on the above discussion, Zhang and coworkers [74] proposed a possible crystallization process for the asymmetric PLA sample, as shown in Figure 1.26. In the first step, owing to the strong hydrogen-bonding interaction between PLLA and PDLA, the nucleation of the SC occurs much more easily than that of the HC phase. Considering chain entanglement, it is very likely that the already-formed SC acts as a crosslinking point, which may restrict the mobility of the PLLA chain. Next, the superfluous PLLA chains are excluded from the SC domains and start to crystallize on the surface of the SC domains. Finally, the SC and HC crystals grow together to form large spherulites.

1.3.5 3D SERS Imaging Using Chemically Synthesized Highly Symmetric Nanoporous Silver Microparticles

Recently, 3D SERS substrates have been developed by many research groups and have been shown to possess promising signal quality owing to their exponentially

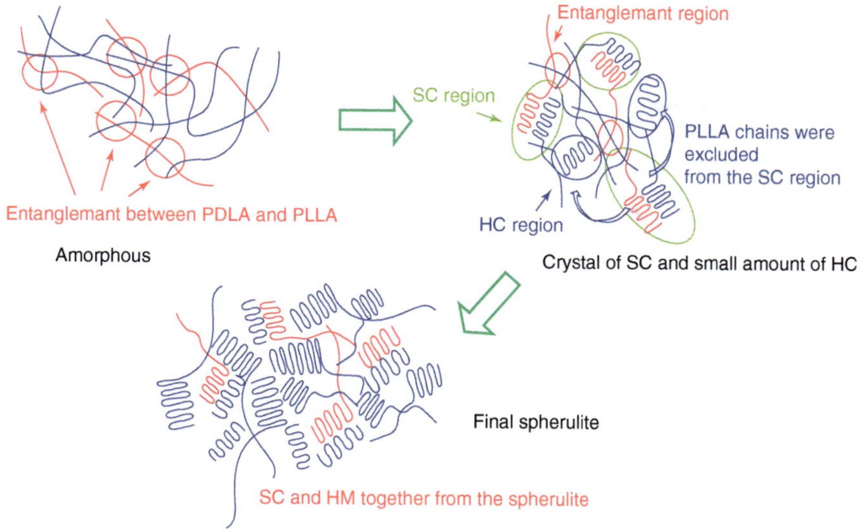

Figure 1.26 Schematic illustrations of isothermal crystallization process of PLLA/PDLA asymmetric blend. Source: Hu et al. [74].

large surface area and substantial number of hotspots from the additional third dimension. However, almost all 3D SERS substrate studies only present point-by-point measurements or 2D SERS imaging, which does not fully utilize the 3D nature of the substrates.

Vantasin et al. [75] developed 3D SERS imaging using highly symmetric 3D silver microparticles as a SERS substrate. The synthesized nanoporous silver microparticles possess a regular hexapodal shape and octahedral symmetry, although the synthetic method is purely chemical (vide infra) and does not involve lithography. The 3D SERS method was applied to detect the 3D inhomogeneity in a polymer blend, which relies on the predictable enhancement pattern of the substrate. 3D SERS imaging using the substrate also provides an improvement in spatial resolution along the *z*-axis, which is a challenge for Raman measurements in polymers, especially layered polymeric systems.

The nanoporous silver microstructures in this study were synthesized by the in-place galvanic reduction of an AgCl template. Figure 1.27a shows SEM images of nanoporous silver microstructures, showing the overall hexapod shape (scale bar: 5 mm). The nanopores, which act as hotspots, are shown in Figure 1.27b, having an average pore size of approximately 60 nm. With this pore size, small molecules such as SERS probe molecules can travel deep inside the cavities of the particles.

Figure 1.28a shows the SERS spectrum of *p*-aminothiophenol (PATP) from the center of the particle in Figure 1.28b. The band at 1074 cm^{-1} in the figure is assigned to the a_1 mode C–S stretching of PATP, which belongs to the C_{2v} point group. Figure 1.28c–e shows 3D SERS images developed using the peak area of the C–S stretching band. The enhancement pattern can be easily predicted because the top-view enhancement mapping in Figure 1.28c is very similar to the particle shape

(a)
(b)

Figure 1.27 SEM images of nanoporous silver microstructures showing (a) the overall hexapodal shape (scale bar: 5 mm) and (b) the nanopores (scale bar: 0.5 mm). Source: Vantasin et al. [75].

in Figure 1.28b. The *xy*-slices shown in Figure 1.28e display an obvious pattern, as the hotspots gradually move from the upper leg positions to the lower leg positions.

A possible application of this 3D substrate was demonstrated by embedding the microparticles into polymers. In this experiment, a 1 : 1 blend of poly(3-hydroxybutyrate) (PHB) and poly(D,L)lactic acid (PDLLA) was used. Owing to the presence of the same functional groups, the interaction between each polymer and silver microparticles should be very similar, and variations in the measured SERS spectra should originate from the polymer distribution.

Figure 1.29a,b shows optical image and a side-view scheme of a silver microparticle in a PHB/PDLLA polymer blend [75]. Figure 1.30c,d shows 3D SERS imaging of the particle based on ratio of the two bands at 870 and 840 cm^{-1} originating from the C–COO stretching modes of PDLLA and PHB, respectively, and 3D Raman imaging showed the same peak ratio, although for the polymer on the outside of the particle. Examples of 2D slices along the *z*-axis shown in Figure 1.30c,d are shown in Figure 1.30e,f, respectively. Only one in five slices (corresponding to 1.0 mm distance) is shown. For example, the upper legs of the silver hexapod in Figure 1.30a are located around 3 mm under the polymer surface, and the particle bottom is still several micrometers above the underlying glass slide (see Figure 1.30b).

The 3D image based on the peak area ratio shows a high value in some microscale areas, which could be interpreted as an inhomogeneity in the blend (in which the area consists of more PDLLA content). In fact, normal 3D Raman imaging can also be used to probe the inhomogeneity in polymers, but it is well known that conventional Raman spectroscopy in polymers suffers from a limitation in spatial resolution owing to the refraction of light at the polymer interface, especially in the *z*-axis,

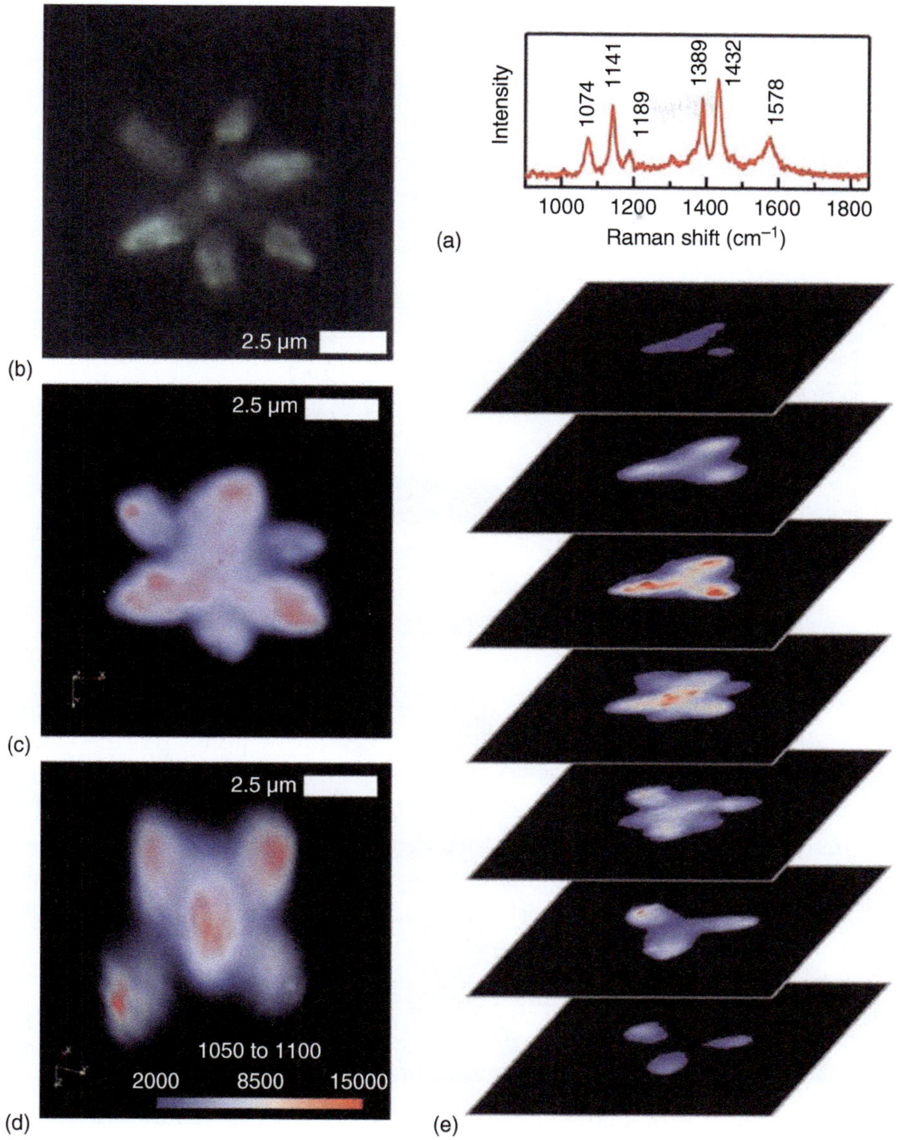

Figure 1.28 (a) SERS spectrum of PATP from the center of a hexapodal silver microstructure. (b) An optical microscopy image of the particle. (c, d) Top and diagonal views of 3D SERS images measured from the particle shown in (b) rendered using the peak area from the PATP a_1 mode at $1074\,cm^{-1}$. (e) Examples of 2D slices along the z-axis of (b). Only one in five slices (corresponding to a 1.5 mm distance) is shown. Note that (c), (d), and (e) use the same color scale. Source: Vantasin et al. [75].

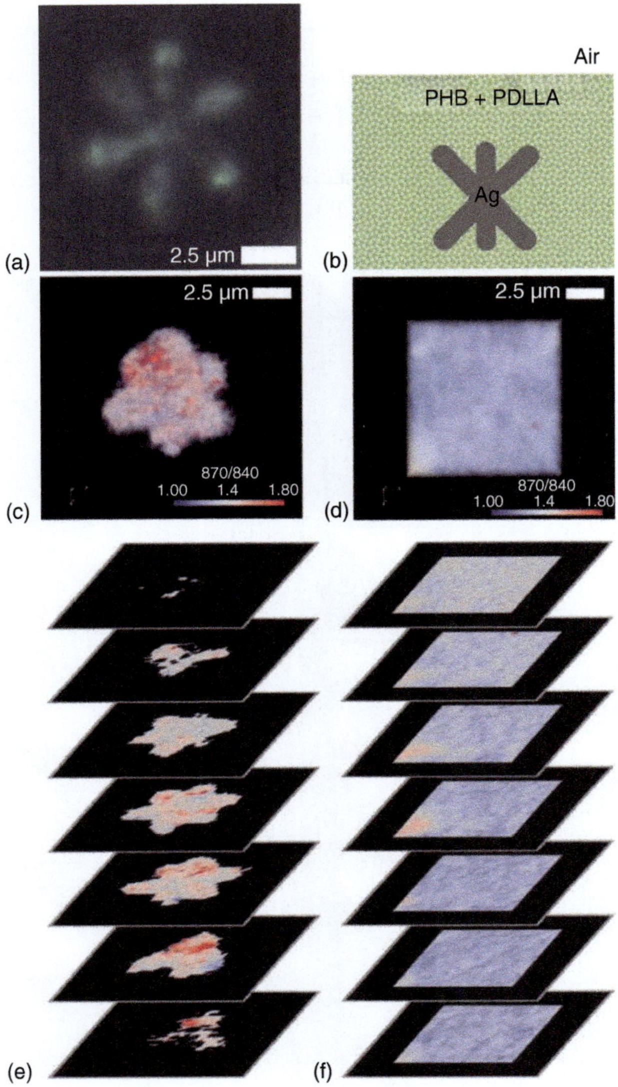

Figure 1.29 A silver microparticle in a PHB/PDLLA polymer blend. (a) An optical image. (b) A side-view scheme. (c) 3D SERS imaging on the particle showing the 870/840 cm^{-1} peak ratio. (d) 3D Raman imaging showing the same peak ratio but on the polymer outside of the particle. (e, f) Examples of 2D slices along the z-axis from (c) and (d), respectively. Only one in five slices (corresponding to 1.0 mm distance) is shown. Source: Vantasin et al. [75].

and this limit could be as low as 20 mm. As shown in the 3D Raman images of the nearby polymer without the silver structure (Figure 1.29d,f), the inhomogeneity is not resolved very well. This is due to the poor resolution in the z-axis, which causes the spectrum from every point to be similar to that of the bulk average. With the symmetric 3D SERS substrate, the probing volume for each measurement point is constrained to a smaller volume near the corresponding surface of the silver particle in the z-axis, thus improving the z-axis spatial resolution. Therefore, the microscale inhomogeneity can be resolved [75].

Double-layered polymeric systems were employed to demonstrate the potential of this method for improving the spatial resolution along the z-axis [75]. The hexapodal silver particles were embedded into the interface between polyvinylpyrrolidone (PVP) and polystyrene (PS; Figure 1.30a,b) [75] used the peak area ratio between the band arising from the PS ring-mode vibration at 1002 cm^{-1} and that arising from

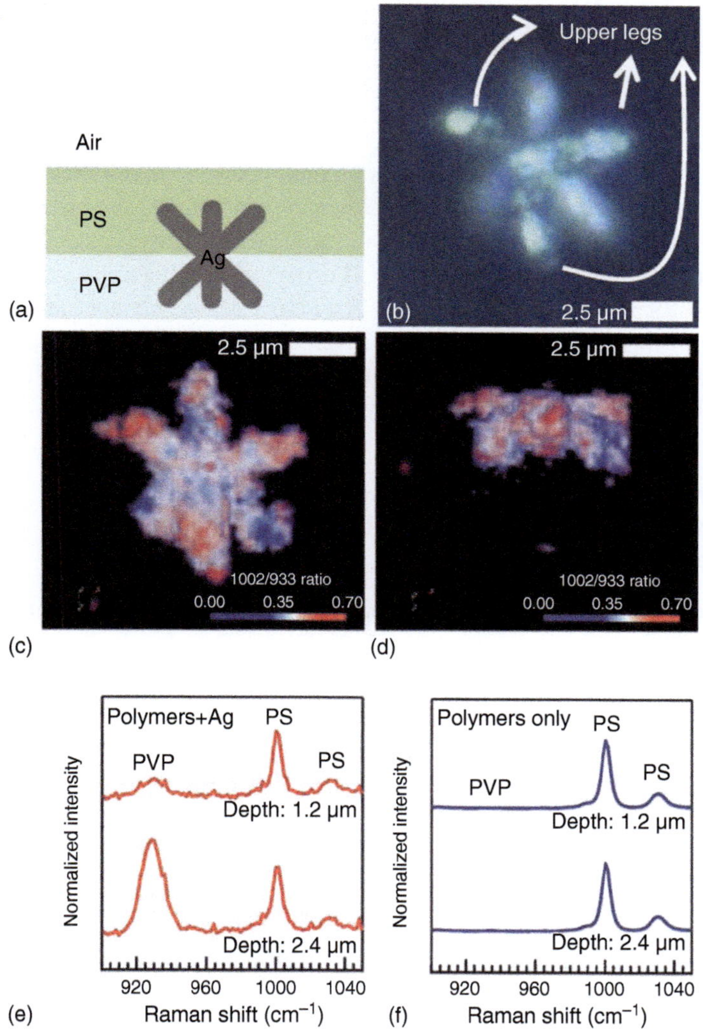

Figure 1.30 (a) Scheme showing a silver particle embedded in double-layered polymers. (b) An optical image of the silver particle in the polymers. (c, d) Top- and side-view 3D SERS images constructed from the peak area ratio between the 1002 cm^{-1} band of PS and the 933 cm^{-1} band of PVP. (e, f) Average spectra at depths 1.2 and 2.4 mm from 3D SERS imaging and normal 3D Raman imaging, respectively. Source: Vantasin et al. [75].

PVP C–C ring breathing at 933 cm^{-1} and clearly observed the partition of the polymers. The top- and side-view 3D SERS images in Figure 1.30c,d show high PS/PVP peak area ratios on the upper legs of the particle and low ratios on the lower legs. This cannot be replicated in normal 3D Raman imaging because the gradient of the peak area ratio is low. More quantitative information can be acquired by considering the averaged spectra at each depth. The particle appears flatter than normal in the side view (Figure 1.30d) because of the refraction (objects in a medium with a high refractive index have a smaller apparent height when viewed from above). Figure 1.30e,f clearly reveals that the contrast between the 1.2- and 2.4-mm depths in 3D SERS imaging is much better than that in normal 3D Raman imaging. However, this resolution improvement in the z-axis overcomes the limitation of refraction at polymer interfaces, and it does not exceed the diffraction limit of light, which is approximately 580 nm in the PS with this setup.

Thus, using 3D SERS imaging, Vantasin et al. [75] succeeded in characterizing highly symmetric nonporous silver microparticles for their SERS properties in three dimensions. The results demonstrate that the substrate can provide volumetric SERS information, and the enhancement pattern of the particles is very predictable because it correlates well with the particle shape. Further, a potential application of the embedded SERS probe was demonstrated: In a PHB/PDLLA polymer blend, 3D SERS imaging of this symmetric nonporous silver structure could resolve the microscale inhomogeneity of the blend. The improved spatial resolution along the z-axis was also illustrated in a double-layered polymeric system.

1.3.6 Tip-Enhanced Raman Scattering Spectroscopy Study of Local Interactions at the Interface of Styrene–Butadiene Rubber/Multiwalled Carbon Nanotube Nanocomposites

This is an example of a TERS study of polymer nanocomposites [76]. TERS is very useful for exploring the structure of nanomaterials [9, 33]. Local molecular interactions at the interface of styrene–butadiene rubber (SBR)/multiwalled carbon nanotube (MWCNTs) nanocomposites in films were explored using TERS spectroscopy [76] (see Figure 1.31 for the structure of SBR). The TERS study revealed that the TERS peaks arising from the phenyl group of SBR were strong when the MWCNT bands were strong, whereas the vinyl-group bands of SBR were strong when the MWCNT bands were weak. These observations indicate that the local distribution

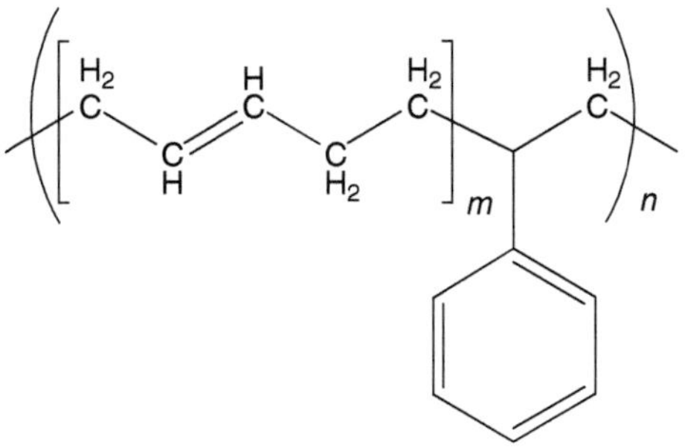

Figure 1.31 Chemical structure of SBR.

of polymer chains is modified by changes in the orientation of the phenyl rings by π–π interactions between the polymer chains and the MWCNTs.

Bokobza investigated SBR polymer nanocomposites containing MWCNTs and carbon black using Raman spectroscopy and discussed the dispersion of the fillers [77]. While this study is important, it could not provide new insights into local molecular interactions at the interface of SBR/MWCNT nanocomposites. Therefore, Suzuki et al. investigated the local molecular interactions at the interface using TERS. Figure 1.32a–c depicts the Raman spectra of pure SBR, pure MWCNTs, and a 1 wt% SBR/MWCNT nanocomposite, respectively [71]. Notably, the Raman spectrum of pure SBR is a normal Raman spectrum but that of pure MWCNTs is a resonance Raman spectrum and that of the 1 wt% SBR/MWCNT nanocomposite is a mixture of normal Raman and resonance Raman. One can easily assign bands at 2698, 1604, and 1354 cm^{-1} in the Raman spectrum of SBR/MWCNTs to G, G, and D bands, respectively (where G is the ordered carbon band and D is the disordered carbon band) of MWCNTs [76]. The Raman bands at 2990, 1668, and 1641 cm^{-1} in the same spectrum are due to the vinyl C–H stretching and *trans* and *cis* C = C stretching modes of SBR, respectively. The Raman bands at 3064 cm^{-1} and those at 1032 and 1004 cm^{-1} were derived from the aromatic C–H stretching and phenyl ring modes of SBR, respectively, and those at 2906 cm^{-1} arise from aliphatic C–H groups. Notably, the G band of SBR/MWCNT shifted to a higher wavenumber compared with that of the pure MWCNT spectrum. It is known that the shift of the

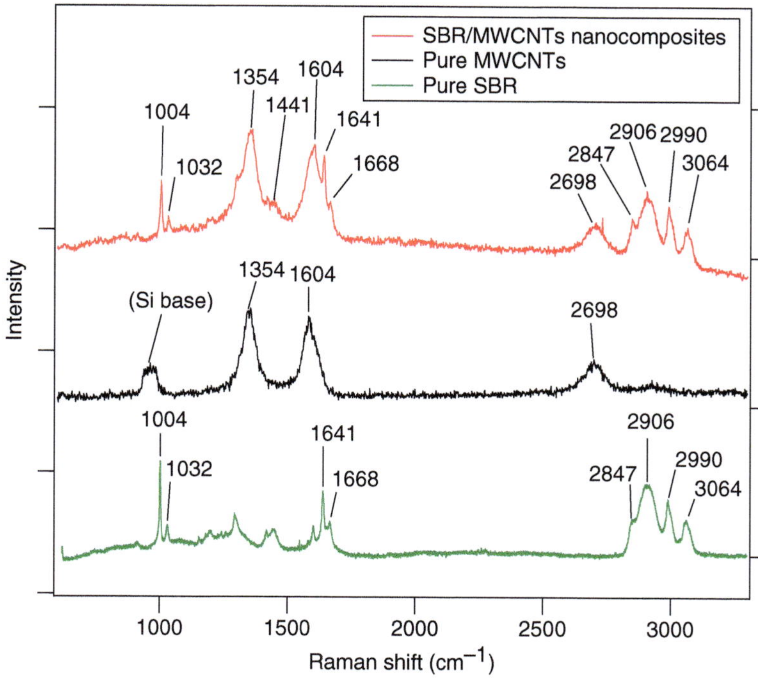

Figure 1.32 Raman spectra of (a) pure SBR, (b) pure MWCNTs, and (c) 1 parts per hundred rubber (phr) SBR/MWCNT nanocomposites. Source: Suzuki et al. [76].

G band in the Raman spectrum of carbon nanotubes (CNTs) is due to the shrinkage of the CNTs induced by the pressure from the polymer matrix [76].

The Raman and TERS spectra of 1 wt% SBR/MWCNT nanocomposite measured at the same eight points, positions (1), (2), (3), (4), (5), (6), (7), and (8), are compared in Figure 1.33a,b [76]. All Raman spectra in Figure 1.33a are similar. The TERS spectra collected at the eight points showed remarkable point-dependent changes. For example, in the spectrum obtained at position 7, the TERS signals assigned to SBR were identified at 2990, 2904, 2847, 1668, and 1641 cm^{-1}; however, the TERS signal at 1354 cm^{-1} arising from MWCNTs (D band) is considerably weaker than the corresponding peak in the Raman spectrum. Moreover, the G bands at 1604 and 2698 cm^{-1} were not observed in the spectrum collected at position 7. On the other

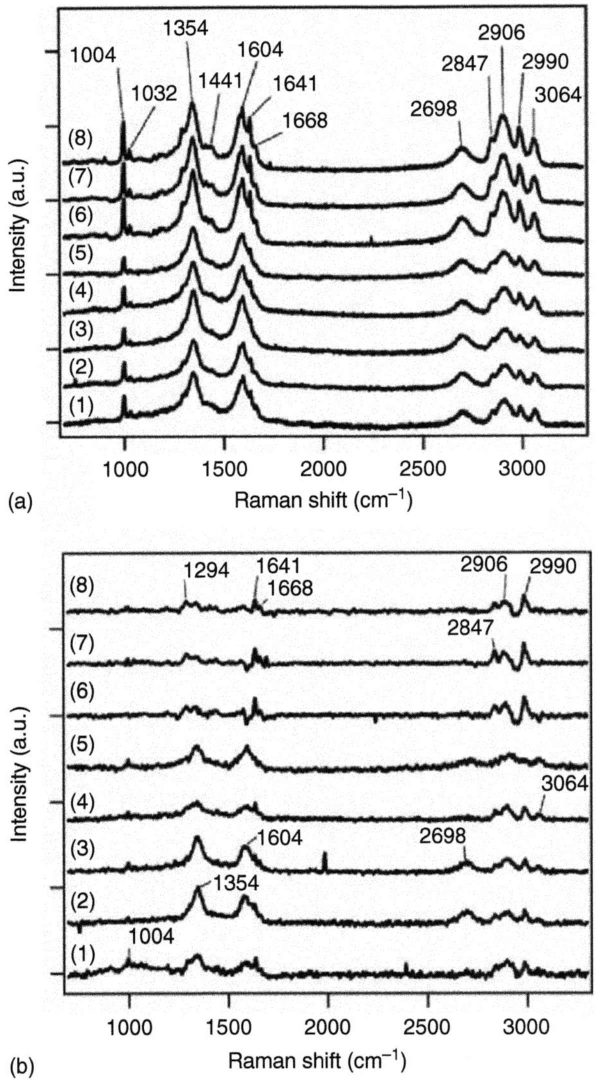

Figure 1.33 (a) Raman and (b) TERS spectra of 1 phr SBR/MWCNT nanocomposites measured at the eight points. Source: Suzuki et al. [76].

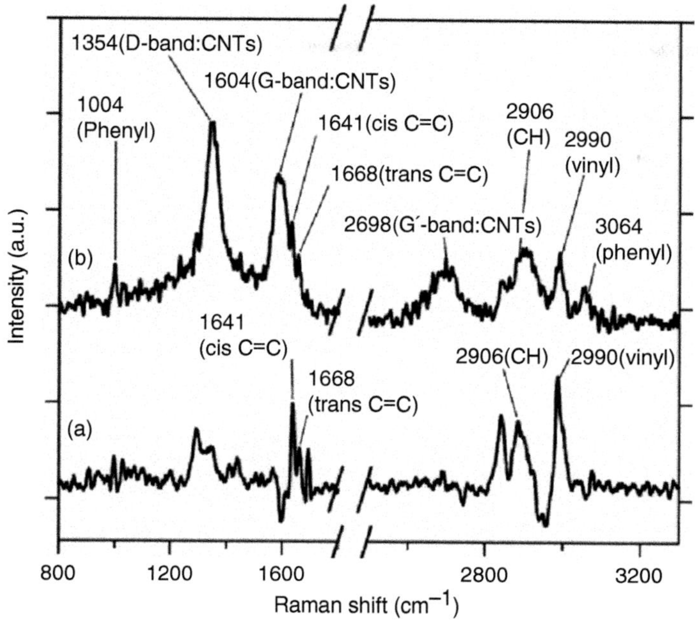

Figure 1.34 Enlarged TERS spectra of 1 phr SBR/MWCNTs from Figure 1.33: (a) spectrum 7 and (b) spectrum 3. Source: Suzuki et al. [76].

hand, in the TERS spectrum collected at position 3, signals originating from SBR were observed at 3064, 2990, 2904, and 2847 cm^{-1}, and those derived from MWCNTs were found at 2698, 1604, and 1354 cm^{-1}. The MWCNT content at 1 wt% was much lower than the polymer content in the sample. Hence, it is very likely that there are areas containing only SBR and those that include both SBR and MWCNTs. TERS measurements clearly detected their distribution.

Figure 1.34 shows the TERS spectra in the 1800 to 800 and 3200 to 2500 cm^{-1} regions measured at positions 7 and 3, respectively, in Figure 1.33b [76]. In the 3200 to 2500 cm^{-1} region, one can find four SBR peaks at 3064, 2990, 2906, and 2847 cm^{-1}. The former two peaks are assigned to CH (phenyl) and CH (vinyl) stretching modes, and the latter two peaks are antisymmetric and symmetric CH_2 stretching modes of the SBR chain, respectively. It is noted that in the TERS spectrum in Figure 1.34a, the signal at 3064 cm^{-1} is nearly absent, and the relative intensity of the signal at 2990 cm^{-1} became stronger than that of the corresponding signal in the spectrum in Figure 1.34a [76]. In Figure 1.34b, the TERS signals arising from MWCNTs appear strongly at 2698, 1604, and 1354 cm^{-1}, and a peak at 3064 cm^{-1} arising from aromatic C–H is clearly observed.

Figure 1.35 shows a plot of the intensities of the 3064 and 2990 cm^{-1} bands vs. that of the 1604 cm^{-1} band [76]. Figure 1.35a,b is related to the TERS data, whereas the data in Figure 1.35c,d are derived from the Raman spectra. As shown in Figure 1.35a,b, in the TERS spectrum (Figure 1.33b), the 2990 cm^{-1} signal (vinyl) decreases as the 1604 cm^{-1} signal (G band) increases, whereas the signal intensity at 3064 cm^{-1} (phenyl) increases as the G-band signal increase. In the Raman spectra

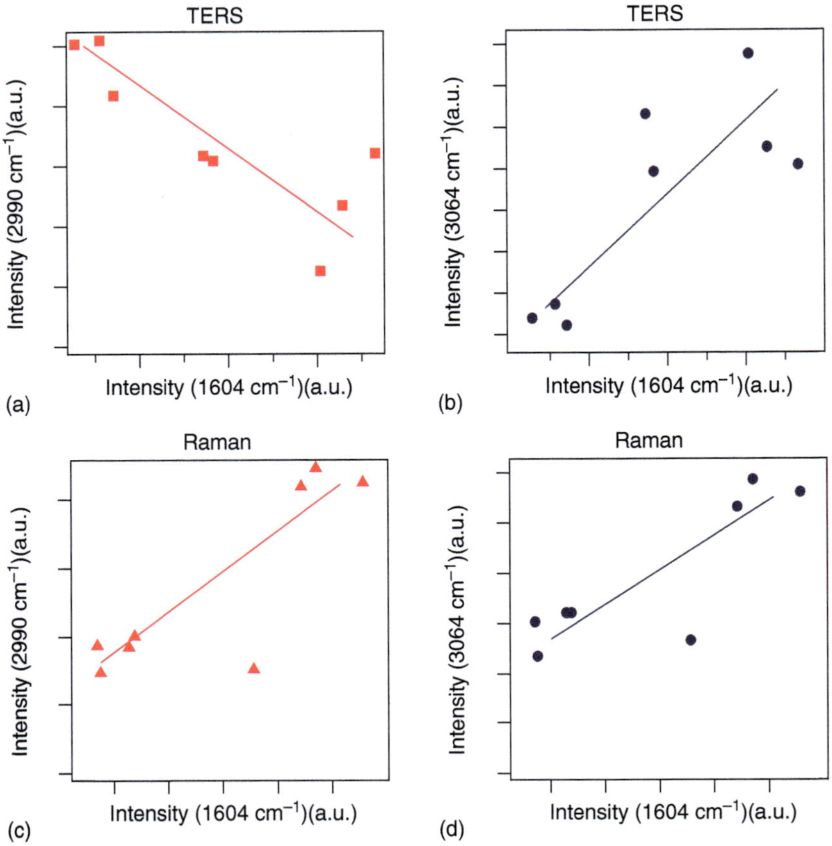

Figure 1.35 Peak intensity plots of (a) the vinyl band (2990 cm^{-1}) vs. the G band (1604 cm^{-1}) in the TERS spectrum, (b) the phenyl band (3064 cm^{-1}) vs. the G band (1604 cm^{-1}) in the TERS spectrum, (c) the vinyl band (2990 cm^{-1}) vs. the G band (1604 cm^{-1}) in the normal Raman spectrum, and (d) the phenyl band (3064 cm^{-1}) vs. the G band (1604 cm^{-1}) in the normal Raman spectrum. Source: Suzuki et al. [76].

(Figure 1.33a), both signals at 2990 (vinyl) and 3064 (phenyl) cm^{-1} increase with increase in the intensity of the 1604 cm^{-1} signal (Figure 1.35c,d). These results suggest that the local distribution of polymer nanocomposites at the interface between the polymer and the filler is different from that of polymer areas without CNTs. The depth resolution of TERS is much smaller than its spatial resolution (approximately 100 nm). It is known that the intensity of the TERS signal decays exponentially from the surface, having a decay length of 2.8 nm. Hence, TERS experiments measure a region several nanometers in depth from the surface of the nanocomposites.

Figure 1.36 illustrates a schematic diagram of the expected surface structure of SBR/MWCNT nanocomposites as proposed by Suzuki et al. [76], specifically, (a) far from the MWCNT area and (b) near the CNT area. In the polymer areas where MWCNTs are not located (a), it is very likely that the C = C groups and phenyl rings

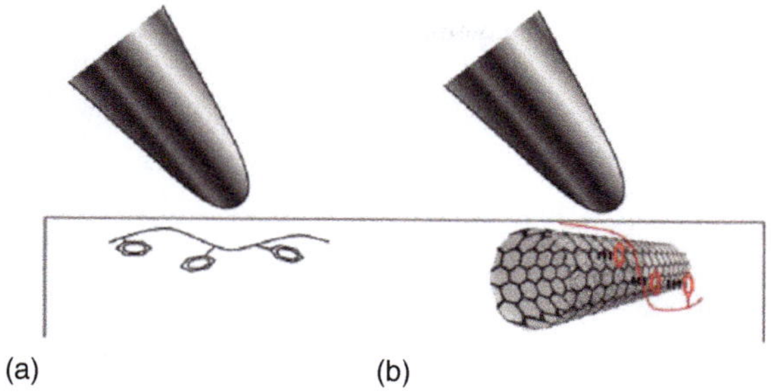

Figure 1.36 Schematic diagram of the surface structure of SBR/MWCNT nanocomposites: (a) far from and (b) near the CNT area. At the interface between the polymer and the filler, the phenyl ring should be oriented perpendicular to the surface through π–π interactions between the CNT and the phenyl ring. Source: Suzuki et al. [76].

are parallel to the polymer surface. In this orientation, the C = C signal is much more enhanced than that of the phenyl ring. On the other hand, at the interface of the polymer and the filler (Figure 1.36b), it is very likely that MWCNTs interact with the phenyl rings of the polymer through π–π interactions. These interactions induce differences in the orientation of the polymer chains and their side residues. The orientation of the phenyl rings at the interface may be nearly perpendicular to the surface because of the π–π interactions between the CNTs and the phenyl rings shown in Figure 1.36b.

1.4 Perspectives for Polymer Spectroscopy

In this chapter, we have reported a brief history of polymer spectroscopy, its roles and present status, and some examples of applications of polymer spectroscopy. In the last section, we discuss the perspectives for polymer spectroscopy.

Surface IR spectroscopy should be developed further for studies on polymer surfaces, interfaces, and thin films [20]. Even today, surface IR spectroscopy is used extensively for the study of polymer surface structure, interaction with substrates, and molecular orientation. ATR, external reflection, and RA spectroscopy are representative of the surface IR spectroscopy of polymers. Spectral analysis methods for surface IR spectroscopy have also been developed. For example, Hasegawa [78] proposed the multi-angle incidence resolution spectrum (MEIRS), which has recently been used extensively for the analysis of surface IR spectra. Surface IR spectroscopy will be further developed in combination with imaging techniques and X-ray techniques such as grazing incident X-ray diffraction (GIXD).

FUV spectroscopy was first applied to polymer studies several years ago (Section 1.2 and Chapter 6). Because FUV spectroscopy is an electronic spectroscopy technique, it provides different information about polymer structure and functions from vibrational spectroscopy. Notably, polymers usually do not absorb in the DUV and UV regions and absorb only in the FUV region. Therefore, FUV spectroscopy is unique for investigating the electronic structure and electronic transitions of polymers. Moreover, it is useful, for example, for studies of polymer nanocomposites, polymers absorb in the FUV region, whereas nanomaterials such as graphene and carbon nanotubes absorb mostly in the DUV and UV regions; thus, one can investigate the electronic structure of a polymer and a carbon nanomaterial independently. Thus, FUV spectroscopy is very promising for the electronic spectroscopy of polymers (Chapter 6).

Currently, imaging techniques are very popular in various fields (Chapters 2 and 5). In polymer science, imaging techniques are used in combination with IR, Raman, NIR, and terahertz spectroscopy. Each imaging spectroscopy technique has different characteristics and advantages for polymer research. However, almost all imaging studies involve 2D imaging. Although we have introduced one example of the 3D imaging of polymers in this chapter, 3D imaging studies are rather rare. As a next-generation technique for polymer spectroscopy, 3D imaging is very important.

Nanospectroscopy, such as TERS and AFM-IR (nano-IR), is also a developing technique. AFM-IR consists of an IR spectrometer and SPM [79]. Originally, the term AFM-IR was used to describe a method that combines a tunable free-electron laser with an AFM equipped with a sharp probe that enables the measurement of the IR absorption of a sample in a nanoscale area. The spatial resolution of AFM-IR reaches 10 nm, which is almost as high as that of TERS. AFM-IR has recently been used for studies of polymers, polymer blends, and polymer nanocomposites, particularly multilayer films and thin films.

Theoretical calculations of electronic and vibrational spectra are particularly important for the development of polymer spectroscopy, and computational molecular spectroscopy is being developed. Of recent note is the development of anharmonic calculations. Because of this, one can reproduce Raman and IR spectra more precisely, and, moreover, one can reproduce NIR spectra that consist of bands arising from overtones and combinations. Moreover, computer power is getting more powerful while computational resources (particularly time and cost) are becoming increasingly smaller, and thus, it is expected that quantum chemical calculations of polymer spectra will become more popular.

Further, it is likely that NIR spectroscopy for polymer studies will advance in two directions. One is industrial applications, and NIR spectroscopy has already been used in many applications in the polymer industry (Chapter 5); for example, the online analysis of polymers, polymer coating, process monitoring, and process control of synthetic textiles. Online analysis and process control will develop further with the introduction of new software technologies, such as artificial intelligence (AI) and machine learning. NIR spectroscopy should also be developed for polymer structural analysis. In this regard, the development of anharmonic calculations will play an important role, which will make the analysis of overtones and combinations easier; thus, more concrete information about the band assignments in the NIR region will be obtained, enabling deeper studies of polymer structures, interactions, and reactions.

References

1 Nyquist, R.A. (1961). *Infrared Spectra of Plastics and Resins*, 2e. Dow Chemical Co.
2 Zbinden, R. (1964). *Infrared Spectroscopy of High Polymers*. Academic Press.
3 Henniker, C.J. (1967). *Infrared Spectrometry of Industrial Polymers*. Academic Press.
4 Siesler, H.W. and Holland-Moritz, K. (1980). *Infrared and Raman Spectroscopy of Polymers*. Dekker.
5 Tadokoro, H. (1990). *Structure of Crystalline Polymers*. Krieger Pub. Co.
6 Coleman, M.M., Graf, J.F., and Painter, P.C. (1991). *Specific Interactions and the Miscibility of Polymer Blends*. Technical Publishing Co.
7 Koenig, J.L. (1999). *Spectroscopy of Polymers*, 2e. Elsevier.
8 Brame, E.G. (2012). *Applications of Polymer Spectroscopy*. Academic Press.
9 Thomas, S. and Rouxel, D. (2016). *Spectroscopy of Polymer Nanocomposites*. Elsevier.
10 Ozaki, Y. and Kawata, S. (eds.) (2015). *Far- and Deep-Ultraviolet Spectroscopy*. Springer.
11 Morisawa, Y., Tanabe, I., and Ozaki, Y. (2018). Advances in far-ultraviolet spectroscopy in the solid and liquid states. In: *Frontiers and Advances in Molecular Spectroscopy* (ed. J. Laane), 251–285. Elsevier.
12 Ozaki, Y., Morisawa, Y., Tanabe, I., and Beć, K.B. (2021). *Spectrochim. Acta A* 253: 119549.
13 (a) Morisawa, Y., Yasunaga, M., Sato, H. et al. (2014). *J. Phys. Chem. B* 118: 11855. (b) Beć, K.B., Morisawa, Y., Kobashi, K. et al. (2018). *Phys. Chem. Chem. Phys.* 20: 8859.
14 Ueno, N., Wakabayashi, T., Sato, H., and Morisawa, Y. (2019). *J. Phys. Chem. A* 123: 10746.
15 Ozaki, Y., Huck, C.W., Tsuchikawa, S., and Elgensen, S.B. (eds.) (2020). *Near-Infrared Spectroscopy*. Springer.
16 Ozaki, Y., McClure, W.F., and Christy, A.A. (eds.) (2007). *Near-Infrared Spectroscopy in Food Science and Technology*. Wiley.
17 Ozaki, Y. (2012). *Anal. Sci.* 28: 545.
18 Griffiths, P.R. and de Haseth, J.A. (1986), 2e. (2007)). *Fourier Transform Infrared Spectroscopy*, 1e. Wiley Interscience.
19 (a) Christy, A.A., Ozaki, Y., and Gregoriou, V.G. (2001). *Modern Fourier-Transform Infrared Spectroscopy*. Elsevier. (b) Gunzler, H. and Gremlich, H.-U. (2002). *IR Spectroscopy*. Wiley-VCH.
20 Hasegawa, T. (2017). *Quantitative Infrared Spectroscopy for Understanding of a Condensed Matter*. Springer.
21 Finch, A., Gates, P.N., Radcliffe, K. et al. (1970). *Chemical Applications of Far-Infrared Spectroscopy*. Academic Press.

22 Moller, K.D. and Rothschild, W.G. (1971). *Far-Infrared Spectroscopy*. Wiley Interscience.

23 Peiponen, K.-E., Zeitler, J.A., and Kuwata-Gonokami, M. (eds.) (2013). *Terahertz Spectroscopy and Imaging*. Springer.

24 Coutaz, J.-L., Garet, F., and Wallace, V. (2018). *Principles of Terahertz Time-Domain Spectroscopy*. Pan Stanford.

25 Baxter, J.B. and Guglietta, G.W. (2011). *Anal. Chem.* 83: 4342.

26 Ozaki, Y. (2019). *Bull. Chem. Soc. Jpn.* 92: 629.

27 Grasselli, J.G. and Bulkin, B.J. (1991). *Analytical Raman Spectroscopy*. Wiley-Interscience.

28 Long, D.A. (2002). *The Raman Effect*. Wiley.

29 McCreery, R.L. (2000). *Raman Spectroscopy for Chemical Analysis*. Wiley Interscience.

30 Smith, E. and Dent, G. (2019). *Modern Raman Spectroscopy*. Wiley.

31 Schlücker, S. (ed.) (2010). *Surface-Enhanced Raman Spectroscopy: Analytical, Biophysical and Life Science Applications*. Wiley-VCH.

32 Ozaki, Y., Kneipp, K., and Aroca, R. (eds.) (2014). *Frontiers of Surface-Enhanced Raman Scattering*. Wiley.

33 Kneipp, K., Ozaki, Y., and Tian, Z.-Q. (eds.) (2018). *Recent Developments in Plasmon-Supported Raman Spectroscopy*. World Scientific.

34 Zoubir, A. (ed.) (2012). *Raman Imaging, Techniques and Applications*. Springer.

35 Kirkwood, J.G. (1939). *J. Chem. Phys.* 7: 506.

36 Whitcomb, S.E., Nielsen, H.H., and Thomas, L.H. (1940). *J. Chem. Phys.* 8: 143.

37 (a) Shimanouchi, T. and Mizushima, S. (1949). *J. Chem. Phys.* 17: 1102. (b) Shimanouchi, T. (1949). *J. Chem. Phys.* 17: 245, 734, and 848.

38 Shimanouchi, T. (1955). *J. Chem. Phys.* 23: 2465. and 25, 660 (1956).

39 (a) Krimm, S. (1954). *J. Chem. Phys.* 22: 567. (b) Krimm, S., Liang, C.Y., and Sutherland, G.B.B.M. (1956). *J. Chem. Phys.* 25: 549.

40 Eliot, A. (1959). The infrared spectra of polymers. In: *Advances in Spectroscopy*, 1e (ed. H.W. Thomson), 214–287. Interscience.

41 Hummel, D.O. (1966). *Infrared Spectra of Polymers in the Medium and Long Wave Regions*. Wiley Interscience.

42 Schachtschneider, J.H. and Snyder, R.G. (1963). *Spectrochim. Acta* 19: 117.

43 Tasumi, M. and Shimanouchi, T. (1965). *J. Chem. Phys.* 43: 1245.

44 (a) Sugeta, H., Miyazawa, T., and Kajiura, T. (1969). *J. Polym. Sci. B Polym. Lett.* 7: 251. (b) Matsuura, H. and Miyazawa, T. (1969). *J. Polym. Sci. A* 2: 1735.

45 Kitagawa, T. and Miyazawa, T. (1972). *Adv. Ploym. Sci.* 9: 336.

46 Kakida, H., Makino, D., Chatani, Y. et al. (1970). *Macromolecules* 3: 569.

47 Tadokoro, H., Takahashi, Y., Chatani, Y., and Kakida, H. (1967). *Makromol. Chem.* 106: 96.

48 Krimm, S. and Bank, M.I. (1965). *J. Chem. Phys.* 42: 4059.

49 Mizushima, S. and Shimanouchi, T. (1949). *J. Am. Chem. Soc.* 71: 1320.

50 Hendra, P.J. (1969). *Adv. Polym. Sci.* 6: 151.

51 (a) Koenig, J.L. and Druesdow, D.J. (1969). *J. Polym. Sci.* A2: 1075. (b) Koenig, J.L. and Angood, A.C. (1970). *J. Polym. Sci.* A8: 1787.
52 Schaufele, R.F. and Shimanouchi, T. (1967). *J. Chem. Phys.* 47: 3605.
53 Stroble, G.R. and Eckel, R. (1976). *J. Polym Sci. Polym Phys. Ed.* 14: 913.
54 Hsu, S.L. and Krimm, S. (1976). *J. Appl. Phys.* 47: 4265. and 48, 4013 (1977).
55 Miller, R.G.J. and Willis, H.A. (1956). *J. Appl. Chem.* 6: 385.
56 (a) Takeuchi, T., Tsuge, S., and Sugimura, Y. (1968). *J. Polym. Sci. A* 1: 3415. (b) Tosi, C. (1968). *Makromol. Chem.* 112: 303.
57 (a) Hoshina, H., Morisawa, Y., Sato, H. et al. (2010). *Appl. Phys. Lett.*: 96. http://www.ncbi.nlm.nih.gov/pubmed/011907. (b) Yamamoto, S., Morisawa, Y., Sato, H. et al. (2013). *J. Phys. Chem. B* 117: 2180. (c) Yamamoto, S., Miyada, M., Sato, H. et al. (2017). *J. Phys. Chem. B* 121: 1128. (d) Yamamoto, S., Onishi, E., Sato, H. et al. (2019). *J. Phys. Chem. B* 123: 536. (e) Funaki, C., Toyouchi, T., Hoshina, H. et al. (2017). *Appl. Spectrosc.* 71: 1537.
58 Bakeev, K.A. (2010). *Process Analytical Technology*, 2e. Wiley.
59 Kessler, R.W. and Kessler, W. (2020). Inline and online process analytical technology with an outlook for the petrochemical industry. In: *Near-Infrared Spectroscopy* (eds. Y. Ozaki, C.W. Huck, S. Tsuchikawa and S.B. Engelsen), 553. Springer.
60 Czarnecki, M.A., Beć, K.B., Grabska, J. et al. (2020). Overview of application of NIR spectroscopy to physical chemistry. In: *Near-Infrared Spectroscopy* (eds. Y. Ozaki, C.W. Huck, S. Tsuchikawa and S.B. Engelsen), 297. Springer.
61 (a) Beć, K.B., Grabska, J., and Hofer, T.S. (2020). Introduction to quantum vibrational spectroscopy. In: *Near-Infrared Spectroscopy* (eds. Y. Ozaki, C.W. Huck, S. Tsuchikawa and S.B. Engelsen), 83. Springer. (b) Beć, K.B., Grabska, J., Huck, C.W., and Ozaki, Y. (2019). Quantum mechanical simulation of near-infrared spectra: application in physical chemistry and analytical chemistry. In: *Molecular Spectroscopy: A Quantum Chemistry Approach* (eds. Y. Ozaki, M.J. Wojcik and J. Popp), 327. Wiley.
62 Dong, J., Nakashima, K., and Ozaki, Y. (1997). *Macromolecules* 30: 1111.
63 Walczak, W.J., Hoagland, D.A., and Hsu, S.L. (1992). *Macromolecules* 25: 7317.
64 Bardet, L., Cassanas-Fabre, G., and Alain, M. (1975). *J. Mol. Struct.* 24: 153.
65 Ng, J.B. and Shurvell, H.F. (1987). *J. Phys. Chem.* 91: 496.
66 Tanaka, N., Kitano, H., and Ise, N. (1990). *J. Phys. Chem.* 94: 6290.
67 Coleman, M.M., Graf, J.F., and Painter, P.C. (1991). *Specific Interactions and the Miscibility of Polymer Blends*, 328. Technomic Publishing Inc.
68 Song, Y.P., Yarwood, J., Tsibouklis, J. et al. (1992). *Langmuir* 8: 262.
69 Sun, L., Kepley, L.J., and Crooks, R.M. (1992). *Langmuir* 8: 2101.
70 Tadokoro, H., Kobayashi, M., Yoshidome, H. et al. (1968). *J. Chem. Phys.* 49: 3359.
71 Malta, V., Cojazzi, G., Fichera, A. et al. (1979). *Eur. Polym. J.* 15: 765.
72 Frank, W.F.X. and Fiedler, H. (1979). *Infrared Phys.* 19: 481.
73 Shimoyama, M., Ninomiya, T., Sano, K. et al. (1998). *J. NIR Spectrosc.* 6: 317.

74 Hu, J., Wang, J., Wang, M. et al. (2019). *Polymer* 172: 1.
75 Vantasin, S., Ji, W., Tanaka, Y. et al. (2016). *Angew. Chem. Int. Ed.* 55: 8391.
76 Suzuki, T., Yan, X., Kitahama, Y. et al. (2013). *J. Phys. Chem. C* 117: 1436.
77 (a) Bokobza, L. (2011). *Macromol. Symp.* 305: 1. (b) Bokobza, L. (2007). *Polymer* 8: 4907.
78 Hasegawa, H. (2007). *J. Phys. Chem. B* 106: 4385. (2002) Anal. Chem. 79, 4112.
79 (a) Hammiche, A., Pollock, H.M., Reading, M. et al. (1999). *Appl. Spectrosc.* 53: 810. (b) Anderson, M.S. (2000). *Appl. Spectrosc.* 54: 349.

2

FTIR Spectroscopy and Spectroscopic Imaging for the Analysis of Polymers and Multicomponent Polymer Systems

Huiqiang Lu, Andrew V. Ewing, and Sergei G. Kazarian

Imperial College London, South Kensington Campus, Department of Chemical Engineering, London SW7 2AZ, UK

2.1 Investigation of Polymers Using FTIR Spectroscopy and Spectroscopic Imaging

Through assessing the molecular vibrations of all functional groups in molecules simultaneously, vibrational spectroscopy can produce a molecular-level insight into the microstructures of polymers and detect the underlying physical basis for polymer processing [1–3]. Fourier transform infrared (FTIR) spectroscopy can be used to monitor the constitution, configuration, conformation, and inter- and intra-molecular interactions of polymers. It is an ideal technique for dynamic studies of polymers because (i) the precise temperature control is feasible since the amount of required sample is small; and (ii) different types of structural and conformational changes of polymers at the molecular level can be detected simultaneously in situ. The obtained information about the molecular-level behavior of polymers through FTIR spectroscopy can improve the macroscopic properties of polymeric materials [4–7].

Attenuated total reflection (ATR)–FTIR spectroscopy depends on the attenuation of internally reflected infrared light at the interface between ATR crystal and the sample. The evanescent wave can measure a thin layer of sample contacted with the ATR crystal. The typical path lengths range from 0.1 micrometre to several micrometres [8, 9]. The advantage of this approach is that little or no sample preparation is needed. However, it only probes the composition of the surface layer instead of the bulk area, which could be problematic for nonhomogeneous samples. The obtained results can be analyzed quantitatively according to the Beer–Lambert law: the absorbance of a certain band is proportional to the concentration of corresponding component [10, 11].

Through combining spectral and spatial information, FTIR spectroscopic imaging can realize the spatial chemical visualization of samples. Compared with the conventional FTIR spectroscopy, which can only provide an average spectrum from the measured area, FTIR spectroscopic imaging can produce a whole image in a single snapshot. Like the common photographic camera, the infrared focal plane array

Spectroscopic Techniques for Polymer Characterization: Methods, Instrumentation, Applications, First Edition. Edited by Yukihiro Ozaki and Harumi Sato.

(FPA) detector can detect thousands of spatially resolved spectra across the illuminated sample area in a single measurement simultaneously. Then the obtained FTIR images can reveal the distribution of specific chemical species in the imaged area [12, 13]. Since the measurement time using an FPA detector is relatively short, it is possible to study dynamic processes in "real time," for example, interdiffusion and phase separation [14, 15].

2.1.1 Investigation of Miscibility in Polymer Blends

Miscibility of polymer blends has a great influence on the properties of the final products [16–19]. A phase diagram comprises a binodal line and a spinodal line as a graphical representation of the miscibility of polymers in a blend under different temperatures and volume fractions. The binodal line indicates whether these two components are miscible under certain conditions (i.e. temperature and volume fraction), and the spinodal line defines the boundary between unstable and metastable mixtures [20–23]. The critical point on these lines reveals the critical temperature and volume fraction where two phases become homogeneous. When the volume fraction moves from zero toward the critical point, the critical temperature will either increase (upper critical solution temperature [UCST] polymer blend) or decrease (lower critical solution temperature [LCST] polymer blend). When the volume fraction moves from the critical point toward 1, the critical temperature shows an opposite trend [20, 24–26]. Briber and Khoury have established the phase diagram for the α-poly(vinylidene fluoride) (α-PVF_2)/poly(ethyl acrylate) (PEA) blend system using cloud-point measurements and differential scanning calorimetry (DSC). It is found that phase separation appears at the temperature above the cloud point indicative of LCST behavior [27].

Koyama and Doi have investigated the effect of different molecular weight poly(lactic acid) (PLA) on the miscibility of poly(3-hydroxybutyrate) (PHB)/PLA blend based on the Flory–Huggins equation:

$$\frac{\Delta G}{RTV} = \frac{\emptyset_1}{N_1}\ln \emptyset_1 + \frac{\emptyset_2}{N_2}\ln \emptyset_2 + \chi_{12}\emptyset_1\emptyset_2 \tag{2.1}$$

where ΔG is the change in free energy by mixing these two polymers, R is the gas constant: 8.314 J (mol K)$^{-1}$, V is the volume of polymer blend, $\emptyset_1$ and $\emptyset_2$ are the volume fractions of these two polymers, respectively. N_1 and N_2 are the molar volumes of these two polymers, respectively. χ_{12} is the Flory–Huggins interaction parameter. In their study, the spinodal curves for the PHB/PLA blends with different molecular weights were obtained and are shown in Figure 2.1 [18, 28].

To improve the physical properties of biopolymer blends, phase separation of PHB/PLA blends and PHB/poly(caprolactone) (PCL) blends was investigated using transmission FTIR spectroscopic imaging. Biopolymers, which are biologically degradable, are of increasing demand as alternatives for fossil raw materials [29]. PHB is stiff but brittle due to its high crystallinity. To overcome its disadvantage, other biopolymers, such as PLA and PCL, are mixed with it to prepare the biopolymer blend [10, 30–33].

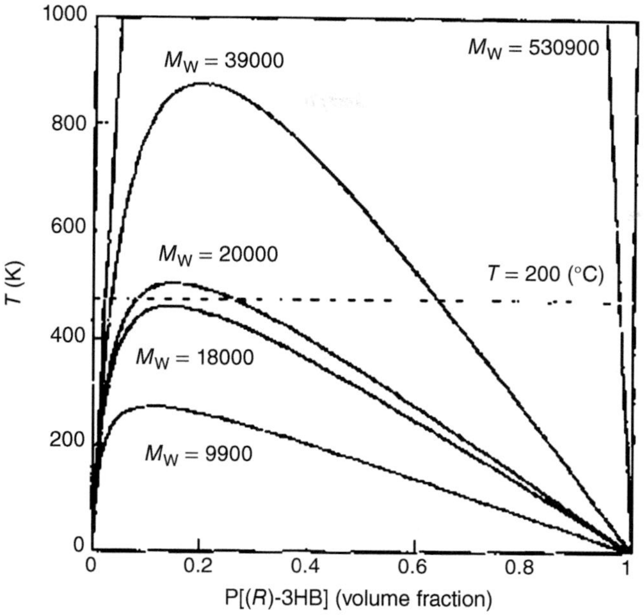

Figure 2.1 Spinodal curves of the PHB/PLA blends with different M_w calculated by the Flory–Huggins equation. Source: Koyama and Doi [18].

As shown in Figure 2.2A, the complementarity of PHB-specific FTIR images and PLA-specific FTIR images revealed the appearance of phase separation in the PHB/PLA (50 wt/50 wt) blend. While the homogeneous distribution of the PHB/PLA (70 wt/30 wt) blend in the FTIR images (see Figure 2.2B) indicated that this blend is a compatible one-phase system. In addition, the glass transition temperature (T_g), measured by DSC, and the elongation-to-break values measured using a miniaturized stretching machine gave more evidence of phase separation. The methods mentioned above were also used to detect the phase separation in PHB/PCL blends, which demonstrated that FTIR spectroscopic imaging is suitable for the analysis of different types of polymer blends. Finally, through detecting the phase separation in polymer blends as a function of blend composition, the miscibility gaps of these two polymer blends were obtained. The miscibility gap of PHB/PLA blends was determined to be between approximately 45 and 65 (wt/wt) and that of PHB/PCL blends is from 55 to 45 (wt/wt) [10].

2.1.2 Investigation of Intermolecular Interactions

FTIR spectroscopy can be used to monitor the samples at the molecular level. Therefore, it is possible to detect intermolecular interactions. For instance, when the carbonyl groups form hydrogen bonds with other functional groups, this leads to a reduction in the wavenumber of ν(C = O) band. When hydrogen bonding forms in compounds with O–H groups, the ν(O–H) band at 3600 cm^{-1} becomes broader. Its intensity increases while its wavenumber reduces [2, 34, 35]. H-bonding enhances

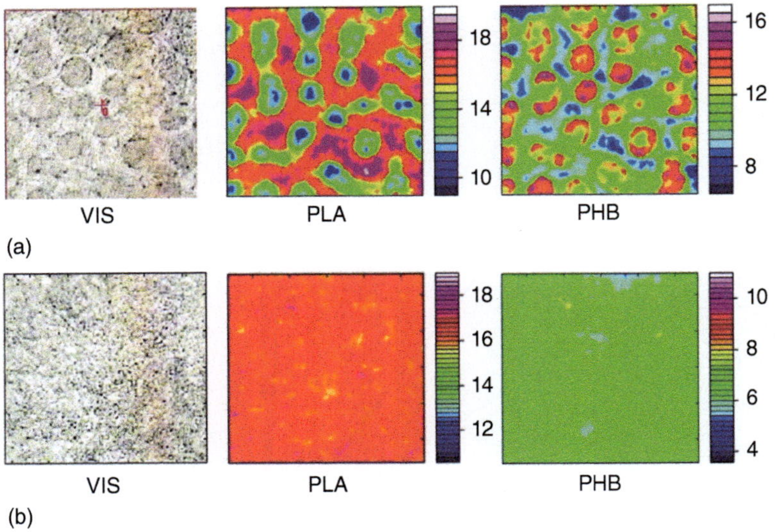

Figure 2.2 (a) Visual image, PLA-specific FTIR image, and PHB-specific FTIR image of the PHB/PLA (50 wt/50 wt) blend. (b) Visual image, PLA-specific FTIR image, and PHB-specific FTIR image of the PHB/PLA (30 wt/70 wt) blend. Source: Vogel et al. [10].

the "molecular friction" in the polymers and thus more attention should be paid to understand it further [36, 37].

2.1.2.1 Investigation of Partially Miscible PMMA–PEG Blends Using Two-Dimensional Disrelation Mapping

Based on multiple perturbation 2D correlation analysis [38, 39], a novel technique named disrelation mapping has been introduced in studies by Shinzawa et al. [40]. It can identify the specific area where coordinated or out-of-phase (dissimilar) variations in the pattern of spectral absorbance appear. As shown in Figure 2.3, the process of constructing the image based on the disrelation intensity is illustrated. Firstly, spectra from different locations within the sample were produced by FTIR spectroscopic imaging in a single measurement, and then a set of FTIR spectroscopic images were produced. The conventional FTIR spectroscopic images were analyzed through visualizing the change of spectral absorbance at a specific wavenumber. In terms of the disrelation mapping, a so-called convolution filter is used to put the spectra in a spatial region into a convolution window. Then, these spectra are subjected to the cross-correlation analysis based on two wavenumbers. The new value of the center pixel in this window is the disrelation intensity between ν_1 and ν_2 within the local area. The disrelation intensity indicates that the spectral absorbance at two wavenumbers changes dissimilarly (asynchronously) with each other. Through moving the window across all of the imaged area, a disrelation map based on the disrelation intensity can be acquired. Unlike the conventional FTIR spectroscopic images, which only show the component-rich domains and component-poor domains, the disrelation map has the capacity to highlight the areas where chemically or physically meaningful variations occur, such as

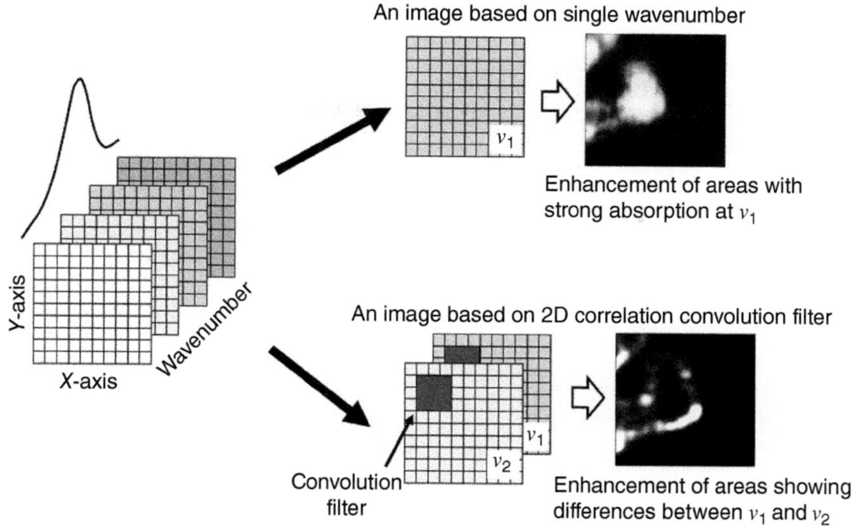

Figure 2.3 A schematic illustration of disrelation mapping based on 2D correlation convolution filter. Source: Shinzawa et al. [40].

the occurrence of molecular interactions or the appearance of intermediate forms between components. This technique can produce key information in the polymeric materials and provide evidence about the interpretation of many different polymer systems [40, 41].

Partially miscible polymer blends of poly(methyl methacrylate) (PMMA) and polyethylene glycol (PEG) were investigated in this study to demonstrate the capacity of this technology to visualize the molecular interactions [40]. The carbonyl groups of PMMA and the terminal hydroxy groups of PEG form hydrogen bonds when PMMA is partially miscible with PEG. Consequently, the preparation of disrelation maps can identify the generation of hydrogen bonds between PMMA and PEG. Figure 2.4 contains ATR–FTIR images of the PMMA/PEG blend, which show the PEG-rich domains and the PMMA-rich domains. However, it cannot provide evidence of interaction between PMMA and PEG, which is expected to be at their interface [40].

To visualize the H-bond interactions, disrelation mapping was applied because the change in spectral absorbance of a certain PMMA band will become different in areas where PMMA forms interactions with PEG. From the disrelation spectra derived from the FTIR spectra of PMMA/PEG blend, a cross-peak (1730, 1714 cm^{-1}) was observed, and it indicated that the variation in the spectral absorbance at these two wavenumbers has different trends. In other words, the spectral absorbance at these two wavenumbers varies differently as a function of spatial positions, which reveals that there is an additional contribution that accelerated or decelerated the absorbance change at 1714 cm^{-1} (PMMA band) [40].

As shown in Figure 2.5, the disrelation maps based on the disrelation intensity between 1730 and 1714 cm^{-1} were prepared, and they reveal that this disrelation

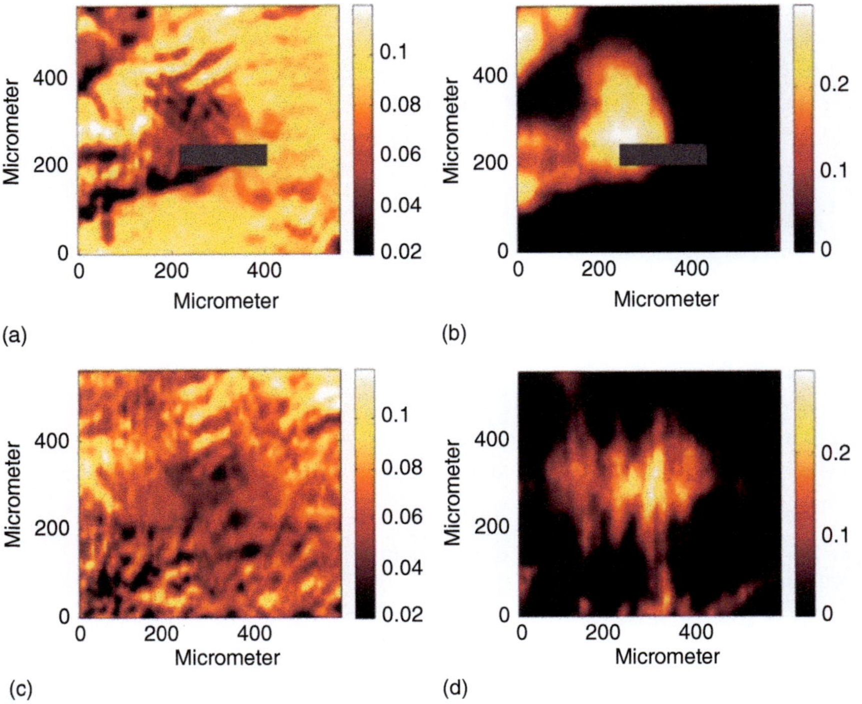

Figure 2.4 ATR–FTIR images of PMMA/PEG-2000000 blends (a, b) and PMMA/PEG-2000 blends (c, d) prepared with the distribution of absorbance of spectral bands at (a) 2882 cm^{-1} (PEG-2000000 band), (b) 1728 cm^{-1} (PMMA band), (c) 2882 cm^{-1} (PEG-2000 band), and (d) 1728 cm^{-1} (PMMA band). Source: Shinzawa et al. [40].

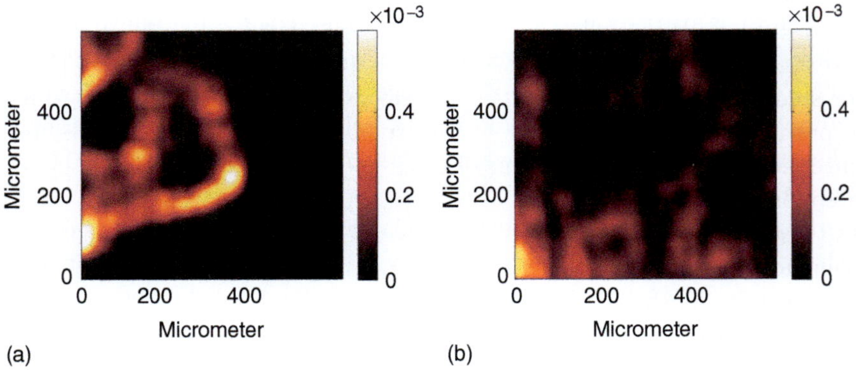

Figure 2.5 Disrelation maps of (a) PMMA/PEG-2000000 and (b) PMMA/PEG-2000 blends calculated with disrelation intensity between 1730 and 1714 cm^{-1}. Source: Shinzawa et al. [40].

intensity is strong only at the interface between PMMA and PEG. The additional factor, which only generates when PMMA and PEG interact, is the C = O stretching mode of the C = O· · ·H—O species. These disrelation maps are the first examples where the intermolecular hydrogen bonding between PMMA and PEG was visualized. Moreover, the level of disrelation intensity indicated the strength of interaction between molecules [40].

2.1.3 Investigation of Crystallization in Polymers

In terms of polymers, some factors, such as the solvent used, temperature, and strain, can induce the crystallization of amorphous polymers. Crystallization is the process of transitioning polymers from entangled melts to an ordered structure, which needs to overcome an energy barrier. The helical chain conformations, chain segment conformations, and chain packing, which are formed during crystallization, can be monitored using FTIR spectroscopy through the absorbance change and peak shift in different spectral bands [36, 42–45].

The isothermal crystallization process can be divided into two types: cold-crystallization and melt-crystallization. In terms of the isothermal cold-crystallization process, temperature is increased above T_c (low critical crystallization temperature) and crystallization occurs. In terms of the isothermal melt-crystallization process, the temperature is reduced below T_{mc} (high critical crystallization temperature) and polymers start to crystallize. Two opposing factors affect the melt-crystallization: increasing thermodynamic driving force with increasing undercooling; and decreasing polymer chain mobility at a lower temperature. These two types of isothermal crystallization process lead to different crystal modifications and intermolecular interactions [16, 36, 46, 47].

2.1.3.1 Investigation of Solvent-Induced Crystallization in Polymers

Due to the scientific interest and technological importance of syndiotactic polystyrene (s-PS), solvent diffusion and solvent-induced crystallization of s-PS were studied simultaneously using transmission FTIR spectroscopic imaging. The previous research using conventional spectroscopy only yields the overall changes about the degree of crystallization but not the crystallinity profile because conventional spectroscopy cannot provide the spatial information. As a result, the s-PS/solvent interface was investigated using transmission FTIR spectroscopic imaging to obtain the spatial distribution of solvent and the polymer morphology simultaneously. As shown in Figure 2.6, the wavenumber regions indicated by gray bars were used to integrate bands representing δ-crystalline s-PS and toluene, respectively [48, 49].

Figure 2.7 shows a visual representation of the movement of solvent into the polymer that induced crystallization. The distribution of crystalline polymer across the measured interface was obtained, and the values of solvent diffusion coefficients and polymer crystallization kinetics were calculated through a series of time-resolved

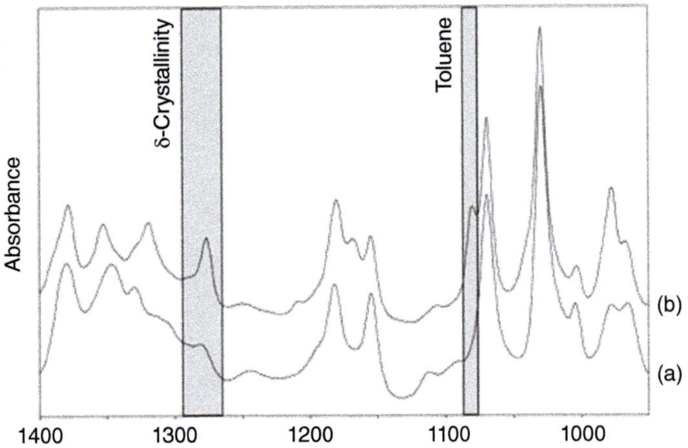

Figure 2.6 FTIR spectra of (a) an amorphous s-PS film and (b) a δ-crystalline s-PS film after exposure to toluene. Source: Gupper et al. [48].

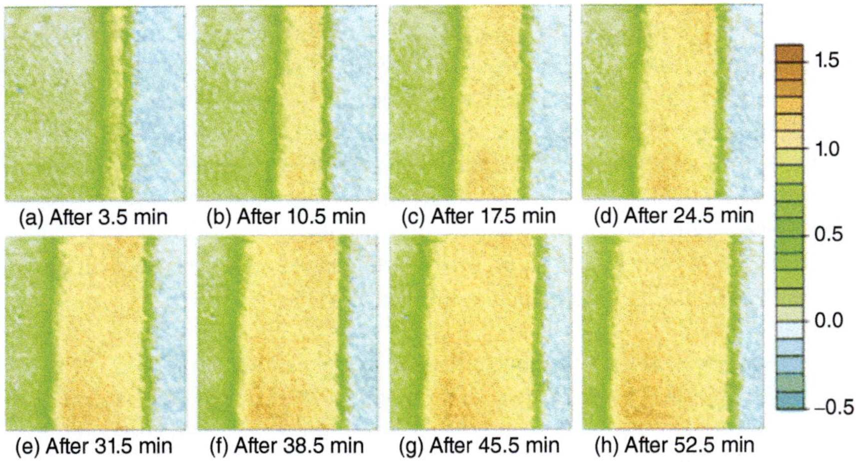

Figure 2.7 FTIR images based on the distribution of the integrated absorbance of the band at 1275 cm^{-1} (crystalline s-PS) as a function of solvent exposure time: (a) after 3.5 minutes, (b) after 10.5 minutes, (c) after 17.5 minutes, (d) after 24.5 minutes, (e) after 31.5 minutes, (f) after 38.5 minutes, (g) after 45.5 minutes, and (h) after 52.5 minutes. Source: Gupper et al. [48].

FTIR images. Evaluation of the diffusion exponent demonstrated that the type of diffusion is Fickian diffusion, which means that the diffusion process obeys Fick's laws. It was also found that the rate of crystallization in this system was much faster than those reported in the previous research. This was found to be a result of a higher solvent concentration being used in this study. Moreover, it could be concluded that the solvent diffusivity, which is affected by the mobility of solvent molecules in the crystalline polymer and the solvent flux on the surface of polymer, is the main limiting factor in the crystallization process. This is because the crystallization process can only be initiated when there is sufficient solvent [48].

2.1.3.2 Investigation of the Crystallization Process of PHB, PLLA, and Their Blends

The details of the dynamic processes during PLLA crystallization have attracted a lot of attention. However, the bands of ester groups in the PLLA spectrum are heavily overlapped, so it is difficult to use FTIR spectroscopy for the detailed analysis. As a powerful method to highlight the subtle spectral changes, 2D correlation spectroscopy was combined with conventional FTIR spectroscopy to investigate the structural changes and crystallization dynamics of PLLA. The generalized 2D correlation spectroscopy can not only enhance the spectral resolution through differentiating highly overlapped bands into single bands, but also reveal the sequence of spectral intensity variations [4, 38].

Through the analysis of the high-wavenumber shift in the $\nu(C=O)$ region during cold-crystallization (see Figure 2.8), it can be concluded that the dipole–dipole interaction between the carbonyl groups is an interchain interaction. Then the normalized peak heights profile for the bands of crystalline PLLA was prepared, and it was observed that the band at 1193 cm^{-1} has the fastest rate of change while the band at 1109 cm^{-1} has the slowest rate of change. It means that the methyl group shows a faster structural adjustment than the ester group during the crystallization of PLLA. To provide more evidence to the above conclusion, 2D asynchronous spectra, which show the sequential changes of spectral intensities at two wavenumbers, were analyzed. When the cross-peak in both the asynchronous spectrum and the corresponding synchronous intensity are positive, it indicates that the intensity change at ν_1 appears ahead of that at ν_2. On the other hand, when the cross-peak in the asynchronous spectrum is negative and the corresponding synchronous intensity is positive, it indicates that the intensity change at ν_1 appears after that at ν_2. Thus, as shown in Figure 2.9, the positive cross-peaks at $\Phi(1193, 1133)$, $\Phi(1193, 1109)$, and $\Phi(1133, 1109)$ in the synchronous spectra, as well as the positive cross-peaks at $\Psi(1193, 1133)$, $\Psi(1193, 1109)$, and $\Psi(1133, 1109)$ in the asynchronous spectra, prove the sequential order of spectral intensity variation: 1193 cm^{-1} > 1133 cm^{-1} > 1109 cm^{-1}, which means that the structural adjustment of methyl group precedes that of ester group [4].

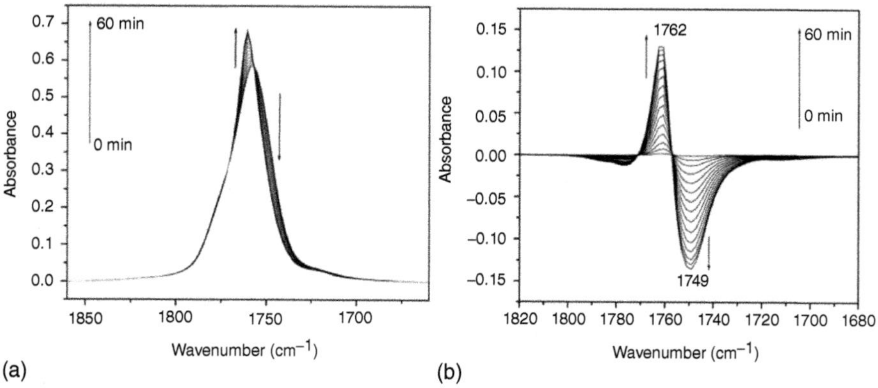

Figure 2.8 (a) FTIR spectra and (b) difference spectra of PLLA during the cold-crystallization process at 78 °C. Source: Zhang et al. [4].

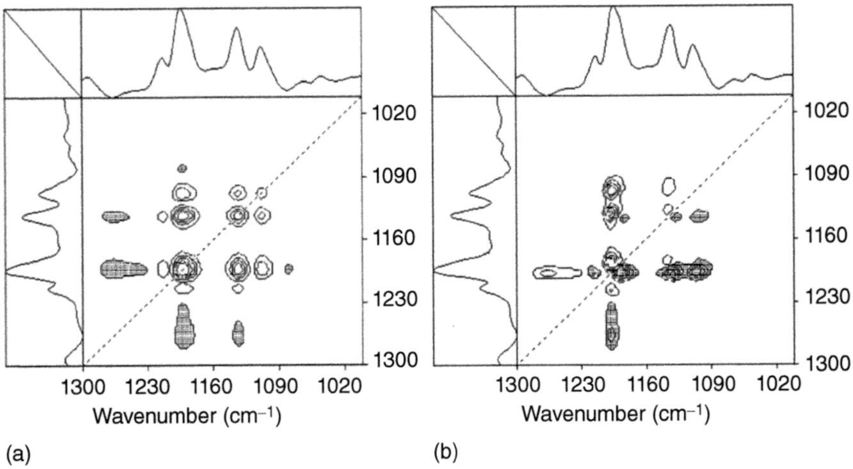

Figure 2.9 Synchronous (a) and asynchronous (b) correlation spectra of PLLA in the region of 1300–1000 cm^{-1} calculated from the spectra obtained during the annealing at 78 °C. Source: Zhang et al. [4].

The combination of these two approaches was also applied to investigate the melt-crystallization of PHB, another important biopolymer. Through detecting the cross-peaks in the asynchronous spectrum, the sequential order of spectral intensity variation in these carbonyl bands was revealed: 1743 cm^{-1} (1747 and 1739 cm^{-1}) > 1731 cm^{-1} > 1722 cm^{-1}. This was not expected because normally the amount of the crystalline polymer increases as the amount of amorphous polymer decreases. The reason for the sequential changes was identified as the existence of an intermediate state, which may be related with the band at 1731 cm^{-1}. Likewise, through plotting the normalized peak heights of crystalline bands of PHB vs. crystallization time, it was found that the rate of change of the absorbance of 1184 cm^{-1} band is faster than that of 825 cm^{-1} band, which means that C–O–C backbone shows a faster structural adjustment than C–C backbone [50].

2D correlation analysis was then used to confirm the above conclusions. The negative cross-peak $\Phi(1184, 825)$ in the synchronous spectrum and the negative cross-peak $\Psi(1184, 825)$ in the asynchronous spectrum verified the more rapid structural rearrangement of the C–O–C backbone compared to the C–C backbone. Through the detection of other cross-peaks in the synchronous spectrum and asynchronous spectrum, it can be concluded that during the melt crystallization of PHB, ester groups adjust their conformation at first, followed by the longer regular 2_1 helix chain sequences and then the regular 2_1 helix chain sequences [50].

To improve the mechanical and processing properties of PHB and PLLA, immiscible and miscible polymer blends were prepared and their crystallization behaviors were investigated via in situ FTIR spectroscopy. When melt crystallization occurs in the immiscible PHB/PLLA blend, the crystallization of PLLA started at a higher temperature and then the crystallization of PHB started at a lower temperature. However, when the T_m of these two blend components are close and melt crystallization occurs in the miscible PHB/PLLA blend, almost simultaneous crystallization is observed at the same isothermal crystallization temperature. Thus, it can be concluded that the miscibility of polymer blend significantly affects the crystallization dynamics of both blend components [42]. The detail of this influence has been further studied in another work [51].

2.2 Investigation of Polymers Subjected to High-Pressure or Supercritical CO_2 Using FTIR Spectroscopy and FTIR Spectroscopic Imaging

As alternatives to the environmentally harmful organic solvents, high-pressure and supercritical CO_2 has a unique and valuable potential to be used for polymer processing. CO_2 is relatively nontoxic, cheap, and abundant in the atmosphere. After the polymer processing, high-pressure CO_2 is easy to remove from the final products because it is gaseous at ambient temperature and pressure. Moreover, it offers fundamental advantages for polymer processing because it can induce plasticization, crystal modification, and temporary swelling [52–54].

The high viscosity of many polymeric materials can be a major obstacle to their processing, especially with high-molecular-weight polymers. The common methods for the viscosity reduction are to increase the temperature or add plasticizing agents. Plasticizing agents work by adding small plasticizers into the polymer matrix and fill its voids; this causes the interchain distance to increase and the polymer segments are more mobile. Both common viscosity reduction techniques have their drawbacks. For example, increasing the temperature consumes more energy and can even lead to thermal degradation. Plasticizer addition usually contaminates the final product if it is not fully removed [55, 56]. As a "molecular lubricant," high-pressure CO_2 has a similar effect for increasing segmental and chain mobility and reducing viscosity as heating the polymer above its T_g, which can facilitate the polymer process due to the lower shear stress [1, 57]. In terms of polymers exposed to high-pressure CO_2, plasticization is the reduction by several tens of degrees in the T_g. Thus, the thermolabile bioactive polymers, which are unobtainable through standard routes, such as some food ingredients and drugs, can be processed at a lower temperature with the help of plasticizing effect of CO_2. However, the effect as a plasticizer is not unlimited for high-pressure CO_2, the T_g no longer reduces when the CO_2 pressure continues to rise beyond a certain point [58–60].

The increase in the temperature of CO_2 decreases the solubility of CO_2 in LCST polymer blends. Consequently, the T_g of LCST polymer blend will increase with increasing temperature at constant CO_2 pressure and the chain and segmental mobility of polymer decreases. Meanwhile, increasing temperature increases this mobility [58, 61]. Thus, the effect of temperature on the chain and segmental mobility of polymer blend, which exhibits lower critical mixing behavior, needs further investigation.

To summarize, the investigation of polymer processing using high-pressure CO_2 has great commercial implications for the overall performance of polymeric materials. With evidence on the molecular level, FTIR spectroscopy can monitor the effects that high-pressure CO_2 induces on polymers [1, 55, 62]. The current state-of-the-art research utilizing FTIR spectroscopy and FTIR spectroscopic imaging will be reported in this chapter, including miscibility, interaction, crystallization, swelling, and CO_2 sorption into pure polymers, polymer blends and interfaces subjected to high-pressure or supercritical CO_2. Understanding these behaviors can benefit the development of more efficient polymer processing routes with this "green" processing technology.

2.2.1 Morphology of Polymeric Materials Under High-Pressure or Supercritical CO_2

Since the visualization of polymer morphology under high-pressure CO_2 has received a lot of attention, ATR–FTIR spectroscopic imaging has been introduced to detect the spatial distribution of different components in the polymer blend/interface under high-pressure CO_2 [63]. The polymer solution or molten polymer was cast directly onto an ATR crystal: diamond or Ge to prepare the film. The thickness of polymer film needed to be higher than the depth of penetration of the evanescent field but not too thick; otherwise, the penetration of CO_2 into the bulk of the film would be difficult. Furthermore, the contact between film and ATR crystal should be sufficient to obtain accurate spectra. After a film had been formed and solvent was totally evaporated, a home-made high-pressure cell was used to cover the polymer film and then high-pressure CO_2 was introduced to the cell. Since the polymer samples needed to reach equilibrium, measurement of spectra was taken several times until the absorbance of spectral bands no longer changed. The home-made miniature high-pressure cell is the key factor for the development of in situ high-pressure ATR–FTIR spectroscopic imaging. It has several advantages to make it safer to use even under very high pressure: (i) there is no fragile window on the cell; (ii) the ATR crystal: diamond is very hard; and (iii) the volume of cell is small [63].

As shown in Figure 2.10, it was observed that the distribution of poly(vinyl methyl ether) (PVME) and polystyrene (PS) was quite homogeneous before exposure to high-pressure CO_2. After subjected to 60 bar CO_2, phase separation occurred and the PS-rich domains and the PVME-rich domains appeared. As shown in the spectra (Figure 2.10e), the absorbance of PVME in the PVME-rich domains is higher than that in the PVME-poor domains (PS-rich domains). Likewise, the absorbance of PS in the PS-rich domains is higher than that in the PS-poor domains (PVME-rich domains). These spectra can also provide evidence of phase separation in the polymer blend [63].

The poly(ethylene oxide) (PEO)/PMMA interface has also been investigated, and the CO_2 sorption into these two polymers was studied simultaneously for the first time. As shown in Figure 2.11, the distribution of PMMA (a–d), PEO (e–h), and CO_2 (i–l) at different conditions is clearly shown. It can be found that the degree of swelling in PMMA increases with increasing CO_2 pressure and decreases with increasing temperature. This is due to the CO_2 sorption into PMMA that increases with increasing CO_2 pressure and decreases with increasing temperature as shown in Figure 2.11i–l. It can also be found that the CO_2 sorption into PEO is lower than

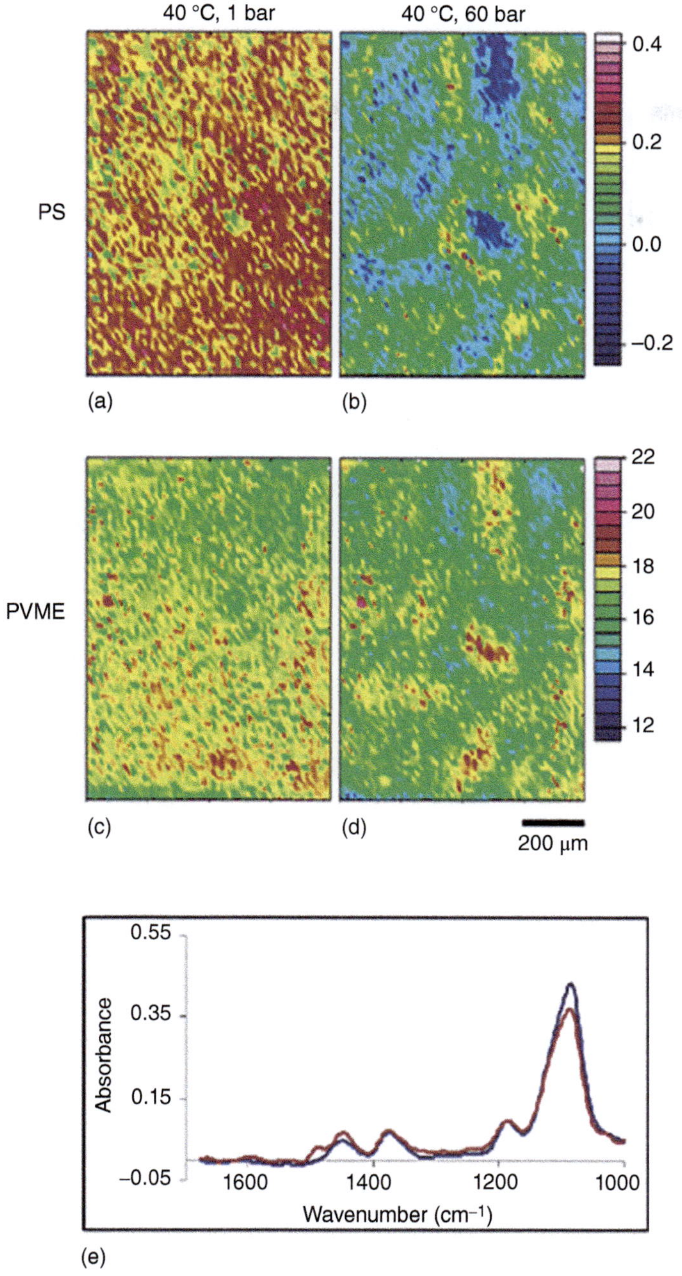

Figure 2.10 ATR–FTIR spectroscopic images of the PS/PVME blend before (a, c) and under 60 bar CO_2 (b, d). Images (a) and (b) are based on the spectral band of PS (1492 cm^{-1}), while images (c) and (d) are based on the spectral band of PVME (1077 cm^{-1}). (e) ATR–FTIR spectra extracted from PVME-rich domains (blue) and PS-rich domains (red). Source: Kazarian and Chan [63].

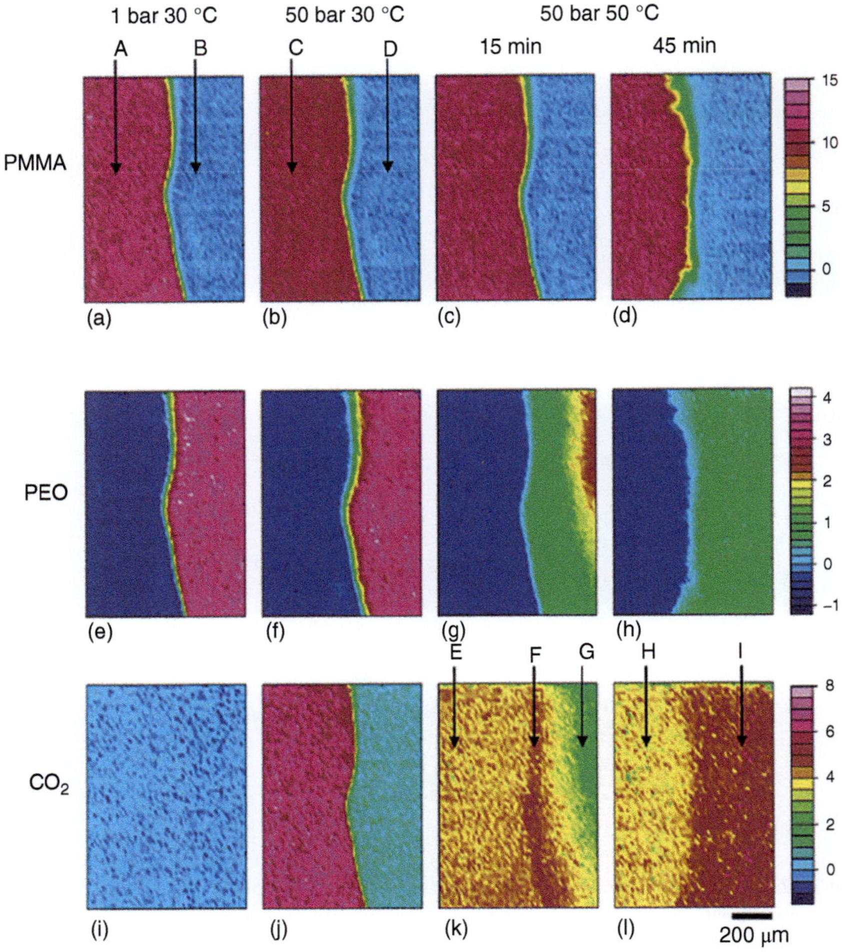

Figure 2.11 ATR–FTIR spectroscopic images of the PMMA/PEO interface. The images (a–d) are based on the spectral band of PMMA, the images (e–h) are based on the spectral band of PEO, and the images (i–l) are based on the spectral band of CO_2. Source: Kazarian and Chan [63].

that into PMMA at 30 °C. However, it increases significantly after the temperature increases. This is due to the high-pressure CO_2 that reduces the T_m of PEO and causes them to melt at 50 °C. ATR–FTIR spectroscopic imaging has been demonstrated as a suitable approach for the investigation of the morphology of polymeric materials under high-pressure CO_2. It has been shown to facilitate faster screening with shorter acquisition times [63].

An LCST polymer blend of PCL/PLA was studied as a further example of a dynamic process of phase separation when it was under high-pressure CO_2. ATR–FTIR spectroscopic imaging was used to demonstrate the gradual separation of PCL-rich domains and PLA-rich domains to reveal the change in the extent of phase separation under different conditions. As shown in Figure 2.12, it was clearly

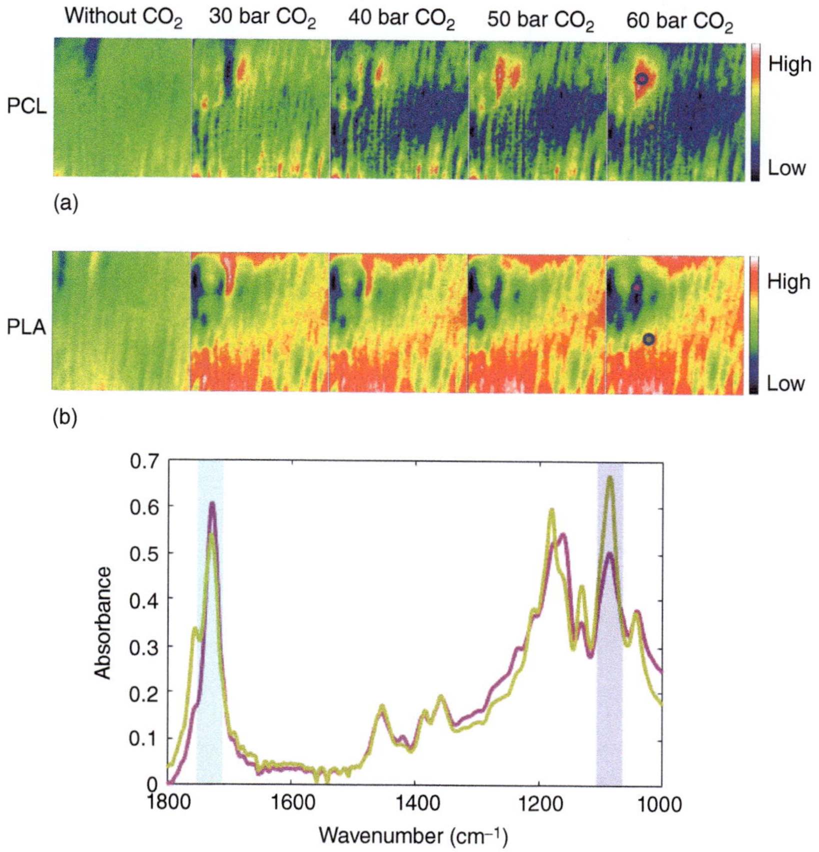

Figure 2.12 (a) ATR–FTIR spectroscopic images of the PCL/PLA blend at 30 °C in the absence of CO_2, with 30, 40, 50, and 60 bar CO_2. The top images are based on the spectral band of PCL, while bottom images are based on that of PLA. (b) Spectra were extracted from the PCL-rich domain indicated by a pink circle in the top right image and the PLA-rich domain indicated by a ginger circle in the bottom right image. Source: Lu and Kazarian [64].

observed that phase separation occurs under high-pressure CO_2 and the extent of phase separation is enhanced with increasing CO_2 pressure. This is because the sorption of CO_2 into this blend increases with increasing CO_2 pressure and then leads to the greater segmental and chain mobility. Through changing other parameters, it can be concluded that the extent of phase separation also becomes greater with increasing temperature and exposure time [64].

2.2.2 Investigation of Interaction in Polymers Under High-Pressure or Supercritical CO_2

FTIR spectroscopy can reveal Lewis acid–base interactions between CO_2 molecules (electron acceptor) and electron donor groups in polymers, such as carbonyl group, by the splitting of ν_2 bending mode of CO_2 at around 660 cm^{-1} [55, 65]. The chemical interaction between CO_2 and polymers has great effects on the physical properties of polymers and consequently has implications on their processing and synthesis [66]. It can disrupt the current inter- and intramolecular interaction in polymers and then increase the distance between polymer chains, which is the plasticization effect of CO_2 [65, 67]. Some factors, such as temperature, CO_2 pressure, and polymer molecular weight, have great effects on the specific interactions between CO_2 and polymers and consequently change the solubility of CO_2 in many glassy polymers [68].

2.2.2.1 Investigation of the Effect of High-Pressure CO_2 on the H-Bonding in PEG–PVP Blends

The hydrogen bond interactions in the poly(ethylene glycol) (PEG)/polyvinylpyrrolidone (PVP) blend are important to pharmaceutical products because they result in the unique elastic and adhesive properties that can be used as transdermal delivery devices. High-pressure CO_2 weakens the hydrogen bond interaction in the polymers and even leads to phase separation. As a result, the effect of high-pressure CO_2 on hydrogen bonding interactions between the terminal OH group of PEG and the C=O group of PVP has been detected through in situ ATR–FTIR spectroscopy [57].

Through processing the second derivative of spectrum in the carbonyl absorption region shown in Figure 2.13, the specific wavenumber and absorbance of these second derivative bands reveal the behavior of interactions under different processing conditions. The ν(C=O) at 1678 cm^{-1} represents the free carbonyl groups, while ν(C=O) at 1656 cm^{-1} represents the interacting carbonyl groups. Since the wavenumber of the band of interacted carbonyl groups increases with increasing PEG concentration, it suggested that a higher PEG concentration promotes the

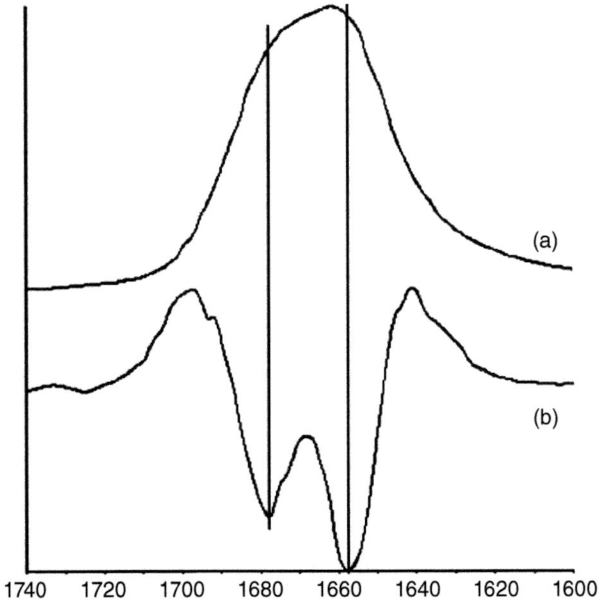

Figure 2.13 (a) FTIR spectrum and (b) its second derivative in the ν(C=O) region of a PVP/PEG blend. Source: Labuschagne et al. [57].

H-bond interactions. Likewise, through monitoring the peak shift and absorbance change of the band of interacted carbonyl groups under different pressures, it was also found that the strength of H-bond interactions is weaker and the number of H-bonds decreases with increasing CO_2 pressure up to 80 bar. This is because more CO_2 molecules interact with the carbonyl groups of PVP and disrupt the current H-bond interactions between PEG and PVP. After depressurization, the shielding effect of high-pressure CO_2 on terminal hydroxyl groups of PEG is reversible. Moreover, choosing a lower-molecular-weight polymer blend or processing at a lower temperature can also enhance this shielding effect [57].

2.2.2.2 Investigation of the Mechanism of Interaction Between CO_2 and Polymers Through the Thermodynamic Parameters Produced from In Situ ATR–FTIR Spectroscopy

In a previous study, it was found that the intermolecular interactions of polymers prevent the polymer swelling and subsequent CO_2 sorption, which demonstrated the importance of understanding these interactions. Thus, there is a need to reveal new physical insights about the interactions of both polymer–CO_2 and polymer–polymer, for optimizing polymer properties. The formation of a "bent" T-shaped complex through the intermolecular interaction of C=O groups in polymers and CO_2 can be monitored via the splitting of ν_2 (CO_2). Then, the change in the shape of ν_2 (CO_2) in polyoxime, polyketone, and polyalcohol samples can be interpreted by the interaction of CO_2 and polymers. Polybutadiene was used as a reference for spectral subtraction because it does not have any functional groups that can form Lewis acid–base interactions with CO_2. Through careful subtraction of the corresponding spectra (i.e. the spectrum of polybutadiene and the spectrum of one of these three samples being studied), the ATR–FTIR spectra of interacting CO_2 were obtained as shown in Figure 2.14 [69].

By changing the temperature at constant pressure, the thermodynamic parameters of CO_2 interactions with polymers were calculated. The concentration of interacted CO_2 is much lower than that of available functional groups, and thus 1 : 1 complexes of CO_2 and the functional groups of polymers were hypothesized, which meant that each functional group could only interact with one CO_2 molecule. Then the equilibrium constant K_c for this Lewis acid–base interaction was calculated. Finally, the van't Hoff equation was used to calculate the thermodynamic parameters: entropy (ΔS) and enthalpy (ΔH). These results revealed that the inter- and intramolecular interaction between the functional groups of polymers are broken by CO_2 and then the CO_2–polymer complexes form, as shown in Figure 2.15. In this work, it was demonstrated that too strong intermolecular interactions within polymers prevent the interaction between polymer and CO_2 and then affect the polymer processing possibilities using high-pressure CO_2 [69].

2.2.3 Investigation of Crystallization in Polymers Under High-Pressure or Supercritical CO_2

The sorption of CO_2 into a polymer will change its glass transition temperature (T_g), crystallization temperature (T_c), and melting temperature (T_m) [58, 70]. Through

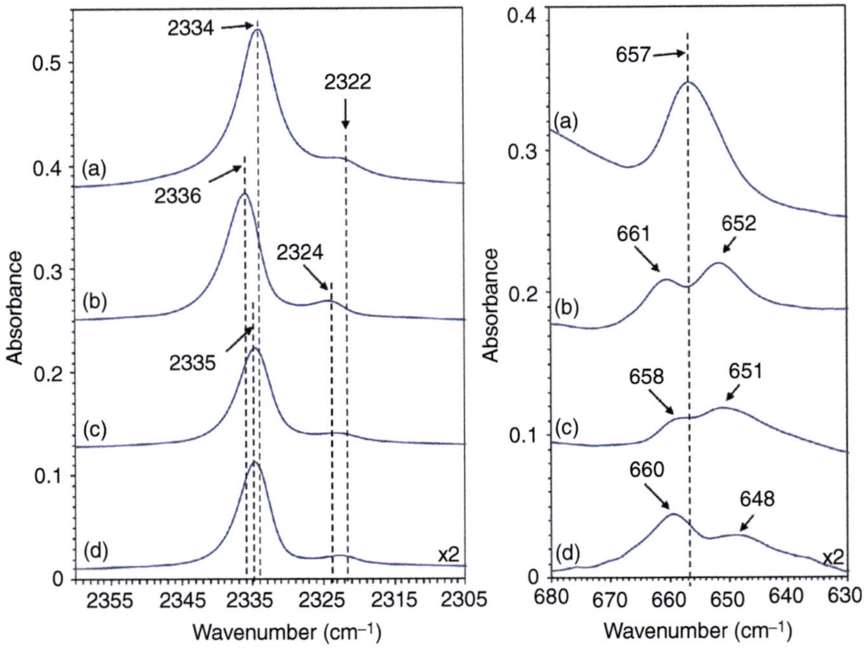

Figure 2.14 ATR–FTIR spectra of noninteracting CO_2 sorbed by polybutadiene (a), interacting CO_2 sorbed by polyketone (b), polyoxime (c), polyalcohol (d) obtained at 300 K and 30 bar CO_2. Source: Gabrienko et al. [69].

the T_g reduction (plasticization), high-pressure CO_2 can induce crystallization in polymers below their original T_g, such as polycarbonate [58] and sPS [59]. Apart from plasticization, the swelling and accelerated movement of polymer chains induced by high-pressure CO_2 can also benefit the formation of more thermodynamically favorable crystalline state. This is because when high-pressure CO_2 dissolves into the polymers, the free volume of the polymers increases. The mobility of polymer chains also increases because of reducing interchain interactions [67, 71]. High-pressure CO_2 can also induce specific types of crystalline forms that cannot be obtained under the same condition without high-pressure CO_2. Handa et al. have used compressed N_2 as a reference and found that the induced unique type of crystalline form was caused by the plasticization effect of CO_2 instead of the hydrostatic pressure of compressed gas. To be more specific, under high-pressure CO_2, the lower critical crystallization temperature reduces so that the energy barrier of crystallization can be overcome at a lower temperature. Moreover, the presence of high-pressure CO_2 can also promote the transition between various types of crystalline forms and bring some new transitions. The dissolved CO_2 also induces morphological modification of polymer crystals and suppresses side reactions [59, 71–73].

Both the isothermal crystallization temperature and CO_2 pressure can affect the crystallization behavior and crystallization kinetic constant. High-pressure CO_2 can accelerate the crystallization at the crystallization temperature in the

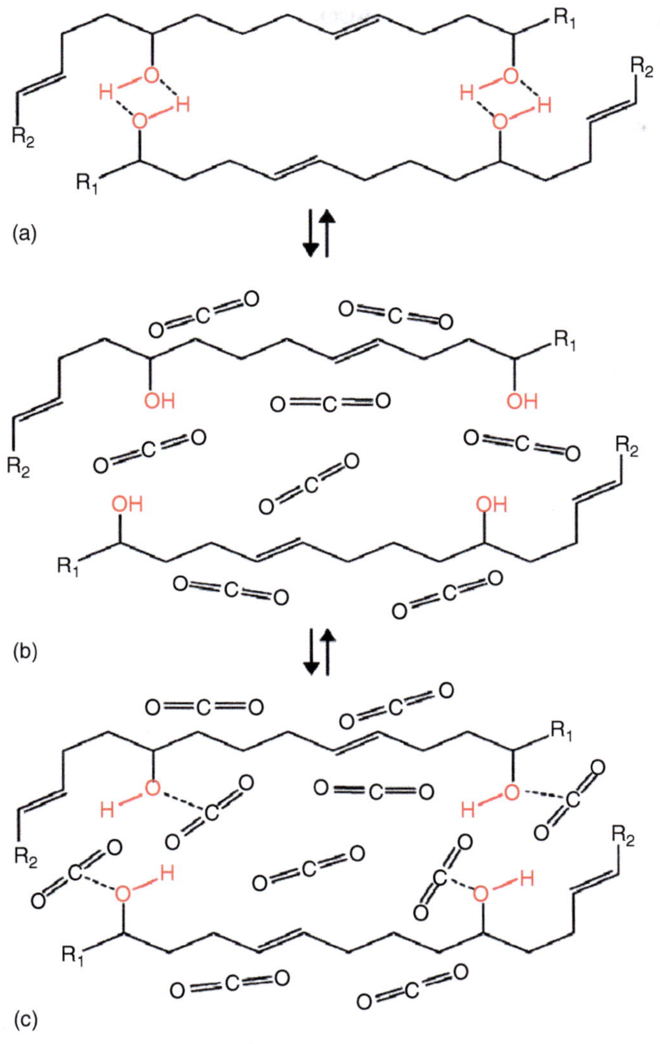

Figure 2.15 The schematic for the formation of CO_2–polymer complexes. Source: Gabrienko et al. [69].

crystal-growth-controlled region and slow down the crystallization at the crystallization temperature in the nucleation-controlled region. At the crystallization temperature in the crystal-growth-controlled region, the effect of CO_2 on self-diffusion of polymer chains (increasing mobility of polymer chains) is dominant. At the crystallization temperature in the nucleation-controlled region, the absorbed CO_2 hinders the nuclei to reach their critical size and then decreases the nucleation rate [72]. To be more specific, high-pressure CO_2 weakens the interchain interaction of polymer molecules and then hinders the transformation of polymer chains to the crystal lattice. Moreover, high-pressure CO_2 sorption reduces the density of polymers and then makes the formation of a steady embryo through density fluctuation more difficult. On the other hand, the molecular chains of polymers are very mobile under high-pressure CO_2 and at high temperatures; decreasing this temperature (increasing thermodynamic driving force) benefits the nucleation of crystals due to the reduced mobility of polymer chains. If the polymer chains are more difficult to move out from the crystal lattice, the nuclei are more likely to reach their critical radius. Therefore, increasing both the CO_2 pressure and temperature can increase the mobility of polymer chains. The increase of mobility, to some extent, can accelerate the crystallization. However, it will slow down the crystallization if it is too high, because the molecular chains are more likely to move out from the crystal lattice [67, 74].

In recent research, the crystallization of polymers after decompression has been observed, which may affect the final product of polymeric foams. This is because the decreasing temperature caused by reduced internal energy benefits the nucleation of polymers. Furthermore, increasing the T_m of polymers enhances the degree of supercooling. The rate of crystallization is faster after depressurization than that under CO_2. The saturation pressure, isothermal crystallization temperature, and the release rate of the pressure can all affect the decompression crystallization [67].

High-pressure DSC and X-ray diffraction (XRD) have been applied to investigate the ScCO_2-induced crystallization of sPS. However, the crystallization under ScCO_2 at a high temperature is usually too fast to measure its kinetics because the process of crystallization is faster than the measurement acquisition times. For example, even at 8 bar, CO_2 can accelerate the sPS crystallization to a great extent and at 30 bar CO_2 can lead to a 13-fold increase in the rate of crystallization [59].

In situ FTIR spectroscopy has been used to monitor crystallization in the polymer–CO_2 system and reveal the change in the polymer morphologies caused by high-pressure CO_2 because of its relatively short acquisitions times. Mizoguchi et al. have used the FTIR spectroscopy to investigate the CO_2-induced crystallization of poly(ethylene terephthalate) (PET). It was found that the crystallization half-life ($t_{1/2}$; the time at which 50% of crystallization has completed) using 50 bar CO_2 at 85 °C is equal to that of crystallization without CO_2 at 118 °C. This research also compared the absorbance of the bands of PET between a heat-induced crystalline sample and a CO_2-induced crystalline sample and found that the density of CO_2-induced crystalline PET is lower than that of heat-induced crystalline PET. This result revealed that high-pressure CO_2 can lead to the formation of voids (foaming) in the fine structure of crystalline PET because the volume change does not fully recover after depressurization [73].

2.2.4 The Investigation of Structural Changes and Crystallization Kinetics of Polymers Exposed to High-Pressure CO_2 Through In Situ High-Pressure FTIR and FT-Raman Spectroscopy

Although the properties of PLLA after isothermal crystallization with high-pressure CO_2 have been investigated by many studies, little attention has been paid to the variation of structure and conformation in PLLA during this process. Thus, in situ FTIR spectroscopy was used in this study to help the comprehension of PLLA crystallization under high-pressure CO_2 [6].

It was found that the crystallization of PLLA occurred under air and high-pressure CO_2, but the resulting crystal structures and crystallization kinetics are different. During the process of crystallization, polymer chains have a more regular (oriented) arrangement in the crystal lattice and become closer so that the chain interaction is enhanced. This can be demonstrated by a change in absorbance and peak shifts in the bands shown in the FTIR spectra of PLLA. Through detecting the $\nu(C=O)$ stretching region of second-derivative spectrum of PLLA, four peaks were observed and they correspond to four distinct conformations: gg (1777 cm^{-1}), tg (1768 cm^{-1}), gt (1759 cm^{-1}), and tt (1749 cm^{-1}). This research has used $t_{1/2}$ to compare the rate of crystallization under different conditions. The results revealed that the rate of PLLA crystallization is accelerated under a higher CO_2 pressure at the crystallization temperature in the crystal-growth-controlled region. The obtained data also suggested that the enhanced mobility of the polymer chain segments caused by high-pressure CO_2 promotes the formation of a stable helix conformation of PLLA during crystallization. On the other hand, the interchain distance in polymers becomes larger because of the plasticization effect and the Lewis acid–base type interactions between CO_2 and basic site of polymers. These two factors weaken the intermolecular interactions between PLLA molecule chains. Finally, it was speculated that the weaker intermolecular interactions and the formation of PLLA helix conformation affect the crystal structures of PLLA [6]. This research supported research that utilized a high-pressure polarized optical microscope (HP-POM), which demonstrated that increasing chain mobility, decreasing molecular chain density, and reducing interchain interactions are three main factors that affect crystallization kinetics and crystallization morphology of PLLA under CO_2 [67].

FTIR and Fourier transform (FT)-Raman spectroscopy were used to investigate the structure and morphology of PET during $ScCO_2$-induced crystallization. This research revealed information about optimizing the processing of PET-based materials with supercritical CO_2. Through the change in the $\nu(C=O)$ band of PET, it was proved that both the type of crystal and the degree of crystallization of $ScCO_2$-treated samples are different from those of heat-treated samples. Compared with FTIR spectroscopy, FT-Raman spectroscopy can be used to detect the subtle aspects of the crystalline PET structures and monitor the dye uptake into polymers [7].

2.2.5 Investigation of Swelling and CO_2 Sorption into the Polymers Under High-Pressure or Supercritical CO_2

High-pressure CO_2 can be used as a solvent to dilute the molecular chain density, which causes an increase in the free volume of polymers and thus induces swelling and increasing mobility of molecular chains [75]. The polymer swelling is defined as a ratio of the polymer volume under high-pressure CO_2 and before exposure to high-pressure CO_2. Thus, it is proportional to polymer concentration before CO_2 treatment (c_0) and polymer concentration after CO_2 treatment (c). According to the Beer–Lambert Law, the absorbance of a spectral band is proportional to the concentration of the corresponding component. Thus, quantitative swelling values can

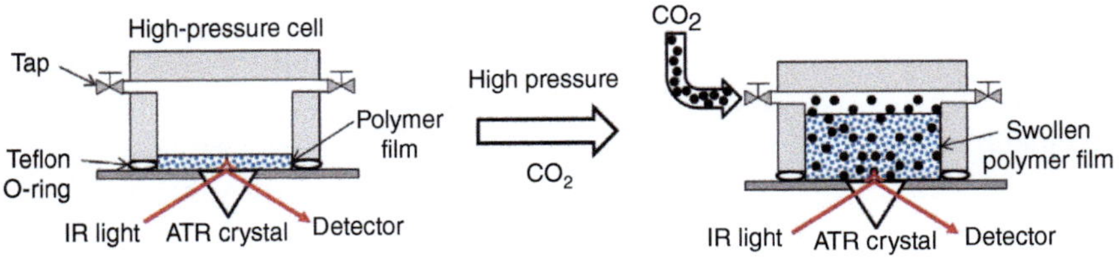

Figure 2.16 Schematic of the in situ ATR–FTIR spectroscopy to study the CO_2 sorption and polymer swelling. Source: Ewing et al. [78].

be obtained through calculating the absorbance of band of polymer before (A_0) and under exposure to high-pressure CO_2 (A) [76, 77].

The simultaneous measurement of CO_2 sorption and the corresponding swelling of polymer in a sample can be challenging, especially when different methods are used. For instance, polymer swelling can be measured by a cathetometer or a video camera, while CO_2 sorption can be measured by the gravimetric method. As shown in Figure 2.16, a novel approach based on ATR–FTIR spectroscopy was introduced to produce the spectrum of both polymers and sorbed CO_2, which reveals the density change in polymer and the concentration of sorbed CO_2 simultaneously [76, 79].

The values of the angles of incidence for diamond and germanium were calibrated using polymers. It was also found that the obtained result of swelling is not affected greatly by the change in the refractive index. Good contact between the sample and ATR crystal was essential to avoid the influence of nonsorbed CO_2 on measured area, which would affect quantitative results [76].

The degree of swelling was obtained via the following equation:

$$S = \frac{A^0}{A} \times \frac{d_{\mathrm{e,u}}}{d^0_{\mathrm{e,u}}} - 1 \tag{2.2}$$

where S is the degree of swelling, A^0 and A are the absorbances of band of polymer before exposure to high-pressure CO_2 and under high-pressure CO_2, respectively. $d^0_{\mathrm{e,u}}$ and $d_{\mathrm{e,\,u}}$ are the average effective thickness in unpolarized light before exposure to high-pressure CO_2 and under high-pressure CO_2, respectively [76].

The concentration of CO_2 sorbed in polymers was obtained via the following equation:

$$c = \frac{A}{\varepsilon \times d_{\mathrm{e,u}}} \tag{2.3}$$

where c is the concentration of CO_2 sorbed in polymers, A is the absorbance of CO_2 band at 2335 cm^{-1}, ε is the molar absorptivity for CO_2: $1 \times 10^6\ \mathrm{cm}^2\ \mathrm{mol}^{-1}$ [76].

The calculated results were found to be consistent with those from the previous research [79], which proved the feasibility of this novel high-pressure in situ spectroscopic approach to reflect the bulk properties of polymers [76].

The in situ ATR–FTIR spectroscopic approach described has also been applied to study the effect of these specific interactions in a systematic manner on a set of polyketone samples [78].

Some improvement was made to the procedure and methodology proposed by Flichy et al. [76]. The effective thickness in Eq. (2.3) was calculated through the following equation:

$$\frac{d_{\mathrm{e,u}}}{\lambda} = \frac{\frac{n_2}{n_1}\cos\theta\left[3\sin^2\theta - 2\left(\frac{n_2}{n_1}\right)^2 + \left(\frac{n_2}{n_1}\right)^2\sin^2\theta\right]}{2\pi\left[1-\left(\frac{n_2}{n_1}\right)^2\right]\left\{\left[1+\left(\frac{n_2}{n_1}\right)^2\right]\sin^2\theta - \left(\frac{n_2}{n_1}\right)^2\right\}\left[\sin^2\theta - \left(\frac{n_2}{n_1}\right)^2\right]^{\frac{1}{2}}} \tag{2.4}$$

where $d_{\mathrm{e,u}}$ is the effective thickness, λ is the wavelength of incident light, θ is the angle of incidence, n_1 and n_2 are the refractive indices of ATR crystal and sample, respectively [78].

Distilled water instead of polymers was used to obtain the angles of incidence for diamond and Ge. Then the refractive indices of polyketones were determined from the spectra measured by two different ATR crystals: diamond and germanium [78]. The higher concentration of electron-donating functional groups results in greater polymer swelling and a higher concentration of CO_2 in polymers, as a result of the Lewis acid–base interactions between CO_2 and functional groups that facilitate the CO_2 sorption. However, it was found that if the concentration of the electron-donating functional groups is too high, the CO_2 sorption capacity of polymers decreases. This was hypothesized to be because too strong intermolecular dipole–dipole interactions between polymers inhibit the access of CO_2 molecules into the bulk area of polymers and thus hinder the functional groups to interact with CO_2 molecules. When the concentration of carbonyl groups in the polymers is the same, polymer molecular weight does not affect the CO_2 sorption and polymer swelling a lot [78].

Through careful spectral subtraction, the bands of noninteracted CO_2 were removed from the spectra and only the bands corresponding to the interacted CO_2 remained. The resulting spectra show that the ratio of the concentration of interacted CO_2 and the concentration of carbonyl groups is 2 : 1. This in situ spectroscopic approach has also been used to measure the polymer swelling and CO_2 sorption in PEG [80] and poly(methylmethacrylate-*co*-ethylhexylacrylate-*co*-ethyleneglycoldimethacrylate) (P(MMA–EHA–EGDMA)) [81].

2.3 Conclusion

This chapter has summarized the recent applications of FTIR spectroscopy and FTIR spectroscopic imaging for the analysis of a range of polymeric materials and systems. The research utilizing FTIR spectroscopic approaches has been demonstrated to be important to reveal a greater understanding of many types of polymers. FTIR spectroscopic approaches have shown to be beneficial for the study of polymer processing applications, such as extraction, drying, impregnation, dyeing, and blending, both with and without high-pressure CO_2. The approaches assessed and reviewed in this book chapter are suitable for the study of pure polymers and multicomponent polymer systems (e.g. blend). The application of FTIR spectroscopic approaches for this type of research is by no means exhausted, and there are still a lot of emerging research fields that can develop novel applications that utilize these methodologies [1, 58]. Thus, as long as the requirement of molecular-level insight of advanced polymer processing exists, the application of in situ FTIR spectroscopy and FTIR spectroscopic imaging has a tremendous analytical potential to aid understanding about the modification and behavior of the polymeric materials and systems.

References

1 Ewing, A.V. and Kazarian, S.G. (2018). Current trends and opportunities for the applications of in situ vibrational spectroscopy to investigate the supercritical fluid processing of polymers. *J. Supercrit. Fluids* 134: 88–95. https://doi.org/10.1016/j.supflu.2017.12.011.

2 Poliakoff, M., Howdle, S.M., and Kazarian, S.G. (1995). Vibrational spectroscopy in supercritical fluids: from analysis and hydrogen bonding to polymers and synthesis. *Angew. Chem. Int. Ed.* 34: 1275–1295. https://doi.org/10.1002/chin.199539328.

3 Grotheer, E., Vogel, C., Kolomiets, O. et al. (2014). FT-IR and NIR spectroscopic imaging: principles, practical aspects, and applications in material and pharmaceutical science. In: *Infrared and Raman Spectroscopic Imaging* (eds. R. Salzer and H.W. Siesler), 341–396. Wiley-VCH.

4 Zhang, J., Tsuji, H., Noda, I., and Ozaki, Y. (2004). Structural changes and crystallization dynamics of poly(L-lactide) during the cold-crystallization process investigated by infrared and two-dimensional infrared correlation spectroscopy. *Macromolecules* 37: 6433–6439. https://doi.org/10.1021/ma049288t.

5 Dougan, J.A., Chan, K.L., and Kazarian, S.G. (2014). FT-IR imaging in ATR and transmission modes: practical considerations and emerging applications. In: *Infrared and Raman Spectroscopic Imaging* (eds. R. Salzer and H.W. Siesler), 397–444. Wiley-VCH.

6 Li, S., He, T., Liao, X. et al. (2015). Structural changes and crystallization kinetics of polylactide under CO_2 investigated using high-pressure Fourier transform infrared spectroscopy. *Polym. Int.* 64: 1762–1769. https://doi.org/10.1002/pi.4977.

7 Kazarian, S.G., Brantley, N.H., and Eckert, C.A. (1999). Applications of vibrational spectroscopy to characterize poly(ethylene terephthalate) processed with supercritical CO_2. *Vib. Spectrosc.* 19: 277–283. https://doi.org/10.1016/S0924-2031(98)00073-3.

8 Salzer, R. and Siesler, H.W. (eds.) (2009). *Infrared and Raman Spectroscopic Imaging*. Wiley-VCH.

9 Iruretagoyena, D., Bikane, K., Sunny, N. et al. (2020). Enhanced selective adsorption desulfurization on CO_2 and steam treated activated carbons: equilibria and kinetics. *Chem. Eng. J.* 379: 122356. https://doi.org/10.1016/j.cej.2019.122356.

10 Vogel, C., Wessel, E., and Siesler, H.W. (2008). FT-IR imaging spectroscopy of phase separation in blends of poly(3-hydroxybutyrate) with poly(L-lactic acid) and poly(ε-caprolactone). *Biomacromolecules* 9: 523–527.

11 Hifumi, H., Ewing, A.V., and Kazarian, S.G. (2016). ATR–FTIR spectroscopic imaging to study the drying and dissolution of pharmaceutical polymer-based films. *Int. J. Pharm.* 515: 57–68. https://doi.org/10.1016/j.ijpharm.2016.09.085.

12 Ewing, A.V. and Kazarian, S.G. (2018). Recent advances in the applications of vibrational spectroscopic imaging and mapping to pharmaceutical formulations. *Spectrochim. Acta, Part A* 197: 10–29. https://doi.org/10.1016/j.saa.2017.12.055.

13 Sroka-Bartnicka, A., Borkowski, L., Ginalska, G. et al. (2017). Structural transformation of synthetic hydroxyapatite under simulated in vivo conditions studied with ATR–FTIR spectroscopic imaging. *Spectrochim. Acta, Part A* 171: 155–161. https://doi.org/10.1016/j.saa.2016.07.051.

14 Kazarian, S.G., Lawrence, C.J., and Briscoe, B.J. (1999). In-situ spectroscopy of polymers processed with supercritical carbon dioxide. *New Trends Atomic Mol. Spectrosc.* 4060: 210. https://doi.org/10.1117/12.375295.

15 Kazarian, S.G. and Chan, K.L.A. (2010). Micro- and macro- attenuated total reflection Fourier transform infrared spectroscopic imaging. *Appl. Spectrosc.* 64: 135A–152A.

16 Blümm, E. and Owen, A.J. (1995). Miscibility, crystallization and melting of poly(3-hydroxybutyrate)/poly(L-lactide) blends. *Polymer (Guildf.)* 36: 4077–4081. https://doi.org/10.1016/0032-3861(95)90987-D.

17 Noda, I., Satkowski, M.M., Dowrey, A.E., and Marcott, C. (2004). Polymer alloys of nodax copolymers and poly(lactic acid). *Macromol. Biosci.* 4: 269–275. https://doi.org/10.1002/mabi.200300093.

18 Koyama, N. and Doi, Y. (1997). Miscibility of binary blends of poly[(*R*)-3-hydroxybutyric acid] and poly[(*S*)-lactic acid]. *Polymer (Guildf.)* 38: 1589–1593. https://doi.org/10.1016/S0032-3861(96)00685-4.

19 Chuang, W.T., Jeng, U.S., Da Hong, P. et al. (2007). Dynamic interplay between phase separation and crystallization in a poly(ε-caprolactone)/poly(ethylene glycol) oligomer blend. *Polymer (Guildf.)* 48: 2919–2927. https://doi.org/10.1016/j.polymer.2007.03.041.

20 Mumby, S.J., Sher, P., and Eichinger, B.E. (1993). Phase diagrams of quasi-binary polymer solutions and blends. *Polymer (Guildf.)* 34: 2540–2545. https://doi.org/10.1016/0032-3861(93)90586-Y.

21 Tanaka, H. and Nishi, T. (1989). Local phase separation at the growth front of a polymer spherulite during crystallization and nonlinear spherulitic growth in a polymer mixture with a phase diagram. *Phys. Rev. A* 39: 783–794. https://doi.org/10.1103/PhysRevA.39.783.

22 Kang, H., Uddin, M.A., Lee, C. et al. (2015). Determining the role of polymer molecular weight for high-performance all-polymer solar cells: its effect on polymer aggregation and phase separation. *J. Am. Chem. Soc.* 137: 2359–2365. https://doi.org/10.1021/ja5123182.

23 Robeson, L.M. (2007). Fundamentals of polymer blends. In: *Polymer Blends: A Comprehensive Review* (ed. L.M. Robeson), 10–23. Hanser.

24 Gelles, R. and Frank, C.W. (1983). Effect of molecular weight on polymer blend phase separation kinetics. *Macromolecules* 16: 1448–1456. https://doi.org/10.1021/ma00243a008.

25 Doi, Y. and Steinbüchel, A. (eds.) (2002). Biopolymers, biology, chemistry, biotechnology, applications. In: *Polyesters III – Applications and Commercial Products*, vol. 4, 410. Wiley Blackwell.

26 Wu, Q.Y., Liu, B.T., Li, M. et al. (2013). Polyacrylonitrile membranes via thermally induced phase separation: effects of polyethylene glycol with different molecular weights. *J. Membr. Sci.* 437: 227–236. https://doi.org/10.1016/j.memsci.2013.03.018.

27 Briber, R.M. and Khoury, F. (1987). The phase diagram and morphology of blends of poly(vinylidene fluoride) and poly(ethyl acrylate). *Polymer (Guildf.)* 28: 38–46. https://doi.org/10.1016/0032-3861(87)90316-8.

28 Lorenzo, A.T., Arnal, L., Albuerne, J., and Mu, A.J. (2007). DSC isothermal polymer crystallization kinetics measurements and the use of the Avrami equation to fit the data: guidelines to avoid common problems. *Polym. Test.* 26: 222–231. https://doi.org/10.1016/j.polymertesting.2006.10.005.

29 George, A., Sanjay, M.R., Srisuk, R. et al. (2020). A comprehensive review on chemical properties and applications of biopolymers and their composites. *Int. J. Biol. Macromol.* 154: 329–338. https://doi.org/10.1016/j.ijbiomac.2020.03.120.

30 Matta, A.K., Rao, R.U., Suman, K.N.S., and Rambabu, V. (2014). Preparation and characterization of biodegradable PLA/PCL polymeric blends, MSPRO. 6: 1266–1270. https://doi.org/10.1016/j.mspro.2014.07.201.

31 Guo, Q. (2016). *Polymer Morphology: Principles, Characterization, and Processing*. Wiley.

32 Patrício, T. and Bártolo, P. (2013). Thermal stability of PCL/PLA blends produced by physical blending process. *Procedia Eng.* 59: 292–297. https://doi.org/10.1016/j.proeng.2013.05.124.

33 Unger, M., Sedlmair, J., Siesler, H.W. et al. (2014). 3D FT-IR imaging spectroscopy of phase-separation in a poly(3-hydroxybutyrate)/poly(L-lactic acid) blend. *Vib. Spectrosc.* 75: 169–172.

34 Marlina, D., Sato, H., Hoshina, H., and Ozaki, Y. (2018). Intermolecular interactions of poly(3-hydroxybutyrate-*co*-3-hydroxyvalerate) (P(HB-*co*-HV)) with PHB-type crystal structure and PHV-type crystal structure studied by low-frequency Raman and terahertz spectroscopy. *Polymer (Guildf.)* 135: 331–337. https://doi.org/10.1016/j.polymer.2017.12.030.

35 Doi, Y. and Steinbüchel, A. (eds.) (2001). *Polyesters II: Properties and Chemical Synthesis (Biopolymers)*, vol. 3b. Weinheim: Wiley-VCH.

36 Zhang, J., Duan, Y., Sato, H. et al. (2005). Crystal modifications and thermal behavior of poly(L-lactic acid) revealed by infrared spectroscopy. *Macromolecules* 38: 8012–8021. https://doi.org/10.1021/ma051232r.

37 Fleming, O.S., Chan, K.L.A., and Kazarian, S.G. (2006). High-pressure CO_2-enhanced polymer interdiffusion and dissolution studied with in situ ATR–FTIR spectroscopic imaging. *Polymer (Guildf.)* 47: 4649–4658. https://doi.org/10.1016/j.polymer.2006.04.059.

38 Lasch, P. and Noda, I. (2019). Two-dimensional correlation spectroscopy (2D-COS) for analysis of spatially resolved vibrational spectra. *Appl. Spectrosc.* 73: 359–379. https://doi.org/10.1177/0003702818819880.

39 Noda, I. (2009). Generalized two-dimensional correlation spectroscopy. *Front. Mol. Spectrosc.*: 367–381.

40 Shinzawa, H., Mizukado, J., and Kazarian, S.G. (2017). Fourier transform infrared (FT-IR) spectroscopic imaging analysis of partially miscible PMMA–PEG blends using two-dimensional disrelation mapping. *Appl. Spectrosc.* 71: 1189–1197. https://doi.org/10.1177/0003702816670917.

41 Sun, B., Lin, Y., Wu, P., and Siesler, H.W. (2008). A FTIR and 2D-IR spectroscopic study on the microdynamics phase separation mechanism of the poly(*N*-isopropylacrylamide) aqueous solution. *Macromolecules* 41: 1512–1520. https://doi.org/10.1021/ma702062h.

42 Zhang, J., Sato, H., Furukawa, T. et al. (2006). Crystallization behaviors of poly(3-hydroxybutyrate) and poly(L-lactic acid) in their immiscible and miscible blends. *J. Phys. Chem. B* 110: 24463–24471. https://doi.org/10.1021/jp065233c.

43 Nogales, A., Hsiao, B.S., Somani, R.H. et al. (2001). Shear-induced crystallization of isotactic polypropylene with different molecular weight distributions: in situ small- and wide-angle X-ray scattering studies. *Polymer (Guildf.)* 42: 5247–5256.

44 Somani, R.H., Hsiao, B.S., Nogales, A. et al. (2000). Structure development during shear flow-induced crystallization of i-PP: in-situ small-angle X-ray scattering study. *Macromolecules* 33: 9385–9394. https://doi.org/10.1021/ma001124z.

45 Lai, W.C., Bin Liau, W., and Lin, T.T. (2004). The effect of end groups of PEG on the crystallization behaviors of binary crystalline polymer blends PEG/PLLA. *Polymer (Guildf.)* 45: 3073–3080. https://doi.org/10.1016/j.polymer.2004.03.003.

46 Chen, X., Hou, G., Chen, Y. et al. (2007). Effect of molecular weight on crystallization, melting behavior and morphology of poly(trimethylene terephalate). *Polym. Test.* 26: 144–153. https://doi.org/10.1016/j.polymertesting.2006.08.011.

47 Xu, J., Zhang, S., and Guo, B. (2017). Insights from polymer crystallization: chirality, recognition and competition. *Chin. Chem. Lett.* 28: 2092–2098. https://doi.org/10.1016/j.cclet.2017.10.012.

48 Gupper, A., Chan, K.L.A., and Kazarian, S.G. (2004). FT-IR imaging of solvent-induced crystallization in polymers. *Macromolecules* 37: 6498–6503. https://doi.org/10.1021/ma049313v.

49 El-Hadi, A., Schnabel, R., Straube, E. et al. (2002). Effect of melt processing on crystallization behavior and rheology of poly(3-hydroxybutyrate) (PHB) and its blends. *Macromol. Mater. Eng.* 287: 363–372. https://doi.org/10.1002/1439-2054(20020501)287:5<363::AID-MAME363>3.0.CO;2-D.

50 Zhang, J., Sato, H., Noda, I., and Ozaki, Y. (2005). Conformation rearrangement and molecular dynamics of poly(3-hydroxybutyrate) during the melt-crystallization process investigated by infrared and two-dimensional infrared correlation spectroscopy. *Macromolecules* 38: 4274–4281. https://doi.org/10.1021/ma0501343.

51 Lu, H., Kazarian, S.G., and Sato, H. (2020). Simultaneous visualization of phase separation and crystallization in PHB/PLLA blends with in situ ATR-FTIR spectroscopic imaging. *Macromolecules* 53 (20): 9074–9085.

52 Handa, Y.P., Zhang, Z., and Roovers, J. (2001). Compressed-gas-induced crystallization in *tert*-butyl poly(ether ether ketone). *J. Polym. Sci., Part B: Polym. Phys.* 39: 1505–1512.

53 Fleming, O.S. (2006). Application of advanced vibrational spectroscopy to polymers processed with supercritical CO_2. PhD thesis. Imperial College London.

54 Kazarian, S.G. (2002). Polymers and supercritical fluids: opportunities for vibrational spectroscopy. *Macromol. Symp.* 184: 215–228. https://doi.org/10.1002/1521-3900(200208)184:1<215::AID-MASY215>3.0.CO;2-X.

55 Ewing, A.V. and Kazarian, S.G. (2015). Interaction of supercritical carbon dioxide with polymers studied by vibrational spectroscopy. In: *Supercritical Fluid Nanotechnology Advances and Applications in Composites and Hybrid Nanomaterials*, 481–516. Jenny Stanford Publishing https://doi.org/10.1201/b19242.

56 Kazarian, S.G., Brantley, N.H., West, B.L. et al. (1997). In situ spectroscopy of polymers subjected to supercritical CO_2: plasticization and dye impregnation. *Appl. Spectrosc.* 51: 491–494. https://doi.org/10.1366/0003702971940765.

57 Labuschagne, P.W., Kazarian, S.G., and Sadiku, R.E. (2011). In situ FTIR spectroscopic study of the effect of CO_2 sorption on H-bonding in PEG–PVP mixtures. *Spectrochim. Acta, Part A* 78: 1500–1506. https://doi.org/10.1016/j.saa.2011.01.040.

58 Beckman, E. and Porter, R.S. (1987). Crystallization of bisphenol a polycarbonate induced by supercritical carbon dioxide. *J. Polym. Sci., Part B: Polym. Phys.* 25: 1511–1517. https://doi.org/10.1002/polb.1987.090250713.

59 Handa, Y.P., Zhang, Z., and Wong, B. (1997). Effect of compressed CO_2 on phase transitions and polymorphism in syndiotactic polystyrene. *Macromolecules* 30: 8499–8504. https://doi.org/10.1021/ma9712209.

60 Parzuchowski, P.G., Gregorowicz, J., Wawrzyńska, E.P. et al. (2016). The phase behavior in supercritical carbon dioxide of hyperbranched copolymers with architectural variations. *J. Supercrit. Fluids* 107: 657–668. https://doi.org/10.1016/j.supflu.2015.07.028.

61 Herek, L.C.S., Oliveira, R.C., Rubira, A.F., and Pinheiro, N. (2006). Impregnation of pet films and PHB granules with curcumin in supercritical CO_2. *Braz. J. Chem. Eng.* 23: 227–234. https://doi.org/10.1590/S0104-66322006000200010.

62 Fleming, O.S. and Kazarian, S.G. (2006). Polymer processing with supercritical fluids. In: *Supercritical Carbon Dioxide: Polymer Reaction. Engineering*, 205–238. https://doi.org/10.1002/3527606726.ch10.

63 Kazarian, S.G. and Chan, K.L.A. (2004). FTIR imaging of polymeric materials under high-pressure carbon dioxide. *Macromolecules* 37: 579–584. https://doi.org/10.1021/ma035420y.

64 Lu, H. and Kazarian, S.G. (2020). How does high-pressure CO_2 affect the morphology of PCL/PLA blends? Visualization of phase separation using in situ ATR–FTIR spectroscopic imaging. *Spectrochim. Acta, Part A* 243: 118760. https://doi.org/10.1016/j.saa.2020.118760.

65 Kazarian, S.G., Vincent, M.F., Bright, F.V. et al. (1996). Specific intermolecular interaction of carbon dioxide with polymers. *J. Am. Chem. Soc.* 118: 1729–1736. https://doi.org/10.1021/ja950416q.

66 Mensitieri, G., Scherillo, G., Panayiotou, C., and Musto, P. (2020). Towards a predictive thermodynamic description of sorption processes in polymers: the synergy between theoretical EoS models and vibrational spectroscopy. *Mater. Sci. Eng., R* 140: 100525. https://doi.org/10.1016/j.mser.2019.100525.

67 Li, S., Chen, T., Liao, X. et al. (2020). Effect of macromolecular chain movement and the interchain interaction on crystalline nucleation and spherulite growth of polylactic acid under high-pressure CO_2. *Macromolecules* 53: 312–322. https://doi.org/10.1021/acs.macromol.9b01601.

68 Kazarian, S.G. (2000). Polymer processing with supercritical fluids. *Polym. Sci. Ser. C* 42: 78–101.

69 Gabrienko, A.A., Ewing, A.V., Chibiryaev, A.M. et al. (2016). New insights into the mechanism of interaction between CO_2 and polymers from thermodynamic parameters obtained by in situ ATR–FTIR spectroscopy. *Phys. Chem. Chem. Phys.* 18: 6465–6475. https://doi.org/10.1039/c5cp06431g.

70 Chiou, J.S., Barlow, J.W., and Paul, D.R. (1985). Polymer crystallization induced by sorption of CO_2 gas. *J. Appl. Polym. Sci.* 30: 3911–3924.

71 Takada, M., Tanigaki, M., and Ohshima, M. (2001). Effects of CO_2 on crystallization kinetics of polypropylene. *Polym. Eng. Sci.* 41: 1938–1946. https://doi.org/10.1002/pen.10890.

72 Takada, M., Hasegawa, S., and Ohshima, M. (2004). Crystallization kinetics of poly(L-lactide) in contact with pressurized CO_2. *Polym. Eng. Sci.* 44: 186–196. https://doi.org/10.1002/pen.20017.

73 Mizoguchi, K., Hirose, T., Naito, Y., and Kamiya, Y. (1987). CO_2-induced crystallization of poly(ethylene terephthalate). *Polymer (Guildf.)* 28: 1298–1302.

74 Gross, S.M., Roberts, G.W., Kiserow, D.J., and Desimone, J.M. (2000). Crystallization and solid-state polymerization of poly(bisphenol A carbonate) facilitated by supercritical CO_2. *Macromolecules*: 40–45. https://doi.org/10.1021/ma990901w.

75 Fleming, O.S., Chan, K.L.A., and Kazarian, S.G. (2004). FT-IR imaging and Raman microscopic study of poly(ethylene terephthalate) film processed with supercritical CO_2. *Vib. Spectrosc.* 35: 3–7. https://doi.org/10.1016/j.vibspec.2003.10.003.

76 Flichy, N.M.B., Kazarian, S.G., Lawrence, C.J., and Briscoe, B.J. (2002). An ATR–IR study of poly(dimethylsiloxane) under high-pressure carbon dioxide: simultaneous measurement of sorption and swelling. *J. Phys. Chem. B*: 754–759.

77 Han, Y., Zheng, H., Jing, X., and Zheng, L. (2018). Swelling behavior of polyester in supercritical carbon dioxide. *J. CO_2 Util.* 26: 45–51. https://doi.org/10.1016/j.jcou.2018.04.017.

78 Ewing, A.V., Gabrienko, A.A., Semikolenov, S.V. et al. (2015). How do intermolecular interactions affect swelling of polyketones with a differing number of carbonyl groups? An in situ ATR–FTIR spectroscopic study of CO_2 sorption in polymers. *J. Phys. Chem. C* 119: 431–440. https://doi.org/10.1021/jp510208e.

79 Sirard, S.M., Green, P.F., and Johnston, K.P. (2001). Spectroscopic ellipsometry investigation of the swelling of poly(dimethylsiloxane) thin films with high pressure carbon dioxide. *J. Phys. Chem. B* 105: 766–772. https://doi.org/10.1021/jp002592d.

80 Pasquali, I., Andanson, J.M., Kazarian, S.G., and Bettini, R. (2008). Measurement of CO_2 sorption and PEG 1500 swelling by ATR–IR spectroscopy. *J. Supercrit. Fluids* 45: 384–390. https://doi.org/10.1016/j.supflu.2008.01.015.

81 Duarte, A.R.C., Anderson, L.E., Duarte, C.M.M., and Kazarian, S.G. (2005). A comparison between gravimetric and in situ spectroscopic methods to measure the sorption of CO_2 in a biocompatible polymer. *J. Supercrit. Fluids* 36: 160–165. https://doi.org/10.1016/j.supflu.2005.04.003.

3

Interfaces in Polymer Nanocomposites Characterized by Spectroscopic Techniques

Liliane Bokobza

Former Professor at ESPCI, 194-196 Boulevard Bineau, 92200 Neuilly-Sur-Seine, France

3.1 Introduction

Mineral fillers are often incorporated into polymeric materials to achieve new physical properties of the resulting composite as, for example, electrical conduction imparted by carbon black (CB) particles to insulating polymeric media [1]. The analysis of the traditional composites has established that the size of the particle, their aspect ratio, their state of dispersion, and essentially their surface characteristics that determine the interface of the polymer–filler system have a strong influence on the properties of the final product [2]. By creating very large polymer–particle interfacial zone, nanoscale fillers with isotropic or anisotropic sheet-like or needle-like morphologies bring much more improved properties than those provided by micron-scale fillers at a same filler content [3]. Therefore, the amount of interfacial region depends on the quality of the filler dispersion also affected by the nanofiller/matrix interfacial interactions [4]. As recently recalled [5], a poor dispersion of the nanosized materials strongly prevents the full realization of their reinforcing capability.

It is admitted that the interfacial region is a critical component in nanocomposite properties, and understanding its role on the macroscale properties of the filled system can allow to tailor interactions with the polymer matrix for the design of advanced materials with specific applications. In particular, the interfacial interactions may cause a change in the polymer chain mobility in the vicinity of the particle surface and impact the mechanical and dielectric behaviors. The formation of a filler network in the host matrix is also highly dependent on the interactions between the two phases. This phenomenon has been widely studied in elastomeric composites through their dynamic mechanical properties by looking at the strain dependence of the storage modulus [6, 7].

The macroscopic information provided by the mechanical analysis can be nicely complemented by molecular spectroscopies able to yield a molecular-level characterization of the polymer–filler interface in filled systems. This chapter is intended

Spectroscopic Techniques for Polymer Characterization: Methods, Instrumentation, Applications,
First Edition. Edited by Yukihiro Ozaki and Harumi Sato.

to illustrate the potential of molecular spectroscopies (fluorescence, solid-state nuclear magnetic resonance (NMR), infrared, and Raman spectroscopies) for the analysis of interfacial effects between polymer chains and the filler surface. Investigations by fluorescence spectroscopy require the use of a fluorescent probe whose emission behavior is sensitive to changes in its surrounding environment. The analysis of the fluorescent response of the incorporated label can provide interesting insight into the polymer dynamics in the interfacial region. One of the major advantages of solid-state NMR is to explore, through the relaxation data, polymer mobility around the filler particles and to estimate the thickness of the interfacial layer formed by polymer–filler interactions. Infrared and Raman spectroscopies provide complementary vibrational information. By the identification of chemical functional groups, they allow the determination of the chemical and physical structure of polymers and are used to probe interactions between different chemical groups. On account of its enhanced sensitivity for the detection of carbon-based materials, the interest in Raman spectroscopy has been revived in the last two decades with the advent of carbon nanofillers materials used to reinforce polymeric systems.

The theoretical backgrounds and the significance of the different spectroscopic techniques in the field of polymer nanocomposites can be found in reference [8] and in the references therein.

3.2 Types of Interactions at the Interface

Interactions between the organic and inorganic phases can take place between functional groups of both polymer chains and filler particles. In some silica-filled systems, interfacial interactions are ensured by hydrogen bonding between the silanols present on the silica surfaces and the oxygen-containing functional groups of polymeric media. Poly(dimethylsiloxane) (PDMS) [9, 10], poly(methyl methacrylate) (PMMA) [11, 12], poly(vinyl acetate) [12, 13], or epoxy resins [14] are examples of polymers able to interact with the silica surface. The polymers are usually filled with fumed silica generally treated with a processing aid to deactivate part of the silanols to decrease filler–filler interactions [15]. It is also of interest to mention that the hydrophilic silica surface can attract water molecules, which hinders a direct contact between the filler and the polymer and thus a more efficient adhesion [14]. The silica–water interface can be characterized by Fourier-transform infrared spectroscopy (FTIR) but more precisely in the near-infrared region (NIR) on account of the strong intensity of the fundamental modes of the water molecule in the mid-IR [16]. The near-infrared spectra of silica-filled PDMS networks (Figure 3.1) display, in addition to combination bands of the fundamental modes of the methyl groups of the PDMS chains, a band of the water molecules present at the water–silica interface. This band located around 5260 cm^{-1} and ascribed to a combination of one of the stretching mode (ν) and the bending mode (δ) of the water molecule, increases with the amount of silica expressed in phr (parts per hundred parts of polymer by weight).

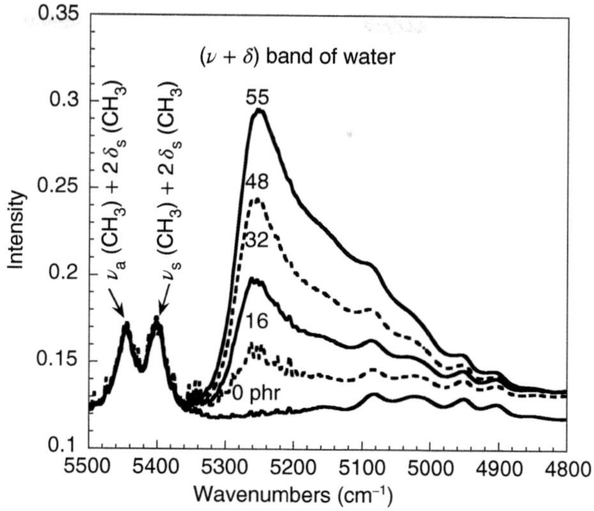

Figure 3.1 Near-infrared spectra of silica-filled PDMS networks. Each curve is labeled with the amount of silica expressed in phr (parts per hundred parts of polymer). *Source:* Dewimille et al. [16].

Despite the possible surface treatment of silica particles, conventional ex situ blending techniques lead to agglomerated filler structures and inhomogeneous dispersions in the resulting composites. The use of a sol–gel route based on the hydrolysis and the condensation, in the presence of a catalyst, of inorganic alkoxides such as tetraethoxysilane (TEOS) for example, has proved to generate small and well-dispersed particles within the host matrix. The morphology of the inorganic structures depends on the hydrolysis and condensation conditions and especially on the nature of the catalyst used to accelerate the gelation process and also on the reactivity of the alkoxide. Moreover, the in situ formation of the filler particles can be carried out before or after the polymerization of macromolecular chains and in the case of elastomeric materials before, after, or during the cross-linking process.

SiO_2 and TiO_2 nanoparticles have been successfully generated in already-preformed PDMS networks to prevent the growth of aggregated particles [16, 17]. The analysis of the surface silanol hydroxyl groups by solid-state ^{29}Si-NMR spectroscopy showed that the silica particles formed by the sol–gel process display a higher content of isolated silanols than a fumed silica [16]. In addition to the increased hydrophilicity of the particle surface, the improved quality of the dispersion results in increased interactions at the polymer–filler interfaces and thus in improved properties. The analysis by Song et al. [18] of in situ and ex situ modification of silica particles in silicone rubber has demonstrated the higher reinforcing capability of the in situ generated particles on account of their high surface area. The appraisal of the sol–gel process for the synthesis of inorganic oxide particles in epoxy and PDMS matrices has been reviewed by Adnan et al. [19].

The sol–gel method has been extended to several other polymer systems that have, contrary to PDMS, a low affinity with silica [20–26]. In that case, to ensure polymer–filler interactions, most of the time, a coupling agent is required to improve adhesion between the two phases by forming bridges at the interface between the matrix and the inorganic particle surface. Coupling agents are generally bifunctional molecules able to react with both the polymer chains and the silanols on the silica surface. A typical example is that of the bis(3-triethoxysilylpropyl)tetrasulfide (TESPT), commonly abbreviated "Si69," that bear ethoxy groups that can interact with silica and a tetrasulfane function that can react with a hydrocarbon rubber as natural rubber (NR).

Nowadays, carbon-based nanofillers such as carbon nanotubes (CNTs), graphene, and graphene oxide (GO) can provide unique nanostructures and superior properties due to their high aspect ratio and their outstanding mechanical properties. But on account of their low solubility and poor dispersibility in polymer matrices, modification of their surface chemistry through functionalization is required to achieve their full potential and exploit their superior properties. The surface chemistry plays a crucial role in promoting a strong covalently bonded interface since the new functional groups introduced on the aromatic structure have to interact with the polymer chains to avoid particle aggregation. Different synthesis methods for functionalizing carbon nanostructures and tailoring surface properties have been reported in the literature [27–30].

Dyke and Tour [31] have shown that sidewall functionalization of surfactant-wrapped single-walled carbon nanotubes (SWCNTs) leads to unbundled entities able to be well dispersed in polymeric systems. A usual measure of the degree of functionalization is a strong reduction of the intensity of the absolute Raman signals of the nanotubes and an increase in the D band. Let us recall that the D band, located between 1300 and 1400 cm^{-1}, is Raman active in defective graphitic lattice owing to defect-induced double-resonance Raman scattering processes [32]. It increases with the amount of disorders or defects or upon functionalization as a result of a loss of resonance enhancement by disruption of the sp^2-hybridized carbon lattice [33]. In Figure 3.2 are displayed the Raman spectra of pristine and functionalized SWCNTs and after annealing at 750 °C in argon [34]. Heating-functionalized nanotubes in an inert atmosphere remove the organic species and restore the pristine nanotube structure [34].

Koval'chuk et al. [35] functionalized multiwall carbon nanotubes (MWCNTs) via attachment of $-(CH_2)_{10}CH_3$ alkyl chains to provide interfacial interactions with polypropylene that have nonpolar chemical structure. The Raman spectrum of the functionalized MWCNTs is not significantly different from that of the pristine material, but FTIR spectroscopy gives evidence of the presence of the aliphatic chains on the MWCNTs. Even if the functionalization leads to an improved dispersion in the host medium, it is concluded that the creation of a stronger interface requires the incorporation of polar functional groups in the polymer chains. Recently, Tamore et al. [36] report the synthesis of functionalized MWCNTs using ethyl-4-amino cinnamate and analyze the effect of their incorporation on the properties of silicone rubber nanocomposites. No clear evidence is given on interactions between the cinnamate group and the polymer chains. Moreover, no significant decrease in the swelling ratio is observed indicating poor interfacial effects. Other studies have mentioned the exceptional affinity of silicon rubber for unfunctionalized MWCNTs attributed to strong CH-π interactions between the PDMS methyl groups and the π-electron-rich surface of MWCNTs [37–39]. Modeling studies based on molecular dynamics have shown that the polymer chain

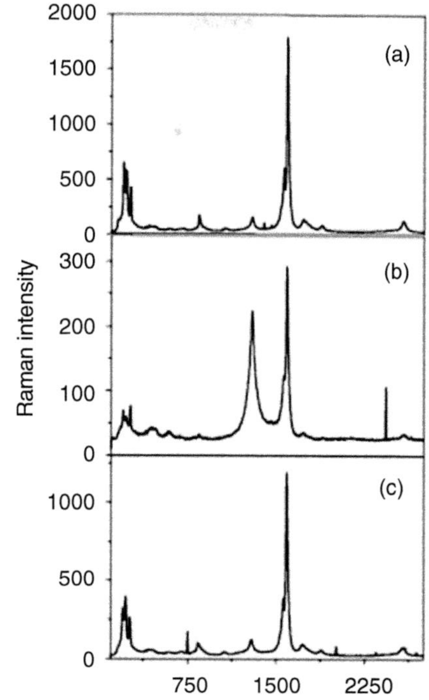

Figure 3.2 Raman spectra (780.6 nm excitation) of (a) pristine SWCNTs, (b) functionalized SWCNTs, (c) after heating at 750 °C in argon. *Source:* Dyke and Tour [34].

completely wraps around the CNT surface [37] with a first layer of methyl groups pointing toward the CNT surface.

The ability to exfoliate graphite into individual or few-layer sheets has attracted an enormous interest in carbon research on account of the extraordinary properties of the two-dimensional sheet of sp^2-bonded carbon atoms called graphene, making it one of the most promising carbon nanostructures for the design of new composite materials [40]. Nevertheless, as in the case of CNTs, graphene nanosheets require chemical modification to improve their processability [28, 41–47]. Different synthetic strategies leading to functionalized graphene sheets are reported in the literature. Organic functionalities can be covalently attached to graphene or few-layer graphenes produced by exfoliation of graphite in organic solvents. Chattopadhyay et al. [41] prepare an exfoliated soluble graphite by adding *n*-dodecyl iodide to the graphite that has been previously treated with lithium in the presence of liquid ammonia. The covalent attachment of the dodecyl radicals was confirmed by the increase in intensity of the D band in the Raman spectrum of the functionalized material.

Other methods of functionalizing graphene nanosheets include nitrene addition of azido-phenylalanine to exfoliated microcrystalline graphite [42], 1,3-dipolar cycloaddition of azomethine ylides [43], functionalization of surfactant wrapped graphene sheets with alkylazides [44], or plasma fluorination treatments [48, 49].

One popular way for the formation of covalently functionalized graphene sheets is the use of GO [50, 51], obtained by an oxidation process with strong acidic media leading to the introduction of oxygen-containing functional groups making GO hydrophilic, thus ensuring better interfacial interactions with organic polymers (Figure 3.3). The chemical oxidation generates structural defects that disrupt the graphene structure and lower its electrical conductivity. A reduction process based essentially on chemical or thermal treatments [52, 53] can partially restore the graphitic network leaving however some oxygen groups and defects. Hydrogen bonding between the hydroxyl and carboxyl groups on the GO surface and functional groups of a polymer is expected to favor the dispersion of the GO sheets [29, 54].

In GO/PDMS composites where hydrogen bonds would have been expected between OH groups of GO and the oxygen atoms of the polymer chains, GO clusters have been observed leading to the belief that the interaction between the GO sheets is stronger than that between GO and PDMS [55]. In a recent work comparing the influence of different carbon-based fillers on electrical and mechanical properties of a polymer blend [56], it is shown that melt-mixed graphene at a concentration of 1 wt% did not allow electrical percolation to be obtained, while only 0.25 wt% of MWCNTs are needed to form a percolated network. This reflects a poor dispersion of graphene and a weak interfacial adhesion with the matrix. Conversely, in the work of Gkourmpis et al. [57], a melt-mixing protocol is also used for the preparation of a composite consisting of a highly isotactic polypropylene and a deagglomerated hierarchical thermally reduced GO. The percolation threshold is obtained around 1 wt%, indicating the existence of a superior dispersion process. These results point out the importance of the processing techniques of the nanocomposites and of the filler treatments devoted to the production of graphene nanosheets.

Another direction mentioned in the review of Singh et al. [30] is the simultaneous reinforcement of epoxy by graphene and CNTs to get synergistic effects between the two different carbon nanostructures. Superior properties are reported by hybrid nanocomposites attributed to better dispersion, the larger aspect ratio of CNTs expected to avoid agglomeration, and restacking of graphene layers.

3.3 Characterization of the Interfaces

Various techniques have been used for the characterization of the interfacial region and on the understanding of its influence on the resulting properties of the nanocomposites.

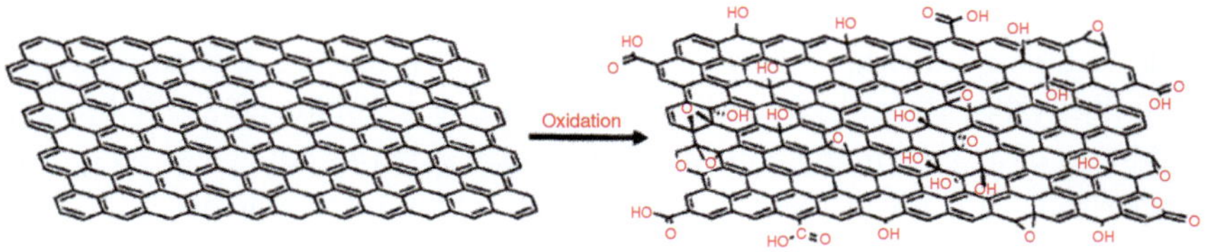

Figure 3.3 Oxidation of graphene sheet to form graphene oxide. *Source:* Georgakilas et al. [50].

The mechanical properties of the composites can serve as a qualitative indicator in evaluating the quality of the interface. The addition of filler particles in a polymeric media is well known to affect the macroscopic mechanical properties of the composites. It leads to an increase in the composite stiffness attributed to the inclusion of rigid particles. This effect depends on the volume fraction and on the aspect ratio of filler according to the Guth equation in the case of filled elastomeric composites [58]. Another contribution arises from interfacial bonding and filler–filler interactions that lead to the formation of a percolated filler network.

The tensile measurements of rubber composites represented in the form of the nominal stress, σ, against strain, ε, can be plotted in the Mooney–Rivlin representation [59, 60] that plots the reduced stress, $[\sigma^*]$ defined by the quantity: $[\sigma^*] = \sigma/(\alpha - \alpha^{-2})$, where α is the extension ratio (ratio of the final length of the sample along the direction of stretch to that of the initial length before deformation, $\varepsilon = \alpha - 1$). The Mooney–Rivlin representation better visualizes specific features of the stress–strain curves and displayed interesting points regarding interfacial adhesion in composite materials.

In highly interacting polymer–filler systems such as silica-filled PDMS networks, an increase in the reduced stress is observed at high elongations as a result of limited chain extensibility of polymer chains (Figure 3.4a). This effect reflects a strong polymer–filler interface due to polymer–filler attachments acting as additional cross-links to the network structure. No upturn in the modulus is observed in the styrene-butadiene rubber (SBR) filled with MWCNTs (Figure 3.4b) indicating a weak interfacial adhesion between the two phases despite a strong reinforcing effect attributed to the high aspect ratio of CNTs [61]. Also associated with the high aspect ratio is the formation of a filler network at a small filler content as evidenced by the strong decrease in stress observed at low deformations. It is also of interest to mention the different tensile behavior of the carbon black (CB)-filled SBR composite that exhibits at high deformations the upturn in the modulus reflecting, as already known, the reinforcing ability of CB particles for hydrocarbon rubbers mainly attributed to interactions occurring at the filler–polymer interface [62]. The strong increase in stress displayed by the unfilled NR (Figure 3.4c) is ascribed to

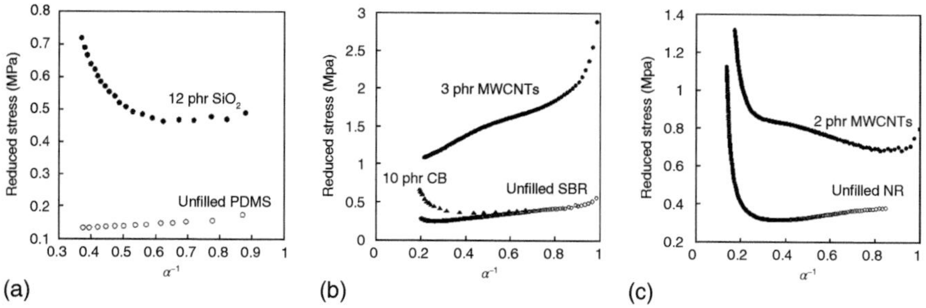

Figure 3.4 Mooney–Rivlin plots for: (a) unfilled poly(dimethylsiloxane) (PDMS) and silica-filled PDMS; (b) unfilled styrene-butadiene rubber (SBR) and SBR filled with multiwall carbon nanotubes (MWCNTs) and carbon black (CB); (c) unfilled natural rubber (NR) and MWCNT-filled NR. *Source:* (b) Bokobza [61]. Licensed under CC BY-4.0.

the strain-induced crystallization phenomenon usually explained by the uniform microstructure (cis-configuration of the macromolecular chains). Interestingly, the onset of crystallization occurs at a lower strain value in the filled sample, which may be explained by the orientation of CNTs along the direction of stretch inducing that of the polymer chains close to the filler surface. In a chapter devoted to the impact of crystallization on the interface in polymer nanocomposites [63], it is concluded that the interface in polymer–clay systems is substantially affected by crystallization of the polymer through crystallites nucleated at the surface of the clay. In polyacrylonitrile/CNT composites, it was shown that the CNTs cause conformation strain in the polymer chains near the CNT surface that depends on the degree of chain preorientation and crystallinity [64, 65].

The above considerations highlight the importance of the polymer–filler interface that largely governs the mechanical performance of the composite material. In particular, a weak interfacial adhesion leads to agglomerated filler structures that act as defects and failure points thus affecting negatively the rupture properties of the resulting composites. In addition to the macroscopic information obtained from the mechanical analysis, a molecular-level characterization provided by fluorescence, solid-state NMR, infrared, and Raman techniques is required to identify the interacting species or to evaluate the mobility of polymer chains at the surface of the filler particle to get a better understanding of the factors that affect the interfacial region. Different examples are given below to illustrate the type of information gained from each type of molecular spectroscopy.

3.3.1 Fluorescence Spectroscopy

Fluorescence spectroscopy has been widely used for investigating at a molecular level various processes in polymer science by incorporating a fluorescent probe molecule able to yield information on the properties of its immediate environment. Polymerization monitoring, dynamical, and relaxation processes of macromolecular chains are examples of topics currently investigated by fluorescence spectroscopy. In particular, fluorescent probes can detect the glass transition temperature T_g, above which the polymer is in the rubbery state and displays a larger free volume. They can be dispersed in the polymeric medium or covalently attached to the macromolecular chains at low concentrations so that no important perturbation is induced in the system under investigation. Ellison and Torkelson [66] have shown that fluorophores intrinsic to the polymer, freely dispersed or covalently linked to the polymer as a label, can be used to sense the glass transition temperature of polymeric systems. A break is usually observed in the temperature-dependent fluorescence intensity upon cooling from the rubbery to the glassy state. The work points out that the probe should be adequately chosen since unexpected temperature effects can be observed as in the case of poly(styrene-*co*-pyrene-labeled methacrylate) for which an interaction of the carbonyl group and the chromophore is expected. In polymer nanocomposites, when interactions are present between the polymer and the filler particles, interface confinement effect can lead to an enhancement of T_g as seen in Figure 3.5 that represents the effect of silica on the

T_g values of poly(2-vinyl pyridine) (P2VP), poly(methyl methacrylate) (PMMA), and polystyrene (PS) doped at trace levels of 1,10-bis(1-pyrene) decane (BPD) [67]. Without attractive interfacial interactions in PS–silica systems, no effect on the T_g of PS is observed upon addition of silica contrary to silica–PMMA and silica–P2VP composites where T_g exhibits significant deviations relative to bulk polymers on account of attractive polymer–nanoparticle hydrogen-bonding interactions.

On account of the impact of the surrounding probe environment on its fluorescence response, fluorescence methods are particularly well suited for gaining important insights into the characterization of the interface and interphase of polymer–filler systems. Important aspects about the use of the fluorescent label (Dansyl label) chemically bonded to specific sites of epoxy filled with silica substrates (silica particles or glass fibers) have been discussed [68, 69]. The filler is first silanized with a silane used as a coupling agent and chemically bonded to the filler/polymer interface or to the epoxy bulk. Differences between the polymer dynamics in the two regions are observed revealing an interfacial region slightly more rigid at the interface than in the bulk. The same approach has been used for the study of the interfaces between silica nanoparticles and thermoplastic polymer matrices by using the fluorescent response from pyrene-1-sulfonamide [70]. In the work of Antonelli et al. [71], the interfacial characterization in epoxy/silica nanocomposites is carried out by using the Dansyl chloride as the fluorescent marker and organosiloxanes and polyetheramines as molecular spacers of different lengths to probe different zones of the interfacial region. Steady-state and fluorescence lifetimes of the chromophore reveal rigidity effects at the interface that extend into the bulk matrix.

Quantum dots are new promising fluorescent nanosystems for the design of high-performance composite materials [72]. Their exceptionally small size of few

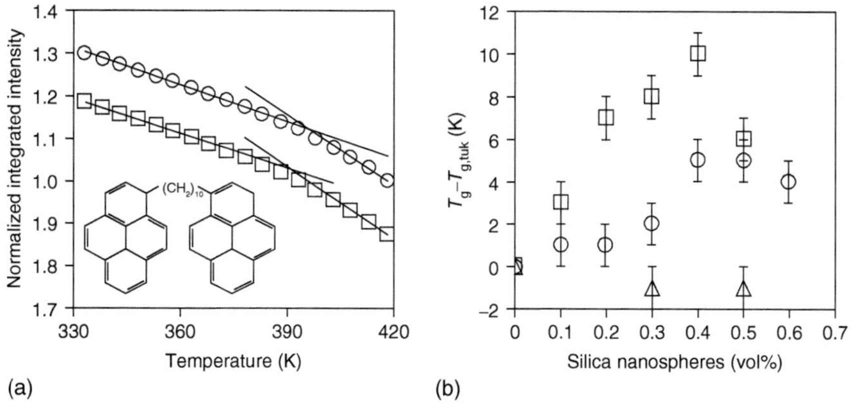

Figure 3.5 T_g data from bulk polymer and polymer nanocomposites determined through fluorescence of 1,10-bis(1-pyrene) decane (BPD). (a) Temperature dependence of the normalized fluorescence of BPD dopant (<0.2 wt%) in a bulk PMMA film (open squares) and in a 0.4 vol% silica–PMMA nanocomposite film (open circles). (b) Deviations from T_g, bulk as a function of silica nanofiller content in three polymer nanocomposites: P2VP (open squares), PMMA (open circles), and PS (open triangles). *Source:* Rittigstein et al. [67].

nm with regard to conventional nanofillers and their photoluminescence properties make them particularly attractive as new additives in polymeric matrices. Inorganic quantum dots are semiconductor nanoparticles made of binary compounds such as zinc sulfide or zinc selenide, cadmium selenide or cadmium telluride, and indium phosphide with size of 2–5 nm [73, 74]. In case of organic quantum dots, size may vary according to the type of nanodots. Organic quantum dots such as carbon, graphene, or GO have emerged as new fascinating nanofillers able to lead to uniform dispersion and strong interfacial interactions when incorporated in polymer matrices even at small loadings [75]. These newcomers in the carbon family can be obtained from different carbon-based precursors, including carbon nanomaterials and more interesting, inexpensive products such as coal, soot, or bio-precursors like banana juice, and citric acid [76–81] by using top-down and bottom-up approaches [82]. In a review devoted to carbon nanodots (CDs), Sciortino et al. [83] consider that these new nanosystems should be regarded as a carbonaceous core whose surface has been passivated by external agents forming a layer a few nanometers thick of functional groups and necessary for carbon nanodots to display a high fluorescence. The authors have showed that the origin of this emission of fluorescence is still a subject of a large debate, but classical interpretations invoke a combination of transitions involving intrinsic and extrinsic states arising from the carbogenic core and from the functional groups on the surface with a core-to-surface migration of the excitation energy.

A state-of-the-art overview has been reported by Kausar [84] on carbon-based quantum dots nanocomposites reinforced in different polymer matrices, including conjugated polymers, epoxy resins, PS, PDMS, and block copolymers. Hyperbranched epoxy/carbon dot nanomaterials have been fabricated using in situ and ex situ methods [85]. It is shown that the in situ technique yields strong interfacial interactions and uniform dispersion of carbon dots because functional groups on the dot surface participate in the formation of the hyperbranched epoxy. This ensures superior mechanical, thermal, and optical properties of the in situ material over the ex situ nanocomposite that only displays physical interactions between carbon dots and the pristine hyperbranched epoxy and also agglomeration of carbon dots. The incorporation of carbon dots in the hyperbranched epoxy through covalent linkages also imparts to the in situ composite a higher fluorescence intensity, exhibiting as already observed [78], excitation wavelength-dependent photoluminescence behavior (Figure 3.6). These exceptional photoluminescence properties can be used for advanced optical applications. PDMS/carbon dot nanocomposite films have been obtained using one-pot synthesis where carbon dots are prepared simultaneously with the polymer host matrix, a strategy that avoids aggregation processes and yields good interfacial effects [86]. Typically, distinct or mixtures of carbon dot precursors, generating different emission properties, are mixed with the silicone elastomer in the presence of the curing agent as depicted in Figure 3.7. This new strategy, through a simple one-pot process and the use of "green" carbon precursors, opens the way for the construction of tunable light emitters. Zeng et al. [87] have studied the interfacial behaviors of luminescent hydrophobic-functionalized graphene quantum dots that enable the synthesis of graphene quantum dot/PS composites.

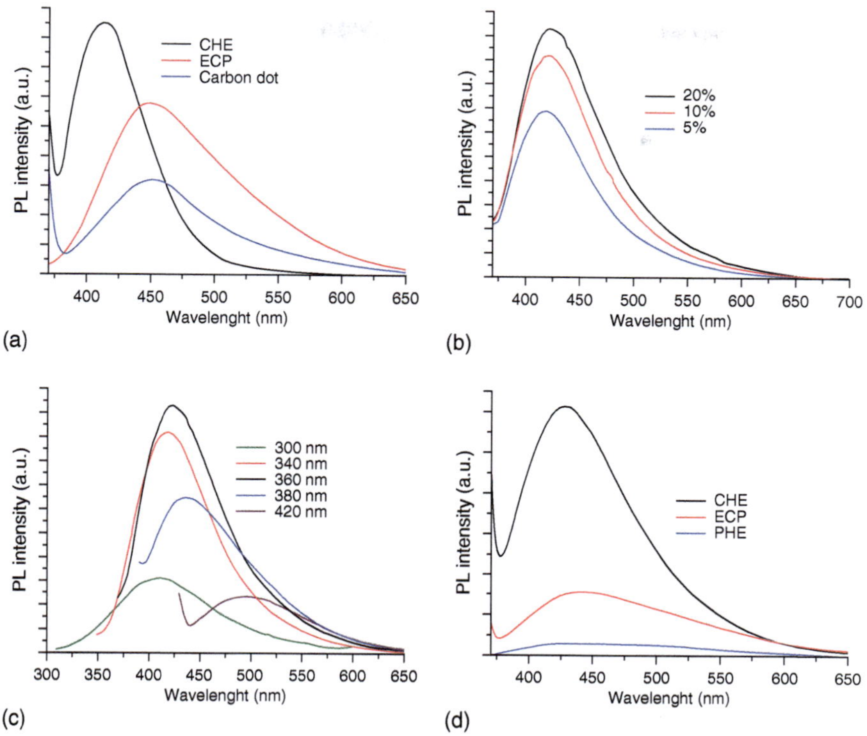

Figure 3.6 Photoluminescence (PL) spectra of (a) CHE (in situ nanocomposite), ECP (ex situ nanocomposite), and carbon dot in ethanolic solution at same concentration (0.001%); (b) CHE (in ethanolic solution) with variation of concentration; and (c) variation of excitation wavelength; and (d) CHE, ECP, and PHE (pristine hyperbranched epoxy) thermosets. *Source:* De et al. [85].

3.3.2 Solid-State NMR Spectroscopy

Solid-state NMR is particularly well suited for the analysis of polymer–filler interfaces and also interphases defined as the volumes of materials affected by the interactions taking place at the interfaces. Polymer–filler interactions usually contribute to the formation of bound polymer where the mobility of macromolecular chains may be strongly hindered in the vicinity of the filler surface. Depending on polymer–filler system, this effect can extend in the interphase region from several nanometers to hundreds of nanometers from the confining surface [88].

The relaxation processes determined by solid-state NMR provide detailed information on polymer dynamics that strongly affect various macroscopic properties of composite materials [89]. The most relevant parameter used to characterize the effects of filler particles on dynamics of polymer chains is the transverse nuclear magnetic relaxation time, T_2, obtained from ^{1}H NMR relaxation methods using spin-echo techniques. The proton relaxation curves are usually fit to a sum of two or three functions reflecting different mobilities [90]. The shorter decaying component is ascribed to the fraction of polymer of restricted mobility, the longer decaying component is related to polymer segments in the bulk phase, while an intermediate

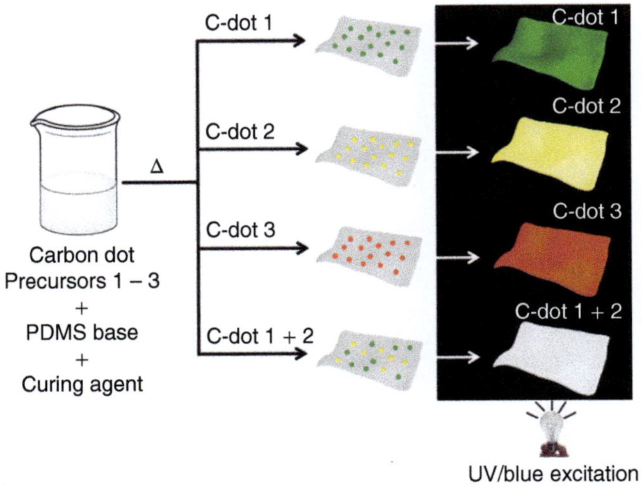

Figure 3.7 Synthesis procedure of luminescent PDMS/carbon dot nanocomposite films. *Source:* Bhunia et al. [86].

decaying component may reflect relaxations in an intermediate proximity to the filler.

Solid-state NMR spectroscopy has been used to characterize the interphase in silica-filled polybutadiene and in carbon black-filled SBR [91]. The surface of the filler was also modified by grafting well-characterized alkyl chains to change the interactions at the polymer–filler interface and monitor the interphase of the filled system. For all filled elastomers, two relaxation times are observed and attributed to tightly and loosely bound rubber components. The surface modification of the filler affects the distribution of the two bound rubbers due to a change in the surface active sites. To get a better understanding of the structure and properties of interphases, Mansencal et al. [92] have analyzed thermal properties and NMR relaxation characteristics of polybutadienes of identical molecular weight but with different microstructures all filled with 50 phr of carbon black. The main idea was to investigate if some groups of the polymer backbone have a specific affinity toward the surface of carbon black. Two spin–spin relaxation times were observed for all the samples making clear the presence of species with different mobilities. Moreover, it was concluded that the vinyl groups have a particular affinity for the filler surface, thus pointing out the existence of specific interactions at the polymer–filler interface.

In silica-filled PDMS networks, specific interactions via hydrogen bonds between the silanols on the filler surface and the oxygen atoms of the polymer chains contribute to the formation of an adsorption layer whose thickness has been estimated around 1.5 nm [16] and resulting as in the above results to two different spin–spin relaxation times ascribed to the polymer in the bulk and to the polymer at the interface.

NMR studies of poly(ethylene glycol) (PEG) macromolecules, well known to form stronger hydrogen bonds with surface hydroxyls of silica [93, 94], confirm the existence of physical adsorption of polymer segments to the particle surface producing a glassy polymer layer [95]. Measurement of the free induction decay (FID) at 70 °C for a PEG of molecular weight 20 000 filled at a silica volume fraction $\phi_c = 0.4$ reveals rigid and mobile signal contributions [95]. A good fit of the FID signal is obtained with the use of an intermediate component with a segmental relaxation time between the glassy and the bulk phases (Figure 3.8).

ten Brinke et al. [96] have studied the interactions between NR and pure silica as well as silicas grafted with various coupling agents also by using ^{1}H NMR T_2 relaxation. Surprisingly, the highest amount of immobilized rubber chains on the silica surface is observed for pure silica. Three separate regions with different mobility are detected in swollen samples: the lowest mobility is attributed to rubber chains and coupling agent molecules tightly bound to the silica surface, and the intermediate and highest mobilities originate from loosely bound and free rubber chains (Figure 3.9).

Borsacchi et al. [97] also use solid-state NMR techniques to investigate polyisoprene samples filled with in situ generated silica. As in the work of ten Brinke et al. [96], three different spin–spin relaxation times T_2 were identified indicating three fractions with different mobilities ascribed to bound rubber, polymer bulk, and free chain ends. The shortest T_2 has been attributed to protons belonging to the bound rubber, but the associated value of 70 μs is higher than that reported in the case of tightly bound rubber indicating, according to the authors, loosely bound rubber only. This is probably the result of a weak interface due to the poor affinity of hydrophilic silica for hydrophobic rubber.

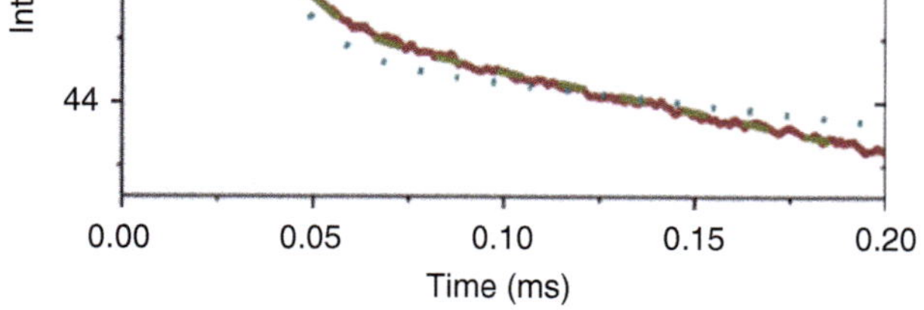

Figure 3.8 Fits to the free induction decay with two or three components for a PEG 20000 at a silica volume fraction of 0.4 and 70 °C. *Source:* Kim et al. [95].

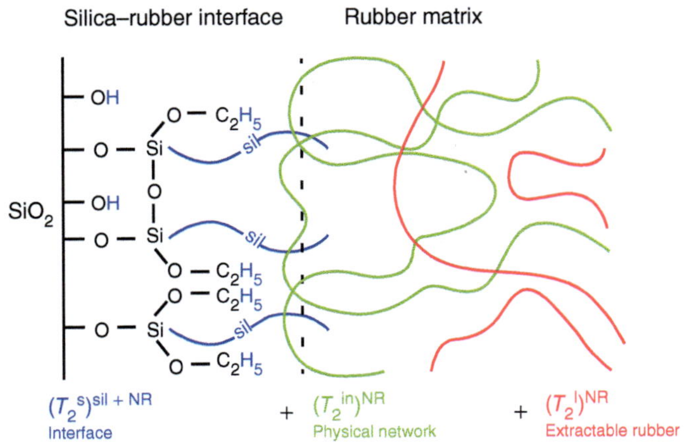

Figure 3.9 Different relaxation components for rubber filled with grafted silica. *Source:* Brinke et al. [96].

3.3.3 Vibrational Spectroscopy

Vibrational spectroscopy, including infrared and Raman spectroscopies, has been widely applied to polymers for a basic identification of their chemical structure, their stereoregularity, the chain configurations and conformations, or the crystallinity. The monitoring of polymerization reactions, the analysis of the interactions between chemical groups of different phases in multicomponent systems, or the orientation of polymer chains under strain are also common applications of vibrational spectroscopy in polymer science.

Infrared absorption and Raman scattering provide complementary information on the vibrational states of a molecule. Infrared activity requires a change in the molecular dipole moment making it particularly sensitive to the presence of polar groups, while Raman activity that requires a change in the polarizability of the molecule during the vibrational mode considered is more sensitive to highly polarizable groups such as C—C, C=C, C≡C, or sulfur species like C—S and S—S, which are important groups for the vulcanization process of rubber compounds. Regarding polymer–filler interfaces, mid- and NIR spectroscopies are particularly well suited for the analysis of silanol groups on the silica surface and their modification by silane coupling agents. For its sensitivity to carbon—carbon bonds, Raman spectroscopy has become an essential tool for the characterization of carbon-based materials exhibiting different nanostructures from diamond, graphene, to fullerenes or CNTs. Moreover, these materials, even if dispersed in a polymer matrix at a very low content, exhibit strong and well-defined bands on account of resonance-enhanced Raman scattering effects.

The emergence of new techniques such as atomic force microscopy-infrared spectroscopy (AFM-IR) or tip-enhanced Raman spectroscopy (TERS) has provided significant progress for the analysis of the structures of interfaces at much smaller scales than the two conventional vibrational spectroscopies [98, 99].

3.3.3.1 Infrared Spectroscopy

Many applications of infrared spectroscopy concern the analysis of the surface species of silica (silanols, physisorbed water) or the adsorption behavior of the silane coupling agents onto the filler surface. The mechanisms of adsorption of bis(3-triethoxysilylpropyl)tetrasulfide "Si69" commonly used to promote adhesion between silica and hydrocarbon rubbers have been investigated by Chirachanchai et al. [100]. A quantitative evaluation of the amount of chemisorbed silane was carried out by FTIR spectroscopy. It was essentially demonstrated that chemisorption is dominant in polar solvent, while physisorption is higher in nonpolar solvents. As water can also be present on the silica surface, infrared bands associated with the fundamental modes of the silanol groups interfere with those of adsorbed water between 3000 and 3800 cm^{-1}. Krysztafkiewicz and Rager [101] used the NIR range for separation of silanol bands at 7326 cm^{-1} from mixed water band at 5270 cm^{-1}, which allows a quantitative estimation of the reactive silanol groups and their degree of condensation at the silica surface in reaction with coupling agents.

A better understanding of the improvement imparted by fillers in the mechanical properties of the host polymer can be obtained by a rheo-optical characterization that combines infrared spectroscopy and mechanical analysis. A molecular-level observation during the deformation processes can provide a better insight into the reinforcement effect significantly affected by the polymer–filler interface. Watanabe et al. [102] studied functionalized polypropylene with hydroxyl groups (PPOH) filled with silica nanospheres by rheo-optical near-infrared spectroscopy. Nanocomposites were prepared with unmodified silica and with silica modified with a cationic surfactant (cetyltrimethyl ammonium bromide, CTAB) to change the chemical structure of the polymer–filler interface. The rheo-optical NIR measurements were combined with two-dimensional correlation spectroscopy to highlight spectral features not identified in conventional one-dimensional spectra often complicated by the presence of overlapped absorptions. Pioneered by Noda [103–105], the 2D correlation analysis examines changes occurring in a series of spectra when the sample is submitted to a perturbation such as temperature, concentration, or mechanical strain. Synchronous and asynchronous correlation spectra taken at two independent wavenumbers can be generated from a set of rheo-optical NIR spectra. According to Noda et al. [106], the synchronous 2D correlation intensity represents the overall similarity between two intensity changes measured at the two wavenumbers, while the asynchronous 2D correlation intensity reflects dissimilarity of the spectral intensity changes. They represent IR signals occurring in phase and out of phase with each other, respectively. More details can be found in the papers of Noda et al. [103–106]. In the work of Watanabe et al. [102], an autopeak in the diagonal position is observed in the synchronous 2D spectrum indicating that the spectral intensities in this region change to the same direction during the tensile deformation of the sample. Cross-peaks at off-diagonal positions appear in the asynchronous spectrum, which reflect changes out of phase (delayed or accelerated) of the spectral features and shading of the peak indicates negative correlation (Figure 3.10). The authors conclude to the presence of different deformation mechanisms associated with the amorphous and crystalline structures of PPOH. A decrease in the correlation

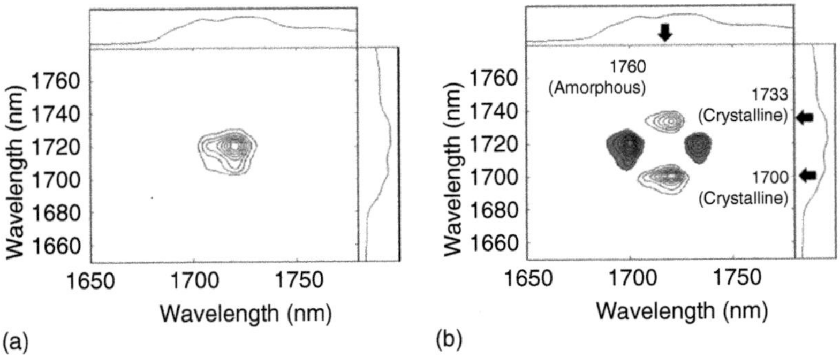

Figure 3.10 Synchronous (a) and asynchronous (b) correlation spectra derived from strain-dependent spectra of functionalized polypropylene with hydroxyl groups (PPOH). *Source:* Watanabe et al. [102].

intensity of the cross peak at the coordinate (1720, 1700 nm) is observed for the PPOH composites and is more significant for the polymer filled with unmodified silica attributed to interfacial adhesion that is reduced for the silica modified with CTAB that acts as a spacer between the polymer and the filler.

Mid- or NIR analysis of polymer composites under mechanical deformation can also be used to measure the orientation of polymer chains along the direction of strain. It is based on the determination of the dichroic behavior of a selected absorption band of the oriented sample. In the case of an uniaxial deformation, the absorbances of the investigated band are measured with radiation polarized parallel and perpendicular to the stretching direction to form the dichroic ratio R ($R = A_{\parallel}/A_{\perp}$) or the dichroic difference ΔA ($\Delta A = A_{\parallel} - A_{\perp}$), respectively, and determine the second Legendre polynomial $\langle P_2 \rangle$. Details on the determination of chain orientation by polarized FTIR spectroscopy either in mid- or NIR can be found in the literature [107–110]. The analysis of orientational behavior in filled elastomeric networks has been used to evaluate the degree of bonding between the polymer and the filler [2, 111, 112]. A filled elastomeric network may be regarded as a two-phase system of rigid particles surrounded by an elastomeric network formed by flexible chains permanently linked together by chemical junctions during the cross-linking process. Additional cross-links created by polymer–filler interactions increase the effective degree of cross-linking. Measurements of chain orientation have been shown to provide a direct estimation of the total network chain density arising from chemical junctions and also from polymer–filler attachments since $\langle P_2 \rangle$ depends on the reciprocal of the apparent molecular weight between cross-links. The number of additional cross-links increases with the amount of filler or with the interface area of the polymer–filler system. In silica-filled PDMS networks, the number of additional cross-links created by hydrogen bonding between the silanols on the silica surface and the oxygen atoms of the PDMS chains has been found to be 0.13 nm^{-2}, which corresponds to a bonding around every 6 nm^2 of silica surface [113].

Atomic force microscopy has proved to be a powerful technique for the analysis of filled polymers by the determination of the particle size, their state of dispersion, and more interestingly the deformation of their structure in the stretched composites, thus allowing the understanding of the phenomena occurring under strain [114, 115]. The coupling of AFM-IR provides nanoscale measurements in polymer composites especially for the characterization of the interphase region between the components with a spatial resolution of tens of nanometers far beyond the diffraction limit. AFM-IR uses the atomic force microscopy (AFM) tip to locally detect thermal expansion in the sample resulting from optical absorption. The thermal expansion causes an oscillation of the AFM cantilever in contact with the material. The oscillation can be measured by a deflection laser reflected from the cantilever onto a photodiode detector. The reader can find details of the principle of this technique elsewhere [116–118], and a recent review reports the recent applications in polymer science [119]. Marcott et al. [117] applied AFM-IR spectroscopy for the analysis of composite materials. The study carried out on silica- or carbon-filled polymers yields information on the filler dispersion and also in the chemical nature of the interphase region between the carbon fiber and the surrounding epoxy matrix. AFM-IR was also used by Mikhalchan et al. [120] to visualize the heterogeneous chemical structure of epoxy nanocomposites reinforced with CNT fibers. The AFM-IR spectra collected at several points in the AFM image (Figure 3.11B) reveal the presence of the aliphatic amine hardener at the CNT fiber/matrix interface (Figure 3.11C) and of the epoxy resin at locations #3 and #4. AFM-IR spectroscopy allows the authors from the comparison with the FTIR spectra of the reference materials (cured epoxy resin and hardener) (Figure 3.11A) to point out the chemical heterogeneity of the polymer–filler interface with resolution of sub-30 nm.

3.3.3.2 Raman Spectroscopy

One of the most important uses of Raman spectroscopy in the field of polymer composites concerns those filled with carbon-based materials that exhibit strong resonance-enhanced Raman scattering effects. Since the work of Tuinstra and Koenig [121] on single crystals of graphite, a tremendous interest has been devoted to Raman spectroscopy that becomes an essential tool for the characterization of carbon structures. The first-order Raman spectrum of single-crystal graphite (excited at 488 nm) exhibits a single band at 1575 cm^{-1} [121] designated as G band (after graphite). The second-order spectrum displays bands around 2700 cm^{-1} (split into two bands and designated G′ or 2D) and around 3200 cm^{-1} (2D′) assigned to overtones of bands of the first-order spectrum. If structural defects are present, new bands appear in the first-order Raman spectrum, in particular, the so-called D band, (D for "disorder" or "defect") and the D′ band visible as a shoulder in the high-wavenumber side of the G band. Figure 3.12 shows, as an example, the Raman spectrum recorded with the 633 nm exciting line of a powder of disorder (defective) graphite nanoplatelets (GNPs) with the assignment of the bands generally accepted in the literature [122–126].

The prominent feature in the Raman spectra of carbon-based materials is the shift of certain Raman bands, especially the D band and its overtone the G′ band to higher frequencies with increasing laser excitation energy, called the dispersive behavior. Their wavenumbers upshift linearly with an increase in the incident energy with a slope around 50 cm^{-1}/eV for the D band and around 100 cm^{-1} for the G′ one [127, 128]. Since this dispersive behavior arises from a coupling between electrons and

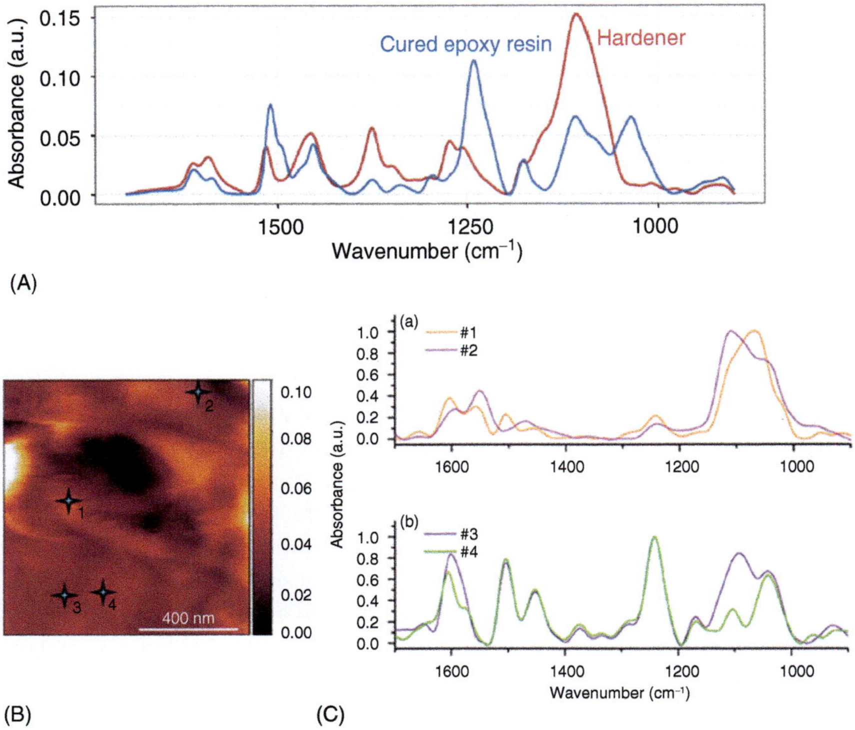

Figure 3.11 (A) FTIR spectra of hardener (red) and cured epoxy resin (blue); (B) Contact mode AFM height image at the CNT fiber/matrix interface; (C) AFM-IR at points #1 and #2 (Panel a) and at points #3 and #4 (Panel b). *Source:* (A) adapted from Mikhalchan et al. [120], Supporting information, (B, C) Mikhalchan et al. [120].

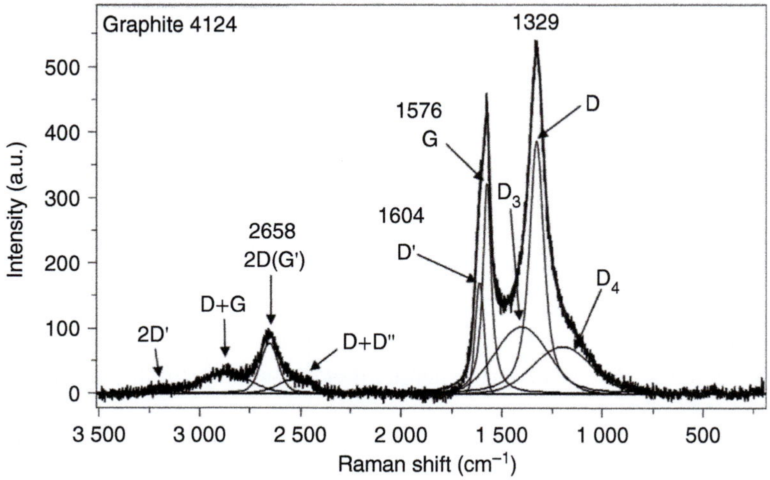

Figure 3.12 Deconvoluted Raman spectrum of a graphite of grade 4124 from Asbury Carbons.

phonons [32, 122, 129, 130], one would expect the excitation-energy dependencies of the D and G′ bands to be affected by interfacial interactions between the polymer and the filler particles in polymer nanocomposites. Figure 3.13 compares the dispersive behavior of the D band of a GNPs and MWCNTs in the pure state and embedded in PDMS. While the slope of the composite filled with graphite is similar to that of the pure filler, the PDMS/MWCNT composite displays a higher slope than that of MWCNTs in the unembedded state [127, 131]. The results reflect a poor interface between graphite and PDMS, while the electronic structure of MWCNTs seems to be affected by the surrounding polymer chains. The unexpected affinity of PDMS for MWCNTs revealed by outstanding electrical and mechanical properties has been connected to strong CH-π interactions between the two phases leading to a wrapping of polymer chains around the tube surface [36–39].

Another point of interest regarding the Raman spectra of carbon materials is the sensitivity of some bands to mechanical strain. Galiotis and Batchelder [132] have shown that the wavenumbers of the D, G, and G′ bands decrease linearly with increasing uniaxial strain. The slope of the plot of wavenumber shift versus strain is higher (almost double) for the G′ band, thus making this band particularly useful to get a better insight into the mechanical behavior of composites essentially into the efficiency of the stress transfer from the matrix to the filler. The strain dependence of the G′ band of carbon fibers was selected for the characterization of the fiber–matrix interface of as-received and plasma-treated carbon fiber-epoxy composites [133, 134]. It was demonstrated that the stress transfer from the matrix to the fiber is more effective in the plasma-oxidized carbon fiber/epoxy composites than in the untreated counterpart due to a better fiber/matrix adhesion.

The analysis of the stress-induced Raman shifts developed for the deformation of carbon fibers was extended to other carbon-based materials such as CNTs [135] and graphene [136]. Since the shift rate (in cm^{-1}/% strain) has been related to Young's modulus of the carbon material [137], its value in the bulk composite reflects the effective modulus of the reinforcing filler. The initial shift rate of SWCNTs in an

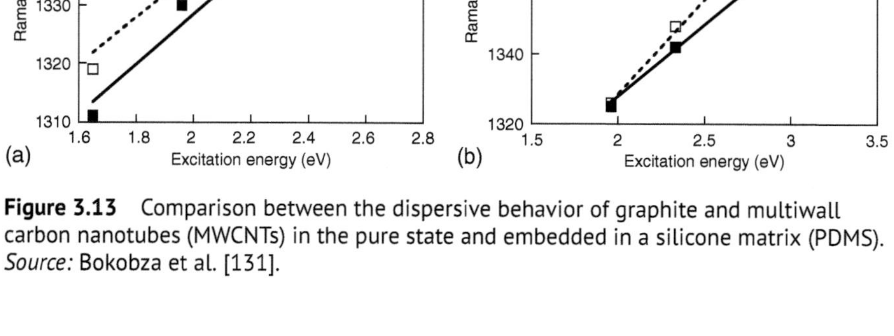

Figure 3.13 Comparison between the dispersive behavior of graphite and multiwall carbon nanotubes (MWCNTs) in the pure state and embedded in a silicone matrix (PDMS). *Source:* Bokobza et al. [131].

epoxy matrix ($-14.1\,cm^{-1}$/% strain) was found to be much higher than that of MWCNTs ($-3.4\,cm^{-1}$/% strain) in the same matrix [135] (Figure 3.14). Therefore, the two values are much lower than those of $-40\,cm^{-1}$/% strain found for the deformation of individual isolated SWCNTs [138], which may be ascribed to a poor interfacial adhesion to the matrix essentially in the case of MWCNTs. It is also interesting to note the nonlinearity of the behavior observed above around 0.5% strain for the SWCNTs attributed by the authors to a slippage at the polymer–filler interface [135].

Similar to AFM-IR, TERS that combined scanning probe microscopy and Raman spectroscopy has a much higher spatial resolution and sensitivity than conventional Raman spectroscopy. This technique has been applied to the analysis, at a nanometer length scale, of a wide variety of nanostructured systems, including carbon nanomaterials, biological specimens, polymer blends, or semiconductors [139–142]. In a basic TERS experiment, a laser beam is focused on the sample. A sharp metal or metal-coated tip is placed in the center of the laser focus and can be positioned at different points on the surface sample. A strong confinement of the electromagnetic field at the tip-apex arising from a combination of localized surface plasmon resonance and antenna effect enhances the Raman signal from the molecules in the vicinity of the tip-apex of the scanning probe. More details on this technique as well as achievements in TERS applications in different fields can be found in the literature [139–142].

With the use of TERS, Ghislandi et al. [143] obtained for graphene, an enhancement of the Raman signal of more than 400% for the G and 2D bands and an enhancement of more than 500% for the D band with regard to conventional confocal Raman spectroscopy. TERS was applied by Suzuki et al. [144] to study local interactions at

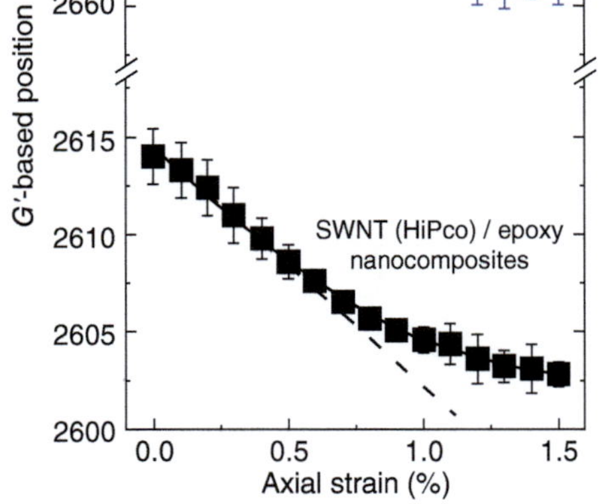

Figure 3.14 Dependence of the 2D Raman band position upon strain during the deformation of single-walled carbon nanotubes (SWNTs) and multiwall carbon nanotubes (MWNTs) in an epoxy resin matrix composite (0.1 wt%). *Source:* Young et al. [135].

the interface of SBR/MWCNT composites. Interesting features not observable in normal Raman spectra are revealed especially the occurrence of bands assigned to the phenyl rings when MWCNT bands are present. This leads the authors to suggest π–π interactions between the two species. The potential of TERS for the analysis of interfacial effects in polymer composites was also demonstrated by the same group [145] in a study of PS filled with SWCNTs. It is shown that the wavenumber of the two sub-bands G^- and G^+ of the G band of SWCNTs in the composite varies significantly with points where the TERS spectra are collected, while negligible shift is observed in the normal Raman spectra. The shift of the G band (up to $13\,cm^{-1}$) is attributed by the authors to a disentanglement of the tubes due to the penetration of PS chains during the mixing process and the mechanical compression distribution from PS.

3.4 Conclusions

This chapter is devoted to show the potential of spectroscopic techniques for the characterization of polymer–filler interface well known to play a critical role in the physical performance of the composite materials. A better understanding of the interface characteristics can guide the processing conditions for the design of filled systems with tailor-made interfacial structures able to yield specific properties. Polymer–filler interactions determine the properties of the interfacial area that may significantly differ from those of the bulk material. These interactions can take place between chemically active sites present or grafted on the filler surface and functional groups of polymer chains. Bifunctional molecules known as coupling agents can also be used in the case of low affinity between the two phases.

Among the different techniques used to understand the structure and properties of polymer–filler interfaces, spectroscopic methods have a great importance because they yield information, at a molecular scale, on both structural and functional aspects of materials. In this chapter, different examples illustrate the information on interfacial effects provided by the common spectroscopic techniques such as fluorescence, solid-state NMR, infrared, and Raman spectroscopies.

The fluorescence response of chemically bonded labels can probe changes appearing in specific sites in composite materials and especially in the interfacial region where a slowing down of polymer chain mobility is often observed. Recently, carbon dots have attracted considerable interest as fluorescent nanofillers on account of their exceptionally small size and the increased possibility of their large-scale industrial production from inexpensive organic sources.

Solid-state NMR is well known to provide an approach of the interface through measurements of spin–spin relaxation times that yield specific information on the molecular dynamics of polymer chains at the surface of the filler particle. Assessment of the degree of polymer with restricted mobility and the thickness of the adsorption layer are important parameters that can be deduced from NMR spectroscopic techniques.

Infrared and Raman spectroscopies are probably the most commonly utilized techniques for the characterization of interactions developed at interfaces of composite materials. Infrared spectroscopy, in the mid- and near-infrared ranges, has proved to be of particular significance for the analysis of silica particles by identifying the silanol groups and water adsorbed on their surface or the new functional groups grafted to decrease their hydrophilic properties. Polarized FTIR spectroscopy of composites submitted to an uniaxial deformation can be used to measure polymer orientation that allows, in filled elastomeric networks, an estimation the number of polymer–filler attachments. Raman spectroscopy has become one of the most important techniques for the analysis of carbon nanostructures on account of their strong resonance-enhanced scattering effects. It can be applied to all carbon-based nanocomposites for an evaluation of polymer–filler interfaces and the understanding of the stress transfer efficiency from the matrix to the filler particle. Coupling each type of vibrational spectroscopy with AFM or scanning probe microscopy to increase the spatial resolution has a great potential for the characterization of polymer–filler interfaces in the nanometer scale.

References

1 Payne, A.R. (1965). A note on the conductivity and modulus of carbon black-loaded rubbers. *J. Appl. Polym. Sci.* 9: 1073–1082.

2 Bokobza, L. (2004). The reinforcement of elastomeric networks by fillers. *Macromol. Mater. Eng.* 289: 607–621.

3 Bokobza, L. (2013). Elastomeric composites based on nanospherical particles and carbon nanotubes: a comparative study. *Rubber Chem. Technol.* 86: 423–448.

4 Li, Y., Huang, Y., Krentz, T. et al. (2017). Polymer nanocomposite interfaces: the hidden lever for optimizing performance in spherical nanofilled polymers, Chapter 1. In: *Interface/Interphase in Polymer Nanocomposites* (eds. A.N. Netravali and K.L. Mittal), 3–69. Beverly, MA: Scrivener Publishing LLC.

5 Fakirov, S. (2020). Polymer nanocomposites: why their mechanical performance does not justify the expectation and a possible solution to the problem? *eXPRESS Polym. Lett.* 14: 436–466.

6 Payne, A.R. (1962). The dynamic properties of carbon black-loaded natural rubber vulcanizates. Part I. *J. Appl. Polym. Sci.* VI: 57–63.

7 Payne, A.R. and Whittaker, R.E. (1971). Low strain dynamic properties of filled rubbers. *Rubber Chem. Technol.* 44: 440–478.

8 Thomas, S., Rouxel, D., and Ponnamma, D. (eds.) (2016). *Spectroscopy of Polymer Nanocomposites*, 1e. Oxford: Elsevier.

9 Yue, Y., Zhang, H., Zhang, Z., and Chen, Y. (2013). Polymer-filler interaction of fumed silica filled polydimethylsiloxane investigated by bound rubber. *Compos. Sci. Technol.* 86: 1–8.

10 Liu, J., Zong, G., He, L. et al. (2015). Effects of fumed and mesoporous silica nanoparticules on the properties of Sylgard 184 polydimethylsiloxane. *Micromachines* 6: 855–864.

11 Eslami, H., Rahimi, M., and Müller-Plathe, F. (2013). Molecular dynamics simulation of a silica nanoparticle in oligomeric poly(methyl methacrylate): a model system for studying the interphase thickness in a polymer-nanocomposite via different properties. *Macromolecules* 46: 8680–8692.

12 Mortazavian, H., Fennell, C.J., and Blum, F.D. (2016). Surface bonding is stronger for poly(methyl methacrylate) than for poly(vinyl acetate). *Macromolecules* 49: 4211–4219.

13 Mortazavian, H., Fennell, C.J., and Blum, F.D. (2016). Structure of the interfacial region in adsorbed poly(vinyl acetate) on silica. *Macromolecules* 49: 298–307.

14 Higuchi, C., Tanaka, H., and Yoshizawa, K. (2019). Molecular understanding of the adhesive interactions between silica surface and epoxy resin: effects of interfacial water. *J. Comput. Chem.* 40: 164–171.

15 Wang, M.-J., Morris, M.D., and Kutsovsky, Y. (2008). Effect of fumed silica surface area on silicone rubber reinforcement. *Kautsch. Gummi Kunstst.* 61: 107–117.

16 Dewimille, L., Bresson, B., and Bokobza, L. (2005). Synthesis, structure and morphology of poly(dimethylsiloxane) networks filled with in situ generated silica particles. *Polymer* 46: 4135–4143.

17 Bokobza, L. and Diop, A.L. (2010). Reinforcement of poly(dimethylsiloxane) by sol–gel in situ generated silica and titania particles. *eXPRESS Polym. Lett.* 4: 355–363.

18 Song, Y., Yu, J., Dai, D. et al. (2014). Effect of silica particles modified by in-situ and ex-situ methods on the reinforcement of silicone rubber. *Mater. Des.* 64: 687–603.

19 Adnan, M.M., Dalod, A.R.M., Balci, M.H. et al. (2018). In situ synthesis of hybrid inorganic-polymer nanocomposites. *Polymers* 10: 1129.

20 Bokobza, L. and Chauvin, J.P. (2005). Reinforcement of natural rubber: use of in situ generated silicas and nanofibres of sepiolite. *Polymer* 46: 4144–4151.

21 Messori, M. and Fiorino, M. (2011). Isoprene rubber filled with silica generated in situ. *J. Appl. Polym. Sci.* 119: 3422–3428.

22 Scotti, R., Wahba, L., Crippa, M. et al. (2012). Rubber-silica nanocomposites obtained by in situ sol–gel method: particle shape influence on the filler-filler and filler-rubber interactions. *Soft Matter* 8: 2131–2143.

23 Mokhothu, T.H., Luyt, A.S., and Messori, M. (2014). Reinforcement of EPDM rubber with in situ generated silica particles in the presence of a coupling agent via a sol–gel route. *Polym. Test.* 33: 97–106.

24 Tohsan, A., Kishi, R., and Ikeda, Y. (2015). A model filler network in nanocomposites prepared by in situ silica filling and peroxide cross-linking in natural rubber latex. *Colloid. Polym. Sci.* 293: 2083–2093.

25 Ohashi, T., Tohsan, A., and Ikeda, Y. (2017). Role of in situ generated silica for rubber science and technology. *Polym. Int.* 66: 250–259.

26 Lipińska, M. and Zaborski, M. (2018). Reinforcement of rubber by in-situ formed silica. *Int. J. Sci. Eng. Sci.* 1: 91–101.

27 Avilès, F., Cauich-Rodriguez, J.V., Toro-Estay, P. et al. (2018). Improving carbon nanotube/polymer interactions in nanocomposites, Chapter 5. In: *Carbon Nanotube-Reinforced Polymers: From Nanoscale to Macroscale* (ed. R. Rafiee), 83–115. Elsevier.

28 Mosnáčková, K., Kollár, J., Huang, Y.-S. et al. (2019). Synthesis routes of functionalized nanoparticles, Chapter 1. In: *Polymer Composites with Functional Nanoparticles: Synthesis, Properties, and Applications, Micro and Nano Technologies* (eds. K. Pielichowski and T.M. Majka), 1–46. Amsterdam: Elsevier Ltd.

29 Speranza, G. (2019). The role of functionalization in the applications of carbon materials: an overview. *C - J. Carbon Res.* 5: 84.

30 Singh, N.V., Gupta, V.K., and Singh, A.P. (2019). Graphene and carbon nanotube reinforced epoxy nanocomposites: a review. *Polymer* 180: 121724.

31 Dyke, C.A. and Tour, J.M. (2004). Covalent functionalization of single-walled carbon nanotubes for materials applications. *J. Phys. Chem.* 108: 11151–11159.

32 Ferrari, A.C. and Basko, D.M. (2013). Raman spectroscopy as a versatile tool for studying the properties of graphene. *Nat. Nanotechnol.* 8: 235–246.

33 Graupner, R. (2007). Raman spectroscopy of covalently functionalized single-wall carbon nanotubes. *J. Raman Spectrosc.* 38: 673–683.

34 Dyke, C.A. and Tour, J.M. (2003). Solvent-free functionalization of carbon nanotubes. *J. Am. Chem. Soc.* 125: 1156–1157.

35 Koval'chuk, A.A., Shevchenko, V.G., Shchegolikhin, A.N. et al. (2008). Effect of carbon nanotube functionalization on the structural and mechanical properties of polypropylene/MWCNT composites. *Macromolecules* 41: 7536–7542.

36 Tamore, M.S., Ratna, D., Mishra, S., and Shimpi, N.G. (2019). Effect of functionalized multi-walled carbon nanotubes on physicomechanical properties of silicone rubber nanocomposites. *J. Compos. Mater.* https://doi.org/10.1177/0021998319827080.

37 Beigbeder, A., Linares, M., Devalckenaere, M. et al. (2008). CH-π interactions as the driving force for silicone-based nanocomposites with exceptional properties. *Adv. Mater.* 20: 1003–1007.

38 Bokobza, L. (2009). Some issues in rubber nanocomposites: new opportunities for silicone materials. *Silicon* 1: 141–145.

39 Bokobza, L. and Rahmani, M. (2009). Carbon nanotubes: exceptional reinforcing fillers for silicon rubbers. *Kautsch. Gummi Kunstst.* 62: 112–117.

40 Geim, A.K. and Novoselov, K.S. (2007). The rise of graphene. *Nat. Mater.* 6: 183–191.

41 Chattopadhyay, J., Mukherjee, A., Chakraborty, S. et al. (2009). Exfoliated soluble graphite. *Carbon* 47: 2945–2949.

42 Strom, T.A., Dillon, E.P., Hamilton, C.E., and Barron, A.R. (2010). Nitrene addition to exfoliated graphene: a one-step route to highly functionalized graphene. *Chem. Commun.* 46: 4097–4099.

43 Quintana, M., Spyrou, K., Grzelczak, M. et al. (2010). Functionalization of graphene via 1,3-dipolar cycloaddition. *ACS Nano* 4: 3527–3533.

44 Vadukumpully, S., Gupta, J., Zhang, Y. et al. (2011). Functionalization of surfactant wrapped graphene nanosheets with alkylazides for enhanced dispersibility. *Nanoscale* 3: 303–308.

45 Špitalský, Z., Danko, M., and Mosnáček, J. (2011). Preparation of functionalized graphene sheets. *Curr. Org. Chem.* 15: 1133–1150.

46 Georgakilas, V., Otyepka, M., Bourlinos, A.B. et al. (2012). Functionalization of graphene: covalent and non-covalent approaches, derivatives and applications. *Chem. Rev.* 112: 6156–6214.

47 Greenwood, J., Phan, T.H., Fujita, Y. et al. (2015). Covalent modification of graphene and graphite using diazonium chemistry: tunable grafting and nano-manipulation. *ACS Nano* 9: 5520–5535.

48 Bon, S.B., Valentini, L., Verdejo, R. et al. (2009). Plasma fluorination of chemically derived graphene sheets and subsequent modification with butylamine. *Chem. Mater.* 21: 3433–3438.

49 Chen, M., Zhou, H., Qiu, C. et al. (2012). Layer-dependent fluorination and doping of graphene via plasma treatment. *Nanotechnology* 23: 115706.

50 Georgakilas, V., Tiwari, J.N., Kemp, K.C. et al. (2016). Noncovalent functionalization of graphene and graphene oxide for energy materials, biosensing, catalytic, and biomedical applications. *Chem. Rev.* 116: 5464–5519.

51 Yu, W., Sisi, L., Haiyan, Y., and Jie, L. (2020). Progress in the functional modification of graphene/graphene oxide: a review. *RSC Adv.* 10: 15328–15345.

52 Acik, M. and Chabal, Y.J. (2013). A review on thermal exfoliation of graphene oxide. *J. Mater. Sci. Res.* 2: 101–112.

53 Sengupta, I., Chakraborty, S., Talukdar, M. et al. (2018). Thermal reduction of graphene oxide: how temperature influences purity. *J. Mater. Res.* 33: 4113–4122.

54 Bhyan, M.S.A., Uddin, M.N., Islam, M.M. et al. (2016). Synthesis of graphene. *Int. Nano Lett.* 6: 65–83.

55 Niu, R., Gong, J., Xu, D. et al. (2014). Influence of molecular weight of polymer matrix on the structure and rheological properties of graphene oxide/polydimethylsiloxane composites. *Polymer* 55: 5445–5453.

56 Dal Lago, E., Cagnin, E., Boaretti, C. et al. (2020). Influence of different carbon-based fillers on electrical and mechanical properties of a PC/ABS blend. *Polymers* 12: 29.

57 Gkourmpis, T., Gaska, K., Tranchida, D. et al. (2019). Melt-mixed 3D hierarchical graphene/polypropylene nanocomposites with low electrical percolation threshold. *Nanomaterials* 9: 1766.

58 Guth, E. (1945). Theory of filler reinforcement. *J. Appl. Phys.* 16: 20–25.

59 Mooney, M. (1940). A theory of large elastic deformation. *J. Appl. Phys.* 11: 582–592.

60 Rivlin, R.S. and Saunders, D.W. (1951). Large elastic deformations of isotropic materials. VII. Experiments on the deformation of rubber. *Philos. Trans. R. Soc. London, Ser. A* 243: 251–288.

61 Bokobza, L. (2017). Mechanical and electrical properties of elastomer nanocomposites based on different carbon nanomaterials. *C - J. Carbon Res.* 3: 10.

62 Bokobza, L., Bruneel, J.-L., and Couzi, M. (2013). Raman spectroscopic investigation of carbon-based materials and their composites. Comparison between carbon nanotubes and carbon black. *Chem. Phys. Lett.* 590: 153–159.

63 D'Souza, N., Pendse, S., Sahu, L. et al. (2017). Impact of crystallization on the interface in polymer nanocomposites, Chapter 4. In: *Interface/Interphase in Polymer Nanocomposites* (eds. A.N. Netravali and K.L. Mittal), 139–170. Beverly, MA: Scrivener Publishing LLC.

64 Gissinger, J.R., Pramanik, C., Newcomb, B. et al. (2018). Nanoscale structure-property relationships of polyacrylonitrile/CNT composites as a function of polymer crystallinity and CNT diameter. *ACS Appl. Mater. Interfaces* 10: 1017–1027.

65 Pramanik, C., Nepal, D., Nathanson, M. et al. (2018). Molecular engineering of interphases in polymer/carbon nanotube composites to reach the limits of mechanical performance. *Compos. Sci. Technol.* 166: 86–94.

66 Ellison, C.J. and Torkelson, J.M. (2002). Sensing the glass transition in thin and ultrathin polymer films via fluorescence probes and labels. *J. Polym. Sci., Part B: Polym. Phys.* 40: 2745–2758.

67 Rittigstein, P., Priestley, R.D., Broadbelt, L.J., and Torkelson, J.M. (2007). Model polymer nanocomposites provide an understanding of confinement effects in real nanocomposites. *Nat. Mater.* 6: 278–282.

68 Albalá, R., Olmos, D., Aznar, A.J. et al. (2004). Fluorescent labels to study thermal transitions in epoxy/silica composites. *J. Colloid Interface Sci.* 277: 71–78.

69 Gonzalez-Benito, J. (2008). Composite materials studied at a molecular scale. Use of fluorescent labels, Chapter 7. In: *Leading-Edge Composite Material Research* (ed. T.G. Wouters), 313–329. New York: Nova Science Publishers, Inc.

70 Barroso, I. and González-Benito, J. (2011). Steady state fluorescence to study nanocomposites based on silica nanoparticles and thermoplastic polymer matrices. *Polym. Compos.* 32: 1874–1887.

71 Antonelli, C., Serrano, B., Baselga, J. et al. (2015). Interfacial characterization of epoxy/silica nanocomposites measured by fluorescence. *Eur. Polym. J.* 62: 31–42.

72 Kausar, A. (2019). Epoxy and quantum dots-based nanocomposites: achievements and applications. *Mater. Res. Innovations* https://doi.org/10.1080/14328917.2019.1636175.

73 Gaddam, R.R., Narayan, R., and Raju, K.V.S.N. (2016). Fluorescence spectroscopy of nanofillers and their polymer nanocomposites, Chapter 7. In: *Spectroscopy of Polymer Nanocomposites* (eds. S. Thomas, D. Rouxel and D. Ponnamma), 158–180. Oxford: Elsevier.

74 Chebrolu, V.T. and Kim, H.-J. (2019). Recent progress in quantum dot sensitized solar cells: an inclusive review of photoanode, sensitizer, electrolyte, and the counter electrode. *J. Mater. Chem. C* 7: 4911–4933.

75 De, B. (2019). Carbon dots and their polymeric nanocomposites, Chapter 7. In: *Nanomaterials and Polymer Nanocomposites, Raw Materials to Applications* (ed. N. Karak), 217–260. Elsevier.

76 Ye, R., Xiang, C., Lin, J. et al. (2013). Coal as an abundant source of graphene quantum dots. *Nat. Commun.* 4: 2943. https://doi.org/10.1038/ncomms3943.

77 Kovalchuk, A., Huang, K., Xiang, C. et al. (2015). Luminescent polymer composite films containing coal-derived graphene quantum dots. *ACS Appl. Mater. Interfaces* 7: 26063–26068.

78 Sreenath, P.R., Singh, S., Satyanarayana, M.S. et al. (2017). Carbon dot – Uning reinforcing filler for polymer with special reference to physico-mechanical properties. *Polymer* 112: 189–200.

79 Karimi, B. and Ramezanzadeh, B. (2017). A comparative study on the effects of ultrathin luminescent graphene oxide quantum dot (GOQD) and graphene oxide (GO) nanosheets on the interfacial interactions and mechanical properties of an epoxy composite. *J. Colloid Interface Sci.* 493: 62–76.

80 De, B. and Karak, N. (2013). A green and facile approach for the synthesis of water soluble fluorescent carbon dots from banana juice. *RSC Adv.* 3: 8286–8290.

81 De, B., Voit, B., and Karak, N. (2013). Transparent luminescent hyperbranched epoxy/carbon oxide dot nanocomposites with outstanding toughness and ductility. *ACS Appl. Mater. Interfaces* 5: 10027–10034.

82 De, B. and Karak, N. (2017). Recent progress on carbon dot-metal based nanohybrids for photochemical and electrochemical applications. *J. Mater. Chem. A* 5: 1826–1859.

83 Sciortino, A., Cannizzo, A., and Messina, F. (2018). Carbon nanodots: a review – from the current understanding of the fundamental photophysics to the full control of the optical response. *C - J. Carbon Res.* 4: 67.

84 Kausar, A. (2019). Polymer/carbon-based quantum dot nanocomposite: forthcoming materials for technical application. *J. Macromol. Sci. Part A Pure Appl. Chem.* 56: 341–356.

85 De, B., Kumar, M., Mandal, B.B., and Karak, N. (2015). An in situ prepared photo-luminescent transparent biocompatible hyperbranched epoxy/carbon dot nanocomposite. *RSC Adv.* 5: 74692–74704.

86 Bhunia, S.K., Nandi, S., Shikler, R., and Jelinek, R. (2016). Tuneable light-emitting carbon-dot/polymer flexible films prepared through one-pot synthesis. *Nanoscale* 8: 3400–3406.

87 Zeng, M., Wang, X., Yu, Y.-H. et al. (2016, Art. No: 6490383). The synthesis of amphiphilic luminescent graphene quantum dot and its application in miniemulsion polymerization. *J. Nanomater.*: 8. https://doi.org/10.1155/2016/6490383.

88 Wang, Y., Zhang, Y., Zhao, H. et al. (2018). Identifying interphase properties in polymer nanocomposites using adaptive optimization. *Compos. Sci. Technol.* 162: 146–155.

89 Bayley, E.J. and Winey, K.I. (2020). Dynamics of polymer segments, polymer chains, and nanoparticles in polymer nanocomposite melts: a review. *Prog. Polym. Sci.* 105: 101242.

90 Robertson, C.G. and Roland, C.M. (2008). Glass transition and interfacial segmental dynamics in polymer-particle composites. *Rubber Chem. Technol.* 81: 506–522.

91 Legrand, A.P., Lecomte, N., Vidal, A. et al. (1992). Application of NMR spectroscopy to the characterization of elastomer/filler interactions. *J. Appl. Polym. Sci.* 46: 2223–2232.

92 Mansencal, R., Haidar, B., Vidal, A. et al. (2001). High-resolution solid-state NMR investigation of the filler-rubber interaction: 2. High-speed [^{1}H] magic-angle spinning NMR spectroscopy in carbon-black-filled polybutadiene. *Polym. Int.* 50: 387–394.

93 Klonos, P., Pissis, P., Gun'ko, V.M. et al. (2010). Interaction of poly(ethylene glycol) with fumed silica and alumina/silica/titania. *Colloids Surf., A* 360: 220–231.

94 Kim, S.Y. and Zukoski, C.F. (2011). Role of polymer segment-particle surface interactions in controlling nanoparticle dispersions in concentrated polymer solutions. *Langmuir* 27: 10455–10463.

95 Kim, S.Y., Meyer, H.W., Saalwächter, K., and Zukoski, C.F. (2012). Polymer dynamics in PEG-silica nanocomposites: effects of molecular weight, temperature and solvent dilution. *Macromolecules* 45: 4225–4237.

96 ten Brinke, J.W., Litvinov, V.M., Wijnhoven, J.E.G.J., and Noordermeer, J.W.M. (2002). Interactions of Stöber silica with natural rubber under the influence of coupling agents, studied by ^{1}HNMR T_2 relaxation analysis. *Macromolecules* 35: 10026–10037.

97 Borsacchi, S., Sudhakaran, U.P., Calucci, L. et al. (2018). Rubber-filler interactions in polyisoprene filled with in situ generated silica: a solid state NMR study. *Polymers* 10: 822.

98 Chen, J., Gao, X., and Xu, D. (2019). Recent advances in characterizing techniques for the interface in carbon nanotube-reinforced polymer nanocomposites. *Hindawi Adv. Mater. Sci. Eng.*: 1–24.

99 Zhou, H., Tang, Y., and Zhang, S. (2020). Advanced spectroscopic technique for the study of nanocontainers: atomic force microscopy-infrared spectroscopy (AFM-IR), Chapter 2. In: *Smart Nanocontainers, Micro and Nano Technologies* (eds. P. Nguyen-Tri, D. Trong-On and T.A. Nguyen), 7–17. Elsevier.

100 Chirachanchai, S., Chungchamroenkit, S., and Ishida, H. (1999). Adsorption of tetrasuldide-functional silane on high surface area silica treated with aqueous and non-aqueous solutions. *Compos. Interfaces* 6: 155–167.

101 Krysztafkiewicz, A. and Rager, B. (1988). NIR studies of the surface modification in silica fillers. *Colloid. Polym. Sci.* 266: 485–493.

102 Watanabe, R., Shinzawa, H., Kunioka, M. et al. (2017). Reinforcement mechanism of functionalized polypropylene containing hydroxyl group nanocomposites studied by rheo-optical near-infrared spectroscopy. *Eur. Polym. Spectrosc.* 92: 86–96.

103 Noda, I., Dowrey, A.E., and Marcott, C. (1983). Dynamic infrared linear dichroism of polymer films under oscillatory deformation. *J. Polym. Sci., Part C: Polym. Lett.* 21: 99–103.

104 Marcott, C., Noda, I., and Dowrey, A.E. (1991). Enhancing the information content of vibrational spectra through sample perturbation. *Anal. Chim. Acta* 250: 131–143.

105 Marcott, C., Dowrey, A.E., and Noda, I. (1994). Dynamic two-dimensional IR spectroscopy. In: *Analytical Chemistry*, vol. 66, 1065 A–1075 A. ACS Publications.

106 Noda, I., Dowrey, A.E., Marcott, C. et al. (2000). Generalized two-dimensional correlation spectroscopy. *Appl. Spectrosc.* 54: 236A–248A.

107 Bokobza, L. (2002). Molecular orientation of polymers by infrared spectroscopy. In: *The Handbook of Vibrational Spectroscopy*, vol. 4 (eds. J.M. Chalmers and P.R. Griffiths), 2496–2513. Hoboken, NJ: Wiley.

108 Buffeteau, T. and Pézolet, M. (2007). Infrared linear dichroism of polymers. In: Vibrational Spectroscopy of Polymers : Principles and Practice (eds. N.J. Everall, J.M. Chalmers and P.R. Griffiths), 255–281. Wiley.

109 Buffeteau, T., Desbat, B., and Bokobza, L. (1995). The use of near-infra-red spectroscopy coupled to the polarization modulation technique to investigate molecular orientation in unaxially stretched polymers. *Polym. Commun.* 36: 4339–4343.

110 Bokobza, L., Buffeteau, T., and Desbat, B. (2000). Mid- and near-infrared investigation of molecular orientation in elastomeric networks. *Appl. Spectrosc.* 54: 360–365.

111 Bokobza, L. (2001). Filled elastomers: a new approach based on measurements of chain orientation. *Polymer* 42: 5415–5423.

112 Bokobza, L. (2005). Infrared analysis of elastomeric composites under uniaxial extension. *Macromol. Symp.* 220: 45–59.

113 Clément, F. (1999). Etude des mécanismes de renforcement dans les réseaux de polydimethylsiloxane chargés silice. PhD thesis, University of Paris VI.

114 Clément, F., Lapra, A., Bokobza, L., Monnerie, L., and Ménez, P. (2001). Atomic force microscopy investigation of filled elastomers and comparison with transmission electron microscopy – application to silica-filled silicone elastomers. *Polymer* 42: 6259–6270.

115 Lapra, A., Clément, F., Bokobza, L., and Monnerie, L. (2003). Straining effects in silica-filled elastomers investigated by atomic force microscopy: from macroscopic to nanoscale strainfield. *Rubber Chem. Technol.* 76: 60–81.

116 Dazzi, A., Glotin, F., and Carminati, R. (2010). Theory of infrared nanospectroscopy by photothermal induced resonance. *J. Appl. Phys.* 107: 124519.

117 Marcott, C., Lo, M., Dillon, E. et al. (2015). Interface analysis of composites using AFM-based nanoscale IR and mechanical spectroscopy. *Microsc. Today* 23: 38–45.

118 Dazzi, A. and Prater, C.B. (2017). AFM-IR: technology and applications in nanoscale infrared spectroscopy and chemical imaging. *Chem. Rev.* 117: 5146–5173.

119 Nguyen-Tri, P., Ghassemi, P., Carriere, P. et al. (2020). Recent applications of advanced atomic force microscopy in polymer science/a review. *Polymers* 12: 1142.

120 Mikhalchan, A., Banas, A.M., Banas, K. et al. (2018). Revealing chemical heterogeneity of CNT fiber nanocomposites via nanoscale chemical imaging. *Chem. Mater.* 30: 1856–1864.

121 Tuinstra, F. and Koenig, J.L. (1970). Raman spectrum of graphite. *J. Chem. Phys.* 53: 1126–1130.

122 Jorio, A. (2012). Raman spectroscopy in graphene-based systems: prototypes for nanoscience and nanometrology. *Int. Scholarly Res. Not.* https://doi.org/10.5402/2012/234216.

123 May, P., Lazzeri, M., Venezuela, P. et al. (2013). Signature of the two-dimensional dispersion in graphene probed by double-resonant Raman scattering. *Phys. Rev. B* 87: 075402.

124 Sadezky, A., Muckenhuber, H., Grothe, H. et al. (2005). Raman spectroscopy of soot and related carbonaceous materials: spectral analysis and structural information. *Carbon* 43: 1731–1742.

125 Pawlyta, M., Rouzaud, J.N., and Duber, S. (2015). Raman microspectroscopy characterization of carbon black: spectral analysis and structural information. *Carbon* 84: 479–490.

126 Couzi, M., Bruneel, J.-L., Talaga, D., and Bokobza, L. (2016). A multi wavelength Raman scattering of defective graphitic materials: the first order Raman spectra revisited. *Carbon* 107: 388–394.

127 Bokobza, L., Bruneel, J.-L., and Couzi, M. (2015). Raman spectra of carbon-based materials (from graphite to carbon black) and of some silicone composites. *C - J. Carbon Res.* 1: 77–94.

128 Bokobza, L., Couzi, M., and Bruneel, J.-L. (2017). Raman spectroscopy of polymer-carbon nanomaterial composites. *Rubber Chem. Technol.* 90: 37–59.

129 Pimenta, M.A., Dresselhaus, G., Dresselhaus, M.S. et al. (2007). Studying disorder in graphite-based systems by Raman spectroscopy. *Phys. Chem. Chem. Phys.* 9: 1276–1291.

130 Dresselhaus, M.S., Dresselhaus, G., Saito, R., and Jorio, A. (2005). Raman spectroscopy of carbon nanotubes. *Phys. Rep.* 409: 47–99.

131 Bokobza, L., Bruneel, J.-L., and Couzi, M. (2014). Raman spectroscopy as a tool for the analysis of carbon-based materials (highly oriented pyrolitic graphite, multilayer graphene and multiwall carbon nanotubes) and of some of their elastomeric composites. *Vib. Spectrosc.* 74: 57–63.

132 Galiotis, C. and Batchelder, D.N. (1988). Strain dependences of the first- and second-order Raman spectra of carbon fibres. *J. Mater. Sci. Lett.* 7: 545–547.

133 Montes-Morán, M.A., Martínez-Alonso, A., Tascón, J.M.D., and Young, R.J. (2001). Effect of plasma oxidation on the surface and interfacial properties of ultra-high modulus carbon fibres. *Composites Part A* 32: 361–371.

134 Montes-Morán, M.A. and Young, R.J. (2002). Raman spectroscopy study of high-modulus carbon fibres: effect of plasma-treatment on the interfacial properties of single-fibre-epoxy composites. Part II: Characterisation of the fibre-matrix interface. *Carbon* 40: 857–875.

135 Young, R.J., Deng, L., Wafy, T.Z., and Kinloch, I.A. (2016). Interfacial and internal stress transfer in carbon nanotube based nanocomposites. *J. Mater. Sci.* 51: 344–352.

136 Papageorgiou, D.G., Kinloch, I.A., and Young, R.J. (2017). Mechanical properties of graphene-based nanocomposites. *Prog. Mater Sci.* 90: 75–127.

137 Deng, L., Eichhorn, S.J., Kao, C.-C., and Young, R.J. (2011). The effective Young's modulus of carbon nanotubes in composites. *ACS Appl. Mater. Interfaces* 3: 433–440.

138 Kannan, P., Eichhorn, S.J., and Young, R.J. (2007). Deformation of isolated single-wall carbon nanotubes in electrospun polymer nanofibres. *Nanotechnology* 18: 235707.

139 Kumar, N., Mignuzzi, S., Su, W., and Roy, D. (2015). Tip-enhanced Raman spectroscopy: principles and applications. *EPJ Tech. Instrum.* 2 (9) https://doi.org/10.1140/epjti/s40485-015-0019-5.

140 Verma, P. (2017). Tip-enhanced Raman spectroscopy: technique and recent advances. *Chem. Rev.* 117: 6447–6466.

141 Wang, X., Huang, S.-C., Huang, T.-X. et al. (2017). Tip-enhanced Raman spectroscopy for surfaces and interfaces. *Chem. Soc. Rev.* 46: 4020–4041.

142 Kumar, N., Weckhuysen, B.M., Wain, A.J., and Pollard, A.J. (2019). Nanoscale chemical imaging using tip-enhanced Raman spectroscopy. *Nat. Protoc.* 14: 1169–1193.

143 Ghislandi, M., Hoffmann, G.G., Tkalya, E. et al. (2012). Tip-enhanced Raman spectroscopy and mapping of graphene sheets. *Appl. Spectrosc. Rev.* 47: 371–381.

144 Suzuki, T., Yan, X., Kitihama, Y. et al. (2013). Tip-enhanced Raman spectroscopy study of local interactions at the interface of styrene-butadiene rubber/multiwalled carbon nanotube nanocomposites. *J. Phys. Chem. C* 117: 1436–1440.

145 Yan, X., Suzuki, T., Kitahama, Y. et al. (2013). A study on the interaction of single-walled carbon nanotubes (SWCNTs) and polystyrene (PS) at the interface in SWCNT-PS nanocomposites using tip-enhanced Raman spectroscopy. *Phys. Chem. Chem. Phys.* 15: 20618–20624.

4

Far-Infrared/Terahertz and Low-Frequency Raman Spectroscopies in Polymers

Harumi Sato[1,2]

[1] *Kobe University, Graduate School of Human Development and Environment, Tsurukabuto 3-11 Nada, Kobe 657-8501, Japan*
[2] *Kobe University, Molecular Photoscience Research Center, Rokkodaicho 1-1, Nada, Kobe 657-8501, Japan*

4.1 Introduction

The far-infrared (FIR, 400 to 10 cm^{-1}) and terahertz waves (THz, 0.1–10 THz, or 333 to 33 cm^{-1}) are electromagnetic waves that are located in the boundary area between radio and light waves [1–12]. The photon energy of FIR and THz waves is 2 orders of magnitude lower than that of infrared (IR) waves; therefore, FIR/THz spectroscopy can be used to study vibrations in the long-range structure of polymers directly. The spectra in the low-frequency region obtained by FIR/THz spectroscopy and low-frequency Raman spectroscopy are sensitive to the long-range structure of molecules, lattice expansion, weak interactions, and intermolecular vibrations [7–19]. Therefore, FIR, THz, or low-frequency Raman spectroscopy of a polymeric material provides information regarding the higher-order structure of polymers and intermolecular hydrogen bonds in the lamellar structures of semicrystalline polymers. Furthermore, by combining the temperature-dependent spectra, polarization spectra, and quantum chemical calculations (QCCs), the band assignments in the low-frequency region, which are difficult to determine, can be obtained. Consequently, the behavior of inter- and intramolecular hydrogen bonds in semicrystalline polymers can be investigated directly by FIR/THz and low-frequency Raman spectroscopy.

In this chapter, we present the following five topics as examples of polymer research using FIR/THz spectroscopy and low-frequency Raman spectroscopy. (i) Intermolecular hydrogen bonds in the low-frequency region of poly(3-hydroxybutyrate) (PHB) by QCCs [7–9], (ii) visualization of different types of intermolecular hydrogen bonds of polycaprolactone (PCL) [12], (iii) stress-induced crystal transition and its associated changes in the inter- and intramolecular hydrogen bonds of polybutylene succinate (PBS) [14], (iv) differences in intermolecular hydrogen bonding between polyethylene terephthalate (PET) and polybutylene terephthalate (PBT) and the flexibility of these molecules [15], and (v) THz imaging [17]. In Chapter 1, an FIR and QCCs study of nylon 6 is described.

Spectroscopic Techniques for Polymer Characterization: Methods, Instrumentation, Applications, First Edition. Edited by Yukihiro Ozaki and Harumi Sato.

4.2 Intermolecular Hydrogen Bonds in the Low-Frequency Region of PHB by QCCs

PHB is a semicrystalline biodegradable polyester with hydrogen bonds between the CH_3 and C=O groups of adjacent parallel chains, confirmed by IR spectroscopy and wide-angle X-ray diffraction (WAXD) [20–32]. More detail concerning the CH· · ·O hydrogen bonds is reported in Chapter 16. We investigated the weak hydrogen bonding in the crystalline structure of PHB using THz spectroscopy and low-frequency Raman spectroscopy, as they are extremely sensitive to the higher-order structure and intermolecular interactions of polymers [7–9] (Figure 4.1).

Figure 4.2 shows the polarized THz absorption spectra of PHB at room temperature, measured using a highly stretched PHB film. The band at 2.49 THz appears in both the parallel and perpendicular polarization spectra of PHB, whereas the band at 2.92 THz is only present in the parallel polarization spectrum. These results indicate that the band at 2.92 THz is attributed to the spring-like motion of the helical

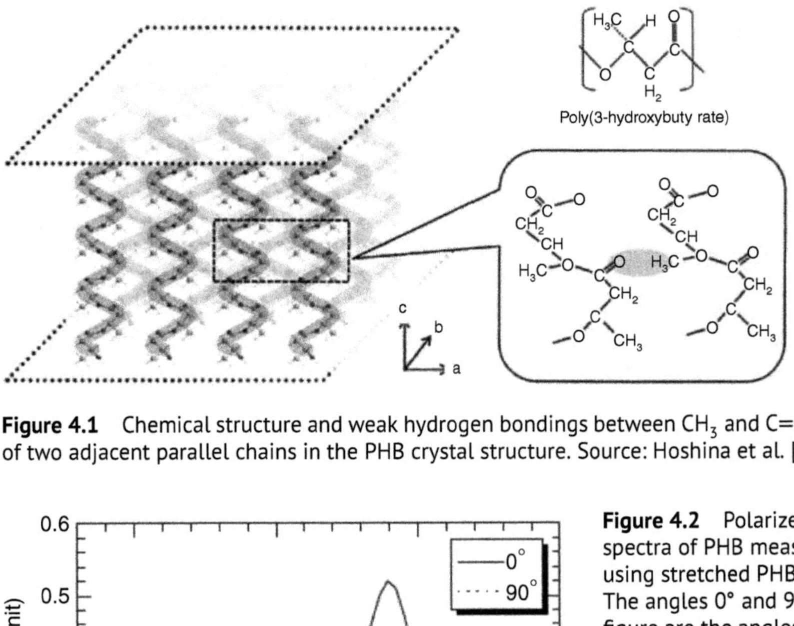

Figure 4.1 Chemical structure and weak hydrogen bondings between CH_3 and C=O groups of two adjacent parallel chains in the PHB crystal structure. Source: Hoshina et al. [8b].

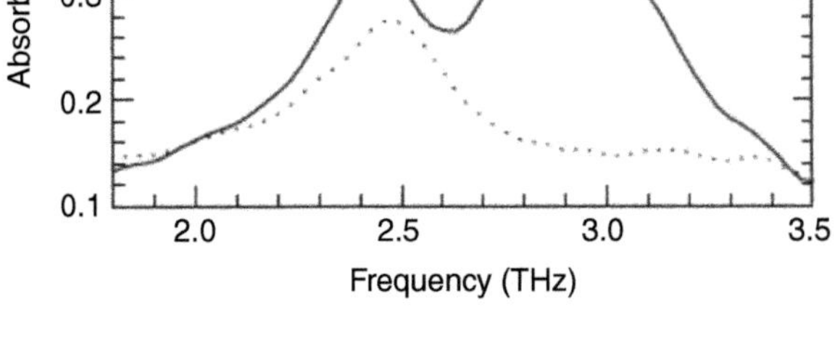

Figure 4.2 Polarized THz spectra of PHB measured using stretched PHB film. The angles 0° and 90° in the figure are the angles between the stretching direction of the film and the electric field of the THz wave. Source: Hoshina et al. [7a].

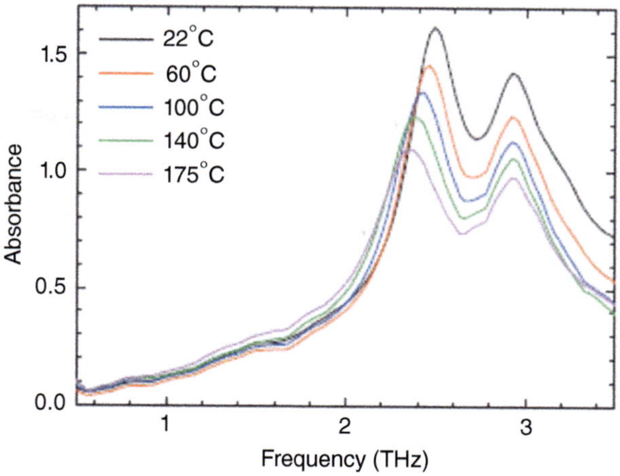

Figure 4.3 Temperature-dependent THz absorption spectra of PHB. Source: Hoshina et al. [7a].

structure, while the band at 2.49 THz is related to the intermolecular hydrogen bonding between the CH_3 and C=O groups of adjacent parallel chains in PHB.

Figure 4.3 shows the temperature-dependent THz spectra of PHB. The band at 2.49 THz shifts to a lower frequency (has a downward shift) as the temperature increases, whereas the band at 2.92 THz is unchanged. The frequency shift of the peak at 2.49 THz can be attributed to the thermal expansion of the PHB crystal lattice. The substantial thermal expansion of the *a*-axis of the PHB crystal lattice suggests that the distance between the $CH_3\cdots O{=}C$ hydrogen bonds along the *a*-axis increases and the hydrogen bonds become weaker. In other words, the peak shift at 2.49 THz represents a change in the $CH_3\cdots O{=}C$ hydrogen bonds.

4.3 Several Types of Intermolecular Hydrogen Bonds in PCL

PCL (Mn 80 000 g mol^{-1}) is a biodegradable polyester with excellent properties for polymer processing. PCL is an innocuous thermoplastic polyester with a low melting point (~60 °C) and high crystallinity, and exhibits good compatibility with other polymers. It is used for several different purposes, including agricultural (mulch film) and medical applications.

PCL maintains high crystallinity despite its low melting point, which indicates the possibility of hydrogen bonding in the crystal structure. As mentioned in Chapter 16 "CH···O hydrogen bonding," several bands are observed in the CH_2 stretching region along with a hydrogen-bonded C=O stretching band. PCL contains CH_2 groups but no CH_3 groups; therefore, the spectrum should exhibit a maximum of four bands corresponding to the CH_2 antisymmetric and symmetric stretching modes of the crystalline and amorphous modes of PCL. However, at least seven bands are observed in the CH stretching vibration region of the PCL IR spectra. In addition, the C=O stretching band in the PCL spectra appeared at 1724 cm^{-1}, which is close to the hydrogen-bonded band in the PHB spectra. Unfortunately, marker bands for hydrogen bonding, such as the band at 3009 cm^{-1} in the PHB spectra, were not observed in PCL. We investigated the intermolecular interactions in PCL using THz spectroscopy as it is a powerful tool for monitoring intermolecular interactions and higher-order structures through low-frequency vibrations exhibited by large molecules [12]. In a previous study, two major peaks were observed in the low-frequency region of the PCL spectra. However, the calculations used only a single oligomer model and did not report detailed band assignments for the low-frequency vibrational modes of PCL [33]. It is important to note that the bands in the low-frequency range are harder to assign than those in the IR spectra. Therefore, we combined THz spectroscopy and QCCs to investigate the intermolecular interactions in PCL [12]. The THz measurements were conducted using a TAS7400 terahertz spectroscopy system (Advantest Corp.; resolution: 7.6 GHz, wavenumber region: 150 to 30 cm^{-1}, no. of scans: 1024).

Figure 4.4a presents the temperature-dependent THz spectra of PCL in the temperature range from −100 to 65 °C, and Figure 4.4b displays the changes in the wavenumber of PCL at 47 cm^{-1} (1.42 THz) and 67 cm^{-1} (2.03 THz). These two bands shift to a lower frequency as the temperature increases. In the range of −100 to 65 °C, the bands at 47 and 67 cm^{-1} exhibit a downward shift of 4 and 9 cm^{-1}, respectively (Figure 4.4b). The disappearance of these bands above the melting point of PCL suggests that they are associated with the crystal structure. Furthermore, it was observed from the polarization THz spectra of PCL that the band at 47 cm^{-1} (1.42 THz) is indicative of perpendicular polarization and the one at 67 cm^{-1} (2.03 THz) consists of overlapped bands, owing to perpendicular and parallel polarization.

Figure 4.5 shows the model structures of PCL used for the natural bond orbital (NBO) calculations [12]. The upper figure shows the arrangement of the two molecules viewed from the direction of the *c*-axis. A, B, and C in Figure 4.5 depict possible intermolecular interactions between the C=O and CH_2 groups in different

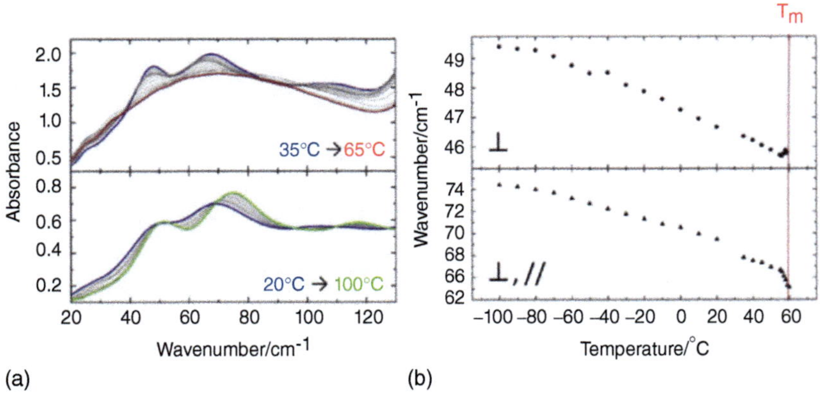

Figure 4.4 (a) Temperature-dependent THz spectra of PCL in the temperature range of 35–65 °C (top) and 20 to −100 °C (bottom). (b) Plots of the wavenumber of the band at 47 cm^{-1} (top) and the band at 67 cm^{-1} (bottom) versus temperature. Source: Funaki et al. [12].

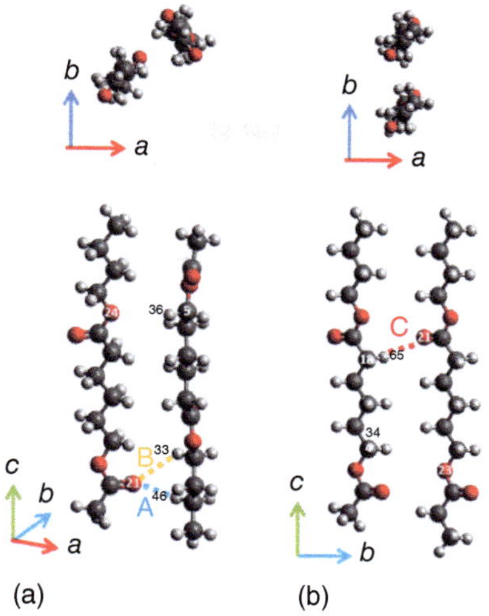

Figure 4.5 Model structures of PCL used for the NBO calculations. The structures were extracted from reported crystal structure [34]. (a) two chains are aligned perpendicularly along the (110) direction and (b) *b*-axis. A, B, and C are different types of hydrogen bonding between C=O and three types of CH_2 group. Source: Funaki et al. [12].

environments and correspond to A, B, and C, respectively, in the chemical structure of PCL in Figure 16.15 in Chapter 16 "CH· · ·O Hydrogen Bonding." In other words, there are several types of CH_2· · ·O=C hydrogen bonds, with different directions. Analysis of the IR spectra allows us to determine that several interactions are possible; however, it is difficult to distinguish them. In contrast, analysis of the THz spectra makes it possible to distinguish several types of hydrogen bonds in PCL. This is due to the combination of measurements from the polarized THz spectra and NBO calculations. Table 4.1 shows the donor–acceptor interaction energies for the model in Figure 4.5. NBO calculations were performed using CAM-B3LYP-GD3BJ/6-31++G**. The donor–acceptor interaction energy between the carbonyl O atom and the methylene H—C bond is different in positions A, B, and C. The stabilization energy can be classified into three patterns (A, B, and C) according to the CH_2 group, as shown in Figure 4.5 and Table 4.1. The three patterns had stabilization energies ranging from weakest to strongest determined by the NBO calculations. It is assumed that the strength of these patterns is due to the difference in the neighboring groups of CH_2.

Subsequently, we discuss the assignment of spectra in the low-frequency region by QCCs. In general, calculations are limited by the size of the molecule as well as the capabilities of the computer. It is necessary to use a model of molecular chains with lamellar thickness when calculating the entire crystal structure of a polymer. However, because it is difficult to perform direct calculations on a large molecular model, the Cartesian coordinate tensor transfer (CCT) method [35, 36] was used to calculate the molecular model of PCL. The CCT method has been used to determine the molecular models of other polymers, including PHB [9], polyglycolic acid (PGA) [10], and nylon 6 [11]. The CCT method performs calculations on fragments of the molecules and transfers them into the original molecular model. Calculations can thus be performed in a reasonable amount of time using the CCT method compared to conventional calculations, especially for large molecular systems.

Table 4.1 Donor–acceptor interaction energies for the model in Figure 4.5a,b calculated at CAM-B3LYP-GD3BJ/6-31++G** based on the crystal structure of PCL [34].

Donor	Acceptor	E (kcal mol^{-1})
(a)		
$n_1(O_{21})$	$\sigma^*(H_{33}–C_4)$	0.41 } B
$n_2(O_{21})$	$\sigma^*(H_{33}–C_4)$	0.60 } B
$n_1(O_{21})$	$\sigma^*(H_{46}–C_{10})$	0.51 } A
$n_2(O_{21})$	$\sigma^*(H_{46}–C_{10})$	0.12 } A
$n_1(O_{214})$	$\sigma^*(H_{36}–C_5)$	0.12
(b)		
$n_1(O_{21})$	$\sigma^*(H_{65}–C_{18})$	1.15 } C
$n_2(O_{21})$	$\sigma^*(H_{65}–C_{18})$	0.37 } C
$n_1(O_{23})$	$\sigma^*(H_{34}–C_4)$	0.07

Source: Funaki et al. [12].

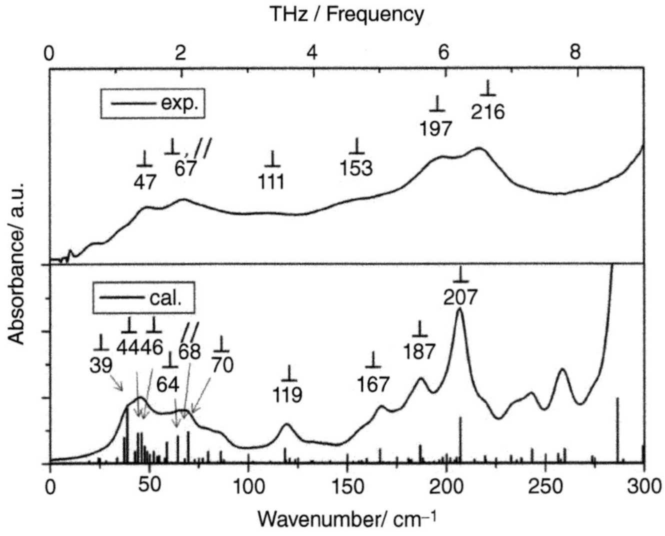

Figure 4.6 The calculated (bottom) and experimental (top) THz spectra of PCL. Source: Funaki et al. [12].

Figure 4.6 shows the calculated and experimental THz spectra of PCL. The calculations were performed with intermolecular interactions at the CAM-B3LYP-GD3BJ/6-31++G** level of theory. The results of the CCT calculations are in agreement with the experimental results, including those regarding the polarization properties and spectral features.

The results of QCCs confirmed that the band in the THz region was formed by the overlapping of several bands. In addition, it was found that the peaks below 100 cm^{-1} originate from the $CH_2\cdots O{=}C$ interaction and those above 100 cm^{-1} originate from the skeletal vibration. A detailed investigation of the vibrational modes in the THz region revealed that the bands at 47 and 67 cm^{-1} are due to the $CH_2\cdots O{=}C$ interaction of pattern A and the $CH_2\cdots O{=}C$ interaction of pattern B, respectively. The temperature-dependent THz spectra showed a decrease in intensity of the band at 47 cm^{-1}, confirming that pattern A is a weak hydrogen bond. This result suggests that PCL forms several types of weak hydrogen bonds, because the THz spectra reflect small differences in the strength of hydrogen bonds and are thus extremely sensitive. The temperature-dependent THz spectra in combination with the QCCs of PCL support the existence of several types of hydrogen bonding in the crystal structure [12].

4.4 Stress-Induced Crystal Transition of Polybutylene Succinate (PBS)

The crystal structure of PBS changes due to stretching, which results in a crystal phase transition. Generally, it is an orientated α crystal with a helical structure; however, it changes to a β crystal with a zigzag structure due to stretching [37–39]. However, this change in crystal structure is reversible and the sample returns to the α crystal helical structure when the stress is released [37–39]. In this study, we investigate the state of the crystal phase transition of PBS using THz spectroscopy, which reflects inter- and intramolecular interactions and higher-order structures of polymers due to its high sensitivity to changes in crystal structure. Previous studies have determined the stretch-induced crystalline phase transition of PBS using Raman spectroscopy or X-ray diffraction. Therefore, we investigated the crystalline phase transition in the low-frequency region using THz and low-frequency Raman spectroscopy [14]. To investigate the higher-order structure and crystal transition of PBS, polarized THz spectra of PBS were measured during the stretching process.

Figure 4.7 shows the THz and low-frequency Raman spectra during the heating of PBS from 30 °C to the molten state [14]. Four major bands can be observed at 58, 76, 90, and 100 cm^{-1} in the 50–150 cm^{-1} range in the THz spectra, and one major band at 88 cm^{-1} was observed in the region from 70 to 140 cm^{-1} in the low-frequency Raman spectra. The band at 100 cm^{-1} in the THz spectra and the band at 88 cm^{-1} in the low-frequency Raman spectra shifted to a lower frequency with increasing temperature. QCCs using the CCT method were performed to assign the bands in the low-frequency region (THz and low-frequency Raman spectra) [35, 36], similar to that of PCL. QCCs indicated that the band at 76 cm^{-1} in the THz spectra was due to the twisting motion of the main chain, while the bands at approximately 100 cm^{-1} in the THz spectra and 88 cm^{-1} in the Raman spectra were due to the intermolecular interaction between the C=O and CH_2 groups perpendicular to the molecular chain. In PHB, PGA, and PCL, the bands in the low-frequency region corresponding to hydrogen bonding showed a downward shift in frequency with

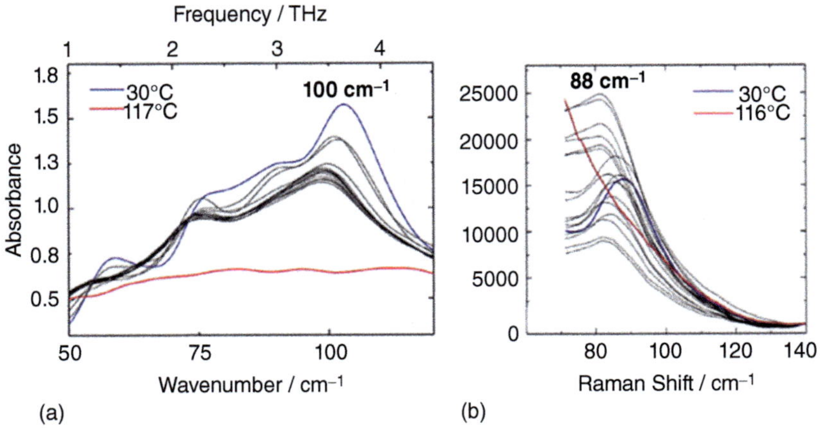

Figure 4.7 (a) THz and (b) low-frequency Raman spectra of PBS during heating process. Source: Tatsuoka and Sato [14].

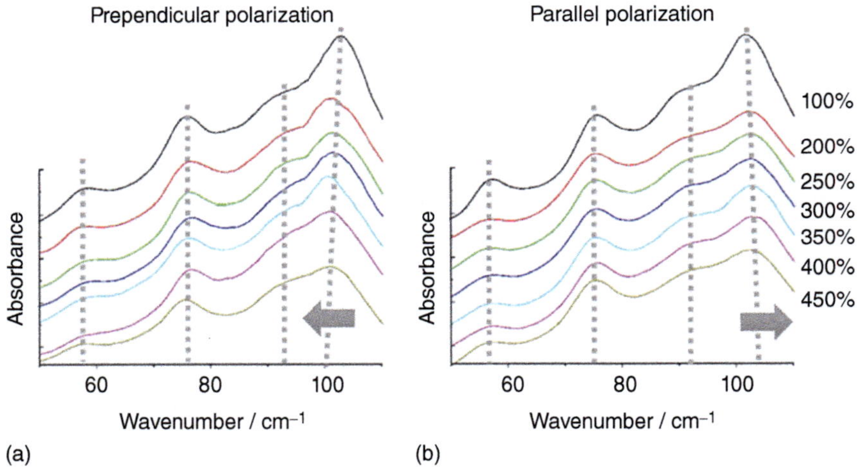

Figure 4.8 THz polarization spectra of PBS at room temperature with several drawn ratios. (a) Perpendicular polarization. (b) Parallel polarization. Source: Tatsuoka and Sato [14].

increasing temperature, indicative of the weakening of the hydrogen bonds. Therefore, the downward shift of the bands at 100 cm^{-1} in the THz spectra and 88 cm^{-1} in the Raman spectra due to the change in temperature also indicates the weakening of the hydrogen bonds.

Figure 4.8 shows THz spectra with perpendicular and parallel polarization spectra of PBS at room temperature with several draw ratios [14]. The THz band at 100 cm^{-1} shifts to lower frequency as the temperature increases. However, in the case of stretching, the band at 100 cm^{-1} shows contradictory behavior in the perpendicular and parallel polarization THz spectra. The band at 100 cm^{-1} has a downward shift as the stretching in the perpendicular THz spectra of PBS, while the band at 100 cm^{-1} has an upward shift in the parallel polarization spectra. The peaks at 58, 76, and 90 cm^{-1} did not exhibit any significant shift during stretching in both the parallel and perpendicular THz spectra. As mentioned above, PBS exhibits a crystal phase transition due to stretching. Therefore, the peak shift due to the stretching in the parallel and perpendicular THz spectra can be attributed to the crystalline phase transition of PBS. It has been suggested that the peak at approximately 100 cm^{-1} in QCCs reflects the intermolecular hydrogen bonding between perpendicular molecular chains. The intermolecular distance between H atoms in the CH_2 group and O atoms in the C=O group in the intermolecular CH···O=C bond increases from 2.62 to 3.04 Å as the crystal structure changes from an α to β crystal. The intermolecular hydrogen bonding between the adjacent parallel chains is weakened with stretching due to the change in the crystal structure from α to β and the increase in interatomic distance. The downward shift of the peak in the perpendicular polarization THz spectra is due to the weakening of the intermolecular interactions. In the temperature-dependent spectra, the peak at 100 cm^{-1} shifted downward as the temperature increased, supporting the result that intermolecular hydrogen bonding is weakened. In contrast, the parallel terahertz spectra showed an upward shift during the stretching process. The calculated THz spectrum of the β crystal exhibited bands at 149 and 166 cm^{-1} that were not present in the α crystal spectrum of PBS. This suggests that a new band in the high-frequency region appears due to the transformation of PBS into a β crystal structure. The QCCs infer that this band reflects the intramolecular O (ester) and H atom interactions in the β crystal structure. The distance between the O (ester) and H atoms decreased from 2.51 Å in the α crystal to 2.09 Å in the β crystal. Therefore, the shift to higher frequency was caused by the crystal transition from the α to β crystal, which strengthens the intramolecular interaction between the O (ester) and H atoms, resulting in the appearance of the β crystal-derived peak at a higher wavenumber than 100 cm^{-1} in the parallel THz spectra [14].

4.5 The Differences in Intermolecular Hydrogen Bonding Between PET and PBT

Poly(ethylene terephthalate) (PET), widely used in plastic bottles, has a slow rate of crystallization. However, poly(butylene terephthalate) (PBT), which contains two more CH_2 groups than PET, has a high crystallization rate and excellent moldability, and is used in a wide range of applications as an engineering plastic. PET and PBT possess a similar crystal structure, as investigated by X-ray diffraction [40–42]. The crystal structure of PET has a triclinic crystal system with lattice parameters of $a = 4.56$ Å, $b = 5.94$ Å, c (chain axis) $= 10.75$ Å, $\alpha = 98.5°$, $\beta = 118°$, and $\gamma = 112°$ [40, 41]. PBT comprises both α- and β-crystalline forms that are reversible via stretching. The α form of PBT is triclinic, with lattice parameters of $a = 4.83$ Å, $b = 5.94$ Å, c (chain axis) $= 11.59$ Å, $\alpha = 99.7°$, $\beta = 115.2°$, and $\gamma = 110.8°$ [42].

Figure 4.9 shows the temperature-dependent THz spectra and the second derivative spectra of PET and PBT during the heating process in the range from 30 to 280 °C [15]. The band at 79 cm^{-1} has shifts downward as the temperature increases. This

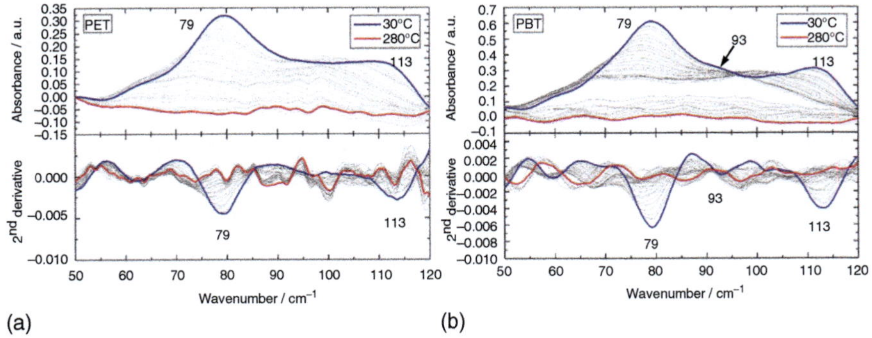

Figure 4.9 The temperature-dependent THz spectra and their second derivatives spectra of (a) PET and (b) PBT in the temperature range from 30 to 280 °C. Source: Yamamoto et al. [15].

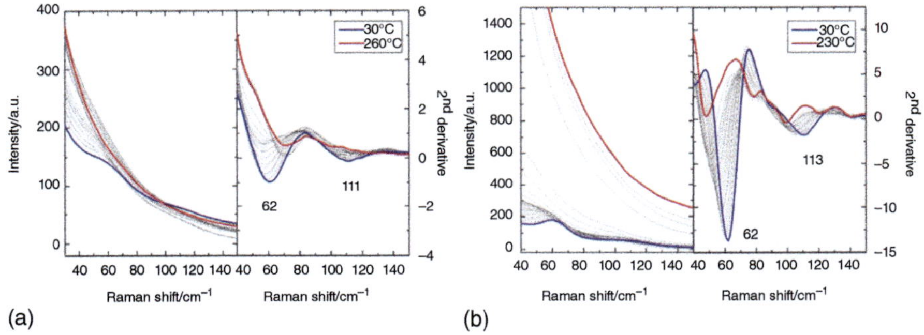

Figure 4.10 The temperature-dependent low-frequency Raman spectra of (a) PET and (b) PBT as the temperature increased. Source: Yamamoto et al. [15].

band is derived from the combination of the out-of-plane bending mode of C=O and the out-of-plane vibration of the benzene ring, and may be related to the π–π stacking of the crystal structure of PET and PBT [43]. Since π–π interactions stabilize the crystal structure [43], the shift of the band at 79 cm^{-1} reflects the thermal expansion of the crystal lattice and accompanying changes in the intermolecular hydrogen bonding.

Figure 4.10 presents the temperature-dependent low-frequency Raman spectra of the PET and PBT structures [15]. The bands at 60 and 111 cm^{-1} in the spectra of PET and bands at 62, 90, 111 cm^{-1} in spectra of PBT appeared in the region below 150 cm^{-1}. The bands at approximately 60 cm^{-1} of PET and PBT show a downward shift as the temperature increases. From the QCCs, this band was assigned to the out-of-plane bending mode of C=O and the out-of-plane vibration of the benzene ring, which corresponds with the band at 79 cm^{-1} in the THz spectra of both PET and PBT. The band at 113 cm^{-1} showed a downward shift due to an increase in temperature in the PBT spectra. However, no shift was observed in the PET spectra. In addition, the small band at 90 cm^{-1} was only present in the PBT spectra. Comparing the QCCs results and the THz spectra indicated that the band at 90 cm^{-1} may correspond to that at 93 cm^{-1} in the THz spectra of PBT [15].

Combining the results of THz and low-frequency Raman spectra, it can be observed that the thermal behavior of the bands at 111 cm^{-1} in the THz spectra and at 113 cm^{-1} in the low-frequency Raman spectra differs between PET and PBT. According to QCCs, the difference in the assignment of those bands relates to whether the CH_2 bending mode is included or not. Therefore, these differences may be due to the differences in intermolecular hydrogen bonding and the mobility of molecular chains between PET and PBT. The difference in the number of CH_2 groups between PET and PBT may be responsible for the difference in molecular chain mobility and intermolecular hydrogen bonding. We concluded that THz and low-frequency Raman spectroscopy could successfully detect the difference in molecular chain mobility between PET and PBT [15].

4.6 THz Imaging of Polymer Film

Herein, we present an example of the application of THz spectroscopy to polymer materials. Imaging spectroscopy is an attractive analytical technique for use in industry due to the visual understandability. THz spectra are sensitive to the intermolecular hydrogen bonding and higher-order structure of polymers. Therefore, THz imaging has attracted attention from academic fields and industry [17, 44–51]. The spatial resolution of THz imaging, which can reach the order of a few hundred microns at most, is lower than that of IR or Raman spectroscopy. Furthermore, it can be used for the imaging of large-sized polymers. To study the physical properties of polymer materials, it is necessary to observe not only the microsize region, as observed by IR or Raman spectroscopy, but also the larger size region. THz imaging reflects the physical properties of polymers because it can provide images of large regions with low spatial resolution.

In this chapter, imaging measurements of biodegradable polyesters in THz-time domain spectroscopy (THz-TDS) were carried out to evaluate its physical properties, including the distribution of crystallinity and progression of secondary crystallization. In the THz spectra of PCL in the frequency range from 0.5 to 4.5 THz, two major bands were found at 1.42 and 2.03 THz, which were ascribed to the crystal structure of PCL [12, 33]. The peaks at 2.03 and 1.42 THz were assigned to the vibration of the zigzag structure along the fiber axis, and the vibration due to the intermolecular interaction between the zigzag structures, respectively. The crystallinity of PCL correlated to the ratio of the intensity of these two bands at 1.42 and 2.03 THz, compared to those of the temperature-dependent THz spectra and temperature-dependent WAXD of PCL. The distribution of crystallinity in the PCL film was investigated using these intensity ratios. Although the film was homogeneous, the distribution of crystallinity was visible in the THz imaging (Figure 4.11).

The intensity ratio of $I_{1.42THz}/I_{2.03THz}$ regarding the positions of A and B is different in the THz imaging of PCL (Figure 4.11). The crystallinity of positions A and B was

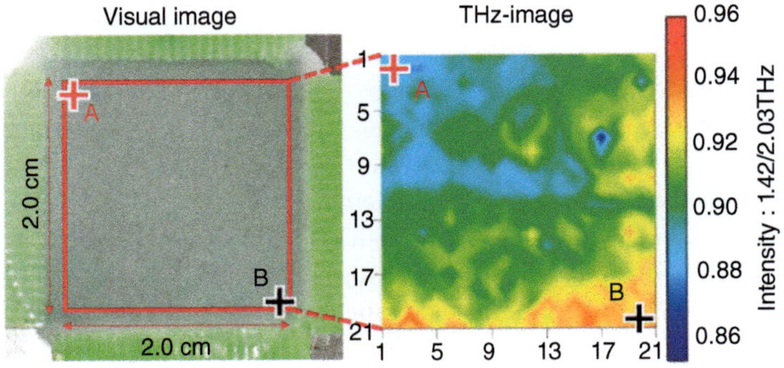

Figure 4.11 A visible image 2 cm^2 PCL film and its corresponding THz imaging. Source: Funaki et al. [12].

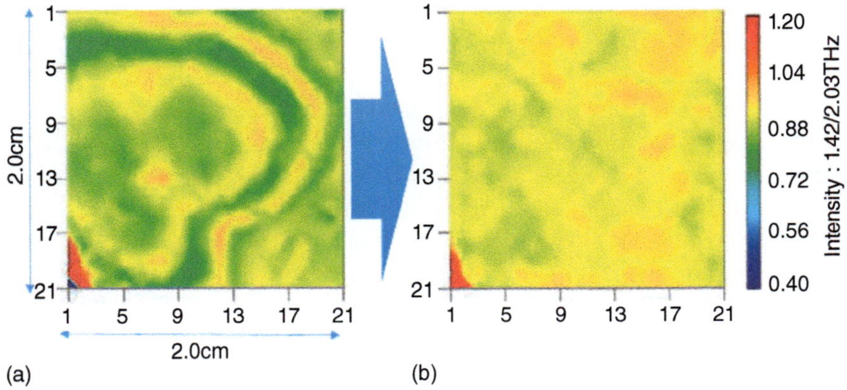

Figure 4.12 The THz images of (a) as-pressed PCL film and (b) after melt removed the stress. Source: Funaki et al. [12].

measured using both WAXD and differential scanning calorimetry (DSC) to confirm the degree of the crystallinity in these positions. The WAXD results indicate that the crystallinity of positions A and B was approximately 33% and 45%, respectively. The heat of fusion experiments determined that A and B were approximately 47 and 64 J g^{-1}, respectively. It was confirmed that the THz imaging using the intensity ratio $I_{1.42THz}/I_{2.03THz}$ accurately reflected the distribution of the crystallinity.

Figure 4.12 shows the annealing effect on THz imaging. Figure 4.12a shows the THz imaging of the as-pressed PCL film. The film was pressed so that the residual stress remained. Furthermore, by annealing the as-pressed PCL film, the inhomogeneous crystallinity of the PCL film clearly disappeared in the THz imaging. The band at 1.42 THz was assigned to the perpendicular polarization of the molecular chain, while the band at 2.03 THz was assigned to the parallel and perpendicular polarizations. Therefore, the intensity ratio is also involved in the orientation of the crystal structure of PCL. Since the intensity ratio is related to the crystallinity and polarization of the inhomogeneous pattern, stripe pattern and concentric rings in the as-pressed PCL film are caused by different crystallinity and/or crystal orientation [12].

The band at 1.42 THz shows only perpendicular polarization, while the band appearing at 2.03 THz shows both parallel and perpendicular polarization. Therefore, THz polarization imaging was conducted only on the band at 1.42 THz. Figure 4.13 shows the THz polarization imaging using the intensity of the band at 1.42 THz (Figure 4.13a,b) and its intensity ratio (Figure 4.13c) when the polarization direction of the THz wave was 90° and 0° to the vertical direction to the sheet. The imaging of the sample film was conducted immediately after pressing, and residual stress due to the pressing remained on the film. The striped lines in Figure 4.13a,b most likely represent the crystal orientation caused by the residual stress. THz imaging was successfully used to visualize the distribution of crystallinity and crystal orientation. The band that appears in the THz region reflects the higher-order structure and intermolecular interaction, and THz imaging constructed by these bands can be used to monitor a variety of properties in the processing of polymers [12].

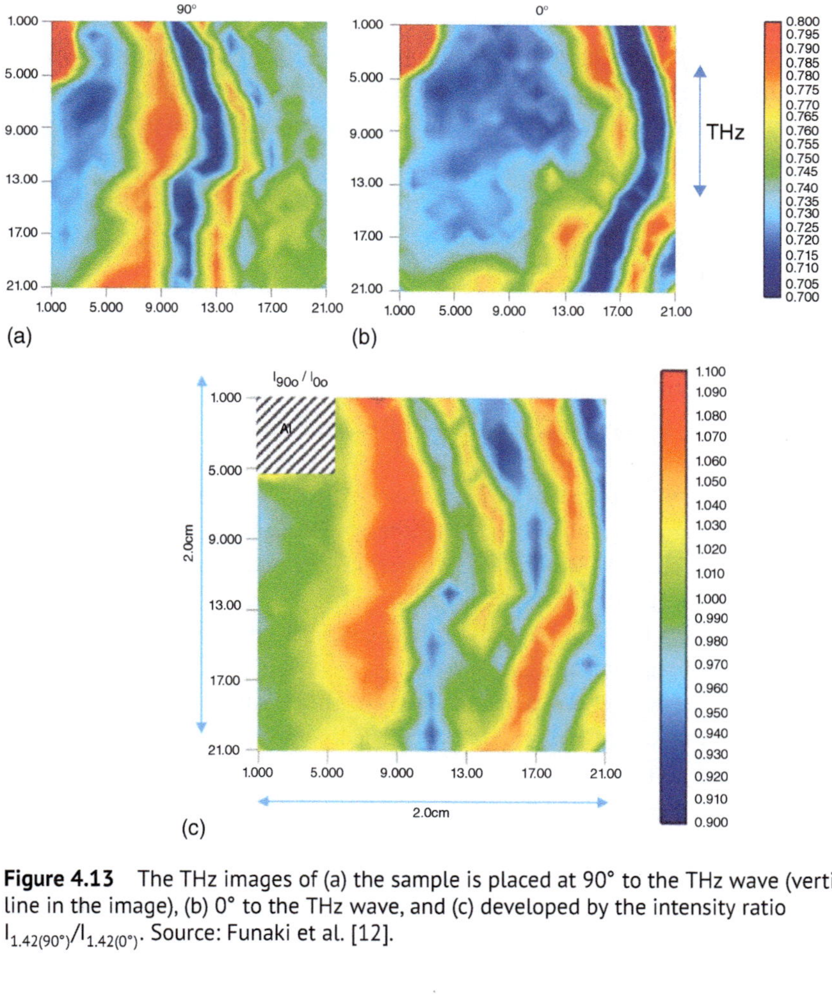

Figure 4.13 The THz images of (a) the sample is placed at 90° to the THz wave (vertical line in the image), (b) 0° to the THz wave, and (c) developed by the intensity ratio $I_{1.42(90°)}/I_{1.42(0°)}$. Source: Funaki et al. [12].

4.7 Conclusions

Low-frequency spectroscopic methods, such as FIR/THz and low-frequency Raman spectroscopy, have been successfully used to measure and investigate higher-order structures, thermal expansion, and intermolecular interactions, including CH···O hydrogen bonds in polymers. The combination of FIR, THz, and low-frequency Raman spectroscopy and QCCs is extremely useful for band assignments in the low-frequency region, which had previously been difficult to determine. In addition, THz imaging has been successfully used to visualize the distribution of crystallinity and crystal orientation in polymer films. In future, THz imaging and low-frequency Raman spectroscopy are expected to be used for the visualization of intermolecular interactions in polymers during crystallization processes. FIR/THz and low-frequency Raman spectroscopy can help in the understanding of intermolecular interactions, including hydrogen bonding and the higher-order structures of polymers. Chemometrics analysis, including multivariate curve resolution (MCR) and two-dimensional correlation spectroscopy (2DCOS), can be extensively used to analyze spectra in the low-frequency regions. In future, depth-profile analysis using THz-TDS, attenuated total reflection-far infrared (ATR-FIR) imaging, as well as 3D-imaging using low-frequency Raman spectroscopy, will be very useful for the characterization of higher-order structures and physical properties of polymers.

References

1 (a) Moller, K.D. and Rothschield, W.G. (1971). *Far-Infrared Spectroscopy*. New York: Wiley Interscience; (b) Finch, A. (1970). *Chemical Applications of Far Infrared Spectroscopy*. New York: Academic Press.

2 Griffiths, P.R. (2002). Far-infrared spectroscopy. In: *Handbook of Vibrational Spectroscopy*, vol. 1 (eds. J.M. Charmers and P.R. Griffiths), 229. Chichester: Wiley.

3 (a) Lee, Y.-S. (2009). *Principle of Terahertz Science and Technology*. New York: Springer; (b) Dexheimer, S.L. (2008). *Terahertz Spectroscopy: Principles and Applications*. Boca Raton, FL: CRC Press; (c) Baxter, J.B. and Guglietta, G.W. (2011). *Anal. Chem.* 83: 4342.

4 Kawano, Y. (2013). Terahertz waves: a tool for condensed matter, the life sciences and astronomy. *Contemp. Phys.* 54: 143–165.

5 Frank, W.F.X., Strohmeier, W., and Hallensleben, M.L. (1981). Far infrared absorption spectra of undeuterated and deuterated poly(ethylene terephthalate). *Polymer* 22: 615–618.

6 Tonouchi, M. (2007). Cutting-edge terahertz technology. *Nat. Photonics* 1 (2): 97–105.

7 (a) Hoshina, H., Morisawa, Y., Sato, H. et al. (2010). Higher order conformation of poly(3-hydroxyalkanoates) studied by terahertz timedomain spectroscopy. *Appl. Phys. Lett.* 96: 101904; (b) Hoshina, H., Morisawa, Y., Sato, H. et al. (2011).

Polarization and temperature dependent spectra of poly(3-hydroxyalkanoate)s measured at terahertz frequencies. *Phys. Chem. Chem. Phys.* 13 (20): 9173–9179.
8 (a) Hoshina, H., Ishii, S., Morisawa, Y. et al. (2012). Isothermal crystallization of poly(3-hydroxybutyrate) studied by terahertz twodimensional correlation spectroscopy. *Appl. Phys. Lett.* 100 (1): 011907; (b) Hoshina, H., Ishii, S., Yamamoto, S. et al. (2013). Terahertz spectroscopy in polymer research: assignment of intermolecular vibrational modes and structural characterization of poly(3-hydroxybutyrate). *IEEE Trans. Terahertz Sci. Technol.* 3: 248–258.
9 Yamamoto, S., Morisawa, Y., Sato, H. et al. (2013). Quantum mechanical interpretation of intermolecular vibrational modes of crystalline poly-(*R*)-3-hydroxydutyrate observed in low-frequency Raman and terahertz spectra. *J. Phys. Chem. B* 117: 2180–2187.
10 Yamamoto, S., Miyada, M., Sato, H. et al. (2017). Low-frequency vibrational modes of poly(glycolic acid) and thermal expansion of crystal lattice assigned on the basis of DFT-spectral simulation aided with a fragment method. *J. Phys. Chem. B* 121: 1128–1138.
11 Yamamoto, S., Ohnishi, E., Sato, H. et al. (2019). Low-frequency vibrational modes of nylon 6 studied by using infrared and Raman spectroscopies and density functional theory calculations. *J. Phys. Chem. B* 123: 5368–5376.
12 Funaki, C., Yamamoto, S., Hoshina, H. et al. (2018). Three different kinds of weak C—H···O=C inter- and intramolecular interactions in poly(ϵ-caprolactone) studied by using terahertz spectroscopy, infrared spectroscopy and quantum chemical calculations. *Polymer* 137: 245–254.
13 Nishimura, F., Hoshina, H., Ozaki, Y., and Sato, H. (2019). Isothermal crystallization of poly(glycolic acid) studied by terahertz and infrared spectroscopy and SAXS/WAXD simultaneous measurements. *Polym. J.* 51: 237–245.
14 Tatsuoka, S. and Sato, H. (2018). Stress-induced crystal transition of poly(butylene succinate) studied by terahertz and low-frequency Raman spectroscopy and quantum chemical calculation. *Spectrochim. Acta Part A* 197: 95–102.
15 Yamamoto, Y., Hoshina, H., and Sato, H. (2021). Differences in intermolecular interactions and flexibility between poly(ethylene terephthalate) and poly(butylene terephthalate) studied by far-infrared/terahertz and low-frequency Raman spectroscopy. *Macromolecules* 54: 1052–1062.
16 Marlinaa, D., Hoshina, H., Ozaki, Y., and Sato, H. (2019). Crystallization and crystalline dynamics of poly(3-hydroxybutyrate)/poly(4-vinylphenol) polymer blends studied by low-frequency vibrational spectroscopy. *Polymer* 181: 121790.
17 Funaki, C., Toyouchi, T., Hoshina, H. et al. (2017). Terahertz imaging of the distribution of crystallinity and crystalline orientation in a poly(ε-caprolactone) film. *Appl. Spectrosc.* 71: 1537–1542.
18 Suzuki, H., Ishii, S., Sato, H. et al. (2013). *Chem. Phys. Lett.* 575: 36–39.
19 Suzuki, H., Ishii, S., Otani, C., and Hoshina, H. (2015). Low-frequency vibrations of polyamide-6 as a function of temperature and thermal history investigated by terahertz absorption spectroscopy. *Eur. Polym. J.* 67: 284–291.

20 Sato, H., Nakamura, M., Padermshoke, A. et al. (2004). Thermal behavior and molecular interaction of poly(3-hydroxybutyrate-*co*-3-hydroxyhexanoate) studied by wide-angle X-ray diffraction. *Macromolecules* 37: 3763–3769.

21 Sato, H., Murakami, R., Padermshoke, A. et al. (2004). Infrared spectroscopy studies of CH···O hydrogen bondings and thermal behavior of biodegradable poly(hydroxyalkanoate). *Macromolecules* 37: 7203–7213.

22 Sato, H., Mori, K., Murakami, R. et al. (2006). Crystal and lamella structure and C—H···O=C hydrogen bonding of poly(3-hydroxyalkanoate) studied by X-ray diffraction and infrared spectroscopy. *Macromolecules* 39: 1525–1531.

23 Sato, H., Murakami, R., Zhang, J. et al. (2005). Infrared spectroscopy and X-ray diffraction studies of C—H···O hydrogen bonding and thermal behavior of biodegradable poly(hydroxyalkanoate). *Macromol. Symp.* 230: 158–166.

24 Sato, H., Dybal, J., Murakami, R. et al. (2005). Infrared and Raman spectroscopy and quantum chemistry calculation studies of C—H/O hydrogen bondings and thermal behavior of biodegradable polyhydroxyalkanoate. *J. Mol. Struct.* 744–747: 35–46.

25 Sato, H., Ando, Y., Dybal, J. et al. (2008). Crystal structures, thermal behaviors, and C—H···O=C hydrogen bondings of poly(3-hydroxyvalerate) and poly(3-hydroxybutyrate) studied by infrared spectroscopy and X-ray diffraction. *Macromolecules* 41: 4305–4312.

26 Sato, H., Murakami, R., Mori, K. et al. (2009). Specific crystal structure of poly(3-hydroxybutyrate) thin films studied by infrared reflection–absorption spectroscopy. *Vib. Spectrosc.* 51: 132–135.

27 Guo, L., Sato, H., Hashimoto, T., and Ozaki, Y. (2010). FTIR study on hydrogen-bonding interactions in biodegradable polymer blends of poly(3-hydroxybutyrate) and poly(4-vinylphenol). *Macromolecules* 43: 3897–3902.

28 Guo, L., Sato, H., Hashimoto, T., and Ozaki, Y. (2011). Thermally induced exchanges of hydrogen bonding interactions and their effects on phase structures of poly(3-hydroxybutyrate) and poly(4-vinylphenol) blends. *Macromolecules* 44: 2229–2239.

29 Guo, L., Spegazzini, N., Sato, H. et al. (2012). Multistep crystallization process involving sequential formations of density fluctuations, "intermediate structures", and lamellar crystallites: poly(3-hydroxybutyrate) as investigated by time-resolved synchrotron SAXS and WAXD. *Macromolecules* 45: 313–328.

30 Suttiwijitpukdee, N., Sato, H., Zhang, J., and Hashimoto, T. (2011). Effects of intermolecular hydrogen bondings on isothermal crystallization behavior of polymer blends of cellulose acetate butyrate and poly(3-hydroxybutyrate). *Macromolecules* 44: 3467–3477.

31 Sato, H., Suttiwijitpukdee, N., Hashimoto, T., and Ozaki, Y. (2012). Simultaneous synchrotron SAXS/WAXD study of composition fluctuations, cold-crystallization, and melting in biodegradable polymer blends of cellulose acetate butyrate and poly(3-hydroxybutyrate). *Macromolecules* 45: 2783–2795.

32 Suttiwijitpukdee, N., Sato, H., Zhang, J. et al. (2011). Intermolecular interactions and crystallization behaviors of biodegradable polymer blends between poly(3-hydroxybutyrate) and cellulose acetate butyrate studied by DSC, FT-IR, and WAXD. *Polymer* 52: 461–471.

33 Komatsu, M., Mizuno, M., Saito, S. et al. (2015). Terahertz spectral change associated with glass transition of poly-ε-caprolactone. *J. Appl. Phys.* 117 (13): 133102.

34 Chatani, Y., Okita, Y., Tadokoro, H., and Yamashita, Y. (1970). Structure studies of polyesters. III. Crystal structure of poly-ε-caprolactone. *Polym. J.* 1 (5): 555–562.

35 Bouř, P., Sopková, J., Bednárová, L. et al. (1997). Transfer of molecular property tensors in Cartesian coordinates: a new algorithm for simulation of vibrational spectra. *J. Comput. Chem.* 18: 646–659.

36 Yamamoto, S., Li, X., Ruud, K., and Bouř, P. (2012). Transferability of various molecular property tensors in vibrational spectroscopy. *J. Chem. Theory Comput.* 8: 977–985.

37 Ichikawa, Y., Suzuki, J., Washiyama, J. et al. (1994). Strain-induced crystal modification in poly(tetramethylene succinate). *Polymer* 35: 3338–3339.

38 Ichikawa, Y., Kondo, H., Igarashi, Y. et al. (2000). Crystal structures of α and β forms of poly(tetramethylene succinate). *Polymer* 41: 4719–4727.

39 Ichikawa, Y., Washiyama, J., Moteki, Y. et al. (1995). Crystal transition mechanisms in poly(tetramethylene succinate). *Polym. J.* 27: 1230–1238.

40 Daubeny, R.D.P., Bunn, C.W., and Brown, C.J. (1954). The crystal structure of polyethylene terephthalate. *Proc. R. Soc. London, Ser. A* 226: 531–542.

41 Fu, Y., Busing, W.R., Jin, Y. et al. (1993). Poly(ethylene terephthalate) fibers. 1. Crystal structure and morphology studies with full-pattern X-ray diffraction refinement. *Macromolecules* 26: 2187–2193.

42 Yokouchi, M., Sakakibara, Y., Chatani, Y. et al. (1976). Structures of two crystalline forms of poly(butylene terephthalate) and reversible transition between them by mechanical deformation. *Macromolecules* 9: 266–273.

43 Kurita, T., Fukuda, Y., Takahashi, M., and Sasanuma, Y. (2018). Crystalline moduli of polymers, evaluated from density functional theory calculations under periodic boundary conditions. *ACS Omega* 3: 4824–4835.

44 Herrmann, M., Tani, M., Sakai, K., and Fukasawa, R. (2002). Terahertz imaging of silicon wafers. *J. Appl. Phys.* 91 (3): 1247–1250.

45 Ajito, K. and Ueno, Y. (2011). THz chemical imaging for biological applications. *IEEE Trans. Terahertz Sci. Technol.* 1 (1): 293–300.

46 Shen, Y.C. (2011). Terahertz pulsed spectroscopy and imaging for pharmaceutical applications: a review. *Int. J. Pharm.* 417 (1): 48–60.

47 Nagatsuma, T., Nishii, H., and Ikeo, T. (2014). Terahertz imaging based on optical coherence tomography. *Photonics Res.* 2 (4): B64–B69.

48 Jansen, C., Wietzke, S., Peters, O. et al. (2010). Terahertz imaging: applications and perspectives. *Appl. Opt.* 49 (19): E48–E57.

49 Rutz, F., Koch, M., Khare, S. et al. (2006). Terahertz quality control of polymeric products. *Int. J. Infrared Millimeter Waves* 27 (4): 547–556.

50 Mittleman, D.M., Guptal, M., Neelamani, R. et al. (1999). Recent advances in terahertz imaging. *Appl. Phys. B: Lasers Opt.* 68 (6): 1085–1094.

51 Wietzke, S., Jördens, C., Krumbholz, N. et al. (2007). Terahertz imaging: a new non-destructive technique for the quality control of plastic weld joints. *J. Eur. Opt. Soc. -Rapid Publ.* 2: 07013.

5

Near-Infrared Spectroscopy and Imaging of Polymers

Daitaro Ishikawa[1], Yuta Hikima[2], and Yukihiro Ozaki[3]

[1] *Fukushima University, Faculty of Food and Agricultural Sciences, 1 Kanayagawa, Fukushima, 960-1296, Japan*
[2] *Kyoto University, Katsura Campus, Department of Chemical Engineering, Nishikyo-ku, Kyoto, 615-8510, Japan*
[3] *Kwansei Gakuin University, School of Biological and Environmental Sciences, 2-1 Gakuen, Sanda, Hyogo, 669-1337, Japan*

5.1 Introduction to NIR Spectroscopy

As already described in Chapter 1, near-infrared (NIR) spectroscopy was performed in the region of 12 500 to 4000 cm^{-1} (800–2500 nm), which lies between the visible and infrared (IR) regions [1–5]. NIR spectroscopy deals with the absorption, emission, reflection, and diffuse reflection of light in the NIR region. The purpose of this chapter is to describe the characteristics and advantages of NIR spectroscopy and NIR imaging in polymer research in session 3 [3–5]. The principles, general characteristics, and advantages of NIR spectroscopy are outlined in Chapter 1. They are described briefly here.

5.1.1 Principles of NIR Spectroscopy

NIR spectroscopy is a type of electronic and vibrational spectroscopy. It is well known that many molecules show absorption bands due to electronic transitions in the NIR region and that electronic NIR spectroscopy has many applications [6, 7]. However, to the best of our knowledge, the application of NIR electronic spectroscopy is rare; therefore, in this chapter, we deal with NIR spectroscopy as a type of vibrational spectroscopy. NIR vibrational spectroscopy involves overtones and combination modes related to prohibited transitions. The prohibited transitions yield very weak bands owing to the anharmonicity of vibrations. Therefore, NIR spectroscopy has high permeability, allowing nondestructive and in situ analyses. Almost all the characteristics and advantages of NIR spectroscopy originate from the fact that NIR spectroscopy involves anharmonicity.

Spectroscopic Techniques for Polymer Characterization: Methods, Instrumentation, Applications, First Edition. Edited by Yukihiro Ozaki and Harumi Sato.

5.1.2 Characteristics and Advantages of NIR Spectroscopy

The characteristics of NIR bands are summarized as follows [1–3]:

(1) NIR bands are, in general, much weaker than IR bands. The intensity of bands due to the first, second, and third overtones of XH stretching bands is less than 1/10, 1/100, and 1/1000, of the intensity of the corresponding band, respectively.
(2) A number of bands due to from overtones and combination modes overlap each other, and multicollinearity is high in the NIR region, and thus, the assignment of the NIR bands is not straightforward.
(3) The NIR region is dominated by bands ascribed to functional groups containing a hydrogen atom (e.g. OH, CH, NH). This is partly because an anharmonic constant of an XH bond is large, and partly because a fundamental of an XH stretching vibration has high frequency. NIR spectroscopy is often called "an XH spectroscopy."
(4) A hydrogen bond and a molecular interaction induce a lower wavenumber shift for XH and C=O bands in a NIR spectrum. The shift is far larger than that of the corresponding IR band.
(5) OH and NH stretching bands of monomeric and polymeric species of a molecule, for example, alcohol, are better separated in the NIR region than in the IR region. A band ascribed to terminal free OH or NH group of polymeric species may be clearly differentiated from a band due to free OH or NH group in the NIR region.
(6) Because of the larger anharmonicity, bands ascribed to the first overtones of OH and NH stretching modes of monomeric species are enhanced compared with the corresponding bands arising from polymeric species.

Let us now consider the advantages of NIR spectroscopy [1–5]. First, NIR spectroscopy allows the nondestructive or in situ analysis of a sample. It can be applied to samples of various states, shapes, and thicknesses. It is also suitable for noncontact analysis and analysis using optical fibers. Therefore, NIR spectroscopy is appropriate for online analyses. Moreover, NIR spectroscopy makes it much easier to study and analyze aqueous solutions than IR spectroscopy. Another advantage of NIR spectroscopy is that one can select an optical path length freely. NIR spectroscopy allows the use of a 1-mm, 1-cm, or even 10-cm cell and facilitates the measurement of a thick film.

5.1.3 Analysis of NIR Spectra

Spectral analysis using vibrational spectroscopy is described in detail in Chapter 1. Some aspects related to spectral analysis using NIR spectroscopy are outlined. As in the case of IR and Raman spectroscopy, band assignment is the basis for spectral analysis using NIR spectroscopy. However, band assignment in the case of NIR spectra is sometimes too complicated because these spectra consist of several bands due to overtones and combinations, and multicollinearity is often severe. In NIR spectra, some bands originate from ternary combinations, for example, the combination of an overtone mode and a fundamental mode ($2\nu_1$ and $+\nu_3$) and that of three fundamental modes ($\nu_1 + \nu_2 + \nu_3$), making precise band assignments difficult. In addition, Fermi resonance may cause complicated spectral patterns, making band assignments almost impossible. However, even in such cases, the functional group to which a band belongs should be known. Conventional band assignment and spectral analysis methods are not always sufficient to analyze complicated NIR spectra. Thus, chemometrics are often employed in NIR spectroscopy as a powerful tool for extracting rich information from NIR spectra [1–5, 8, 9]. An important aspect of chemometrics is multivariate data analysis methods, such as principal component analysis/principal component regression (PCA/PCR) and partial least squares regression (PLSR) [8, 9]. Self-modeling curve resolution (SMCR), which is used to predict pure component spectra and pure component concentration profiles from a set of NIR spectra, is also attracting attention [8, 9].

In addition to conventional spectral analysis methods and chemometrics, two-dimensional (2D) correlation spectroscopy has been used in NIR spectroscopy [10]. In this method, spectral peaks are spread over a second dimension to simplify the visualization of complex spectra consisting of several overlapped bands and to explore any correlation between the bands. Moreover, quantum chemical calculations, such as density functional theory (DFT) calculations, can be used nowadays, even in NIR spectral analysis [11–13]. One can calculate the intensities and frequencies of overtones and combination bands using quantum chemistry calculations, and the spectra can be reproduced. Figure 5.1 shows the experimental (5×10^{-3} M CCl_4) and simulated NIR spectra of diluted methanol; the simulated spectrum was obtained using anharmonic calculations [13]. Note that there is a

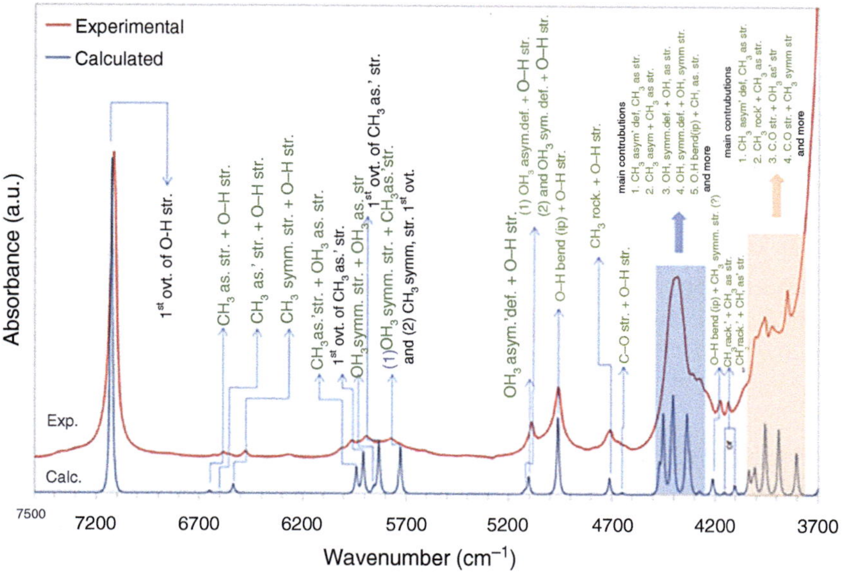

Figure 5.1 Experimental and calculated NIR spectra of low-concentration (5×10^{-3} M) solutions of methanol in CCl_4 with their band assignments. GVPT2 calculations on CPCM-B2PLYP-D/SNST level of theory were used here. Source: Bec et al. [13].

good agreement between the experimental and calculated spectra, including the reproduction of minor bands. The NIR spectra of more complicated compounds, such as long-chain fatty acids, were successfully obtained [13].

Thus, we can use various types of spectral analysis methods to analyze NIR spectra, depending on the purpose. Notably, in NIR spectroscopy, the original spectra are often subjected to pretreatment or data transformation before conducting spectral analysis [1–5]. Various pretreatment methods are applied to the experimental data to reduce any noise, correct baseline variations, enhance apparent spectral resolution, and normalize the data. Pretreatment methods have been described in several references [1–5].

5.2 Applications to Polymer Science and Engineering of NIR Spectroscopy

The application of NIR spectroscopy began in the 1960s [1–5]. NIR spectroscopy has been used extensively in polymer science and engineering [1, 2, 14–27]. It is a powerful tool to investigate the constitution, configuration, conformation, structure, molecular interactions (e.g. hydrogen bonding), and physical properties (e.g. crystallinity) of materials [15–17]. NIR spectroscopy can also be employed to monitor polymer reactions. NIR polymer applications include phase separation studies of polymer blends (Chapter 1), rheo-optical NIR spectroscopy of polymers (Chapter 3), online monitoring of polymers [18, 19], and NIR imaging studies (Sections 5.3 in this chapter). In the following sections, three NIR studies are introduced: an NIR-IR spectroscopy study of molecular orientation (Section 5.2.1) [16, 17, 23], the isothermal crystallization kinetics of poly(3-hydroxybutyrate) (PHB) (Section 5.2.2), and the crystallization of poly(3-hydroxybutyrate-*co*-3-hydroxyhexanoate) (PHBH) during the melt extrusion promoted by residual crystals [25]. In Chapter 1, the NIR spectroscopic and chemometric analyses of linear low-density polyethylene (LLDPE) were reported. Other examples of NIR spectroscopy studies of polymers can be found in Chapter 1. Moreover, recently, some interesting reviews on the NIR spectroscopic analysis of polymers have been published; for example, Scherzer reported applications of NIR techniques to polymer coatings and synthetic textiles in [22]. Watari published a review on the online monitoring of polymers using NIR spectroscopy [20]. Machado et al. outlined an NIR study of the characterization of dispersion in polymer–clay nanocomposites [21].

5.2.1 Polarized NIR Spectroscopy Studies of Molecular Orientation of Polymers

Bokobza et al. [16, 17, 23] conducted an NIR-IR spectroscopy study on the precise determination of dichroic effects by using the polarization modulation of an incident electromagnetic field. NIR spectroscopy enabled the analysis of samples of thickness several millimeters owing to the reduced intensity of absorptions. Bokobza and coworkers [23] investigated the segmental orientation in uniaxially stretched poly(dimethylsiloxane) (PDMS) elastomeric networks using NIR and IR dichroism. They revealed that the dichroic effects measured in the NIR region were stronger than those obtained in the IR region. They also observed that vibrations were less coupled at high energies. They revealed that the introduction of reinforcing fillers, such as silica in PDMS networks, leads to an increase in orientation, which is attributed to the additional cross-links produced by the filler.

Various spectroscopic techniques, including NIR dichroism, have been extensively employed to explore the segmental orientation in deformed polymeric networks [16, 17, 23]. These spectroscopic methods can directly study the orientational behavior of network chains at the molecular level. The detailed analysis of segmental orientation in elastomeric networks leads to an understanding of the mechanisms of deformation and physical properties, particularly mechanical properties.

Figure 5.2a,b shows the IR spectrum of a PDMS film of thickness 125 μm and the NIR spectrum of a PDMS film of thickness 2 mm, respectively [23]. Notably, most of the IR bands are very intense and cannot be used to explore the orientation of the PDMS networks. In a previous study, Bokobza and coworkers [23] investigated the dichroic behavior of the band at 2500 cm^{-1} due to the overtone of the symmetric bending mode of the CH_3 group of the PDMS network. Because of the very weak extinction coefficient of this mode, they could analyze samples up to 2 mm in thickness.

Figure 5.3a,b illustrates the positions of the local chain axis and transition moment with respect to the stretching direction and the proposed direction of the chain segment with respect to the transition moment of A_1 modes, respectively [23]. Note that the transition moment associated with the vibrational mode of A_1 symmetry lies along the CH_3—Si bond, which is the symmetry axis of the methyl group. The determination of the orientation function requires the definition of the angle β between the transition moment of the investigated band and the directional vector characteristic of a given chain segment.

The much weaker intensity of the NIR bands (Figure 5.2a,b) enables their use for the evaluation of anisotropy in polymeric materials. The NIR bands at 5917, 5873, 5447, 5402, and 4164 cm^{-1} in Figure 5.2b can be used for this purpose. Table 5.1 summarizes the frequency, assignment, local symmetry, and dichroism of the characteristic absorption bands of PDMS in the IR and NIR spectra [23]. Figure 5.4a–c depicts the strain dependence of the dichroic difference spectra for a silica-filled PDMS network (2 mm thick) in (a) the 2400–2600 cm^{-1} region, (b) the 4000–4500 cm^{-1} region, and (c) the 5300–6000 cm^{-1} region [23]. As shown in Figure 5.4, the absolute value of ΔA for each studied band, and thus, the anisotropy of the sample, increases with the draw ratio λ. One of the advantages of the polarization modulation method is the possible separation of the bands by their different dichroic behaviors. Notably, depending on the symmetry of the considered vibrations, the bands appear to be positive (E modes) or negative (A_1 mode) with respect to the baseline. A precise assignment of the observed absorptions can be provided, and the two spectral regions, NIR and IR, can be correlated.

Figure 5.5a,b plots the dichroic functions versus the strain function for the unfilled and silica-filled PDMS networks, respectively [23]. An evident proportionality

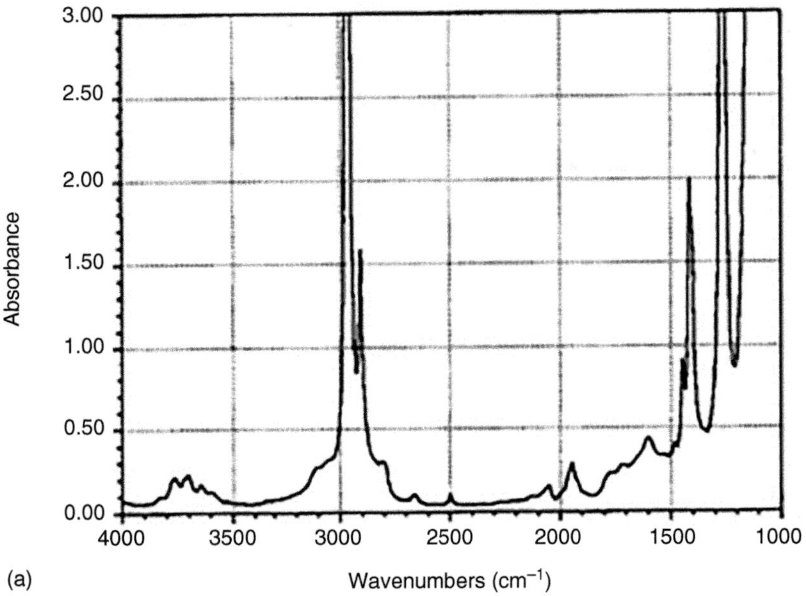

(a)

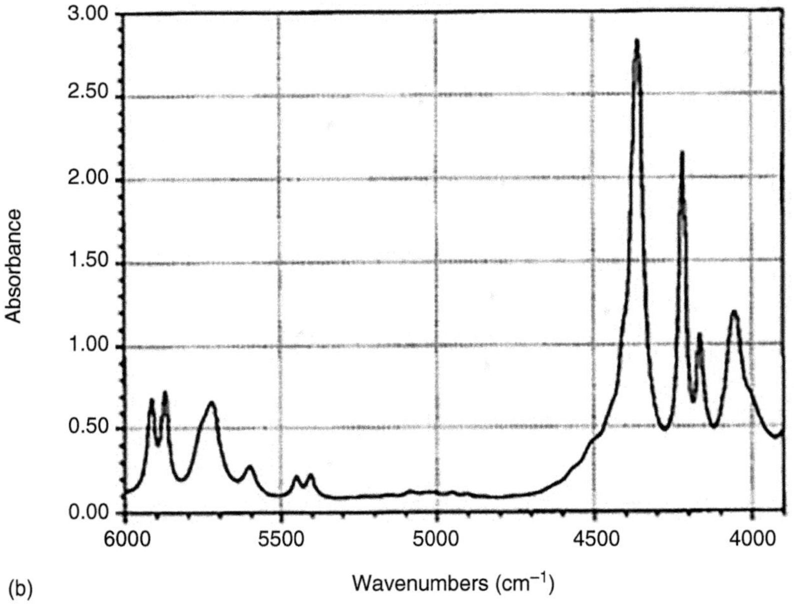

(b)

Figure 5.2 (a) The IR spectrum of a PDMS film of 125 μm in thickness and (b) the NIR spectrum of a PDMS film of 2 mm in thickness. Source: Aoyama, Sato, and Ozaki [26].

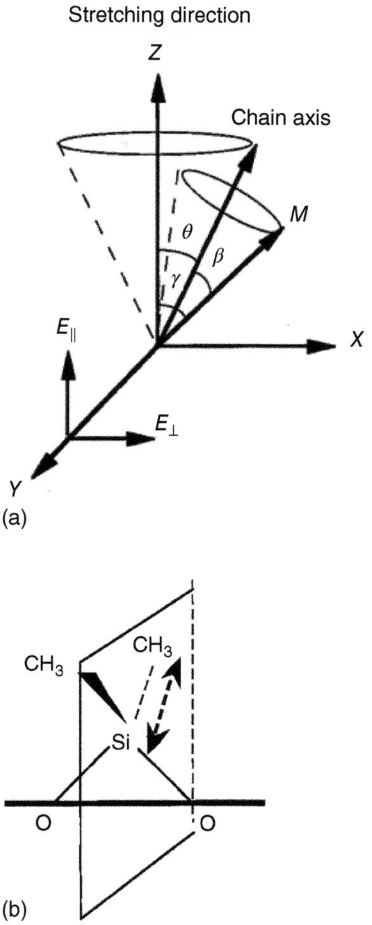

Figure 5.3 (a) Positions of local chain axis and transition moment with respect to the stretching direction. (b) Proposed direction of the chain segment with respect to the transition moment of A_1 modes, respectively. Source: Based on Aoyama, Sato, and Ozaki [26].

between the orientational function and the strain function can be observed. The data in Figure 5.5 are related to the vibrational modes belonging to the two A_1 and E irreducible representations of the C_{3v} point group of the methyl group. The same dichroic function was expected for each irreducible representation. Positive and negative values of the dichroic functions were obtained for the E and A_1 modes, respectively. Notably, for each draw ratio, the E modes yield similar values for the dichroic function, whereas the values obtained for the A_1 modes are more scattered. It can be observed in Figure 5.5 that a wide range of deformations can be studied using the silica-filled PDMS network, and the introduction of reinforcing fillers, such as silica in PDMS networks, increases the orientation due to the additional cross-links provided by the filler.

The important conclusion of this study was that NIR spectroscopy allows not only the analysis of thick samples but also the monitoring of vibrations less coupled than those in the IR region [23].

Table 5.1 Frequency, assignment, local symmetry, and dichroism of the characteristic absorption bands of PDMS in the IR and NIR spectra.

Frequency (cm^{-1})	Assignment[a)]	Local symmetry[b)]	Expected value of β
1260	$\delta_s(CH_3)$	A_1	90°
1398	$\delta_a(CH_3)$	E	(0°, 90°)
1410	$2r_a(CH_3)$	A_1	90°
1445	$2r_a(CH_3)$	E	(0°, 90°)
2500	$2\delta_s(CH_3)$	A_1	90°
2906	$\nu_s(CH_3)$	A_1	90°
2965	$\nu_a(CH_3)$	E	(0°, 90°)
4164	$\nu_s(CH_3)+\delta_s(CH_3)$	A_1	90°
4216	$\nu_a(CH_3)+\delta_s(CH_3)$	E	(0°, 90°)
4357	$\nu_a(CH_3)+\nu_a(CH_3)$	E	(0°, 90°)
5402	$\nu_s(CH_3)+2\delta_s(CH_3)$	A_1	90°
5447	$\nu_a(CH_3)+2\delta_s(CH_3)$	E	(0°, 90°)
5725[c)]	$\nu_s(CH_3)+2\delta_a(CH_3)$	A_1, E	90°, (0° 90°)
	$2\nu_s(CH_3)$	A_1	90°
5873	$2\nu_a^0(CH_3)$	A_1	90°
5917	$2\nu_a^2(CH_3)$	E	(0°, 90°)

a) ν = stretching, δ = bending, r = rocking, s = symmetric, a = antisymmetric.
b) Local symmetry of X–CH_3 group (C_{3v}).
c) The PM-IRLD spectrum shows that this overlapped mode contains three components.
Source: Buffeteau, Desbat, and Bokobza [23].

Mizushima et al. [24] examined the orientation of high-density polyethylene (HDPE), LLDPE, and low-density polyethylene (LDPE) with respect to the stretching direction under uniaxial deformation using polarized NIR spectroscopy. They directly evaluated the orientation function of the polyethylene (PE) chain axis (*c*-axis) from a band due to the first overtone of the CH stretching vibration as a function of the extension time or strain. The following conclusions were drawn. The orientation function of HDPE determined using NIR spectroscopy remained at approximately 0 before the second yield point and markedly increased at the second yield point or neck initiation point. In contrast, the corresponding functions of LLDPE and LDPE remained at approximately zero before the first yield point. It is likely that the destruction of spherulite and the rearrangement of lamellar clusters for the draw axis require a longer time for HDPE owing to its high crystallinity. The spherulite of HDPE may be packed tightly, and the space for the rotation of its lamellar clusters appears to be narrow. Hence, the orientation function fluctuated before the second yield point. On the other hand, the space for rotation is large for LLDPE and LDPE because of their low crystallinities. Thus, the rotation of the lamellar clusters may accrue smoothly. Therefore, the orientation

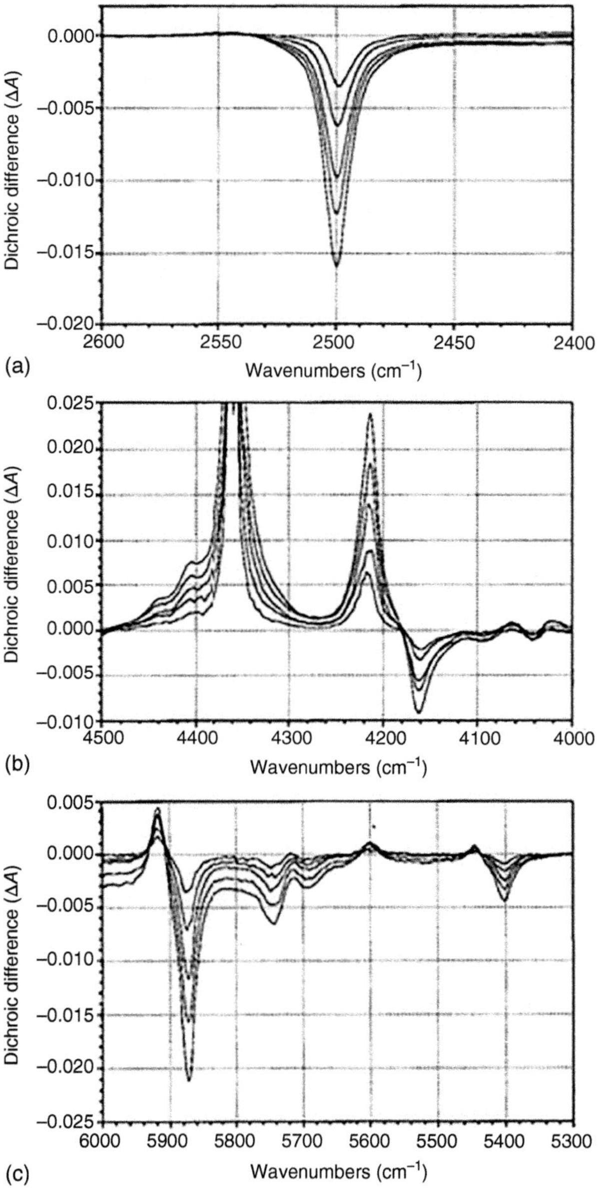

Figure 5.4 Strain dependence of the dichroic difference spectra for a silica-filled PDMS network (2 mm thick) in (a) the 2400–2600 cm^{-1}, (b) the 4000–4500 cm^{-1}, and (c) the 5300–6000 cm^{-1} region. Source: Buffeteau, Desbat, and Bokobza [23].

functions of LLDPE and LDPE increased earlier than that of HDPE. Mizushima et al. compared the NIR method with the IR method, where the orientation function of the PE *c*-axis (chain-axis) was indirectly obtained from the *b*- and *a*-axes, assuming an orthogonal crystal form using the CH_2 rocking vibrations in the IR spectra [24].

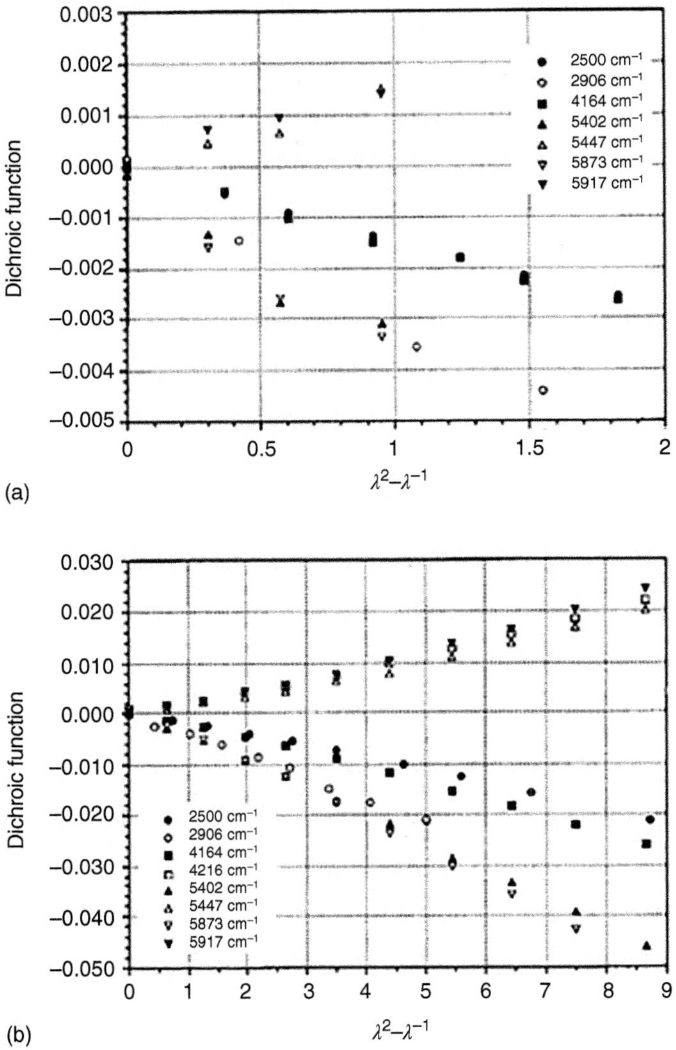

Figure 5.5 (a) and (b) dichroic functions versus strain function for unfilled and silica-filled PDMS network, respectively. Source: Buffeteau, Desbat, and Bokobza [23].

5.2.2 Isothermal Crystallization Kinetics of Poly(3-hydroxybutyrate)

The NIR spectra of PHB have been reported [25]. The relevant band assignments were discussed in comparison with those of the IR spectra of PHB in [25], and the NIR bands were separated into crystalline and amorphous bands. This chapter outlines the isothermal crystallization kinetics of PHB studied using NIR spectroscopy.

Hu et al. [25] analyzed the four spectral regions of PHB: (A) 6200 to 5600, (B) 5200 to 5000, (C) 4800 to 4600, and (D) 4500 to 4000 cm^{-1} regions. These regions correspond to the first overtone and combinations of CH vibration, the second overtone of C=O vibration, the combination of C—H and C=O vibrations, and the combination

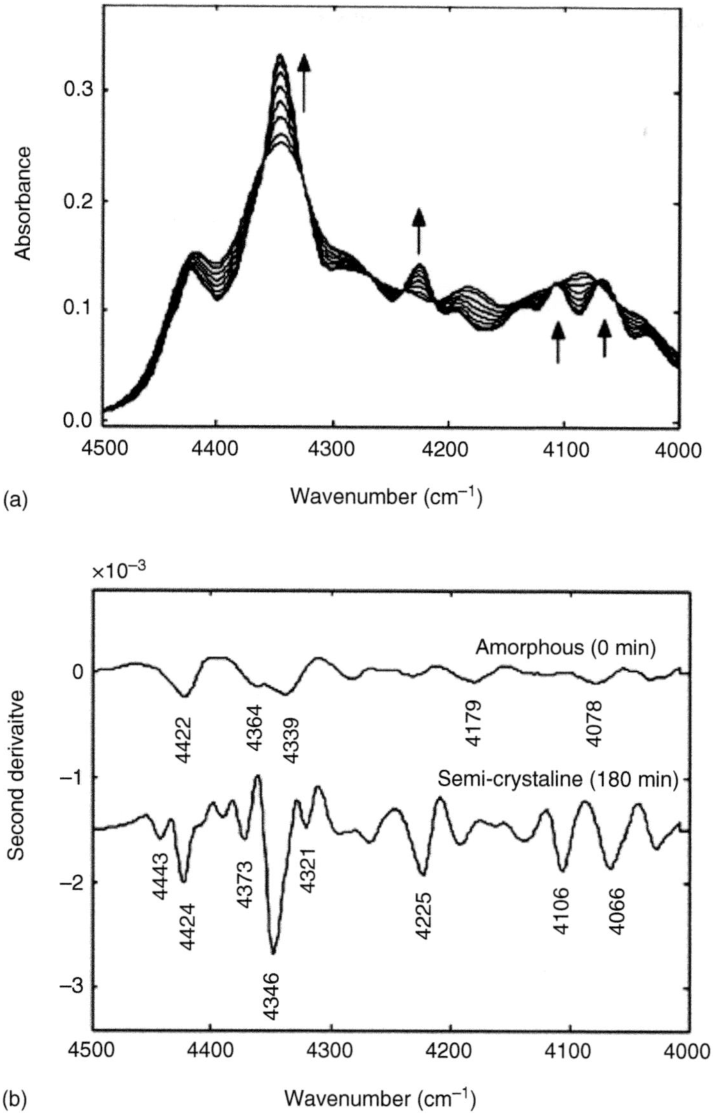

Figure 5.6 (a) and (b) Time-dependent changes of the NIR spectra in the region of 4500 to 4000 cm^{-1} during the melt-crystallization process of PHB at 125 °C and second derivatives of the spectra measured at 0 and 180 minutes shown in (a). Source: Hu et al. [25].

of CH vibrations, respectively. The spectral regions (A), (B), and (C) were described in Chapter 1. Thus, the spectral region (D) is analyzed here. Figure 5.6a,b shows the time-dependent changes of the NIR spectra in the region of 4500 to 4000 cm^{-1} during the melt-crystallization process of PHB at 125 °C, and the second derivatives of the spectra measured at 0 and 180 minutes are shown in Figure 5.6a [25]. The NIR spectra in this region are rather complicated because bands due to various combination modes of CH stretching and deformation vibrations appear. Among them, it can be easily recognized that the bands at 4364 and 4225 cm^{-1} originate from the crystalline state.

PCA was used to analyze the real-time NIR difference spectra measured during the isothermal crystallization process of PHB [25]. PCA extracts the necessary information from a large dataset using only a few factors, known as principal components (PCs), which provide reliable access to the variance of the spectral data. Hence, the spectral analysis using PCA becomes substantially more simplified and pronounced in the pertinent information content.

Figure 5.7a shows the normalized PC1 score plot (*) obtained from the NIR difference spectra in the 6200 to 4000 cm^{-1} region of PHB and an evolving trace of the normalized peak heights of the band at 4346 cm^{-1} (□) with time during the isothermal melt crystallization process of PHB at 125 °C. As shown in Figure 5.7a, the score plot of PC1 and the evolving trace of the band at 4346 cm^{-1} coincide with each other, showing that PC1 reflects the crystallization behavior of PHB with time. The loading plot of PC1, which accounts for 99.932% of the total spectral variance, is shown in Figure 5.7b. Positive bands such as those at 4653, 4346, and 4225 cm^{-1} are strongly correlated with the crystalline state of PHB. As a variance of 99.932% has been accounted for, all the important spectral changes can be explained by PC1. However, the PC2 loading shown in Figure 5.7c indicates that PC2 also provides more significant information than noise, although it accounts for only 0.046% of the total variance. Based on the results of PC1 and PC2, Hu et al. [25] considered that, during the crystallization process, the structural evolution and development of the crystalline structure of PHB may not be a simple transition of the binary mixture system (crystalline and amorphous states). This idea may be consistent with the conclusion based on a previous 2D IR correlation analysis by Zhang et al. [28], which demonstrated that sequential changes in functionalities are also responsible for the structural adjustment of polymer chains, except for the cooperative changes during the crystallization process. Zhang et al. [28] suggested by using an additional weak C=O stretching band of PHB at 1731 cm^{-1} that there is an intermediate state between the ordered crystalline and disordered amorphous states during the phase transition of PHB.

Hu et al. [25] used the above four regions to analyze the kinetics of the isothermal crystallization process of PHB, which is often expressed by the Avrami equation:

$$\frac{A_t - A_\infty}{A_0 - A_\infty} = \exp\left(kt^n\right) \tag{5.1}$$

where A_t is the peak intensity at the crystallization time t, A_0 and A_∞ are the initial and final peak intensities during isothermal crystallization, respectively, k is the overall kinetic constant of crystallization, t is the time of crystallization, and n is the Avrami exponent, which reflects the nature of nucleation and the geometry of the growing crystals. Eq. (5.2) can also be rewritten in the following form:

$$\ln\left[-\ln\left(\frac{A_t - A_\infty}{A_0 - A_\infty}\right)\right] = \ln k + n \ln t \tag{5.2}$$

In the Avrami equation, the normalized score vector of PC1 in different spectral regions was employed. To compare the results from PCA with those from the spectral

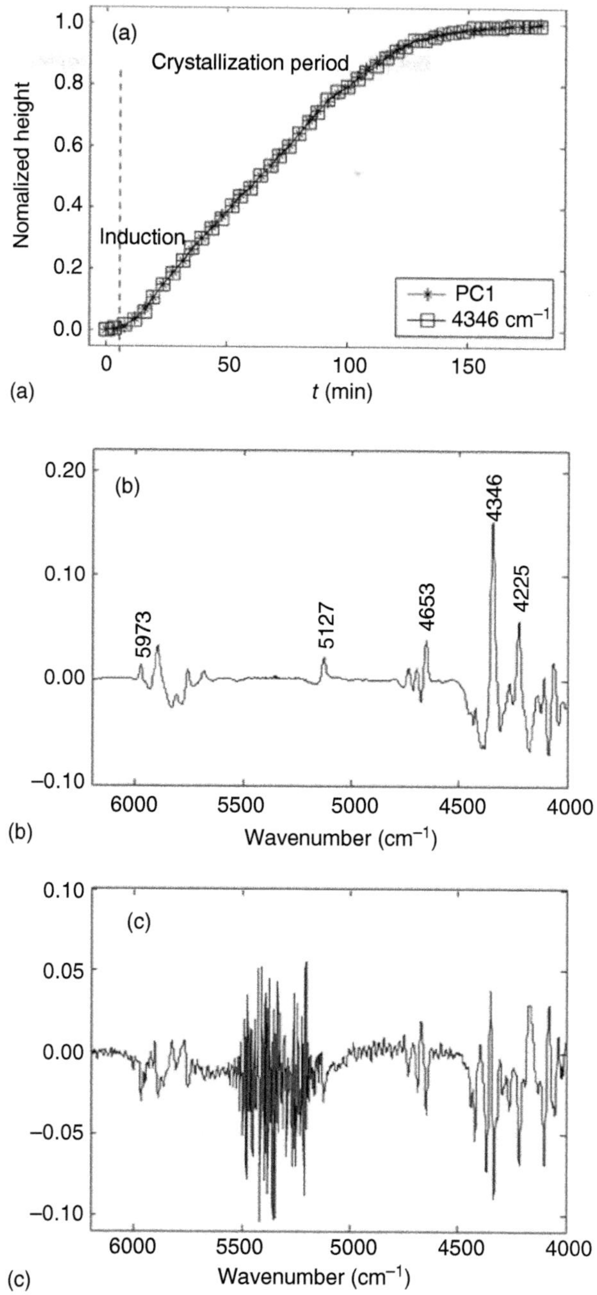

Figure 5.7 (a) Normalized PC1 score plot (*) obtained from the NIR difference spectra in the 6200 to 4000 cm^{-1} region of PHB and an evolving trace of normalized peak heights of the band at 4346 cm^{-1} (□) with time during the isothermal melt crystallization process of PHB at 125 °C. Loading plots of (b) PC1 and (c) PC2. Source: Hu et al. [25].

change of a band, Hu et al. [25] also used the normalized peak height of the band at 4346 cm^{-1} in Eq. (5.2). By plotting the first item vs. ln t, two Avrami parameters, n and k, can be obtained from the slope and intercept, respectively.

The top and bottom panels of Figure 5.8 depict the Avrami plots of the crystallization dynamics of PHB at 125 °C obtained from the normalized peak height of the

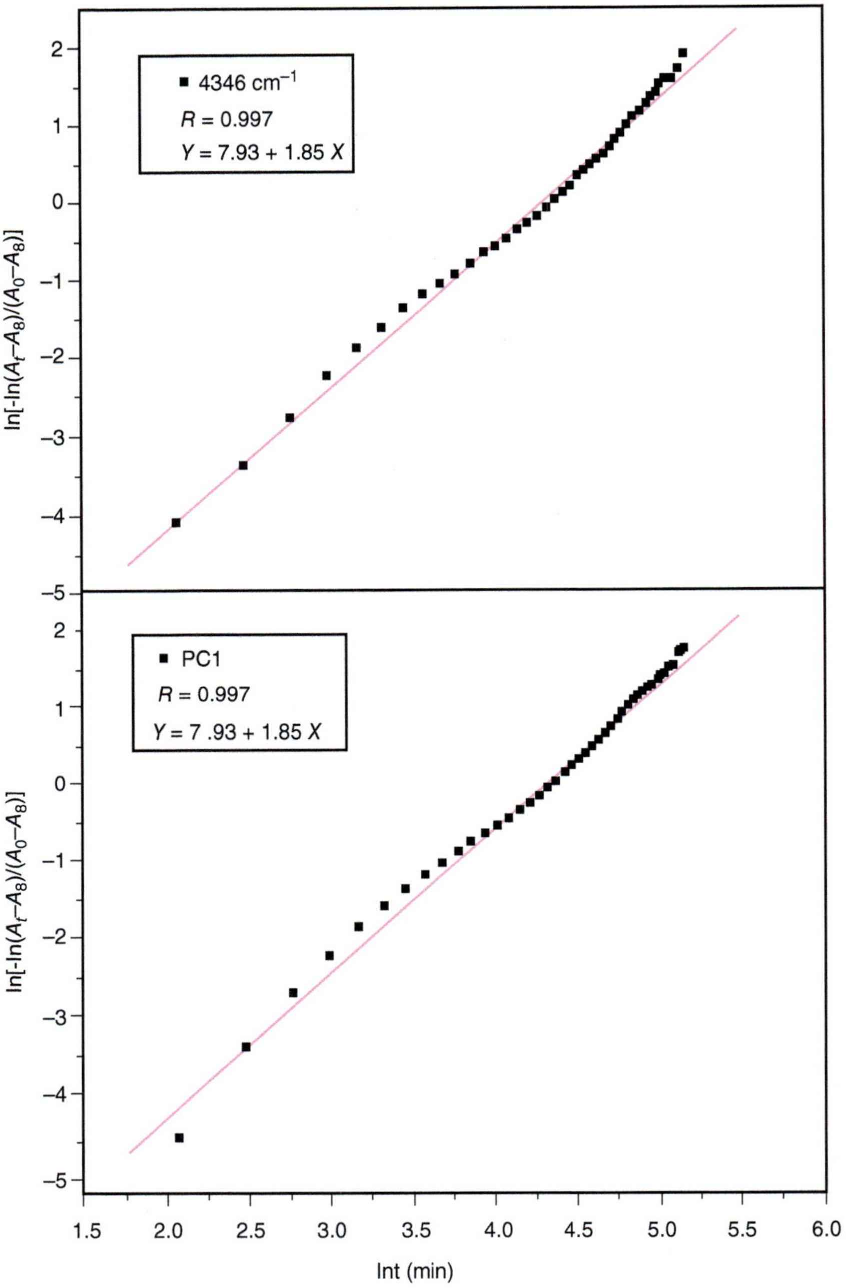

Figure 5.8 Top and bottom panels depict Avrami plots of the crystallization dynamics of PHB at 125 °C obtained from the normalized peak height of the band at 4346 cm^{-1} and from the PC1 score in the 6200 to 4000 cm^{-1} region, respectively. Source: Hu et al. [25].

band at 4346 cm^{-1} and from the PC1 score in the 6200 to 4000 cm^{-1} region, respectively. Notably, these two Avrami plots yield a consistent linear relation in terms of slopes (Avrami parameter $n = 1.85$) and intercepts (−7.93). They both exhibit an initial linear portion that subsequently tends to level off. This deviation may arise from the secondary crystallization caused by the slower crystallization, crystal perfection, or spherulite impingement in the later stage of the crystallization process.

The half-time of crystallization, $t_{1/2}$, is an important parameter in the discussion of the crystallization kinetics and can be calculated from n and k, such that

$$t_{1/2} = \left(\frac{\text{In}2}{k}\right) \tag{5.3}$$

Table 5.2 summarizes the Avrami parameters calculated from the normalized PC1 scores of the five broad spectral regions for the isothermal crystallization dynamic analysis of PHB [25]. As presented in Table 5.2, the crystallization parameters calculated from the normalized PC1 scores obtained from the individual regions are not always consistent with each other. However, Hu et al. [25] suggested that PCA can be performed with certain appropriate spectral regions to monitor the crystallization process of a polymer and to investigate its local crystallization behavior without considering prior information.

As presented in Table 5.2, the average value of the Avrami exponent is $n \approx 2$ for the isothermal melt crystallization of PHB at 125 °C. This result indicates that, during the isothermal process of PHB, the crystallization starts from heterogeneous nucleation and that the primary crystallization stage corresponds to a 2D, circular, diffusion-controlled growth of nucleation. According to a previous IR study, the Avrami exponent n was calculated to be 2.5 using the crystalline sensitive bands of PHB at 1184 and 825 cm^{-1} during the melt crystallization process at 129 °C [25]. Moreover, the differential scanning calorimetry (DSC) measurements of PHB at other crystallization temperatures (80–114 °C) yielded an Avrami exponent n close to 2 [25]. Hence, the Avrami exponent n from the first score vector of the NIR data is consistent with those from the IR and DSC data.

This study demonstrated that crystallization kinetics can be investigated by combining PCA with the Avrami equation. It was observed that PCA can analyze the entire crystallization process of polymers instead of using the intensity change of

Table 5.2 Avrami parameters calculated from the normalized PC1 scores of five broad spectral regions for the isothermal crystallization dynamic analysis of PHB.

Parameters	Spectral region used for PCA calculation (cm^{-1})				
	6200–4000	6200-5600	5200–5000	4800–4600	4500–4000
Rate constant, k ($min^{-n} \times 10^4$)	3.60	2.70	3.90	2.95	3.56
Avrami index, n	1.85	1.93	1.85	1.91	1.85
Half-time, $t_{1/2}$ (min)	60	58	57	58	60

Source: Hu et al. [25].

certain peaks, particularly when these peaks significantly overlap with each other. This study also provided fundamental information for the use of NIR spectra in the online monitoring of the chemical and physical properties of PHB.

5.2.3 Crystallization of Poly(3-hydroxybutyrate-*co*-3-hydroxyhexanoate) During Melt Extrusion Promoted by Residual Crystals

PHBH (Figure 5.9), which is a biodegradable polymer, has promising applications in inflation films, T-die extruded films and sheets, injection-molded and blow-molded articles, etc. However, PHBH has the serious defect of slow crystallization during molding, and no nucleating agent examined to resolve this issue so far has proven successful. The addition of a higher-melting biodegradable polyester may be an effective solution, as the residual crystals preserved during PHBH melting are important for subsequent crystallization. Therefore, Aoyama et al. [26] investigated the relationship between the amount of residual crystals and the subsequent crystal behavior under dynamic conditions, such as practical extrusion molding. They used online NIR spectroscopy to analyze the residual amount of crystals at the extruder outlet and explored the crystallization behavior of the extruded strands in real time. Even under dynamic conditions, they observed a close relationship between the amount of residual crystals and the subsequent crystallization behavior.

5.2.3.1 Outline of Online NIR Analysis and Online NIR Monitoring of the Residual Crystal Amount at the Extruder Outlet Nozzle

Figure 5.10a illustrates a schematic diagram of the first online NIR monitoring system to determine the residual crystal amount at the extruder outlet. Figure 5.10b depicts the second online NIR monitoring system for the extruded strands [26]. For more details on the experimental setup, refer to [26]. Tables 5.3 and 5.4 summarize the detailed extrusion conditions and results of the visual observation of the strands, respectively [26]. Aoyama et al. [26] used direct online NIR analysis to determine the amount of residual crystals in the molten resin after melt kneading it using an extruder screw and before its extrusion into a strand. This analysis was conducted using the NIR data collected from a cell installed in the nozzle of the single-screw extruder (first online spectrum).

Structure:

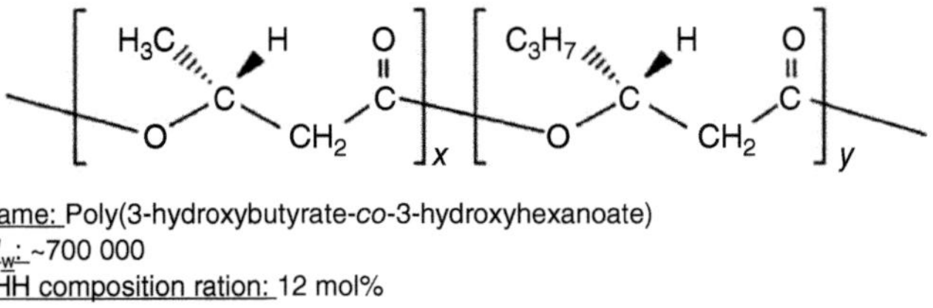

Name: Poly(3-hydroxybutyrate-*co*-3-hydroxyhexanoate)
M_w: ~700 000
3HH composition ration: 12 mol%

Figure 5.9 Molecular structure of poly(3-hydroxybutyrate-*co*-3-hydroxyhexanoate).

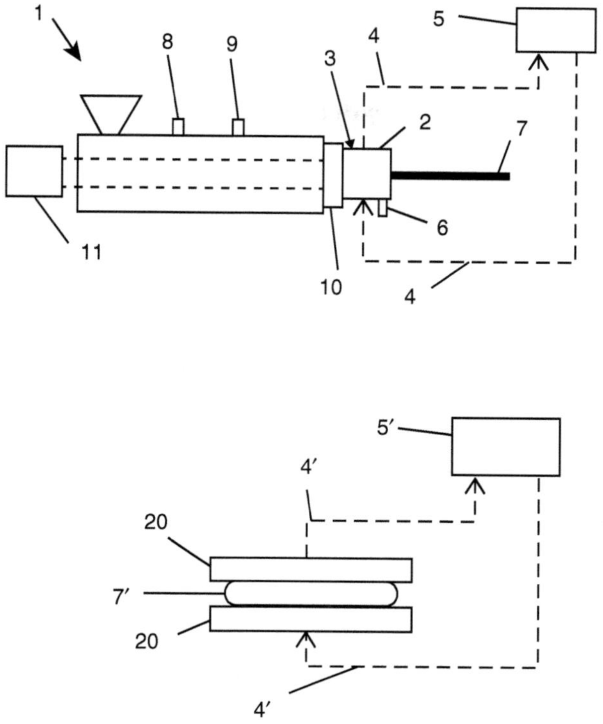

Figure 5.10 Chemical structure of poly(3-hydroxybutyrate-*co*-3-hydroxyhexanoate). (a) A schematic diagram of the first online NIR monitoring system to determine the residual crystal amount at the extruder outlet. (b) That of the second online NIR monitoring system for the extruded strands. Source: Aoyama, Sato, and Ozaki [26].

Table 5.3 Detailed extrusion conditions.

	Temperature of extrusion					Resin temperature (°C)	Motor (rpm)	Discharge rate ($kg\,h^{-1}$)	Atmosphere temperature (°C)
	C_1 (°C)	C_2 (°C)	Flange (°C)	Nozzle (°C)	Cell (°C)				
Exp-1	120	135	135	135	135	145	20.8	1.32	25.4
Exp-2	120	140	140	140	140	147	21.2	1.32	25.4
Exp-3	121	150	150	150	150	161	22.7	1.32	25.4
Exp-4	120	160	160	160	160	172	21.6	1.32	26.2
Exp-5	120	170	170	170	170	181	21.4	1.32	24.3

Source: Aoyama, Sato, and Ozaki [26].

5.2.3.2 Amount of Residual Crystals at the Extruder Outlet

Figure 5.11 shows the NIR spectrum in the 11 000 to 4000 cm^{-1} region of Experiment (Exp)-1 measured at the extruder outlet. According to Hu et al. [25] (Chapter 1), the bands at 5129 and 4670 cm^{-1} are ascribed to the crystalline state, whereas those at

Table 5.4 Results of visual observation of strands.

	Temperature of extrusion (°C)	Resin temperature (°C)	Crystallization time of extruded strand (visual judgment)
Exp-1	135	145	75 s or less
Exp-2	140	147	75 s
Exp-3	150	161	15 min
Exp-4	160	172	75 min or more
Exp-5	170	181	75 min or more

Source: Aoyama, Sato, and Ozaki [26].

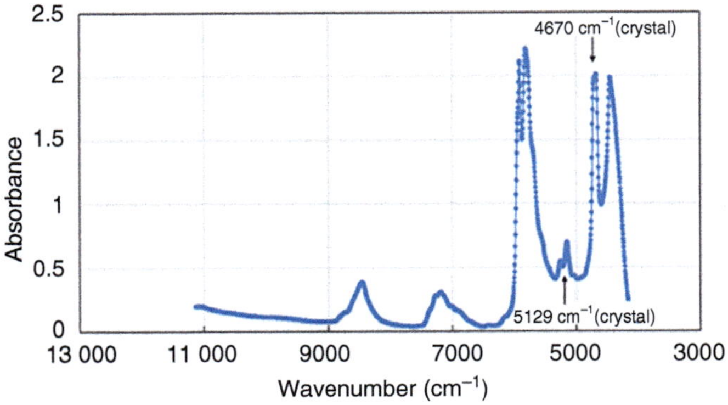

Figure 5.11 A NIR spectrum in the 11 000 to 4000 cm^{-1} region of Experiment (Exp)-1 measured at the extruder outlet. Source: Aoyama, Sato, and Ozaki [26].

5150 cm^{-1} are due to the amorphous state; all the bands are assigned to the combinations of the C=O and CH stretching modes. PHBH crystals consist of PHB components, and the 3HH components of PHBH exist in the amorphous part. The amount of residual crystals at the extruder outlet was determined using the peak intensities of the bands at 4670 and 5129 cm^{-1}. Figure 5.12a,b depicts enlarged online NIR spectra in the 4800 to 4600 and 5300 to 5100 cm^{-1} regions collected at the extruder outlet. In Exp-4 (resin temperature = 172 °C), the extruded strands required more than an hour to crystallize, indicating that the amount of residual crystals was very small. Hence, the amount of residual crystals in the other experiments was determined by calculating the difference spectrum between the spectrum of a given experiment and that of Exp-4. Figure 5.13a,b displays the difference spectra calculated using Exp-4 as the reference in the 4800 to 4600 and 5200 to 5100 cm^{-1} regions, respectively [26]. As shown in Figure 5.7a, the intensity of the crystalline peak at 4670 cm^{-1} decreases with increasing temperature in Exp-1–5, reflecting the decrease in the amount of residual crystals. In Exp-1, 2, and 3, the peak at 4670 cm^{-1} is positive, indicating a greater residual crystalline content in Exp-1, 2, and 3 than in Exp-4. On the other hand, the residual crystalline content was smaller in Exp-5 compared with that in

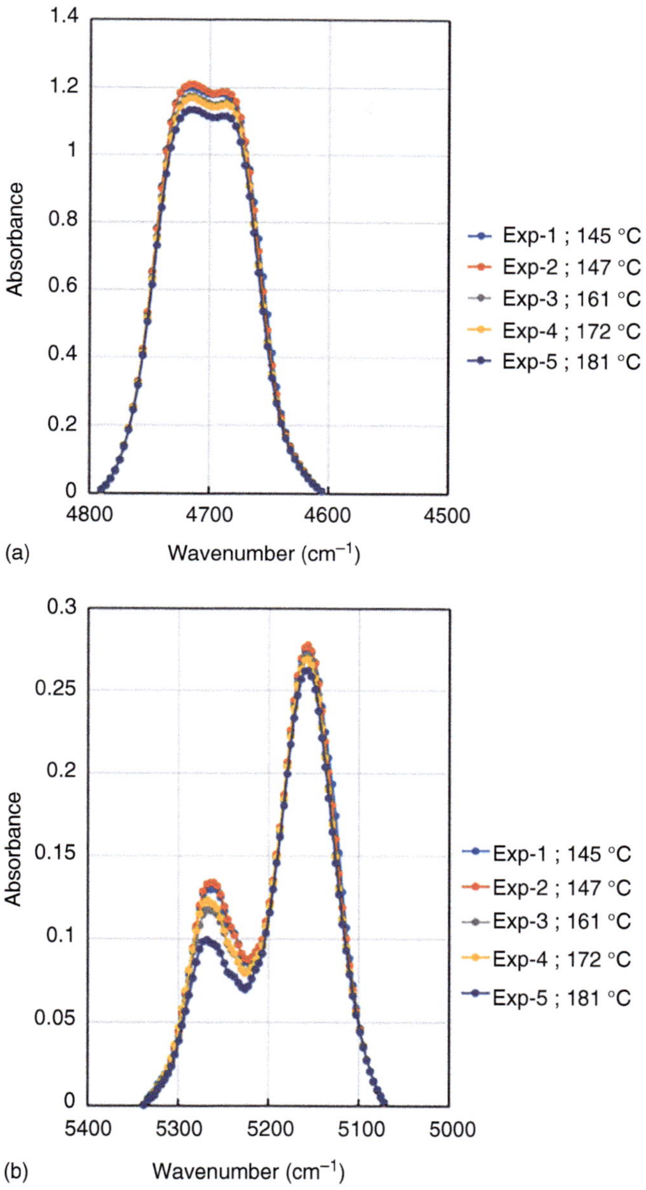

Figure 5.12 Enlargements of online NIR spectra in the (a) 4800 to 4600 cm^{-1} and (b) 5300 to 5100 cm^{-1} regions collected at the extruder outlet. Source: Aoyama, Sato, and Ozaki [26].

Exp-4. As shown in Figure 5.13b, the intensity of the crystalline peak at 5129 cm^{-1} decreases with increasing temperature in Exp-1–5. The decrease in intensity corresponds to a reduction in the amount of residual crystals. Thus, the online NIR monitoring at the extruder outlet was capable of tracking the amount of residual crystals using the intensity variations of the crystalline bands [26].

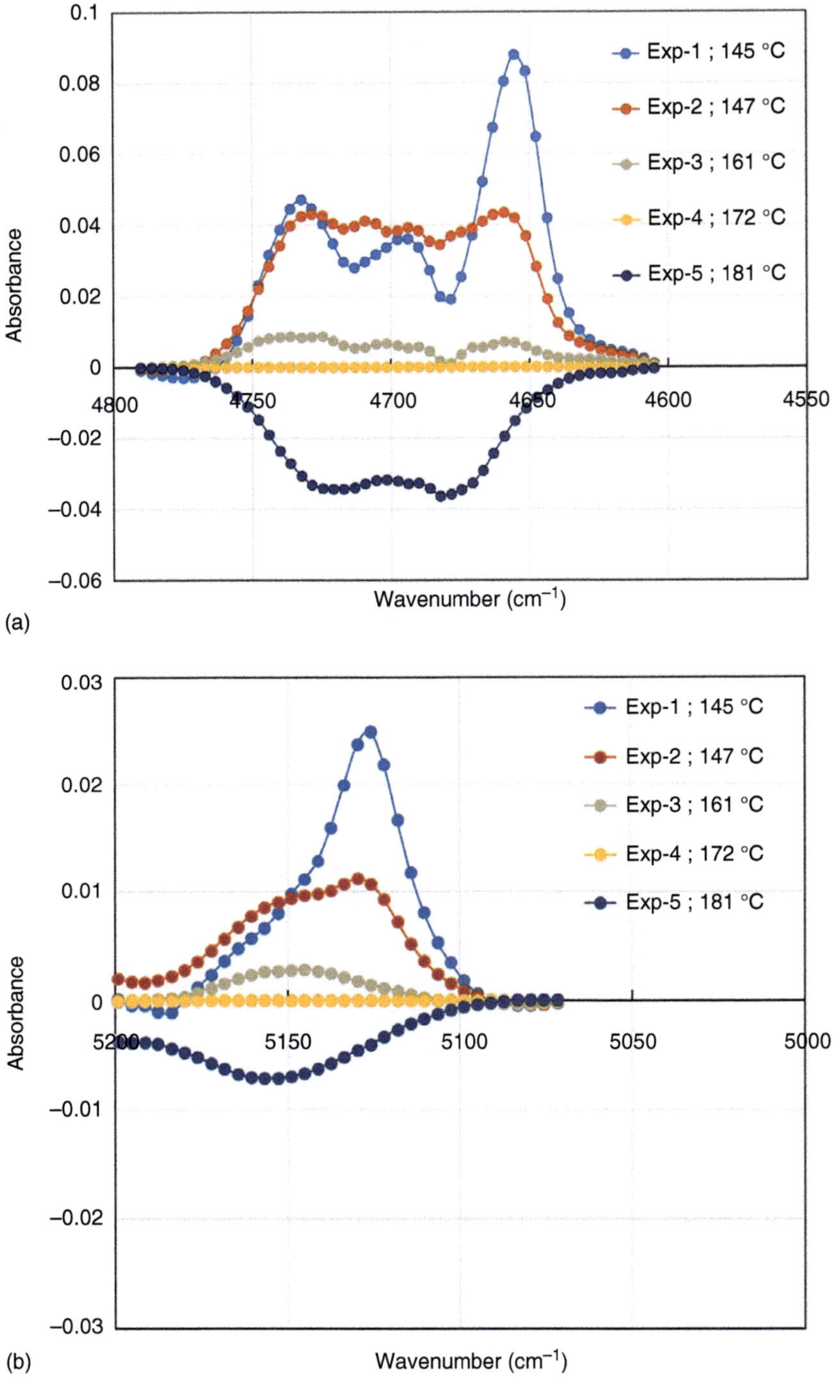

Figure 5.13 Difference spectra calculated using Exp-4 as the reference in the region of (a) 4800 to 4600 cm^{-1} and (b) 5200 to 5100 cm^{-1} regions. Source: Aoyama, Sato, and Ozaki [26].

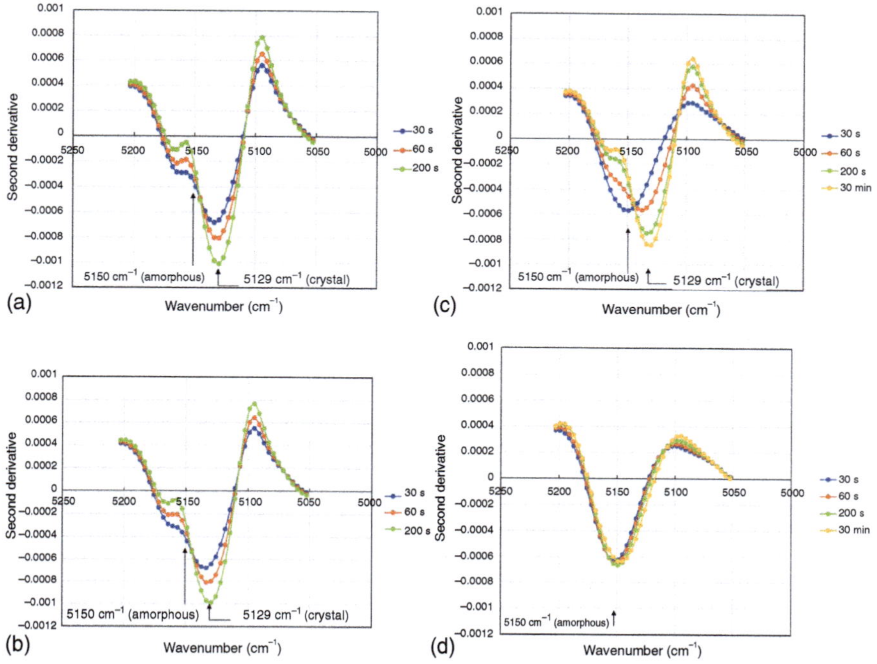

Figure 5.14 Time-dependent variations in the second-derivative spectra measured during online NIR monitoring of extruded strands in Exp-1–4. (a) Exp-1, (b) Exp-2, (c) Exp-3, (d) Exp-4. Source: Aoyama, Sato, and Ozaki [26].

5.2.3.3 Crystallization of Extruded Strands

The crystallization behavior of the extruded strands was also probed using online NIR monitoring (second online monitoring) and correlated with the amount of residual crystals at the extruder outlet. Figure 5.14 shows the time-dependent variations in the second-derivative spectra measured during the online NIR monitoring of extruded strands in Exp-1–4 [26]. The band at 5150 cm^{-1} was used to estimate the noncrystalline (amorphous) parts, whereas that at 5129 cm^{-1} was employed to quantify the amount of crystals (second crystallization peak). As shown in Figure 5.14a,b for Exp-1 and 2, the crystalline peak of the sampled strands at 5129 cm^{-1} was observed after 30 seconds, indicating that the crystallization process was very fast and was consistent with the visual observation. It is very likely that the presence of a large amount of residual crystals at the extruder outlet promotes fast crystallization. In Exp-3 (Figure 5.14c), the amorphous peak appeared at 5150 cm^{-1} after 30 seconds, whereas that of the crystalline peak was observed at 5129 cm^{-1} after 60 seconds; the intensity of the latter peak increased between 200 seconds and 30 minutes. Visual observation revealed that the strands were no longer fused after 15 minutes, and thus, the crystallization was slow compared with that in Exp-1 and 2. In Exp-4, the peak at 5129 cm^{-1} was not observed after 30 minutes (Figure 5.14d), suggesting a very slow crystallization process. Visual observation indicated the absence of crystallization even after 75 minutes. Moreover, the residual crystal at the extruder outlet in Exp-4 was very small, resulting in an extremely slow crystallization of the extruded strand.

From the above experimental results, Aoyama et al. [26] concluded that the crystallization behavior of the extruded strands analyzed via online NIR monitoring was closely related to the residual crystallinity at the extruder outlet. Hence, for an extrusion temperature of 160 °C, a relatively high amount of residual crystals was observed in the faster crystallization, whereas almost no residual crystals were observed at an extrusion temperature of 170 °C, making the subsequent crystallization significantly slower.

5.2.3.4 Analysis of Extruded Strand Crystallization Using the Avrami Equation

The crystallization of extruded strands was investigated using the results of online NIR monitoring and the Avrami equation. The second derivatives of the NIR spectra were used to compare the crystallization behaviors of the extruded strands, whereas the original spectra were employed for the analysis of the Avrami equation. Figure 5.15 presents the time-dependent variations in the NIR spectra of the extruded strand in Exp-1 [26]. Figure 5.16 compares the temporal intensity changes of the peak at 5129 cm^{-1} in the NIR spectra of the extruded strand in Exp-1–4 [26]. Notably, for strands in Exp-1 and 2, the intensity of the peak at 5129 cm^{-1} quickly increased, and the strand exhibited continuous growth. For the strand

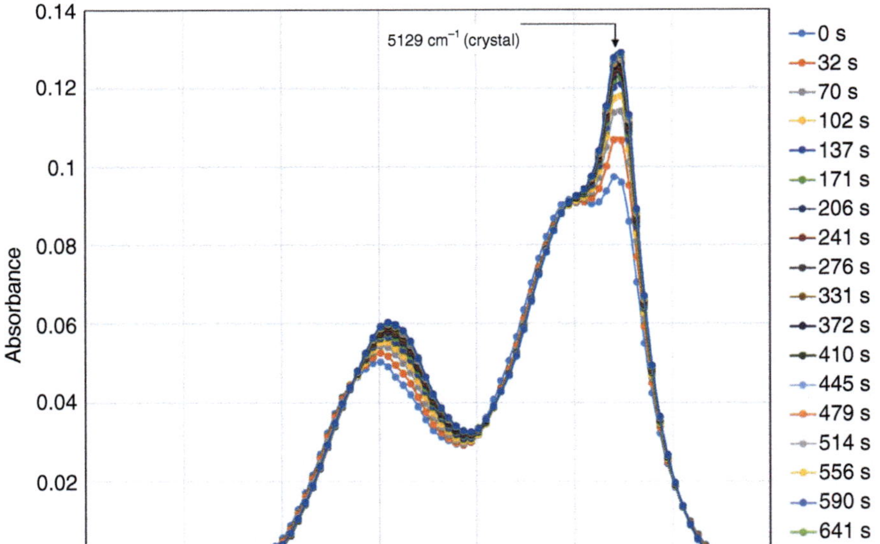

Figure 5.15 Time-dependent variations of the NIR spectra of the extruded strand in Exp-1. Source: Aoyama, Sato, and Ozaki [26].

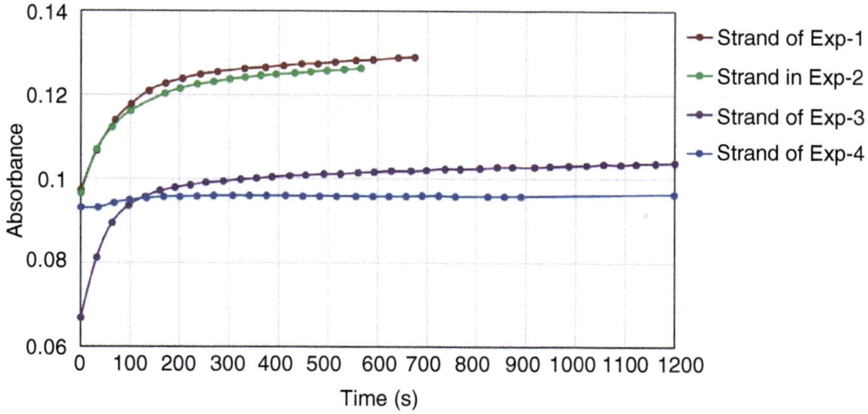

Figure 5.16 Temporal intensity changes of the peak at 5129 cm^{-1} in the NIR spectra of the extruded strand in Exp-1–4. Source: Aoyama, Sato, and Ozaki [26].

Table 5.5 Avrami parameters calculated from the crystallization dynamics analysis of extruded strands in Exp-1 to Exp-3.

	Extruded strand in Exp-1	**Extruded strand in Exp-2**	**Extruded strand in Exp-3**
n	0.81	0.77	0.63
$-\ln k$	3.75	3.50	2.74

Source: Aoyama, Sato, and Ozaki [26].

in Exp-3, the intensity increase was rather slow, and its subsequent growth was more moderate, whereas the strand in Exp-4 showed a very small change in peak intensity.

Aoyama et al. [26] determined the Avrami exponent (n) from the slope of each graph and the crystallization rate constant (k) from the corresponding intercepts. Table 5.5 summarizes the Avrami parameters calculated using the crystallization dynamics analysis of the extruded strands in Exp-1–3. For the strands in Exp-1–3, the reduction in the value of n was accompanied by the decrease in the value of k. It has been reported that the Avrami index of PHB is approximately 1.9 under static conditions [25], and that the nucleation is heterogeneous, and the primary crystal growth is 2D (disc shape). Aoyama et al. [26] observed that the Avrami index of PHBH was approximately 0.8–0.6 under dynamic conditions, and that the nucleation was heterogeneous and the primary crystallization growth showed a more one-dimensional (rod-like) growth. This may be due to the nucleation by residual crystals and the molecular orientation in the extrusion direction of the strands. They considered that, in the extruded strand, the polymer chains were oriented in the extrusion direction and caused one-dimensional crystal growth, as predicted by the Avrami equation.

5.3 NIR Imaging for Polymer Sciences

5.3.1 Introduction

NIR light has high penetration in polymeric materials; thus, NIR spectroscopy is also applicable for polymeric products of various sizes, with the thickness ranging from micrometers to centimeters. In addition, the NIR spectroscopic imaging method provides useful information with a spatial resolution, which makes it possible to assess the morphological and chemical properties of a polymer simultaneously [29–32]. NIR spectroscopic imaging has attracted considerable attention owing to these advantages. For example, spectroscopic imaging techniques can chemically evaluate several specimens simultaneously; thus, they can be used for a comprehensive inspection. Therefore, NIR spectroscopic imaging could be used in industrial applications, such as quality control [33a, 33b], the inspection of polymer films [33c, 33d], and the discrimination of specific components in a large number of parts [34–39]. Herein, we introduce examples of the application of NIR spectroscopic imaging for the melt processing of polylactic acid (PLA).

5.3.2 Theory of NIR Imaging

5.3.2.1 Acquisition of Hypercube

NIR imaging acquires the spectral data of a certain point on the *XY* axes corresponding to each wavelength acquired by a NIR spectrometer at each pixel or multiple pixels by scanning the entire target [30, 31]. Some imaging methods acquire positional and spectral information by combining a planar multielement detector and a spectroscope. Thus, spectral data consisting of three dimensions, the *XY* axes for position and the λ (wavelength) axis, are defined as a hypercube. The NIR imaging obtained from the hypercube has positional and spectral information, the spatial spread of a certain chemical component, and the boundary with multiple components, and their heterogeneity can be visualized and analyzed.

As it is difficult to acquire the above hypercubes simultaneously owing to time constraints, 2D data are usually acquired first, and then the fragments (slices) are scanned to obtain the final hypercube. In NIR imaging, hypercube acquisition is to fix the measurement area, as shown in Figure 5.17, measure the *XY* plane data for a certain wavelength band, and perform this process for all wavelengths [32, 40]. There is another method in which the spectral data in all wavelength regions are acquired for a certain measurement point on the *XY* plane, and they are performed for whole points.

The imaging by the former can acquire hypercubes at a high speed because it can acquire position information all at once, but uneven sensitivity and illuminance unevenness of each element can easily occur. In addition, multielement detectors in the NIR region must be used at extremely low temperatures to reduce noise, which generally results in a high cost and large detector size. In the imaging by the scanning method, position information is acquired by scanning the *XY* axes. As this method acquires spectra for each pixel, the measurement takes a long time when the number of pixels is large. However, a highly accurate hypercube can be obtained since

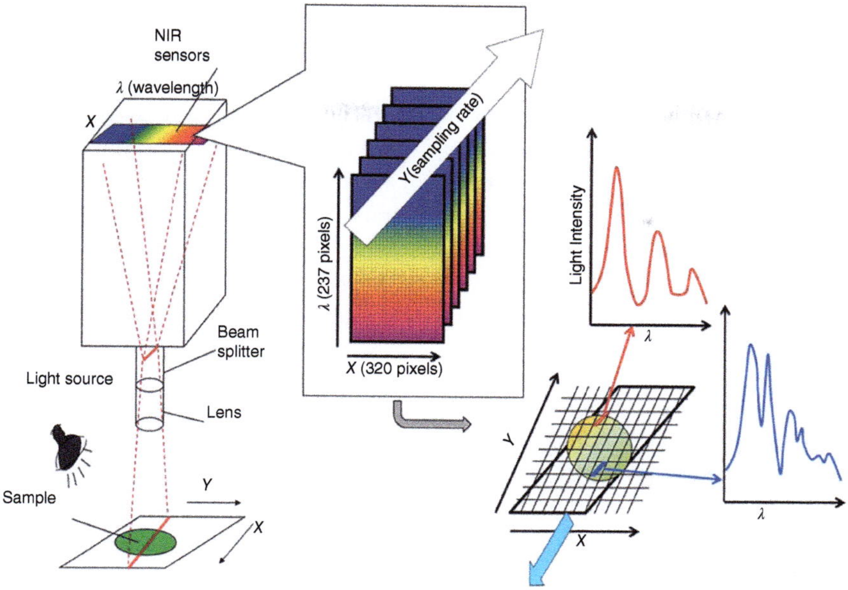

Figure 5.17 Schematic diagram of NIR imaging measurement and acquisition of hypercube. Source: Ishikawa et al. [40].

the occurrence of the measurement error was reduced described above [32]. In addition, the scanning method has the advantage that the size of the imaging device itself can be reduced because the size of the detector can be smaller than that of another imaging methods.

5.3.2.2 Data Transfer and Mapping

The spectral information obtained from each pixel was immediately transferred by the computer. It is resaved as data in a general text format (comma separated value; .csv, .txt); however, in some cases, it may be directly stored in a binary format via dedicated software. Positional information, wavelength (or wavenumber), and spectral values (absorbance, reflectance, etc.) corresponding to each wavelength and position are recorded in the stored data. Note that the amount of information transferred at one time is much larger than the amount of point data; hence, the transfer may be interrupted depending on the performance of the PC [32].

Methods for visualizing (mapping) spectral data with information on location, (i) spectral intensity, (ii) intensity ratio techniques, and (iiii) chemometric techniques have been proposed. Method (i) draws a map using the intensity of the obtained spectrum. By using the absorption peak of a specific component appearing in the NIR region, mapping for each of the various components can be performed. Accordingly, the distribution of specific components and the interactions at the molecular level can be visualized. (ii) For the mapping based on the intensity ratio, the intensity ratio of two wavelengths (wavenumbers) with high absorption intensity, or to normalize the spectrum, the wavelength ratio (wavenumber) band in which usually one has change obviously and the intensity of the other has a small change (or constant) is used. By using the ratio of the two wavelengths, the influence of the sample thickness can be removed. (iii) PCA is mainly used for imaging combined with chemometrics method. The information obtained in the 2D field is merged, and the difference (inhomogeneity of spectra in the case of imaging) within the object can be clarified. In addition, NIR imaging developed by wavelength (wavenumber) shifts has been proposed in recent years. By using the amount of shift in the absorption band for mapping, the changes in intermolecular interactions can be evaluated. However, in the NIR region, shifts often occur because of the overlapping peaks. It is necessary to examine thoroughly whether the shift observed in the spectrum represents a chemical change.

5.3.2.3 Feature of NIR Imaging Devices

A typical NIR imaging device is of a microscope type. The incident light of these types of devices is obtained from a spectrometer, and the output light is introduced to the microscope side after passing through an interferometer inside the spectrometer. The transmitted/reflected light path is usually selected, and the incident light is introduced from the bottom in the case of transmission measurement and from the top in the case of reflection measurement. The light introduced by each measurement method is focused by a Cassegrain mirror, and it finally reaches the detector. Although the lens is used in a general optical microscope, there is no lens that can cover a wide wavelength region from visible to IR with a uniform refractive index. Thus, for NIR imaging devices, even in spectroscopic imaging devices, a focusing method using a reflector is usually used. The detector is equipped with mercury cadmium telluride; HgCdTe (MCT) and/or triglycine sulfate (TGS) [32]. When the sample is measured, the area of interest (AOI) of the sample can be determined using the visible image of the sample taken by the charge coupled device (CCD) camera. The movable sample stage is driven with a three-dimensional space by a motor, and a joystick or mouse controls the minute movement of the region. Typically, the hypercube is detected by a linear array, and a maximum area of approximately $10 \times 10\ mm^2$ can be measured with a spatial resolution of submicron order.

A wide-area investigation of the inhomogeneity of components and/or states with a high speed is important for controlling the process of various objects. Compovision is a wide-area NIR imaging camera developed by Sumitomo Electric Industries Ltd. (Figure 5.18) [40]. The wide-area and high-speed monitoring by Compovision is superior; this instrument can measure 2D NIR spectra in an area of $150 \times 200\ mm^2$ within five seconds. High-speed monitoring with Compovision is achieved using the newly developed InGaAs detector. The detector of Compovision consists of InGaAs/GaAsSb type-II quantum wells (QWs) laminated on an indium phosphide (InP) substrate. This detector also contributes to the acquisition of high-quality spectral data in the wide NIR spectral region (1000–2350 nm). Therefore, Compovision measures an object with a wide area (approximately $150 \times 200\ mm^2$) in the 1000–2350 nm region within approximately five seconds. Commercial NIR imaging instruments obtain data of similar size within approximately 5–10 minutes.

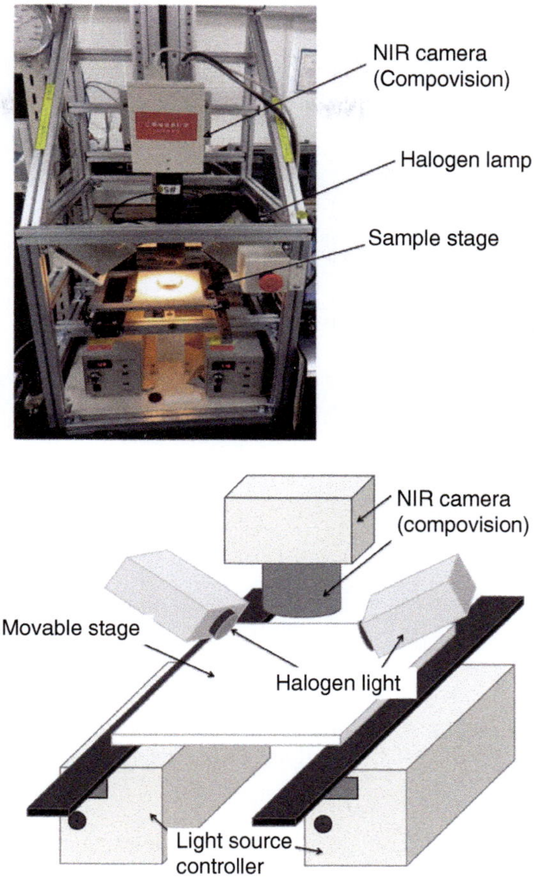

Figure 5.18 Outline view of the imaging instrument, including the developed NIR camera. Source: Ishikawa et al. [40] and Ishikawa et al. [41].

Moreover, the measurement area of Compovision is 1000 times larger than that of the commercial one [40].

5.3.3 Applications of NIR Imaging

5.3.3.1 Monitoring of Crystal Evolution Combined with Chemometrics

Crystalline biodegradable polymers, such as PLA and PHB, have different crystallinities depending on the temperature history, and the mixed state of each component in the sample differs depending on the mixing ratio. Therefore, from the viewpoint of the determination of polymer quality, it is important to investigate the crystallinity of the polymer and the homogeneity of the mixed polymers.

We would like to introduce an emerging study of NIR imaging for polymers by Furukawa et al. [42] Blend polymers with the PLA:PHB mixing ratios of 80 : 20, 60 : 40, 40 : 60, and 20 : 80 were prepared, and NIR spectra were acquired and imaged. In the 1000–2500 nm region, informative bands due to the overtone of the CH_3 stretching vibration mode appeared at 1690 and 1720 nm, and bands due to C=O stretching vibration mode appeared at 1910 and 1950 nm. In such samples, it

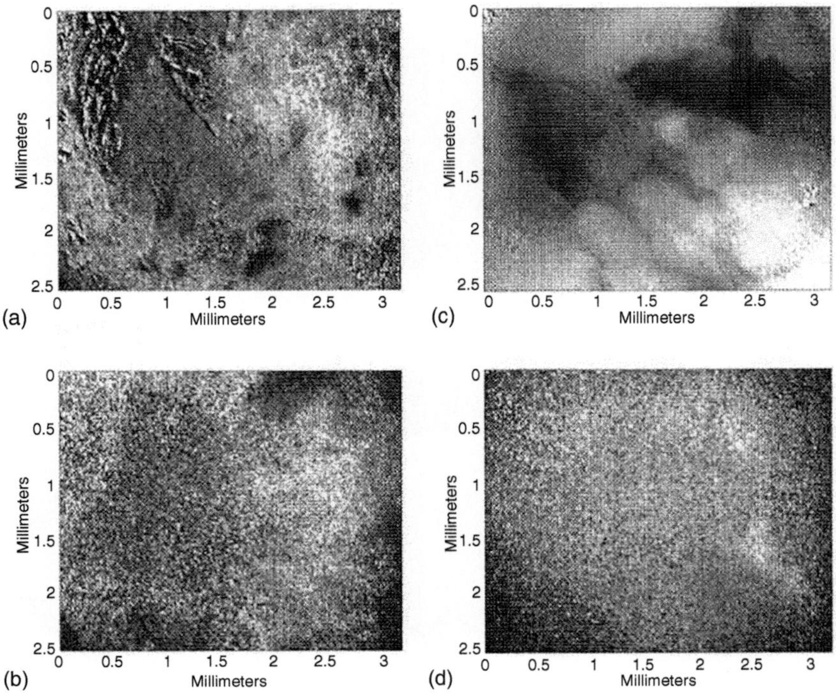

Figure 5.19 Score images of PHB/PLLA blends derived from PLSR. PHB/PLLA (a) 80/20, (b) 60/40, (c) 40/60, (d) 20/80. Source: Reproduced from Ref. [42] with permission.

is difficult to isolate and examine the exact changes in the components; therefore, analytical techniques combined with chemometrics are useful.

Figure 5.19 shows the score images developed by the first component of the PCA of the spectra of PLA/PHB blend polymers [42]. In the case of the mixing ratio of 60 : 40, the roughness of the image was slightly lower than that for the other ratios. The distribution of score slightly differed from a normal distribution. Thus, the difference of homogeneity induced by the mixing ratio was demonstrated. As it is recognized that this may occur due to the evaporation rate of the solvent in the mixed polymer production process, the polymer production process can be applied to NIR imaging data by applying chemometrics, particularly PCA. This is expected to lead to a quality evaluation method.

Suttiwijitpukdee et al. reported the crystal structures and crystallization behaviors of PHB and cellulose acetate butyrate (CAB) blends using IR and NIR imaging [43]. They successfully visualized the evolution of heterogeneous spherulite during isothermal crystallization by using the IR and NIR imaging of the first and second overtones of the C=O stretching vibrations of PHB and CAB. Moreover, the distribution of the different morphologies was classified using IR and NIR spectra with PCA. The first and second components largely indicate crystallinity and amorphous bands. They concluded that the PC1 and PC2 scores of different parts of the spherulite corresponded to areas of low crystallinity in both PHB and CAB.

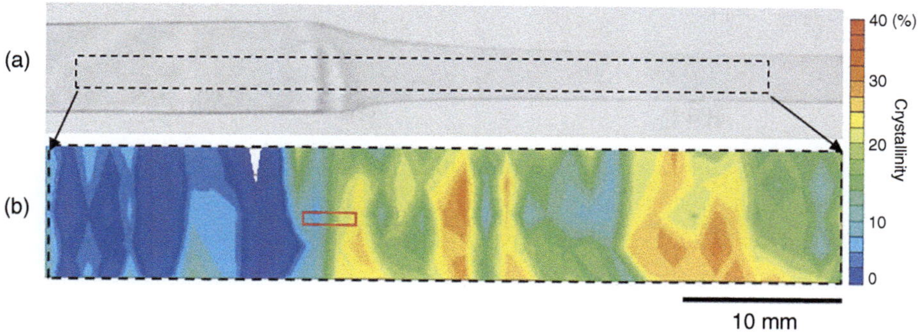

Figure 5.20 (a) Photos of stretched sample of PLA and (b) NIR images for crystallinity developed by the SNV spectra. Dotted and solid lines in figures depict the field of view of Compovision, which can record images in several seconds, and that of a general imaging system. Source: Reproduced from Ref. [40] with permission.

5.3.3.2 Quality Evaluation Potential for Wide Area

Stretched PLA films were investigated using Compovision to demonstrate the potential of wide-area NIR imaging to examine the properties of the polymer [40]. A PLA film was prepared from several pellets and gradually stretched to five times the initial size at 60–70 °C. The crystallinity of the film was determined using an X-ray diffraction (XRD) pattern. The partial least squares (PLS) regression model of the crystallinity of the PLA film was developed using standard normal variate (SNV) spectra and was applied to the data obtained from the stretched sample. Figure 5.20 shows the NIR images developed using SNV spectra for the 500% stretched sample and its visible photo [40]. As depicted in the figure, the field of view (FOV) of Compovision is superior to that of conventional NIR imaging devices. Moreover, the inhomogeneity of the crystallinity distribution in the PLA sample is shown by NIR imaging. Therefore, NIR imaging over a wide area not only provides good quantitative solutions regarding the concentration but also is a powerful tool for the monitoring of physical properties, such as the crystallinity of the polymer.

5.3.3.3 Diffusion Process Monitoring

Unger et al. conducted an experiment on the difference in the diffusion rate of butanol into polyamide during the deuteration process above and below the glass transition temperature of polyamide [44]. In this study, the progress of diffusion was monitored by the changes in the intensity of the NH-specific absorption band of polyamide (PA) 11 due to NH/ND isotope substitution. Figure 5.21 shows the NIR images obtained (a) immediately and (b) after 10.15 hours under 25 °C and (c) immediately and (d) after 1.45 hours at 50 °C [44]. The NIR image was created using the ν(NH) + amide II band around 4935 to 4800 cm^{-1}. There was no diffusion of butanol to the polymer side initially ($t = 0$ minute); hence, the color of the polyamide side was homogeneous. On the other hand, after 10 hours, the edge of butanol moved to the polyamide side. The band of ν(NH) + amide II decreased according to the substitution of NH/ND, as shown in yellow in the figure. The progress of

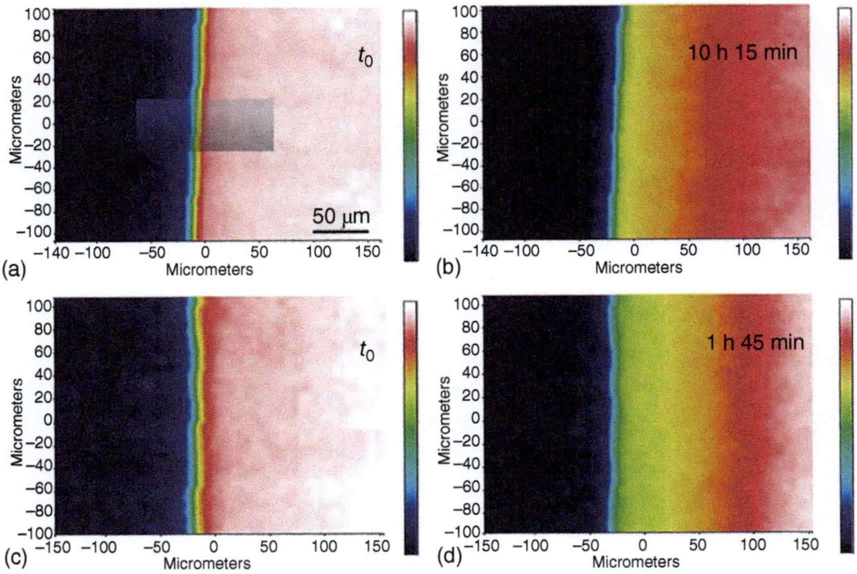

Figure 5.21 FT-NIR images based on the integrated ν(NH) + amide II absorbance for (a) the deuteration times t_0 and (b) 10 hours 15 minutes at 25 °C and (c) t_0 and (d) 1 hour 45 minutes at 50 °C. Source: Reproduced from Ref. [44] with permission.

diffusion at 50 °C was rapid compared with that at 25 °C. The result suggested that the temperature of the glass transition was influenced by the diffusion temperature.

5.3.3.4 Degradable Process Monitoring of Biodegradable Polymer

Shinzawa et al. expanded the potential of NIR imaging by combining it with multiple-perturbation 2D correlation spectroscopy [45]. An example is provided with the NIR imaging data of PLA nanocomposite films undergoing UV-induced variation to show the potential of the multiple-perturbation 2D correlation spectroscopy for applications. The peak at 5810 cm^{-1} due to antisymmetric CH_2 was observed in the average spectra of the PLA nanocomposite samples. Doubly synchronous and double disrelation correlation spectra were calculated from the NIR spectra of the samples subjected to UV irradiation. Interestingly, the correlation map indicated that a cross-peak at the coordinates 5810 and 5940 cm^{-1} generated different trends. Thus, development theory is very attractive for the identification of subtle and/or complicated spectral features. Consequently, the potential of the method was successfully revealed to establish an unambiguous assignment of NIR bands. Figure 5.22 is constructed with the ratio intensities of 5810 and 5940 cm^{-1} for samples subjected to UV irradiation for 0 and 30 hours, respectively. A notable pattern can be observed in Figure 5.22 [45]. The image indicated that the decrease in the crystalline band delayed compared with that of the amorphous band, and it is especially acute in the region where the clay was aggregated. Thus, the incorporation of the clay leads to the cleavage-induced crystallization of PLA by UV irradiation. And it is suggested that the additional development of the crystalline structure occurred due to the delay in the weight loss of the PLA [45].

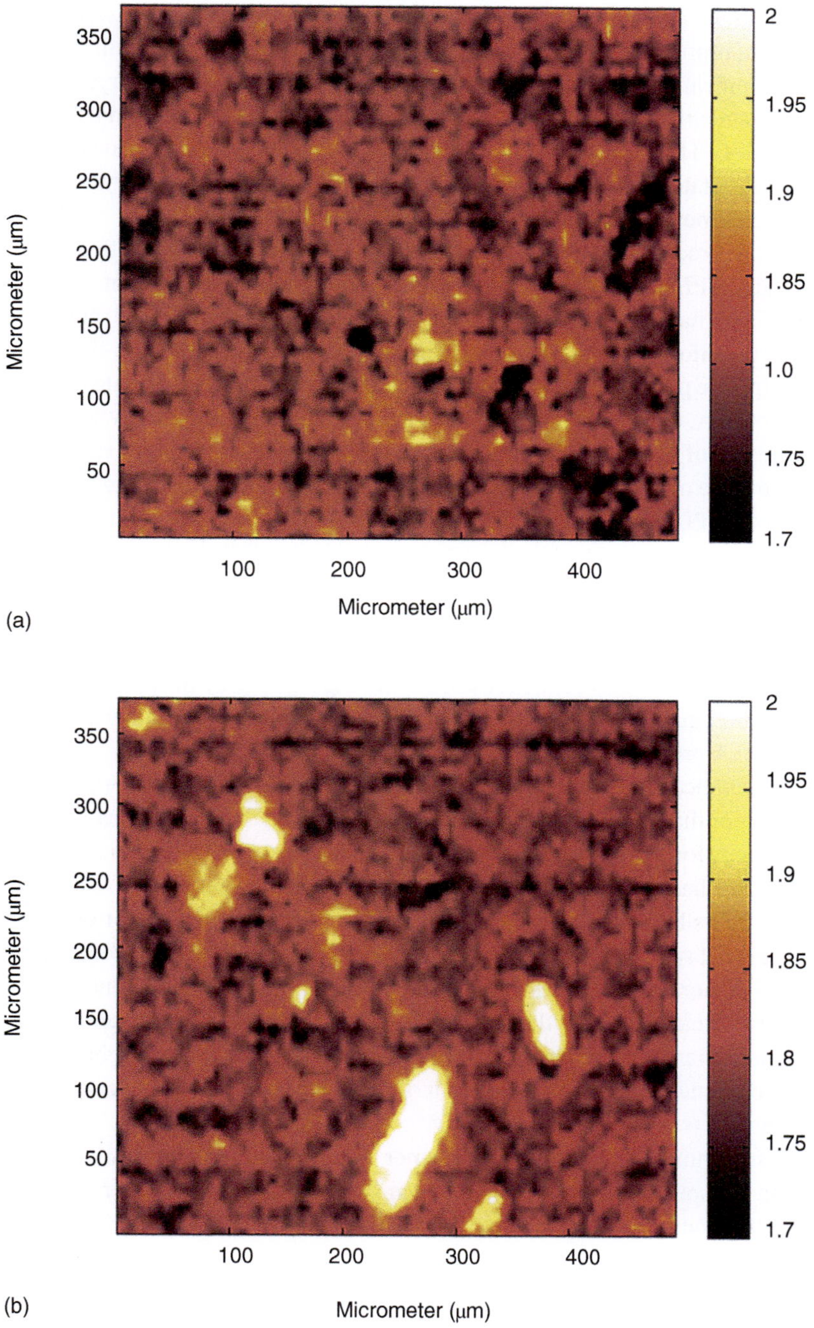

Figure 5.22 Images of samples subjected to UV-irradiation for crystalline/amorphous (a) 0 and (b) 30 hours constructed with ratio of spectral intensities at 5810 and 5940 cm^{-1}. Source: Reproduced from Ref. [45] with permission.

5.3.3.5 Rapid Evaluation of the Water Content in PLA Pellets

PLA is a well-known biomass-based biodegradable polymer. The crystallinity of PLA is an important property for industrial applications. For example, the crystallinity of PLA has a significant effect on its mechanical properties. PLA is a semicrystalline polymer with a melting point of approximately 170 °C and a glass transition temperature of 60–70 °C. Therefore, glassy PLA products can be obtained by rapid cooling from the molten state. The heat resistance of PLA could be improved via crystallization by further annealing above the glass transition temperature.

Polymer melt processing, such as injection molding and extrusion, is often used to mass produce PLA industrial products. In these processes, the PLA materials in granular form, namely pellets, are first heated to approximately 200 °C, and the molten PLA is uniformly mixed in a high-temperature screw barrel. PLA, like several other aliphatic polyesters, is highly hydrolysable. This feature contributes to the importance of PLA, that is, biodegradability. However, high hydrolysability severely affects the quality of PLA products manufactured using melt processes. For example, as the reaction rate of hydrolysis becomes very fast at high temperatures, the degradability of molten PLA must be controlled in the molding machine. A particularly important solution is to decrease the water content of PLA pellets before molding. If the pellets contain moisture, hydrolysis proceeds quickly in the molten PLA in the screw barrel. Consequently, the formability of the material is affected by the change in viscosity, and the mechanical properties of PLA products deteriorate significantly even after molding. Therefore, it is necessary to dry the PLA pellets at approximately 80 °C for several hours prior to melt processing, to prevent the mechanical deterioration of the PLA product.

Herein, an application of NIR spectroscopic imaging to determine whether the drying process of the pellets is sufficient is introduced [46]. Several applications of NIR spectroscopy for evaluating the water content in polymers have previously been reported [47–50] because this method has a good absorption sensitivity to water. Although it is possible to evaluate the water content in a polymer plate or powder with a typical NIR spectrometer, evaluating the water content in each pellet is time consuming. Furthermore, the method to evaluate the average water content of pellets does not sufficiently detect the variation in the water content between pellets or the contamination of inadequately dried pellets. NIR spectroscopic imaging techniques can solve such problems, because a large number of pellets can be evaluated simultaneously.

The NIR spectrum of a hygroscopic polymer indicates the variation of absorption peaks in the spectral range of 8000–6500 cm^{-1} depending on the water content [46, 47]. Multivariate analysis can be applied to estimate the water content in PLA using NIR spectral data. In this experiment, PLSR was employed to estimate the water content. The PLA plates were molded using a hot-press machine. The plates were dried in a hot-air oven or immersed in water to control the water content. The calibration dataset was prepared by measuring samples with known water contents. Diffuse reflection NIR spectroscopic imaging measurements were conducted on the samples placed on a white reflectance plate. Each obtained NIR spectrum was subjected to the Kubelka–Munk (KM) transform [51], second-derivative pretreatment, and SNV transformation [52]. The preprocessed spectra were then used to optimize the number of latent variables in the PLS model with leave-one-out cross-validation. The constructed calibration model was applied to the NIR spectrum at each pixel measured using the NIR spectroscopic imaging technique. Four types of PLA pellets were prepared on a glass Petri dish in dried, undried, and water-absorbed states. The water-absorbed samples were prepared by immersing the PLA pellets in water for 3 and 45 hours. The average water contents calculated from the change in weight were 0.28 and 0.45 wt%, respectively. Figure 5.23 shows the NIR images of the dried, undried, and water-absorbed PLA pellets with the water content estimated using PLSR. The difference in the water content of the PLA pellets was visualized using the NIR spectroscopic imaging technique. As the shape of each pellet is distinguishable, NIR spectroscopic imaging allows the identification of the shape of each pellet for an easy comparison. The variation in the estimated water content between different pellets suggests that this technique is applicable for evaluating the drying process. Therefore, NIR spectroscopic imaging can assess the variation in the drying process of PLA pellets and has the potential to reduce defects in PLA melt processing processes.

5.3.3.6 Nondestructive Detection of Degraded Polylactic Acid Moldings

As stated above, drying PLA pellets is necessary to prevent their deterioration during polymer melt processing. However, the degradation of PLA sometimes occurs during melt processing, even when dried pellets are used [54]. The degradability of PLA is intrinsic; hence, it is necessary to deal with degradation during melt processing.

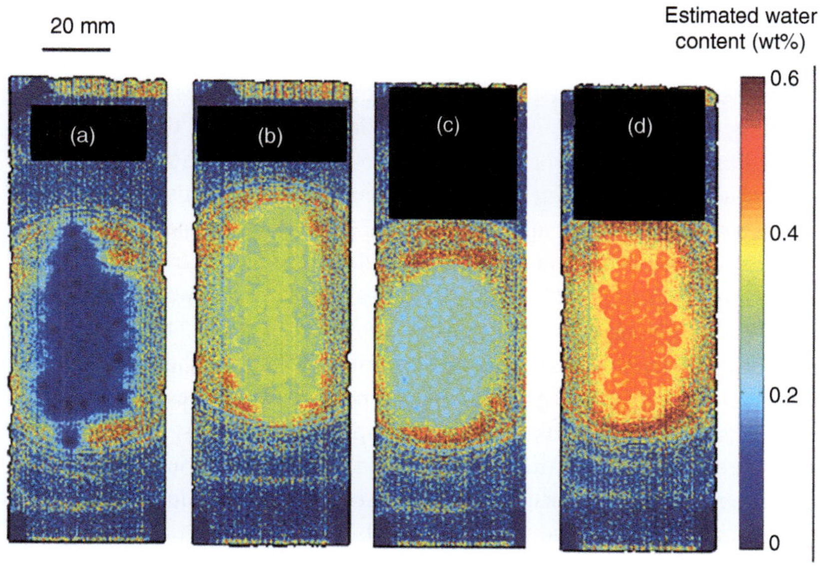

Figure 5.23 NIR images of the water content of (a) dried, (b) undried, and water-absorbed for (c) 3 and (d) 45 hours PLA pellets estimated using a PLS model. The average water contents of the PLA pellets in (c) and (d) were 0.28 and 0.45 wt%, respectively.

As the decomposition rate of PLA is much higher at temperatures above the melting point, an appropriate processing temperature and residence time in the screw barrel should be selected. If the barrel temperature is increased to gain good formability or the materials are left in the screw barrel for a relatively long time, PLA will deteriorate, and the mechanical properties will be significantly degraded. Evident molding defects, such as yellowing and embrittlement, can be detected visually via visual inspection or image analysis. However, it is difficult to distinguish a product whose mechanical properties only are considerably degraded. For such cases, NIR spectroscopic imaging is an effective means of detecting chemical changes in polymers. As the thermal degradation of PLA proceeds, the hydroxyl group increases as a result of the breaking of the ester bond. Therefore, the progress of degradation can be determined by focusing on the changes in the absorption bands of the hydroxyl band.

In this section, an application of the NIR spectroscopic imaging technique to distinguish the deterioration of the mechanical properties of PLA products is presented [53]. The aim of this study was to estimate the change in the mechanical properties of PLA using the NIR absorption spectrum. Notably, both degradation and crystallization affect the mechanical properties and the shape of the NIR absorption spectrum. For example, the flexural modulus increases, and simultaneously, the flexural strain decreases with an increase in crystallinity. Simultaneously, the growth of the crystalline band is observed in the NIR spectrum. Thus, the variation in PLA crystallinity also needs to be considered in estimating the change in the mechanical properties of PLA products from the hydroxyl bands. Both thermal degradation and crystallinity were controlled to prepare the PLA samples. The combination of melting time (t_m) at 180 °C and annealing time (t_a) at 110 °C was varied during the molding of PLA using a hot-press machine [53].

First, the results of NIR spectroscopy are demonstrated. Figure 5.24a shows the variation in the KM-transformed spectra of the quenched samples ($t_a = 0$ minute) as the melting time increased. A significant difference between the NIR spectra was observed in the wavelength range of 1200–1600 nm. To eliminate the effects of the baseline fluctuation and optical path deviation, second-derivative pretreatment and SNV transformation were applied to the KM-transformed spectra, as shown in Figure 5.24b. The development of the second-derivative spectrum at 1457 nm was observed. In contrast, the NIR absorption peaks in the same wavelength range increased as the annealing time increased, as shown in Figure 5.24c. Figure 5.24d shows the second-derivative spectra of the annealed PLA. The shape of the second-derivative spectrum at 1419 nm changed notably after annealing.

To prepare the calibration data set, the three-point flexural test and DSC measurements were performed for the PLA products molded with various melting and annealing times, after NIR spectroscopic measurements. The flexural strength, flexural strain, and flexural modulus were calculated from the stress–strain curves obtained from the flexural test. The degree of crystallization was also determined using DSC measurements. PLS regression was applied to the second-derivative spectra to construct models for estimating these four response variables. The response variables of each PLA sample were estimated by applying the PLS models to the NIR

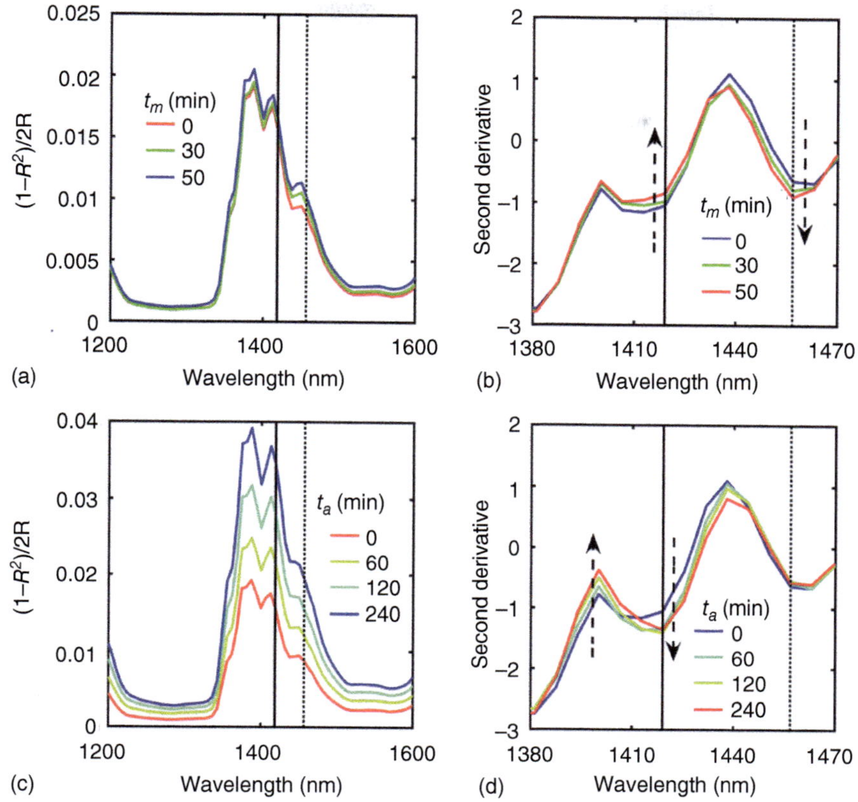

Figure 5.24 NIR spectra of PLA with different molding conditions. (a) KM-transformed and (b) second-derivative and SNV-transformed spectra of the quenched samples ($t_a = 0$ minutes) with different melting times (t_m). (c) KM-transformed and (d) second-derivative and SNV-transformed spectra with different annealing times t_a at the melting time t_m of 0 minutes. The black solid lines show the wavelength of 1419 nm. The black dotted lines indicate the wavelength of 1457 nm. Reproduced with permission from Ref. [53]; ©2018, John Wiley and Sons.

spectrum at each pixel in the imaging data. Figure 5.25 shows the 2D distributions of the estimated response variables: (i) flexural strength, (ii) flexural strain, (iii) flexural modulus, and (iv) crystallinity. The flexural strength and flexural strain indicated a remarkable deterioration as the melting time increased. On the other hand, the variation in flexural modulus was strongly correlated with that of crystallinity.

In conclusion, the deterioration of mechanical properties due to thermal degradation was visualized for PLA products with different degrees of crystallization by combining multivariate analysis with the NIR spectroscopic imaging technique. This technique would be applicable for assessing changes in the mechanical properties of polymers induced by changes in chemical or higher-order structures. Therefore, NIR spectroscopic imaging can be used as a nondestructive quality control method for polymer products.

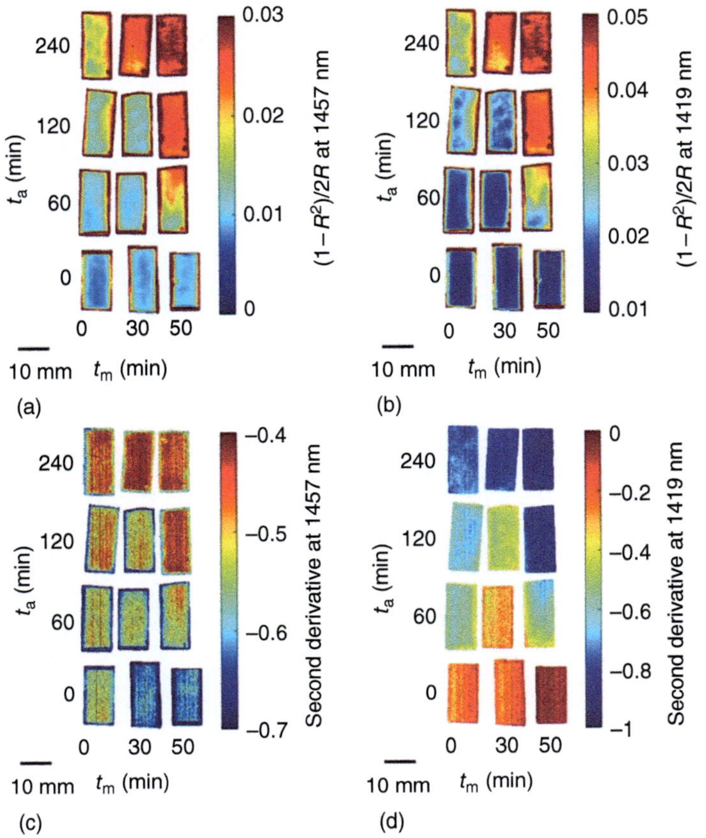

Figure 5.25 NIR images of the (a) flexural strength, (b) flexural strain, (c) flexural modulus, and (d) crystallinity of PLA estimated using PLS models. The PLA samples were molded with different melting time (t_m) at 180 °C and annealing time (t_a) at 110 °C. Reproduced with permission from Ref. [53]; ©2018, John Wiley and Sons.

References

1 Ozaki, Y., Huck, C.W., Tsuchikawa, S., and Engelsen, S.B. (eds.) (2020). *Near-Infrared Spectroscopy, Theory, Spectral Analysis, Instrumentation, and Applications*. Springer.

2 Siesler, H.W., Ozaki, Y., Kawata, S., and Heise, H.M. (eds.) (2002). *Near-Infrared Spectroscopy, Principles, Instruments, Applications*. Wiley-VCH.

3 Ozaki, Y., McClure, W.F., and Christy, A.A. (eds.) (2006). *Near Infrared Spectroscopy in Food Science and Technology*. Wiley Interscience.

4 Workman, J. and Weyer, L. (2012). *Practical Guide and Spectral Atlas for Interpretive Near-Infrared Spectroscopy*, 2e. Boca Raton, FL: CRC Press.

5 Burns, D.A. and Ciurczak, E.W. (eds.) (2008). *Handbook of Near-Infrared Spectroscopy*, 3e. Boca Raton, FL: CRC Press.

6 Martens, H. and Næs, T. (1989). *Multivariate Calibration*. Chichester: Wiley.

7 Jue, T. and Masuda, K. (2013). *Application of near Infrared Spectroscopy in Biomedicine*. Springer.

8 National Institutes of Health (ed.) (2020). Near-infrared imaging of adoptive immune cell therapy in breast cancer model using cell membrane labeling.

9 Mark, H. and Workman, J. Jr. (2007). *Chemometrics in Spectroscopy*. Amsterdam: Elsevier.

10 Varmuza, K. and Filzmoser, P. (2009). *Introduction to Multivariate Statistical Analysis in Chemometrics*. Boca Raton, FL: CRC Press.

11 Noda, I. and Ozaki, Y. (2004). *Two Dimensional Correlation Spectroscopy-Applications in Vibrational and Optical Spectroscopy*. Wiley.

12 Czarnecki, M.A., Morisawa, Y., Futami, Y., and Ozaki, Y. (2015). Advances in molecular structure and interaction studies using near-infrared spectroscopy. *Chem. Rev.* 115: 9707–9744.

13 Bec, K.B., Futami, Y., Wojcik, M.J., and Ozaki, Y. (2016). Spectroscopic and theoretical study in the near-infrared region of low concentration aliphatic alcohols. *Phys. Chem. Chem. Phys.* 18: 13666–13682.

14 Bec, K.B., Grabska, J., Huck, C.W., and Ozaki, Y. (2016). Quantum mechanical simulation of near-infrared spectra applications in physical and analytical chemistry. In: *Molecular Spectroscopy; A Quantum Chemistry Approach* (eds. Y. Ozaki, M.J. Wojcik and J. Popp), 353–388. Wiley-VCH.

15 Siesler, H.W. and Holland-Moritz, K. (1980). *Infrared and Raman Spectroscopy of Polymers*. Marcel Dekker.

16 Siesler, H.W. Applications to polymers and textiles. In: *Near-Infrared Spectroscopy, Principles, instruments, applications* (eds. H.W. Siesler, Y. Ozaki, S. Kawata and H.M. Heise), 213–246. Wiley-VCH.

17 Bokobza, L. (1998). Near infrared spectroscopy. *J. Near Infrared Spectrosc.* 6: 3–17.

18 Bokobza, L. (2019). Some applications of vibrational spectroscopy for the analysis of polymers and polymer composites. *Polymers* 11: 1159–1171.

19 Ozaki, Y. and Amari, T. (2000). Near-infrared spectroscopy in chemical process analysis. In: *Spectroscopy in Process Analysis* (ed. J.M. Chalmers), 53–95. Scheffield Academic Press/CRC Press.

20 Watari, M. (2014). A review of online real-time process analyses of melt-state polymer using the near-infrared spectroscopy and chemometrics. *Appl. Spectrosc. Rev.* 49: 462–491.

21 Machado, A.V., Barbas, J.M., and Covas, J.A. (2015). Near IR spectroscopy for the characterization of dispersion in polymer-clay nanocomposites. In: *Synthesis Techniques for Polymer Nanocomposites* (ed. V. Mittal), 241–266. Wiley-VCH.

22 Scherzer, T. (2020). Applications of NIR techniques in polymer coatings and synthetic textiles. In: *Near-Infrared Spectroscopy, Theory, Spectral Analysis, Instrumentation, and Applications* (eds. Y. Ozaki, C.W. Huck, S. Tsuchikawa and S.B. Engelsen), 475–516. Springer.

23 Buffeteau, T., Desbat, B., and Bokobza, L. (1995). The use of near-infra-red spectroscopy coupled to the polarization modulation technique to investigate molecular orientation in uniaxiary stretched polymers. *Polym. Commun.* 36: 4339–4343.

24 Mizushima, M. Kawamura, T., Takahashi, K., and Nitta, K. (2012). E-Polymers no. 068.

25 Hu, Y., Zhang, J., Sato, H. et al. (2006). C–H· · ·O=C hydrogen bonding and isothermal crystallization kinetics of poly(3-hydroxybutyrate) investigated by near-infrared spectroscopy. *Macromolecules* 39: 3841–3847.

26 Aoyama, T., Sato, H., and Ozaki, Y. (2019). Crystallization of poly(3-hydroxybutyrate-*co*-3-hydroxyhexanoate) during melt extrusion promoted by residual crystals. *Polym. Cryst.* https://doi.org/10.1002/pcr2.10076.

27 Watanabe, R., Shinzawa, H., Kunioka, M. et al. (2017). Reinforcement mechanism of functionalized polypropylene containing hydroxyl group nanocomposites studied by reo-optical near-infrared spectroscopy. *Eur. Polym. Spectrosc.* 92: 86–96.

28 Zhang, J., Sato, H., Noda, I., and Ozaki, Y. (2005). Conformation rearrangement and molecular dynamics of poly(3-hydroxybutyrate) during the melt-crystallization process investigated by infrared and two-dimensional infrared correlation spectroscopy. *Macromolecules* 38: 4274–4281.

29 Šašić, S. and Ozaki, Y. (eds.) (2011). *Raman, Infrared and Near Infrared Chemical Imaging*. Willy.

30 Salzer, R. and Siesler, H.W. (eds.) (2014). *Infrared and Raman Spectroscopic Imaging*. Wiley-VCH.

31 Ishikawa, D., Shinzawa, H., Genkawa, T. et al. (2014). Recent progress of near-infrared imaging–development of novel instruments and their applicability for practical situations. *Anal. Sci.* 30: 143–150.

32 Ishikawa, D., Ishigaki, M., and Gaowen, A.A. (2020). NIR imaging (Chapter 22). In: *Near-Infrared Spectroscopy* (eds. Y. Ozaki, C. Huck, S. Tsuchikawa and S.B. Engelsen), 517–551. Singapore: Springer.

33 (a) Gosselin, R., Rodrigue, D., González-Núñez, R., and Duchesne, C. (2009). Potential of hyperspectral imaging for quality control of polymer blend films. *Ind. Eng. Chem. Res.* 48 (6): 3033–3042. (b) Gosselin, R., Rodrigue, D., and Duchesne, C. (2011). A hyperspectral imaging sensor for on-line quality control of extruded polymer composite products. *Comput. Chem. Eng.* 35 (2): 296–306. (c) Gosselin, R., Rodrigue, D., and Duchesne, C. (2008). On-line prediction of crystallinity spatial distribution across polymer films using NIR spectral imaging and chemometrics methods. *Can. J. Chem. Eng.* 86 (5): 869–878. (d) Ghasemzadeh-Barvarz, M., Rodrigue, D., and Duchesne, C. (2014). Multivariate image analysis for inspection of multilayer films. *Polym. Test.* 40: 196–206.

34 Serranti, S., Gargiulo, A., and Bonifazi, G. (2011). Characterization of post-consumer polyolefin wastes by hyperspectral imaging for quality control in recycling processes. *Waste Manage.* 31 (11): 2217–2227.

35 Serranti, S., Gargiulo, A., and Bonifazi, G. (2012). Classification of polyolefins from building and construction waste using NIR hyperspectral imaging system. *Resour. Conserv. Recycl.* 61: 52–58.

36 Ulrici, A., Serranti, S., Ferrari, C. et al. (2013). Efficient chemometric strategies for PET–PLA discrimination in recycling plants using hyperspectral imaging. *Chemom. Intell. Lab. Syst.* 122: 31–39.

37 Bonifazi, G., Serranti, S., Potenza, F. et al. (2017). Gravity packaging final waste recovery based on gravity separation and chemical imaging control. *Waste Manage.* 60: 50–55.

38 Zheng, Y., Bai, J., Xu, J. et al. (2018). A discrimination model in waste plastics sorting using NIR hyperspectral imaging system. *Waste Manage.* 72: 87–98.

39 Caballero, D., Bevilacqua, M., and Amigo, J. (2019). Application of hyperspectral imaging and chemometrics for classifying plastics with brominated flame retardants. *J. Spectral Imaging.*

40 Ishikawa, D., Nishii, T., Mizuno, F. et al. (2013). Development of a high-speed monitoring NIR hyperspectral camera (Compovision) with wide area and its applications. *NIR News* 24 (5): 6–11.

41 Ishikawa, D., Nishii, T., Mizuno, F. et al. (2013). Potential of a newly developed high speed near-infra red (NIR) camera (Compovision) in polymer industrial analyses – monitoring of Crystallinity and crystal evolution of poly lactic acid (PLA) and concentration of PLA in PLA/poly-(*R*)-3-hydroxybutyrate blends. *Appl. Spectrosc.* 67 (12): 1441–1446.

42 Furukawa, T., Sato, H., Shinzawa, H. et al. (2007). Evaluation of homogeneity of binary blends of poly(3-hydroxybutyrate) and poly(L-lactic acid) studied by near infrared chemical imaging (NIRCI). *Anal. Sci.* 23: 871–876.

43 Suttiwijitpukdee, N., Sato, H., Unger, M., and Ozaki, Y. (2012). Effects of hydrogen bond intermolecular interactions on the crystal spherulite of poly(3-hydroxybutyrate) and cellulose acetate butyrate blends: studied by FT-IR and FT-NIR imaging spectroscopy. *Macromolecules* 45 (6): 2738–2748.

44 Unger, M., Ozaki, Y., Siesler, W.H. (2011). Variable-temperature Fourier transform near-infrared (FT-NIR) imaging spectroscopy of the diffusion process of butanol(OD) into polyamide 11. *Appl. Spectrosc.* 65: 1051–1055.

45 Shinzawa, H., Murakami, T., Nishida, M. et al. (2014). Near-infrared (NIR) imaging analysis of polylactic acid (PLA) nanocomposite by multiple-perturbation two-dimensional (2D) correlation spectroscopy. *J. Mol. Struct.* 1069: 171–175.

46 Muroga, S., Hikima, Y., and Ohshima, M. (2017). Near-infrared spectroscopic evaluation of the water content of molded polylactide under the effect of crystallization. *Appl. Spectrosc.* 71 (6): 1300–1309.

47 Camacho, W., Hedenqvist, M., and Karlsson, S. (2002). Near infrared (NIR) spectroscopy compared with thermogravimetric analysis as a tool for on-line prediction of water diffusion in polyamide 6,6. *Polym. Int.* 51 (12): 1366–1370.

48 Camacho, W., Vallés-Lluch, A., Ribes-Greus, A., and Karlsson, S. (2003). Determination of moisture content in nylon 6,6 by near-infrared spectroscopy and chemometrics: nylon 6,6 moisture content. *J. Appl. Polym. Sci.* 87 (13): 2165–2170.

49 Zhang, X. and Wyeth, P. (2007). Moisture sorption as a potential condition marker for historic silks: noninvasive determination by near-infrared spectroscopy. *Appl. Spectrosc.* 61 (2): 218–222.

50 Kuda-Malwathumullage, C.P.S. and Small, G.W. (2014). Determination of moisture content of polyamide 66 directly from combination region near-infrared spectra. *J. Appl. Polym. Sci.* 131 (16): 40645.

51 Kubelka, P. (1948). New contributions to the optics of intensely light-scattering materials. Part I. *J. Opt. Soc. Am.* 38: 448.

52 Barnes, R.J., Dhanoa, M.S., and Lister, S.J. (1988). Standard normal variate transformation and de-trending of near-infrared diffuse reflectance spectra. *Appl. Spectrosc.* 43: 772–777.

53 Muroga, S., Hikima, Y., and Ohshima, M. (2018). Visualization of hydrolysis in polylactide using near-infrared hyperspectral imaging and chemometrics. *J. Appl. Polym. Sci.* 135 (8): 45898.

54 Taubner, V. and Shishoo, R. (2001). Influence of processing parameters on the degradation of poly(L-lactide) during extrusion. *J. Appl. Polym. Sci.* 79: 2128–2135.

6

Far Ultraviolet Spectroscopy for Polymers

Yusuke Morisawa[1] *and Nami Ueno*[2]

[1] *Kindai University, School of Science and Engineering, Department of Chemistry, 3-4-1 Kowakae, Higashi-Osaka 577-8502, Japan*
[2] *University of Innsbruck, Institute of Analytical Chemistry and Radiochemistry, Innrain 80-82, 6020 Innsbruck, Austria*

6.1 Introduction

Light with a wavelength shorter than that of visible rays (>380 nm) and longer than that of X-rays (<10 nm) is considered as ultraviolet (UV) light [1–6]. Conventional UV spectroscopy in the range of 190–380 nm has been used together with visible spectroscopy as UV–vis spectroscopy for various qualitative analyses of compounds with “chromophore,” complex of transition metals, etc. UV spectroscopy at wavelengths shorter than 190 nm has been primarily utilized for gas molecules because of the extremely large absorption coefficients of allowed electronic transitions observed below 190 nm, and it is difficult to measure the UV spectra of solid and liquid samples via absorption spectroscopy [7, 8]. Although no clear definition exists, the UV region in the 10–380 nm range may be segmented into four regions: extreme ultraviolet (EUV, 10–120 nm), far ultraviolet (FUV, 120–200 nm), deep ultraviolet (DUV, 200–300 nm), and UV (UV, 300–380 nm) regions. The term “vacuum UV region” is occasionally used in the 10–200 nm region, which is inappropriate for the 120–200 nm region because most recent spectrometers used in this region are not necessarily equipped with a vacuum evaporation system; instead, they are equipped with a nitrogen-purged system or those of other inert gases (such as He and Ar). This section describes recent advances in the FUV spectroscopy of polymers [9–13]. Electronic transitions from valence electrons were observed in both the FUV (120–200 nm) and DUV (200–300 nm) regions.

In 2007, Higashi et al. introduced an attenuated total reflection (ATR) spectrometer in the 140–200 nm range to measure the FUV spectra of solid and liquid materials under nitrogen gas purge [14, 15]. At the interface between the internal reflection element (IRE) and a sample, an evanescent wave is created through total reflection when light is passed through, and it interacts with the sample. Hence, the ATR technique, which is typically used in the IR region, can yield absorption information corresponding to those of transmittance (Tr) spectra with an extremely short optical

Spectroscopic Techniques for Polymer Characterization: Methods, Instrumentation, Applications, First Edition. Edited by Yukihiro Ozaki and Harumi Sato.

path length via appropriate spectral corrections such as the Kramers–Kronig transformation (KKT) [16–18]. Without the KKT, the repeatability of the ATR spectra is extremely high because the interaction length between light and samples depends on the refractive indices of the IRE and samples. Consequently, the ATR method is advantageous in terms of both quantitative measurements and sample management in the condensed phase with strong absorption compared to Tr spectroscopy. Currently, ATR–FUV spectroscopy can be used to measure the electronic spectra of almost all types of solid and liquid samples.

FUV spectroscopy offers the following features [7] compared with those in the DUV region. (i) Because of the absorption of oxygen, the FUV region requires a vacuum evaporation system or a nitrogen gas-purging system, in contrast to the DUV region. (ii) Absorption bands appear owing to lower energy electronic transitions from π and nonbonding electrons such as π–π* and n–π* in the DUV region, whereas in the FUV region, not only π–π* and n–π* transitions, but also n–σ* transitions and σ-Rydberg transitions can be observed. Water, alkanes, alcohols, and other molecules that have only a single bond do not exhibit any peaks in the DUV region but result in extremely strong bands in the FUV region. (iii) Compared with DUV spectroscopy, the progress of FUV spectroscopy is slow because of three major reasons. One is the instrumental issue mentioned above. The other is the extremely high absorptivity of the molecules in the FUV region. The third reason is the lack of practical applications of FUV spectroscopy.

To interpret the measured ATR–FUV spectra, quantum chemical calculations of electronic excitation for various types of molecules were performed to investigate the band assignments, electronic transitions, and electronic structure using time-dependent density functional theory (TD–DFT) and symmetry-adapted cluster-configuration interaction (SAC–CI) calculations [9–12, 19, 20].

6.2 Measurement of ATR–FUV Spectra of Polymer

Details of the ATR–FUV spectrometer for solid samples have been reported in [14]. Briefly, it comprised an FUV monochromator (KV-200, Bunkoh-Keiki, Japan) and a sample compartment. To prevent the absorption of oxygen, both the monochromator and the sample compartment were separately purged with nitrogen gas (99.999%). A 30 W D_2 lamp was used as the light source. The FUV light from the monochromator was split into a reference beam and a sample beam using a MgF2 splitter. The sample beam was incident on an ATR probe equipped with an IRE made of vacuum-UV-grade sapphire. Both the reference and the sample beams finally passed through a synthetic quartz plate coated with sodium salicylic acid, which fluoresced. Subsequently, the fluorescence of each beam was detected using a photomultiplier. First, we measured the reflection spectrum through the prism as the background intensity, I_0. Next, we placed a sample on the IRE and measured its ATR intensity, I. The ATR absorption A is defined as follows:

$$A = -\log\left(\frac{I}{I_0}\right) \tag{6.1}$$

The observed ATR spectra were not identical to the corresponding Tr spectra, as the ATR technique changed the peak intensity and peak position. None of the spectra reported herein required ATR correction. The ATR spectra of polymers cannot be subjected to the KKT because the referring refractive indices of polymers are difficult to measure.

All spectra were obtained at room temperature (24–25 °C). Cast films of nylons were directly formed on the probed surface of the IRE made of sapphire. The elimination of the solvent was confirmed using the observed spectra. Solid samples were clamped to the IRE to improve the adhesion between the samples and IRE surface.

6.3 ATR–FUV Spectra of Nylons

Nylons are synthetic polymers in which alkyl chains and amide groups combine with each other. In the FUV spectra of the nylon, these groups show an absorption band due to transitions that originate from *t* *n*-alkanes and amide moieties. The Tr absorption spectra of nylon-3, -4, -6, -6/6, and -6/10 were originally reported by Onari [21]. The π–π* transition of the amide was observed at approximately 190 nm, and the absorbance of the band in the wavelength region shorter than 165 nm increased as the number of methylene units increased. However, the assignment of the shorter wavelength band has not been elucidated for polymers. In relation to nylon, the absorption spectra of α-polypeptides are of interest. These spectra exhibited a charge-transfer (CT) transition between the interpeptides at 165–155 nm [22]. The authors theoretically predicted that this CT transition would weaken rapidly in β-peptides. Therefore, the absence of CT transition in β-peptides and polyamides with longer methylene units has recently been confirmed [23]. However, most theoretical studies regarding solid amides and polyamides do not focus on the Rydberg transitions in the condensed phase/solid state. However, recent studies using ATR–FUV for liquids revealed that Rydberg transitions occurred even in the liquid or condensed phase, as mentioned above.

In this section, we describe the FUV spectra of nylon-6, -11, -12, -6/6, and -6/12, whose chemical structures are shown in Figure 6.1 [9]. For nylon-6 and -6/6 in particular, quantum chemical calculations for the crystalline models involve TD–DFT, including Rydberg transitions [9].

Figure 6.2a shows the ATR–FUV spectra in the 150–260 nm region of cast films of these nylons and liquid *N*-methylacetamide (NMA), which is an amide with a methyl group in the nitrogen atom. All nylons had two intense bands with peaks near 150 and 200 nm. Liquid NMA and other amides exhibited one band at approximately 190 nm [24], whereas liquid *n*-alkanes ($m = 5$–14) exhibited only the 160 nm band [19, 25]. Hence, the 150 and 200 nm bands of nylons originated from the alkane and amide parts of the main chain of nylons, respectively. Figure 6.2b shows the spectra normalized to the peak intensity near 200 nm shown in Figure 6.2a. As shown in Figure 6.2b, the intensity ratio between the absorbance at 150 nm and that at 200 nm increased with the number of carbon atoms in the alkane part, as nylon-6 < nylon-6/6 < nylon-6/12 < nylon-11 < nylon-12. It was observed that this

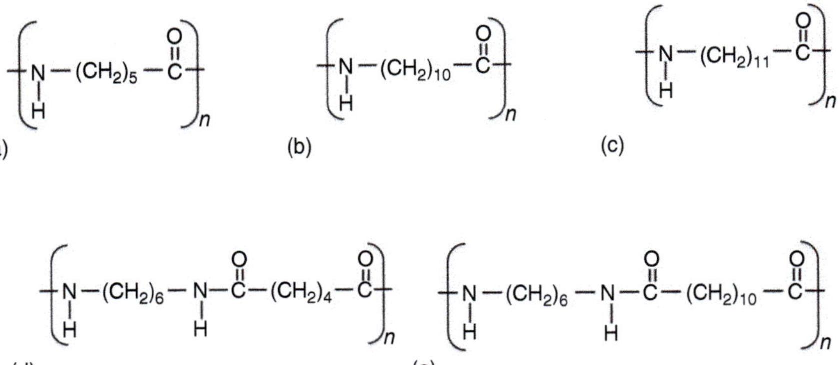

Figure 6.1 Chemical structures of (a) nylon-6, (b) nylon-11, (c) nylon-12, (d) nylon-6/6, and (e) nylon-6/12. Source: Morisawaet al. [9].

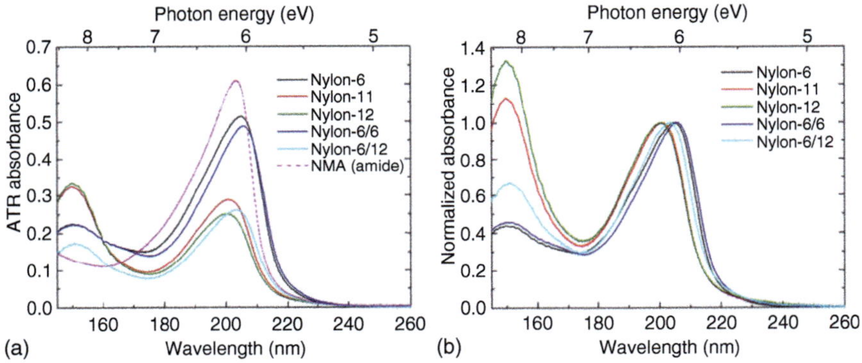

Figure 6.2 (a) ATR–FUV spectra of nylon-6, -11, -12, -6/6, and -6/12, and liquid NMA in cast film; and (b) their normalized spectra with band near 200 nm. Source: Adapted from Beć et al. [10].

intensity trend was the same as that of the 150 nm band in the *n*-alkane spectra, i.e. it increased with the number of carbon atoms. One can estimate the relationship between the peak ratio of the 150–200 nm bands and the average carbon number of nylon. Figure 6.3 shows the relationship between the number of carbon chains and the intensity ratio of the 150–200 nm bands. The intensity ratio was proportional to the average number of alkyl chains of nylon [8].

Figure 6.4a shows the Tr spectra of the observed nylons in the cast film. The intensity of the Tr spectra is affected by the thickness and density of the samples. Similar to Figures 6.2b and 6.4b shows the spectra normalized by the intensity of the 190 nm band. The peak wavelength of the bands near 190 nm was observed at a shorter wavelength by ~10 nm in the Tr spectra compared with that of the corresponding band in the ATR–FUV spectra. The 150 nm band, which was clearly observed in the ATR–FUV spectra, appeared as only as a tail on the longer wavelength side in the Tr spectra. These differences were due to the real part of the refractive index,

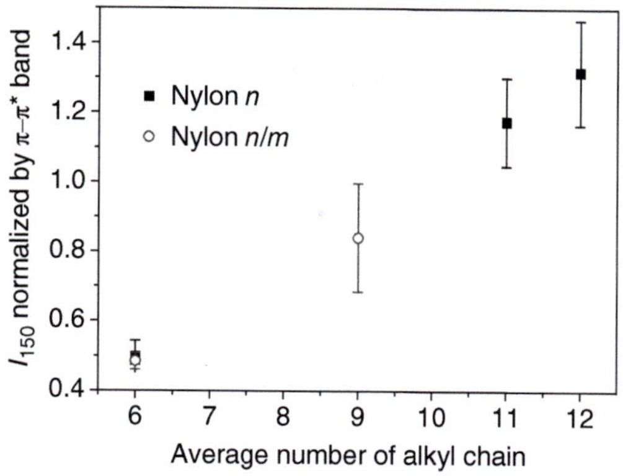

Figure 6.3 Correlation between average alkyl chain and intensity ratio of $I_{150}/I_{\pi-\pi^*}$. Source: Ozaki et al. [8].

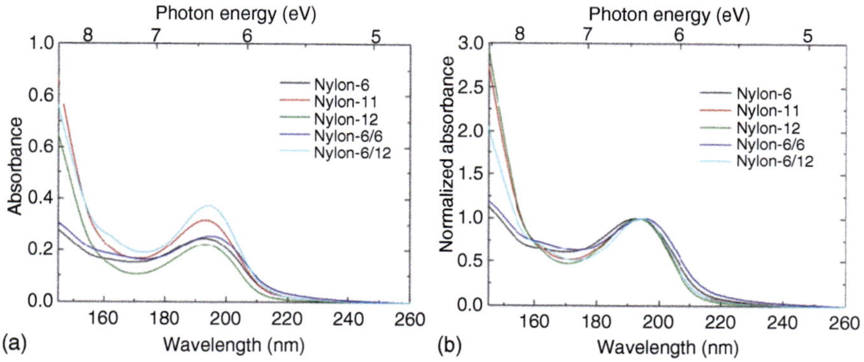

Figure 6.4 (a) Tr–FUV spectra of nylon-6, -11, -12, -6/6, and -6/12 in cast film; and (b) their normalized spectra with band near 200 nm. Source: Morisawa et al. [9].

which exhibited a peak in the absorption band on the longer wavelength side in the ATR–FUV spectrum instead of in the Tr spectrum.

Figure 6.5 shows the relation between the peak positions of the π–π* band in the Tr spectra and those in the ATR spectra. Nylon-*m*/*n* (nylon-6/6 and -6/12) exhibited a peak in the longer wavelength region for both the Tr and ATR spectra compared with nylon-*m* (nylon-6, -11, and -12), except for the ATR spectrum of nylon-6. It is a fact that nylon-*m*/*n* has a stronger hydrogen bonding than nylon-*m*. This systematic difference between nylon-*m*/*n* and nylon-*m* is due to hydrogen bonding, which will be discussed later via quantum chemical calculations. The longer wavelength shift in the ATR spectrum than that in the Tr spectrum is attributable to two reasons. First, nylon-6 has a larger reflective index, yielding a more prominent effect on its real part. Second, the surface of nylon-6 might exhibit a stronger hydrogen bonding than the interior of the polymer.

To reveal the assignments of the transitions and the effects of the hydrogen bond in the ATR–FUV spectra of the nylons, we performed TD–DFT calculations for monomer units of nylon-6, -11, and -12. Figure 6.6 shows the simulation spectra convoluted by Gaussian functions using calculated excitation energies and oscillator strengths with a width of 0.3 eV. The simulation spectra show bands near 170 and

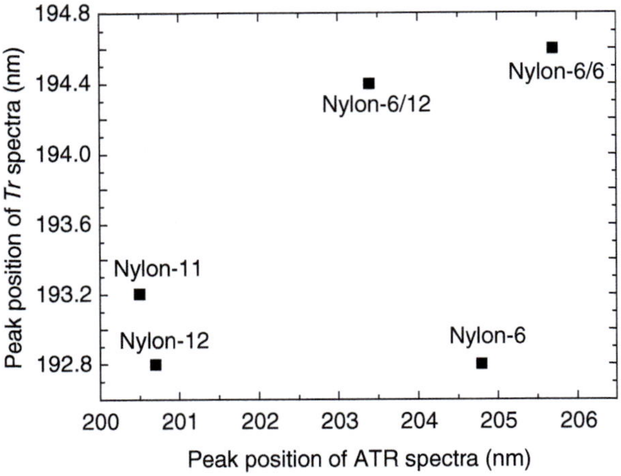

Figure 6.5 Peak position of band near 195 nm of Tr spectra vs. that of band near 205 nm of ATR–FUV spectra for five types of nylon. Source: Morisawa et al. [9].

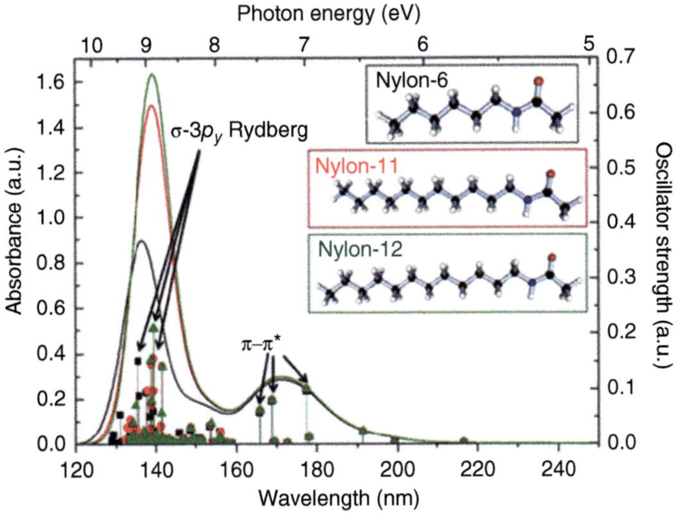

Figure 6.6 Calculated spectra of model compounds of nylon-6, -11, and -12 in vacuum. Source: Morisawa et al. [9].

140 nm, whose assignments were π–π* and σ-3p Rydberg transitions. The calculated bands at 170 and 140 nm corresponded to the observed bands near 200 and 150 nm, respectively.

As discussed above, the hydrogen bonding of NH—O=C between neighboring nylon chains resulted in a longer wavelength shift in the π–π* absorption. Therefore, the effects of hydrogen bonding are discussed via quantum chemical calculations using monomer, dimer, and trimer models.

Figure 6.7 shows the calculated transition dipole strengths together with the structures of the nylon-6 and -6/6 trimer models. As shown in Figure 6.7, the nylon-6 model formed hydrogen bonds perfectly by inverting the alternate chain inside. Meanwhile, nylon-6/6 exhibited a progressive shear structure in the α-form. The TD–DFT results of the π–π* and few σ-Rydberg transitions are summarized in [9].

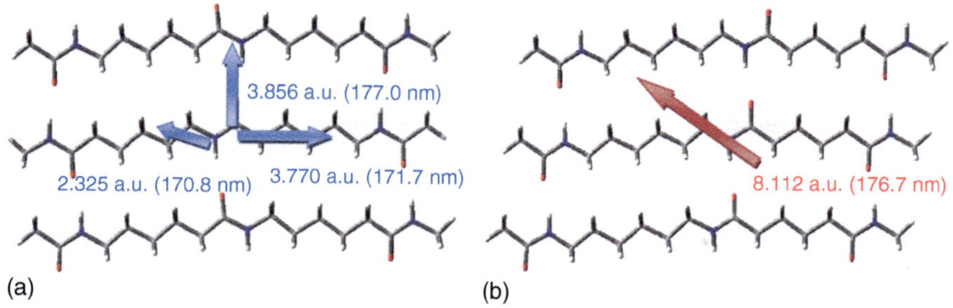

Figure 6.7 Computational trimer models of hydrogen bonded systems for (a) nylon-6 and (b) nylon-6/6 with transition dipole strengths. Source: Morisawa et al. [9].

The peak wavelengths of the π–π* bands were calculated for a band that was convoluted for several transitions by oscillator strengths and excitation wavelengths calculated using a band width of 0.3 eV. Longer wavelength shifts from monomer to trimer of 171–173 nm for nylon-6 and 171–176 nm for nylon-6/6 were observed. The π–π* band of nylons was due to many transitions with significant oscillator strengths and different excitation wavelengths in the dimer and trimer models owing to hydrogen-bonding effects. The difference in the shift calculated between nylon-6 and -6/6 was approximately 3 nm (0.12 eV), which corresponds to the experimental value of 2 nm (0.06 eV). Figure 6.8 shows the distribution of excitation energies of π–π* transitions with significant intensities. As shown in Figure 6.8, the intense transition at 176.7 nm (7.02 eV) enhanced compared with that of the low-energy side for the trimer model of nylon-6/6. For the nylon-6 trimer model, three intense transitions at 170.8 nm (7.26 eV), 171.7 nm (7.22 eV), and 177.0 nm (7.00 eV) with similar oscillator strengths occurred owing to its crystal structure, which is the inverse structure of the alternate chain. This different transition dipole coupling scheme might have caused the difference in the red shifts of nylon-6 and -6/6.

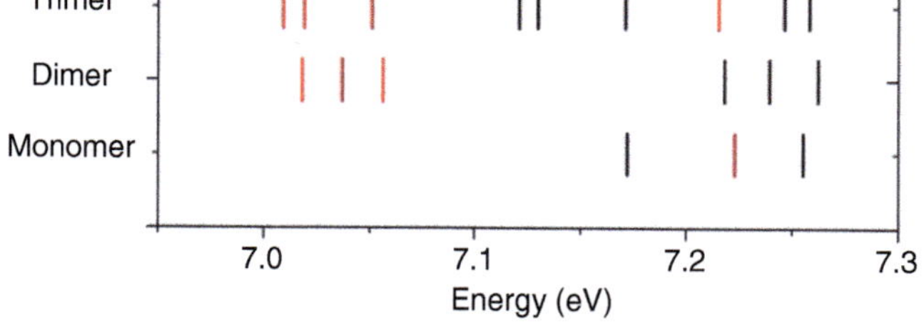

Figure 6.8 Distributions of π–π* transitions with relatively strong ($f \geq 0.1$, red bar) and weak ($f < 0.1$, black bar) intensities calculated using three models of nylon-6 and -6/6. Source: Based on Beć et al. [10].

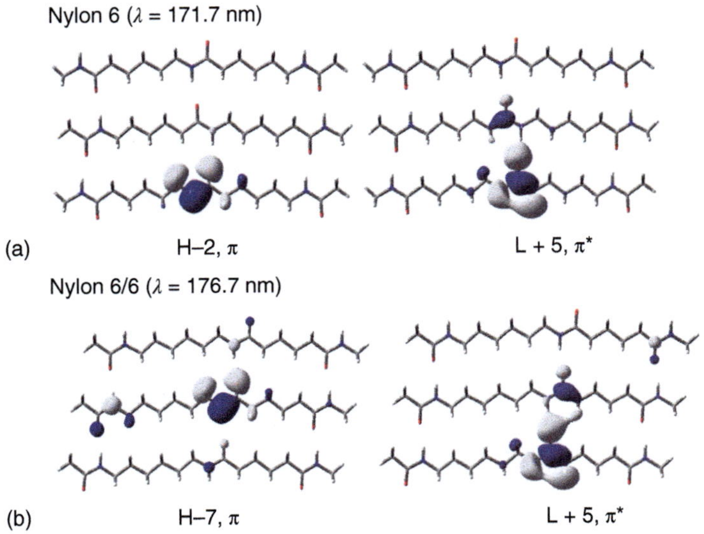

Figure 6.9 Typical MOs relevant to π–π* transition in trimer model of nylon-6 and -6/6 [10]. (a) nylon-6 ($\lambda = 171.7$ nm); (b) nylon-6/6 ($\lambda = 176.7$ nm). Source: Based on Beć et al. [10].

Typical π–π* transition molecular orbitals (MOs) of the trimer models of nylon-6 and -6/6 are shown in Figure 6.9. As shown in Figure 6.9, the first three transitions with relatively large oscillator strengths calculated at approximately 176 nm in the nylon-6/6 model corresponded to the transitions from/to the relatively delocalized MOs. As shown in Figure 6.9b, π and π* appeared in two or more amide groups. The low-energy transitions originated from a linear combination of the delocalized π–π* orbitals. Meanwhile, the valence π orbitals of nylon-6 were well localized in the separated amide group, as shown in Figure 6.9a. In the dimer and trimer of nylon-6, these amide groups generated various π–π* transitions. However, the coupling between the series was weak. Hence, several types of intense π–π* states with different excitation wavelengths appeared in the dimer and trimer of nylon-6.

6.4 ATR–FUV Spectra of Poly(3-hydroxybutyrate) (PHB) and Its Graphene Nanocomposites

Graphene–polymer nanocomposites are garnering broad interest. Recently, nanocomposites of biodegradable polymer poly(3-hydroxybutyrate) (PHB) and graphene nanoparticles were investigated; they exhibited higher melting point (by 10 °C), thermal stability (by 10 °C), tensile strength (by two times) as well as three and two times reduction in oxygen and water vapor permeabilities, respectively [26]. Electronic transitions observed in the FUV region were extremely sensitive to the polymer structure owing to interactions between the polymer chain and nanofillers. Beć et al. obtained new insights regarding high-energy electronic

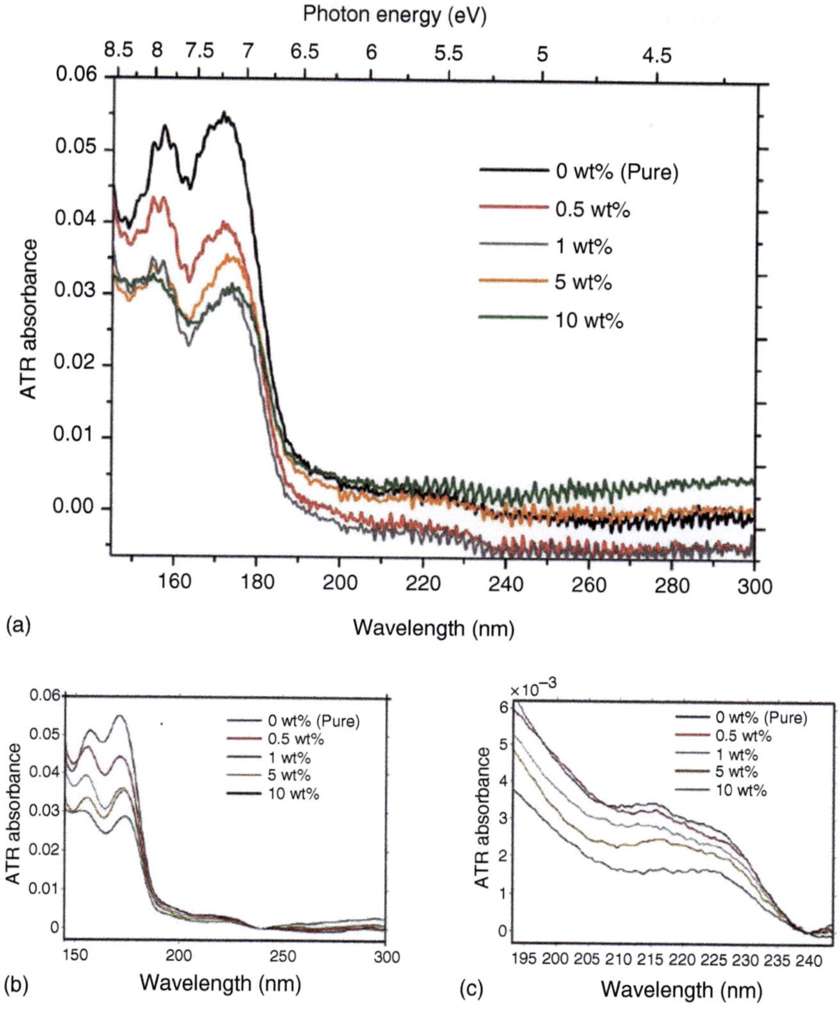

Figure 6.10 Graphene-concentration-dependent ATR–FUV–DUV spectra of PHB and PHB–graphene nanocomposites. Original measured spectra (a); post-processing including smoothing (Savitzky–Golay algorithm, 151 data points/0.01 nm) and vertical offset performed at 240 nm (b); expansion of 190–245 nm region of (b, c). Source: Beć et al. [10].

transitions, in particular Rydberg transitions, and structural information regarding nanocomposites constituting PHB and graphene nanoflakes [10].

The ATR–FUV–DUV spectra of a sample set with gradually varying graphene contents (0.5, 1, 5, and 10 wt%) and pure PHB are shown in Figure 6.10 [10]. In the FUV region, the spectra of the nanocomposites were dominated by the strong bands of the polyester of PHB. Although pure graphene nanoflakes have a weak FUV peak [25], the peak did not appear in the spectra of the nanocomposites. Moreover, the DUV band of the graphene nanoflakes could not be clearly observed. Therefore, PHB bands can be used to elucidate changes in the electronic transitions and the polymer structure upon the formation of a nanocomposite. Hence, the study by Beć et al. [10]

focused primarily on the interpretation of the absorption features in the 145–180 nm region of the PHB polymer.

A domain of strong bands of PHB in the 155–180 nm region (6.89–8.00 eV) was observed in the FUV region, where the peak maximum was 171 nm (7.25 eV). Considering the graphene content dependence of pure PHB, the dominant FUV peak (~171 nm) showed a red shift reaching 2.2 nm for the sample containing 10 wt% of graphene. The shift was accompanied by a significant decrease in the peak intensity (Figure 6.12).

Because of the high crystallinity of the polymer, the main contribution to the FUV spectra of PHB was from the crystalline domains. Therefore, in the TD–DFT calculation, the electronic transitions of PHB to represent the structure of crystalline PHB were first investigated [10]. To perform TD–DFT calculations of the electronic transitions, simplified molecular models representing the structural motif of crystalline PHB are required. Figure 6.11a presents the FUV–DUV spectrum convoluted by applying a Gaussian band profile to the transition energies and oscillator strengths obtained using TD-CAM-B3LYP/aug-cc-pVDZ calculations. The most relevant MOs corresponding to the observed FUV–DUV peaks are shown in Figure 6.11c. The primary contribution to the 162 nm band of PHB is attributable to the transition from an n orbital of the oxygen atoms to the Rydberg 3p orbital (Figure 6.11c), with two primary calculated contributions at 157 and 161 nm. Furthermore, the π and π^* MOs as occupied and unoccupied orbitals, respectively, contributed significantly to the 162 nm transition. The absorption bands in the region around 150 nm were primarily attributed to multiple n–π^* and π–π^* transitions. The n–π^* transitions appeared as minor bands in the region around 220 nm. The two-chain model did not exhibit any significant changes in the simulated spectra. Beć et al. concluded that an elongated model of the polymer chain excluding two adjacent chains improved the consistency between the calculated and experimental spectra. These results confirmed that the FUV–DUV spectra were primarily sensitive to the local structural features of the molecule, and a simple model was sufficient for the study (Figure 6.11). Unlike crystalline PHB (Figure 6.11a), a more sophisticated approach is required to consider the polymer structure relaxed from the confined crystalline state. A systematic method to perform a relaxed scan calculation was performed for the analysis, and the energetically favored structure was subsequently reoptimized using the higher-level method (B3LYP/cc-pVTZ) and then used for the TD–DFT calculations of the electronic transitions (CAM-B3LYP/aug-cc-pVDZ).

This set of calculations improved the interpretation of the alternation in the ATR–FUV–DUV spectra from pure PHB to its nanocomposites with varying graphene flake content (Figure 6.11). The most prominent difference in terms of the simulated FUV–DUV spectra between the relaxed (or "amorphous") and crystalline PHB was the major FUV band calculated at approximately 162 nm in crystalline PHB, i.e. it increased in intensity and blue shifted to 155 nm in the relaxed polymer. Meanwhile, a blue shift in the minor DUV band occurred at 220–209 nm, but its intensity was much lower than that of crystalline PHB. The transitions that contributed significantly to the FUV peak of PHB, either crystalline or relaxed, involved Rydberg states, as presented in Figure 6.11c,d, respectively.

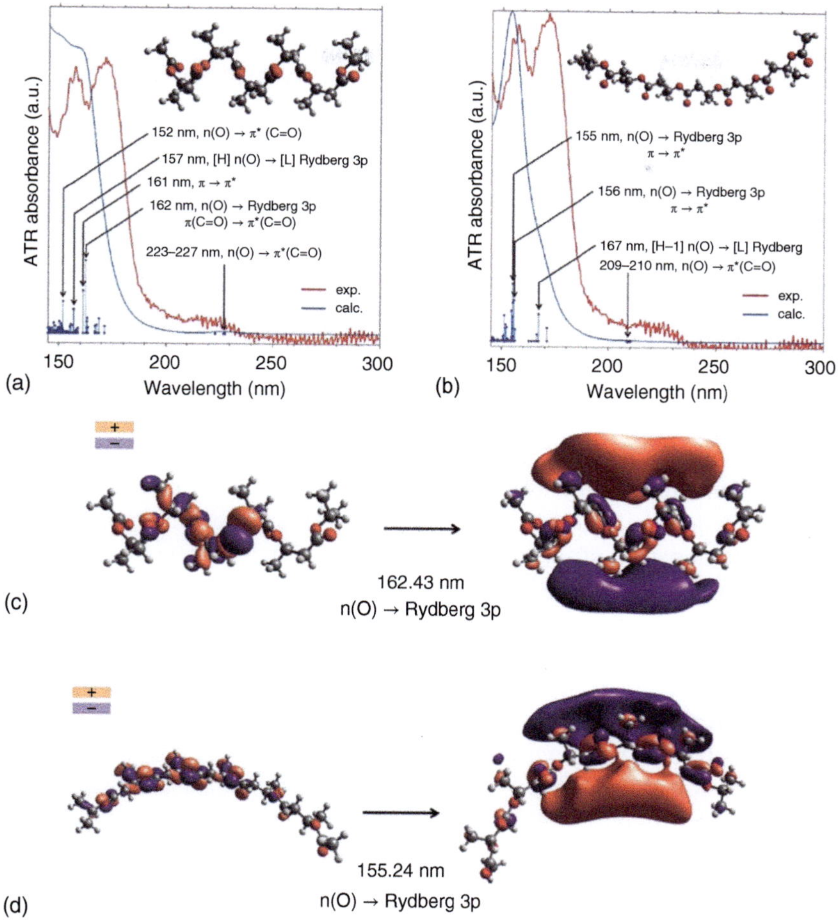

Figure 6.11 Simulated (TD-CAM-B3LYP/aug-cc-pVDZ) FUV–DUV spectra of (a) α-crystalline PHB based on six-unit single-chain model, and (b) model of fully relaxed helix of amorphous PHB (determined by relaxed scan performed on B3LYP/6-31G(d, p) level and reoptimization on B3LYP/cc-pVTZ level), compared with experimental spectrum of PHB. The most relevant MOs for major FUV transition for crystalline (c) and amorphous (d) PHB (TD-CAM-B3LYP/aug-cc-pVDZ) models. Source: Beć et al. [10].

Hence, the primary contributions to the FUV–DUV spectra of PHB and their correspondence with the crystallinity of the polymer were revealed. The crystal structure of PHB had a higher strain than its relaxed structure. The unfolding of the helix significantly affected the environment around the n orbital of oxygen in PHB, where the highest electron density was localized. This resulted in an increase in the energy difference between the ground and Rydberg states, resulting in a blue shift in the FUV band of the disordered PHB (Figure 6.11). Therefore, as concluded by Beć et al. [10], the spectral alternation induced to the local conformation of the polymer by adding graphene increased the strain around the ester fragment in the helix of the crystalline PHB. In other words, the local "crystallinity" of the polymer may have been enhanced because of graphene. The responsiveness of the Rydberg transitions to this effect was extremely high. Therefore, the magnitude of this structural change was not required to be manifested clearly in the ATR–FUV–DUV spectra.

6.5 ATR–FUV Study of Poly(ethylene glycol) (PEG) and Its Complex with Lithium Ion (Li^+)

Because alkali metal complexes, particularly Li^+/PEG complexes, demonstrate characteristic properties such as electronic conductivity, they have been employed for various purposes in electrochemistry, electric devices, and Li-ion batteries. Li^+/PEG complexes derived from a combination of various types of Li salts and ethers have been employed extensively as electrolytes. Because the electronic states of ethers are composed of only single bonds, the electronic structures determining the energies of the ground and excited orbitals in the condensed phase have not been elucidated. Moreover, because alkali metal complexes do not undergo CT transitions, the electronic states of alkali metal complexes have not been investigated. As mentioned above, electronic transitions of single-bond molecules occur in the FUV (120–200 nm) region; therefore, these transitions can be observed using ATR–FUV spectroscopy. Ueno et al. measured the ATR–FUV spectra of poly(ethylene glycol) (PEG) (Figure 6.12) and assigned the electronic transitions of PEG by comparing the experimental and calculated results pertaining to their dependence on the molecular weight [11]. As shown in Figure 6.12, the PEGs demonstrated broad absorption in the 145–200 nm region, with three bands at approximately 155, 163, and 177 nm. The transition was as follows: 155 (*n*(OH)–3p Ryd), 163 (*n*(ether)–3p Ryd), and 177 nm (*n*(ether)–3s Ryd) [11]. Consequently, the n orbitals in PEG can be distinguished between the OH and ether groups in PEG.

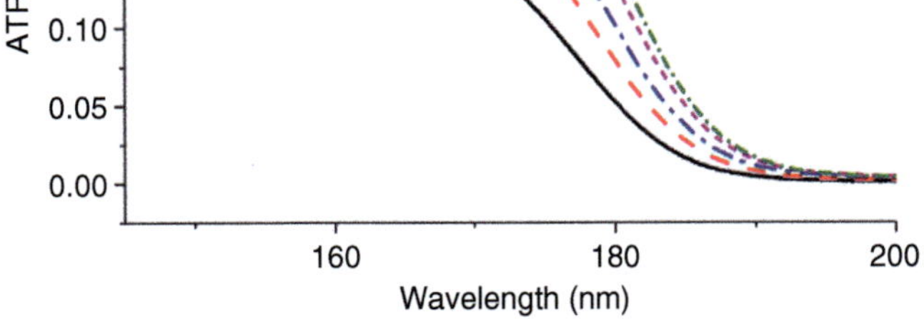

Figure 6.12 ATR–FUV spectra in 145–200 nm wavelength region for diethylene glycol (DG), triethylene glycol (TG), diethylene glycol dimethyl (DGDM), diethylene glycol diethyl (DGDE), and triethylene glycol dimethyl (TGDM) in liquid phase. Source: Ueno et al. [11].

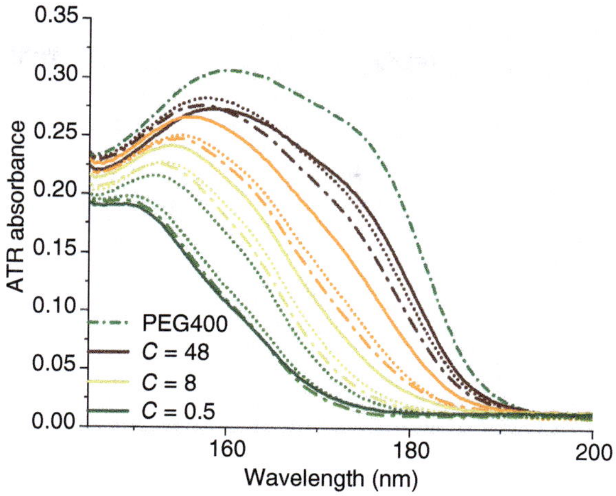

Figure 6.13 Dependence of absorbance on concentration of Li^+ for mixture samples composed of PEG400 and LiTFSI. Source: Ueno et al. [12].

Figure 6.13 shows the Li^+-concentration dependence of the ATR absorbance spectra for mixture samples composed of PEG400 and lithium bis(trifluoromethanesulfonyl)imide (LiTFSI) [12]. The composition of the solution, *C*, denotes the ratio of the ether unit to Li ion, [COC]/[Li^+]; a small value of *C* reflects the high concentration of Li salt in the samples. As shown in Figure 6.14, the ATR absorbance of the solutions decreased with *C* as the concentration of LiTFSI increased. Figure 6.14a,b shows plots of the peak shift and intensity ratio vs. *C*, respectively. On the horizontal axis of these figures, the lowest *C* corresponded to the highest concentration of LiTFSI. Shorter wavelength shifts of 155 and 177 nm increased with the decrease in *C*. The inflection points for the shifts of these bands were at different values of *C*. The 177 nm band showed a significant shift in the high-*C* region ($C > 16$); meanwhile, the 155 nm band began showing a significant shift in the low-*C* region ($C < 8$). The 163 nm band did not indicate any shift in the entire concentration range. As shown in Figure 6.14b, the absorption intensities of all bands decreased with *C*, and the 177 nm band showed the most significant change compared with the others; hence, it was easily changed by coordination.

The value for the composition of the solution, *C*, denotes the ratio of ether units to Li ions, [COC]/[Li^+], and a small value of *C* reflects the high concentration of Li salt in the samples. As shown in Figure 6.14, the ATR absorbance of the solutions decreased with *C* as the concentration of LiTFSI increased. Figure 6.14a,b shows plots of the peak shift and intensity ratio vs. *C*, respectively. On the horizontal axis of these figures, the lowest *C* corresponded to the highest concentration of LiTFSI. Shorter wavelength shifts of 155 and 177 nm increased as *C* decreased, and the inflection points for the shifts of these bands were at different values of *C*. The 177 nm band showed a significant shift in the high-*C* region ($C > 16$); meanwhile, the 155 nm band began showing a significant shift in the low-*C* region ($C < 8$). The 163 nm band did not show a shift in the entire concentration range. As shown in Figure 6.14b, the absorption intensities of all bands decreased with *C*, and this trend was more significant in the 177 nm band compared with those of the others; hence, it was easily changed by coordination.

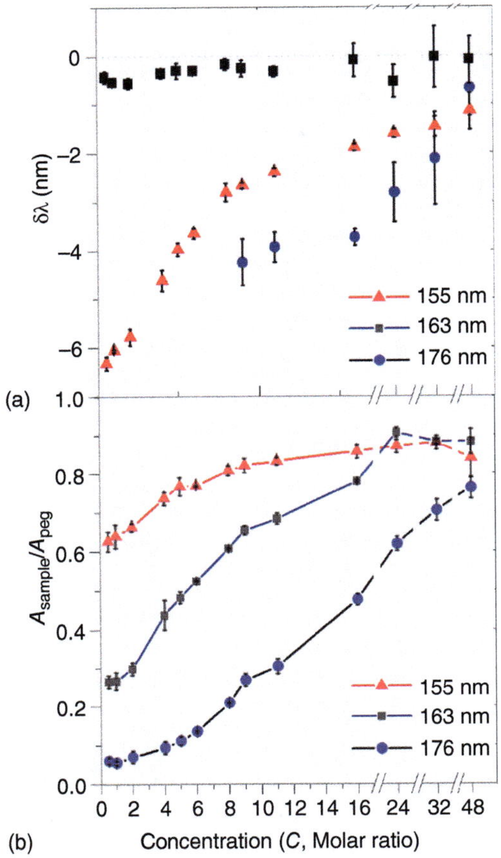

Figure 6.14 (a) Band shift ($\delta\lambda$) of three bands at 155, 163, 177 nm vs. concentration of Li complexes, (b) intensity ratio (sample/neat PEG) of three bands at 155, 163, and 177 nm for binary mixture and PEG vs. concentration. Source: Ueno et al. [12].

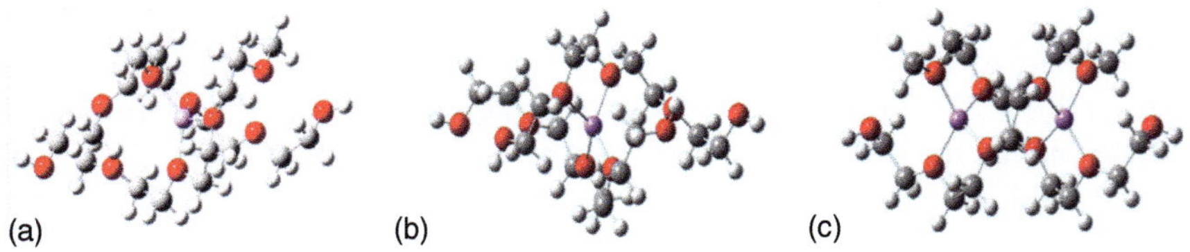

Figure 6.15 Complex structure models of (a) Int 1, (b) Int 2, and (c) Int 3. Source: Ueno et al. [12].

To interpret the experimental variations caused by coordination, a TD–DFT calculation at the level of the CAM-B3LYP/6–311++G (2d,p) was performed for the three models of complexes shown in Figure 6.15 [12]. Those models were intercalation coordination models that can explain the concentration dependences. Int 1 formed coordination between Li^+ ions and two PEGs, each comprising five ether units. The difference between models Int 2 and Int 1 was the end group of the ligands. Int 2 contained OH as the end group of the ligands. In Int 3, two Li^+ ions were present in the ligand of Int 2. The primary difference between these models was the different numbers of noncoordinated (free) ether O atoms that did not coordinate with Li^+,

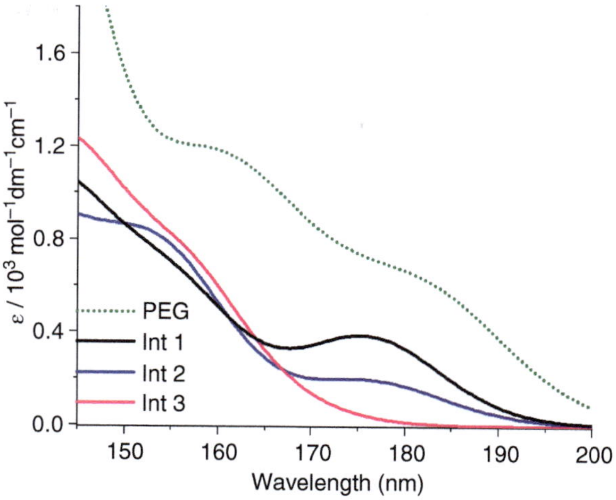

Figure 6.16 Simulation spectra of neat PEG and complexes (Int 1, Int 2, and Int 3) obtained using structure model based on TD–DFT. Source: Ueno et al. [12].

although the total number of O atoms among the models did not change. For Int 1, Int 2, and Int 3, the numbers of ethers without coordination was 4, 3, and 0, respectively. Figure 6.16 shows the simulation spectra of the three models above and pure PEG. Spectral differences between PEG and the Li^+/PEG complex and concentration dependence on absorbance were reproduced in the simulation spectra. The origin of the intensity decreased at approximately 177 nm as C decreased, as shown clearly from the quantum chemical calculations. In the region of the 177 nm band, only one transition occurred, i.e. from the uncoordinated (free) O atom in the ether. We concluded that the absorbance of the 177 nm band depended on the number of free ether O atoms in the solution. In the low C range ($8 > C$), the 155 nm band showed a blue shift. We assigned the band in the 145–155 nm region to the alkali metal complex because we observed a new band at approximately 150 nm in the ATR–FUV spectra in the high-C region. Furthermore, the simulation spectra of the complexes showed a new band at this wavelength. It was speculated that the blue shift around 155 nm was due to the appearance of a new band, and that this was likely caused by the formation of coordinate bonds. This result proves the occurrence of transitions from alkali metal complexes.

Based on the finding above regarding the 177 nm band, we estimated the coordination number of the PEG/Li^+ complex in the solution using ATR–FUV spectra [13]. Although the coordination number in a crystal can be obtained by X-ray structure analysis [27, 28], only limited information is available on the coordination number of Li^+ cations in solutions with different solvents using Raman spectroscopy [29, 30]. The following coordination reaction was assumed to model the change in ATR absorption observed in the spectra shown in Figure 6.14:

$$Li^+ + n(\text{—COC—}) \rightarrow Li(\text{—COC—})n^+, \tag{6.2}$$

where n is the coordination number of the Li-COC complexes. In this study, we set the initial concentrations of LiTFSI and COC-units as $[Li^+]_0$ and $[COC]_0$, respectively. Because of the strong Lewis acidity of Li^+, all the Li^+ ions in the samples formed complexes in the low-concentration regions of LiTFSI ($C = 47 - 16$). In this case, the concentration of uncoordinated COC ([COC]) is expressed as shown in (6.3).

$$[\mathrm{COC}] = [\mathrm{COC}]_0 - n[\mathrm{Li}^+]_0 \tag{6.3}$$

The ATR absorption band at 177 nm was due to the transition from uncoordinated COC, as mentioned above. Therefore, the ratio of uncoordinated COC in the total COC ($[\mathrm{COC}]/[\mathrm{COC}]_0$) can be estimated by the ratio of ATR absorption at 177 nm, i.e. $A_{177}^{\mathrm{sol}}/A_{177}^{0}$. Hence, we derived the following formula:

$$\frac{A_{177}^{\mathrm{sol}}}{A_{177}^{0}} = 1 - n\frac{[\mathrm{Li}^+]_0}{[\mathrm{COC}]_0} = 1 - \frac{n}{C} \tag{6.4}$$

A_{177}^{0} was calculated by multiplying the ATR absorption of the pure solvent by the ratio of the concentration of solvent molecules in the solution to the pure ligand. Figure 6.17 shows plots of the ratio $A_{177}^{\mathrm{sol}}/A_{177}^{0}$ vs. $[\mathrm{Li}^+]_0/[\mathrm{COC}]_0$ for solutions of ethylene glycol dimethyl (G1), diethylene glycol dimethyl (G2), triethylene glycol dimethyl (G3), and PEG400. Based on Eq. (6.3), the coordination number can be derived from the slope of the plots, and the intercept should be 1. The slopes and intercepts are listed in Table 6.1. As shown in Table 6.1, the intercept value of the observed solutions ranged between 1.0 and 0.9, and the data in Figure 6.17 corresponded well to Eq. (6.4). The average coordination number of each solution was

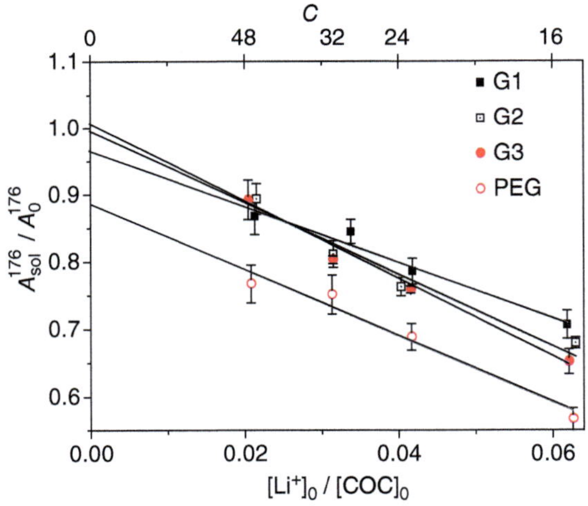

Figure 6.17 Plots of intensity ratio ($\log([A]/[A_0])$) vs. concentration ratio ($\log([\mathrm{Li}^+]/[\mathrm{COC}]_0)$) using ATR–FUV absorption intensity at 177 nm. Source: Modified from Ueno et al. [13].

Table 6.1 Values of slope (coordination numbers) and intercept with error ranges for each ligand, based on Eq. (6.3).

	G1	G2	G3	PEG400
Slope (n)	4.1 ± 0.5	5.3 ± 0.8	5.8 ± 0.3	4.9 ± 0.7
Intercept	0.96 ± 0.02	0.99 ± 0.03	1.01 ± 0.01	0.89 ± 0.03

determined to be the following: G1 = 4, G2 = 5, G3 = 6, and PEG400 = 5. Consequently, the coordination numbers in the solutions estimated from the ATR–FUV spectra corresponded to the coordination numbers in the crystal [27, 28].

6.6 Summary

In this chapter, we provided an overview of recent studies using ATR–FUV spectroscopy applied to polymers that exhibit characteristic electronic transitions in the FUV region. The effectiveness of this spectroscopy was demonstrated for investigating the electronic structures and transitions in the FUV energy region, i.e. functions not afforded by other spectroscopies. Using this spectroscopy with the aid of quantum chemical calculations, we extracted interesting physics and chemistry in polymers, e.g. the existence of Rydberg excited states in polymers; strong valence–Rydberg coupling; effects of intermolecular interactions, such as dispersion force, hydrogen bonding, and coordinate bonds with alkali metal ions on the excited states; and polarization of the surroundings. These interesting phenomena have been systematically investigated via a series of studies pertaining to nylon, polymer composites, and complex polymers with alkali ions. These novel chemical and physical insights with robust experimental techniques enable future developments in the field of chemistry, particularly for investigating the electronic structure and transition characteristics of the FUV region.

References

1 Millikan, R.A. (1916). A direct photoelectric determination of Planck's "h". *Phys. Rev.* 7: 0355.

2 Reontgen, W.C. (1945). Radiology, on a new kind of rays. *Radiology* 45: 428.

3 Mulliken, R.S. and Chem, H. (1935). Electronic structures of polyatomic molecules. VII. Ammonia and water type molecules and their derivatives. *J. Chem. Phys.* 3: 506.

4 Robin, M.B. (1974). *Higher Excited States of Polyatomic Molecules.* New York/London: Academic Press.

5 Damany, N., Romand, J., and Vodar, B. (eds.) (1974). *Some Aspects of Vacuum Ultraviolet Radiation Physcis.* Oxford: Pergamon Press.

6 Sándorfy, C. (ed.) (1999). *The Role of Rydberg States in Spectroscopy and Photochemistry-Low and High Rydberg States.* Dordrecht: Kluwer Academic Publishers.

7 Ozaki, Y. and Kawata, S. (2015). *Far- and Deep-Ultraviolet Spectroscopy.* Springer.

8 Ozaki, Y., Morisawa, Y., Tanabe, I., and Beć, K.B. (2021). *Spectrachim. Acta A* 253: 119549–119549.

9 Morisawa, Y., Yasunaga, M., Sato, H. et al. (2014). *J. Phys. Chem. B* 118: 11855–11861.

10 Beć, K.B., Morisawa, Y., Kobashi, K. et al. (2018). *Phys. Chem. Chem. Phys.* 20: 8859–8873.
11 Ueno, N., Wakabayashi, T., and Morisawa, Y. (2018). *Spectrochim. Acta A* 197: 170–175.
12 Ueno, N., Wakabayashi, T., Sato, H., and Morisawa, Y. (2019). *J. Phys. Chem. A* 123: 10746–10756.
13 Ueno, N., Wakabayashi, T., and Morisawa, Y. (2020). *Anal. Sci.* 36: 723–729.
14 Higashi, N., Ikehata, A., and Ozaki, Y. (2007). *Rev. Sci. Instrum.* 78: 103107.
15 Ozaki, Y., Morisawa, Y., Ikehata, A., and Higashi, N. (2012). *Appl. Spectrosc.* 66: 1–25.
16 Ikehata, A., Higashi, N., and Ozaki, Y. (2008). *J. Chem. Phys.* 129: 234510.
17 Ikehata, A., Mitsuoka, M., Morisawa, Y. et al. (2010). *J. Phys. Chem. A* 114: 8319–8322.
18 Ikehata, A., Goto, T., and Morisawa, Y. (2017). *Appl. Spectrosc.* 71: 1530–1536.
19 Morisawa, Y., Tachibana, S., Ehara, M., and Ozaki, Y. (2012). *J. Phys. Chem. A* 116: 11957–11964.
20 Tanabe, I., Kurawaki, Y., Morsawa, Y., and Ozaki, Y. (2016). *Phys. Chem. Chem. Phys.* 18: 22526–22530.
21 Onari, S. (1970). *Jpn. J. Appl. Phys.* 9: 227.
22 Serrano-Andres, L. and Fulscher, M.P. (1998). *J. Am. Chem. Soc.* 120: 10912–10920.
23 Goto, T., Morisawa, Y., Higashi, N., and Ozaki, Y. (2013). *Anal. Chem.* 85: 4500–4506.
24 Morisawa, Y., Yasunaga, M., Fukuda, R. et al. (2013). *J. Chem. Phys.* 139: 154301.
25 Tachibana, S., Morisawa, Y., Ikehata, A. et al. (2011). *Appl. Spectrosc.* 65: 221–226.
26 Manikandan, N.A., Pakshirajan, K., and Pugazhenthi, G. (2020). *Int. J. Biol. Macromol.* 154: 866–877.
27 Frech, R., Rhodes, C.P., and Khan, M.A. (2002). *Macromol. Symp.* 186: 41.
28 Henderson, W., McKenna, F., Khan, M.A. et al. (2005). *Chem. Mater.* 17: 2284.
29 Oldiges, K., Diddens, D., Ebrahiminia, M. et al. (2018). *Phys. Chem. Chem. Phys.* 20: 16579.
30 Seo, D.M., Boyle, P.D., Sommer, R.D. et al. (2014). *J. Phys. Chem. B* 118: 13601.

7

Synchrotron-Based UV Resonance Raman Spectroscopy for Polymer Characterization

Barbara Rossi[1], Mariagrazia Tortora[1,2], Sara Catalini[3], Alessandro Gessini[1], and Claudio Masciovecchio[1]

[1]*Elettra Sincrotrone Trieste, S.S. 114 km 163.5, Basovizza, Trieste 34149, Italy*
[2]*Area Science Park, Padriciano, 99, 34149 Trieste, Italy*
[3]*European Laboratory for Non-Linear Spectroscopy, LENS, Via Nello Carrara, 1, 50019, Sesto Fiorentino, Firenze, Italy*

7.1 Basic Principles of Raman and UV Resonance Raman Spectroscopy

7.1.1 Molecular Vibrations and Raman Effect

Spectroscopy is a scientific branch that is able to provide information about the chemical–physical properties of the matter by means of interaction with the light. Depending on the molecular transitions that are probed, it is possible to classify the different spectroscopic techniques into three broad groups: rotational, vibrational, and electronic.

The classical vibrational spectroscopies used to investigate molecular vibrations are the Fourier transform infrared spectroscopy (FTIR) and the Raman spectroscopy. Both these techniques probe the vibrational levels of the fundamental electronic state of the molecules. The difference lies in the processes at the basis of the two spectroscopies, i.e. absorption in the case of FTIR and scattering for Raman technique. It follows that the selection rules that make a vibrational motion infrared or Raman active are different. Depending on the molecular symmetry, some vibrations give very intense infrared instead of Raman signals and vice versa.

To give a classical description of the molecular vibration in a simple molecule, we have to consider the relative displacement of the atomic nuclei with respect to the center of mass. The simplest model to describe this motion is that of a harmonic oscillator (Figure 7.1).

A molecular vibration involves the displacement of the atomic nuclei that oscillate around their equilibrium position with deformations of bond angles and bond lengths. Due to the chemical bonds, the nuclei are brought back to the equilibrium position by a force that can be described by Hook's law:

$$F = -K\overline{\mathbf{x}} \tag{7.1}$$

Spectroscopic Techniques for Polymer Characterization: Methods, Instrumentation, Applications, First Edition. Edited by Yukihiro Ozaki and Harumi Sato.

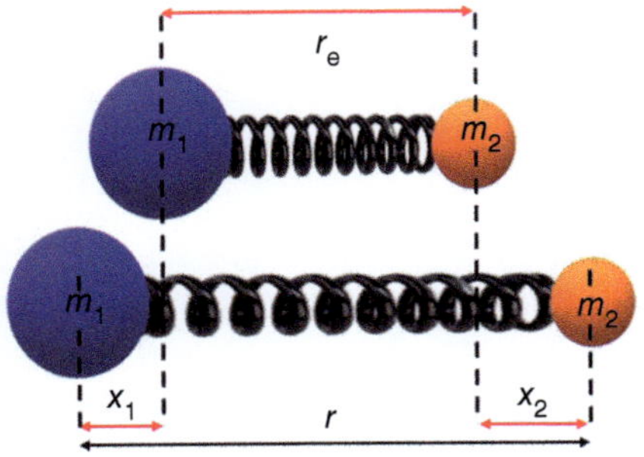

Figure 7.1 Graphical representation of the molecular vibration in terms of the harmonic oscillator for a simple molecule composed of two atoms with masses m_1 and m_2.

where F is the restoring force that acts in the opposite direction to the nuclei motion, K is the harmonic force constant, and $\mathbf{x}$ is the vector that describes the motion of the nucleus.

The vibrational motion corresponds to the quantification of the amplitude of the oscillation in a system where the oscillations are constrained. We can refer to classical mechanics and, in particular, to the Lagrange one to describe this effect (Figure 7.1).

By exploiting the principle of energy conservation, it is possible to relate kinetic (T) and potential (V) energy to each other through the following equation:

$$\frac{\partial}{\partial t}\left(\frac{\partial T}{\partial \dot{x}_{\mathrm{i}}}\right) + \left(\frac{\partial V}{\partial x_{\mathrm{i}}}\right) = 0 \tag{7.2}$$

Since the total energy is redistributed between kinetic and potential contributions:

$$\begin{cases} T = \frac{1}{2} m_1 \dot{x}_1^2 + m_2 \dot{x}_2^2 \\ V = \frac{1}{2} K (x_2 - x_1)^2 \end{cases} \tag{7.3}$$

the equations for the two coordinates x_1 and x_2 that represent the movement of the two mass m_1 and m_2 from the equilibrium position r_e can be written as:

$$\begin{cases} m_1 \ddot{x}_1 - K(x_2 - x_1) = 0 \\ m_2 \ddot{x}_2 - K(x_2 - x_1) = 0 \end{cases} \tag{7.4}$$

Equation (7.4) admits solutions for the coordinate x:

$$\begin{cases} x_1 = A_1 \cos(2\pi v t + \varphi) \\ x_2 = A_2 \cos(2\pi v t + \varphi) \end{cases} \tag{7.5}$$

When $m_1 \neq m_2$, the movement (x_1 and x_2) and the amplitude (A_1 and A_2) in Eq. (7.5) are different. Since the chemical bond is the same, so are the phase (φ) and frequency (v); therefore, the cosine argument is equal in both equations.

In Eq. (7.5), the second derivative of x can be written as:

$$\ddot{x} = -A \cos(2\pi v t + \varphi)(2\pi v)^2 \tag{7.6}$$

Replacing $\ddot{x}$ in Eq. (7.4) and solving the system, we find two solutions.

The first defines a translational motion and is not taken into consideration, while the second relates to the characteristic oscillation frequency v to the chemical bond strength constant (k) and the reduced mass of the system (μ).

$$v = \frac{1}{2\pi}\sqrt{\frac{k}{\mu}} \tag{7.7}$$

Figure 7.2 Representation of the quantized vibrational energy: vibrational eigenstates associated to the vibrational quantum number.

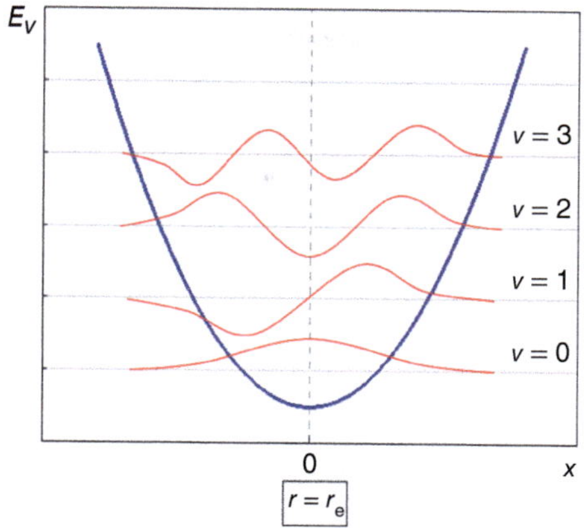

By graphing the value of potential energy as a function of the shift from the equilibrium position (r_e), a parabolic trend is produced, with the origin of the parabola at $x = 0$, i.e. in $r = r_e$.

The limit of the classical model is to predict continuous energy values as a function of the displacements (x), restraining from a discrete spectrum. Quantum mechanics helps in overcoming this drawback. Solving the Schrödinger equation for the nuclear coordinates, the vibrational energy eigenstates and eigenvalues can be found. Equation (7.8) defines the vibrational energy values (eigenvalues) connected with each vibrational function (eigenstates):

$$E_v = \left(v + \frac{1}{2}\right) h\nu \tag{7.8}$$

The energy associated with the oscillation motion depends on the vibrational frequency (ν) and the vibrational quantum number (v) (see Figure 7.2). The latter can assume all integer values ≥0.

The harmonic oscillator model provides a series of levels in which the vibrational ground state ($v = 0$) has nonzero energy ($E_v = \frac{1}{2}h\nu$), and the gap between each level is invariably equal to ($h\nu$).

The general selection rule in vibrational absorption spectroscopy allows transitions between different vibrational levels only when the molecular vibrational motion induces a variation of the dipole moment of the molecule $\left(\frac{\partial \mu}{\partial x}\right)_{x=0} \neq 0$. For instance, the vibration mode of homonuclear diatomic molecules is not infrared active. The special selection rule implies that it is possible to have transitions just between contiguous vibrational levels ($\Delta v \pm 1$).

Molecules do not always behave as harmonic oscillators, especially during vibration, the two atoms are close together with a prevalence of the repulsion forces, or when they are very far from each other, and the recall force cannot result in the chemical bond dissociation. For this reason, it is necessary to correct the potential energy (7.8) introducing the anharmonicity term. A Taylor's series expansion can illustrate the anharmonic potential, giving the following expression for the vibrational energy:

$$E_v = \left(v + \frac{1}{2}\right) h\nu - \left(v + \frac{1}{2}\right)^2 \chi\, h\nu + \ldots \tag{7.9}$$

Table 7.1 Schematic representation of the main molecular vibrational motions.

	Symmetric stretching	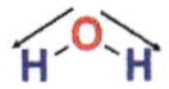
Variations of bond lengths	Asymmetric stretching	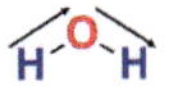
In-plane variation of bond angles	Deformation	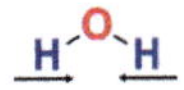
	Rocking	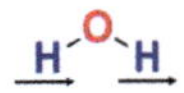
Out-of-plane variation of bond angles	Wagging	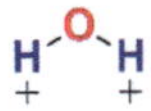
	Twisting	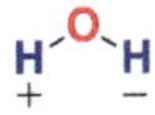

The χ term is the anharmonic constant ($\sim 10^{-3}$) and becomes more relevant as the vibrational quantum number rises. Due to anharmonicity, the interval between levels decreases with increasing v. The introduction of anharmonicity modifies the expression of the vibrational energy and even the selection rules. Consequently, it is possible to have weak transitions also for $\Delta v \pm 2, \pm 3, \pm 4$, etc. known as overtone bands.

A coordinate system integral to the geometry of the molecule, namely internal coordinates, can conveniently disentangle the description of the vibrational motion. The linear combination of the internal coordinates represents the normal coordinates that enable simple expressions for kinetic and potential energy and are used as reference systems in vibrational spectroscopy. A normal coordinate depicts a normal mode of vibration, while complex vibrational motions can be described as a combination of several normal modes. The modes of vibration are reported as deformations from the equilibrium structure-like variations of bond lengths and bond angles, as summarized in Table 7.1.

As a first attempt to simplify the interpretation of the vibrational spectrum, we can consider the molecular groups frequencies constant. This approximation is valid in two cases:

1. When we examine the displacement of atoms that are lighter than those to which they are bound, which can be considered stationary ($m_1 \gg m_2$) and the oscillation frequency only depends on the mass of the light atoms. The vibrational frequencies of groups such as CH, NH, OH, and SH always fall in the same spectral region.
2. In the case of very rigid bonds, whose constant force is very different from the rest of the molecule (e.g. $C\equiv C$, $C\equiv N$, and $C=O$). Table 7.2 summarizes the characteristic frequencies of vibrations of molecular groups for various chemical compounds.

Molecular vibrations are very sensitive to the molecules structural variation. As an example, in Figure 7.3, we report the infrared absorption spectra of formamide and *N,N*-dimethylformamide, two simple molecules that have a similar structure.

Table 7.2 Characteristic vibrational frequencies of some molecular groups for different chemical compounds [1].

Wavenumber (cm^{-1})	Assignment
Alkanes	
2960	Methyl symmetric C — H stretching
2930	Methylene asymmetric C — H stretching
2870	Methyl asymmetric C — H stretching
2850	Methylene symmetric C — H stretching
1470	Methyl asymmetrical C — H bending
1465	Methylene scissoring
1380	Methyl symmetrical C — H bending
1305	Methylene wagging
1300	Methylene twisting
720	Methylene rocking
Alkenes	
3100 to 3000	=C — H stretching
1680 to 1600	C = C stretching
1400	=C — H in-plane bending
1000 to 600	=C — H out-of-plane bending
Alkynes	
3300 to 3250	≡C — H stretching
2260 to 2100	C ≡ C stretching
700 to 600	≡C — H bending
Amines	
3335	N — H stretching
2780	N — CH_2 stretching
1615	NH_2 scissoring, N — H bending
1360 to 1250	Aromatic C — N stretching
1220 to 1020	Aliphatic C — N stretching
850 to 750	NH_2 wagging and twisting
715	N — H wagging
Alcohol and phenols	
3600	Alcohol O — H stretching
3550 to 3500	Phenol O — H stretching
1300 to 1000	C — O stretching

(continued)

Table 7.2 (Continued)

Wavenumber (cm^{-1})	Assignment
Ethers	
1100	C—O—C stretching
Aldehydes and ketones	
2900 to 2700	Aldehyde C—H stretching
1740 to 1720	Aliphatic aldehyde C=O stretching
1730 to 1700	Aliphatic ketone C=O stretching
1720 to 1680	Aromatic aldehyde C=O stretching
1700 to 1680	Aromatic ketone C=O stretching
Esters	
1750 to 1730	Aliphatic C=O stretching
1730 to 1705	Aromatic C=O stretching
1310 to 1250	Aromatic C—O stretching
1300 to 1100	Aliphatic C—O stretching
Carboxylic acids	
3300 to 2500	O—H stretching
1700	C=O stretching
1430	C—O—H in-plane bending
1240	C—O stretching
930	C—O—H out-of-plane bending
Anhydrides	
1840 to 1800	C=O stretching
1780 to 1740	C=O stretching
1300 to 1100	C—O stretching

Source: Based on Lin-Vien et al. [1].

In both spectra, vibrational signals attributable to vibrations involving CH and CO groups can be recognized. Compared to formamide, *N,N*-dimethylformamide has two CH_3 linked to nitrogen that replace the hydrogens, so its spectrum shows also signals associated with CH_3 contributions but not related to NH group.

The information resulting from the interaction between infrared radiation and the molecules permanent dipole moment is like that obtained by studying the effect of electromagnetic radiation on the induced dipole moment ($\overline{\mu}$). In this case, we refer to vibrational Raman spectroscopy, where the electromagnetic radiation brings the system into a virtual state that generates the induced dipole moment, responsible for the light scattering phenomenon. The expression of the induced electric dipole moment can be written as:

$$\overline{\mu} = \alpha\overline{E} \tag{7.10}$$

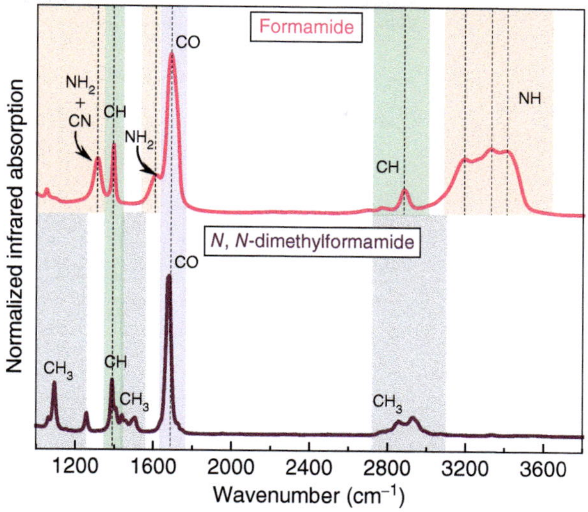

Figure 7.3 Infrared absorption spectra of formamide and *N,N*-dimethylformamide with the assignment of the main vibrational signals. Source: Refs. [2, 3].

where $\overline{E}$ is the incident electric vector of the radiation and α is the molecular polarizability. The latter quantity is a tensor that describes the molecule's tendency to deform its electronic cloud.

The variation of the elements in the dynamic polarizability tensor ($\alpha_{\rho\sigma}$) and the vibrational coordinates (Q_k) can be expressed in a Taylor series expansion:

$$\alpha_{\rho\sigma} = \alpha(0)_{\rho\sigma} + \sum_{k}\left(\frac{\partial \alpha_{\rho\sigma}}{\partial Q_k}\right)_0 Q_k + \frac{1}{2}\sum_{k,l}\left(\frac{\partial^2 \alpha_{\rho\sigma}}{\partial Q_k \partial Q_l}\right)_0 Q_k Q_l + \ldots \tag{7.11}$$

Considering (7.10, 7.11), each component of the induced dipole moment can be defined by the following equation:

$$\mu_\rho(Q_k) = \sum_{\sigma=x}^{z} \alpha(0)_{\rho\sigma}\, E(0)_\sigma \cos 2\pi \upsilon_0 t + \frac{1}{2}\sum_{\sigma=x}^{z} E(0)_\sigma Q(0)_k \left(\frac{\partial \alpha_{\rho\sigma}}{\partial Q_k}\right)_0 [\cos 2\pi t(\upsilon_0 + \upsilon) + \cos 2\pi t(\upsilon_0 - \upsilon)] \tag{7.12}$$

The first term of the Eq. (7.12) represents the elastic collision without energy transfer called Rayleigh (or elastic) scattering. Less efficient collisions with energy transfer could also occur and are mathematically formalized by the second term of (7.12). These photon–molecule inelastic collisions can transfer and subtract the energy necessary to populate high (Stokes Raman scattering) or low (anti-Stokes Raman scattering) energy vibrational levels. Figure 7.4 schematizes the virtual energy diagram of the elastic and inelastic scattering processes.

The intensity of the Rayleigh component is usually 10^{-6} times lower than the exciting radiation, while the Stokes components are 10^{-10} lower. The anti-Stokes bands are even less intense as they originate from the vibrational transitions of excited energy levels. At room temperature, the population of these levels is negligible compared to that of the ground state. Indeed, usually the vibrational Raman spectrum registers just Stokes events. The intensity of the scattered light is reported as a function of the frequency difference between the incident and the scattered radiation.

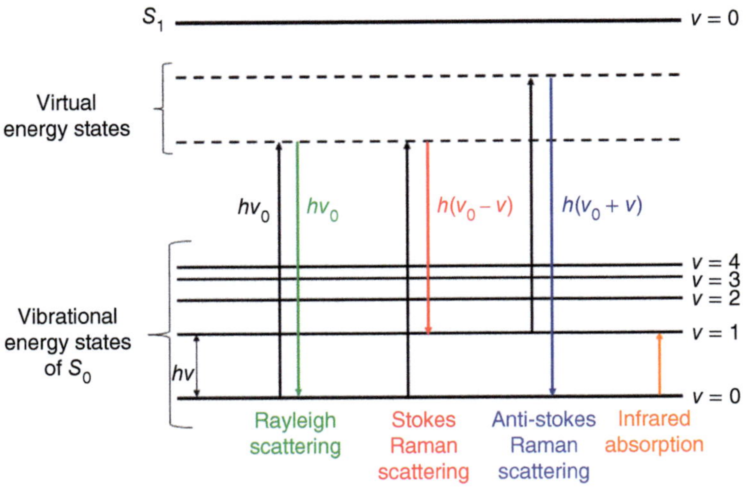

Figure 7.4 Virtual energy diagram representing the elastic and the two inelastic scattering processes, Stokes and anti-Stokes Raman scattering.

A Raman active mode causes a change in the molecular polarizability (general selection rule) and affects a transition between contiguous vibrational levels (specific selection rule). The vibrational motion can modify the molecular polarizability producing a variation of the molecular electronic cloud volume (total-symmetric modes) or changing its shape (non-total-symmetric modes).

Depending on the symmetry of the molecule and of the vibrational motion, it is common practice to employ infrared absorption and Raman scattering techniques in conjunction. Complementarity is due to the different physical quantity used to observe each phenomenon, i.e. the variation of the permanent dipole in infrared (IR) absorption and of the induced dipole (molecular polarizability) in Raman scattering. Molecules with very low or no symmetry have very similar IR and Raman spectra. The more symmetric is the molecule, the smaller is the number of active modes leading to less comparable IR and Raman spectra. Hence, the mutual exclusion rule highlights that molecules with a center of symmetry have total-symmetric vibrations active only in the Raman spectrum, while non-total-symmetric vibrations are visible only in the IR one.

Figure 7.5 shows the comparison between the Raman spectra of formamide and *N,N*-dimethylformamide collected using visible excitation radiation at 633 nm.

Since formamide and *N,N*-dimethylformamide molecules have low symmmetry, IR (see Figure 7.3) and Raman (see Figure 7.5) spectra display similar spectral features at specific frequencies of vibration. However, the signals exhibit different relative intensities in IR and Raman spectra: for example, the band attributable to CO stretching vibration is very intense in the infrared absorption spectrum, but not in the Raman one. On the other hand, the signals related to CH and CH_3 are much more pronounced in the Raman spectrum.

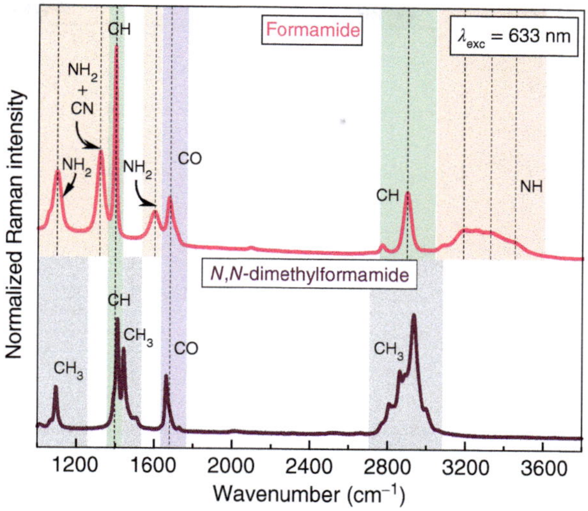

Figure 7.5 Raman spectra of formamide and *N,N*-dimethylformamide obtained using visible excitation radiation at 633 nm and assignment of the main experimental signals to vibrations involving specific groups of the molecules. Source: Based on Park et al. [4].

7.1.2 Resonance Raman (RR) Scattering

In the previous section, we have explained the Raman effect as the induction of a dipole moment in the sample through the interaction with electromagnetic radiation. The displacement of the electronic charge oscillates with the electromagnetic field at the frequency of the incident radiation (ν_0). Even the nuclei vibrational movements cause an oscillation of the induced dipole moment, with characteristic frequencies for the molecular vibrational motions (ν_v). The Raman scattering results from the coupling between the oscillating electronic cloud at ν_0 frequency and that induced by the vibration at ν_ν ($\nu_0 \pm \nu_\nu$). The molecule's electron cloud produces natural oscillation frequencies, which generate the electronic absorption bands (UV-visible absorption spectra). When the energy of the radiation wavelength used to investigate the molecule matches with its natural oscillation frequencies, the Raman experiment is performed in resonance with the electronic transition. Figure 7.6 illustrates the main differences between the mechanism of Raman and resonance Raman scattering. Resonance condition involves an increase in the charge displacement during the oscillation and a rise in the induced dipole moment. Since latter is responsible for the Raman effect, the Raman scattering efficiency enhances approximately 10^8 times more than in conventional Raman.

The resonance effect provides great sensitivity (sample concentrations lower than 10^{-8} M can be detected) and selectivity to the Raman scattering technique. We can observe significant differences in the Raman spectra collected using resonance or non-resonance excitation wavelengths. The extreme selectivity of the resonance Raman derives from various factors. First, the analytes that compose the sample or the specific segments of a molecule (chromophores) can have distinct absorption bands. By changing the excitation wavelength, it is possible to match the electronic transition of each analyte or molecule's portion, increasing their Raman signals by up to 10^8 times. Second, several absorption bands may be associated with

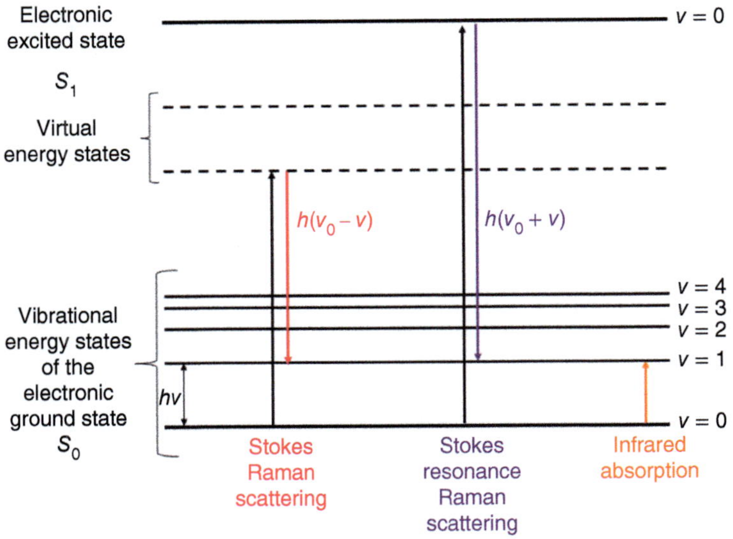

Figure 7.6 Virtual energy diagram representing the Raman and resonance Raman scattering processes.

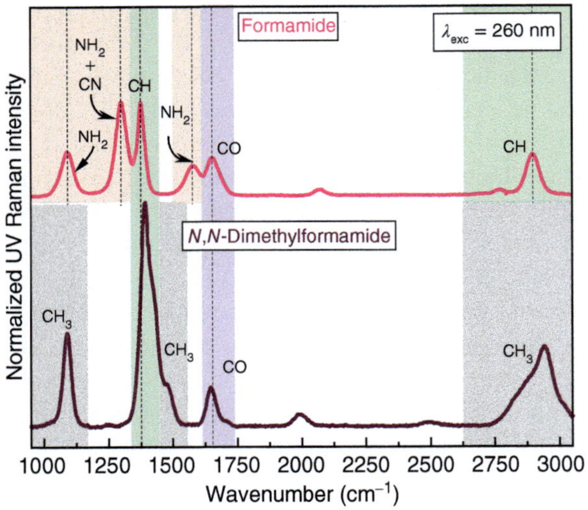

Figure 7.7 Resonance Raman spectra of formamide and *N,N*-dimethylformamide collected using 260 nm as excitation wavelength with the relative assignment of the main vibrational signals. Source: Based on Hildebrandt et al. [6].

the same molecular chromophore. Here, by tuning the excitation wavelength, the excited electronic states change and the resonance Raman spectra will be different [5]. Figure 7.7 shows the resonance Raman spectra of formamide and *N,N*-dimethylformamide collected using ultraviolet excitation radiation at 260 nm.

Comparing IR, Raman, and resonance Raman spectra in Figures 7.3, 7.5, and 7.7, we can notice that the signals attributable to carbonyl group vibrations are at the same frequency in the spectra recorded by using the different techniques, but the relative intensities of these bands change. Raman intensities of $C = 0$ bands depend mainly on resonance effects. Signals enhancement occurs when the Raman excitation wavelength approaches the specific electronic transition frequency of formamide and *N,N*-dimethylformamide that lies in the deep-ultraviolet range [7].

For instance, we observe in Figure 7.7 that C–N vibrational modes are particularly intense in resonance Raman than in Raman spectra because of the increased proximity of excitation wavelength at 260 nm to the $\pi \rightarrow \pi^*$ transition of C–N bond that takes place in the 150–180 nm range. Interestingly, the enhancement effect occurs even in pre-resonance conditions [8].

7.1.3 Fundamental Applications of UV Resonance Raman Spectroscopy

The ability of the Resonance Raman spectroscopic technique to study the vibrational behavior of particular analyte within a complex mixture makes it a powerful technique for analytical applications, especially in the biochemical field [9]. Due to the fact that many electronic transitions associated to several organic functional groups lie in the wavenumber ultraviolet (UV) region, the UV resonance Raman (UVRR) spectroscopy is becoming particularly popular for the study of organic and biological samples. Early in the 1990s, UVRR spectroscopy with excitation wavelengths from 229 to 280 nm had been used to obtain tryptophan and tyrosine vibrational signatures in hemoglobin [10], while ~200 nm laser had proven to resonantly enhance Raman scattering from the amide chromophore, a building block of a polypeptide backbone [11]. Biological macromolecules such as proteins and DNA contain different chromophores in their structure. In the study of peptide of proteins, by appropriately choosing the excitation wavelength, it is possible to distinguish the signals coming from the aromatic amino acids (230 nm) rather than those relating to the peptide bond (below 210 nm). This allows isolating the molecular portion under consideration, simplifying the Raman spectrum, and highlighting some signals not observable with conventional Raman spectroscopy [12].

Similarly, we can selectively increase the Raman signals of each single DNA nucleobase. For example, the 250 nm excitation wavelength matches the energy of an electronic transition of the imidazole ring, satisfying the resonance condition for the guanine residue rather than the adenine one, whose signals can be enhanced through a 270 nm excitation energy [13]. The resonance enhancement reduces the amount of sample required for the experiment allowing to investigate macromolecules and biological samples also in very high diluted conditions that are more similar to the physiological ones. Moreover, the interfering fluorescence background that often is superimposed to the Raman signal is completely suppressed if the excitation wavelength is below 250 nm [14], thus allowing to simplify the analysis of the spectra.

The capability to give a special focus on different portions of a molecule is a general property of the UVRR technique that may apply to all molecules presenting electronic transitions in the UV spectral range. For this reason, applications of UVRR spectroscopy extend to different analytical fields, including polymer science, whether of biological or synthetic origin.

7.2 Synchrotron-Based UV Resonance Raman: Basic Principles and Instrumentation

UVRR spectroscopy enables investigations of complex samples in liquid, solid, or gas phases. As most molecules have strong UV absorption bands, UVRR can supply

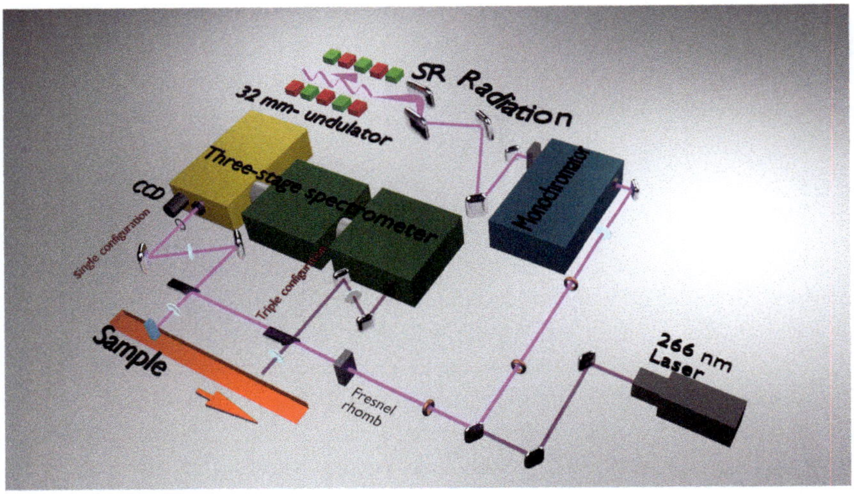

Figure 7.8 Technical layout of the synchrotron-based UVRR setup at BL10.2-IUVS beamline. Source: (Elettra Synchrotron Facility, Trieste, Italy).

analytical data from the resonance excitation of specific analytes and chromophoric portions, resulting in high sensitivity and selectivity [5]. By changing the excitation wavelengths in the UV region, it is possible to study single segments of a molecule or differences in the excited states of the same chromophore. Due to technical limitations, most UVRR spectrometers are designed to be operated in a single-color (monochromatic light source) configuration only. However, tunable excitation sources over a wide UV range have turned out to be of paramount importance since they permit: (i) to select the precise excitation energy for enhancing the Raman cross section of a particular chromophore, (ii) to map the whole resonance range, minimizing the samples' self-absorption that is excitation frequency-dependent; (iii) to explore outer electronic transitions in the matter (i.e. up to ≈10–15 eV) and specific orbitals and bands thanks to the extended UV domain. Several types of UVRR apparatus have been developed over the years by exploiting different technological solutions that ensure the combination of tunability, narrow linewidth, and high repetition rate suitable for UVRR applications [15–20]. Here we report the design, development, and characterization of a synchrotron-based UV resonance Raman (SR-UVRR) spectroscopic apparatus with excitation tunable in the deep-UV range that exploits the wide emission of the synchrotron source.

7.2.1 Synchrotron-Based UVRR Setup on IUVS@Elettra

Figure 7.8 shows the optical setup of the UVRR instrument developed at the BL10.2-Inelastic Ultra-Violet Scattering (IUVS) Beamline (Elettra Synchrotron Facility, Trieste, Italy), which takes advantage of a tunable UV synchrotron radiation (SR) source [21].

The low photon energy undulator source (Figure 7.8) of 32 mm installed in the storage ring produces linear polarized radiation with photon energy from 4.4 to 11 eV

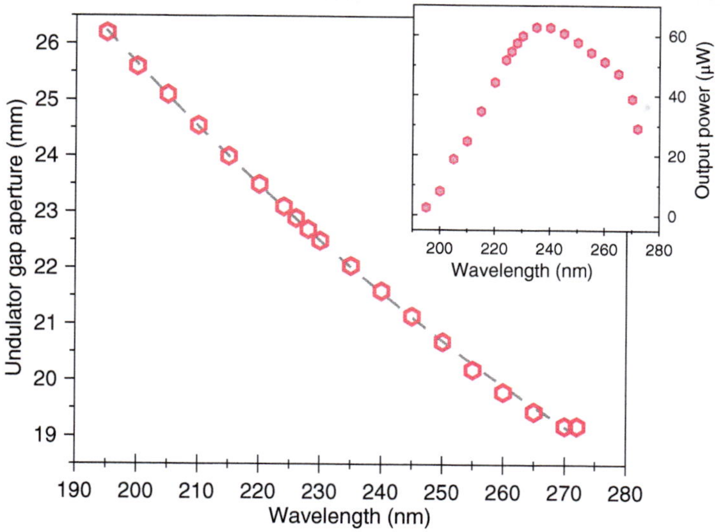

Figure 7.9 Radiation wavelength as a function of the gap aperture for Figure 7.8 32 mm undulator inserted in the IUVS beamline; inset: output power after the monochromator as a function of the selected wavelength.

(corresponding to wavelengths between about 113 and 280 nm). Its particular design [22, 23] reduces the total on-axis power density without affecting the first harmonic flux of the output spectrum. A dedicated optics system can inhibit unwanted higher harmonics. First, a gold-coated GlidCop mirror deviates the beam in the vertical plane and, second, a silicon mirror takes it back parallel to the floor. Direct internal and external cooling of the optical surfaces minimizes the radiation's thermal distortions [24] to grant the best performance.

The SR emission is then addressed toward a monochromator via suitable UV-enhanced coating mirrors with a reflectivity of 90–92% between 185 and 250 nm. The UV radiation collected at the entrance of the monochromator at $\lambda = 270$ nm has a power of ≈10 mW and a typical bandwidth of $\Delta\lambda/\lambda \cong 0.01$ (≈350 cm^{-1}). The Czerny–Turner monochromator (Acton SP2750, Princeton Instruments, USA) in use at IUVS operates with three exchangeable flat holographic gratings with 1800, 2400, and 3600 grooves mm^{-1}.

Figure 7.9 displays the monochromatized radiation as a function of the undulator's gap and the beam power after the monochromator (inset), depending on the selected wavelength.

The low-energy limit of 272 nm in the SR-UVRR apparatus is due to the minimum undulator gap of 19 mm.

The relatively low powers of the emitted UV radiation at the exit of monochromator can be efficiently exploited to analyze organic samples, which usually undergo photodegradation when irradiated with high UV powers. The theoretical resolving power of the monochromator strictly depends on the resolving power of the gratings at a fixed wavelength and the exit slit width. Figure 7.10 shows the CCD image of the elastic peak for the 250 nm excitation wavelength recorded by varying the

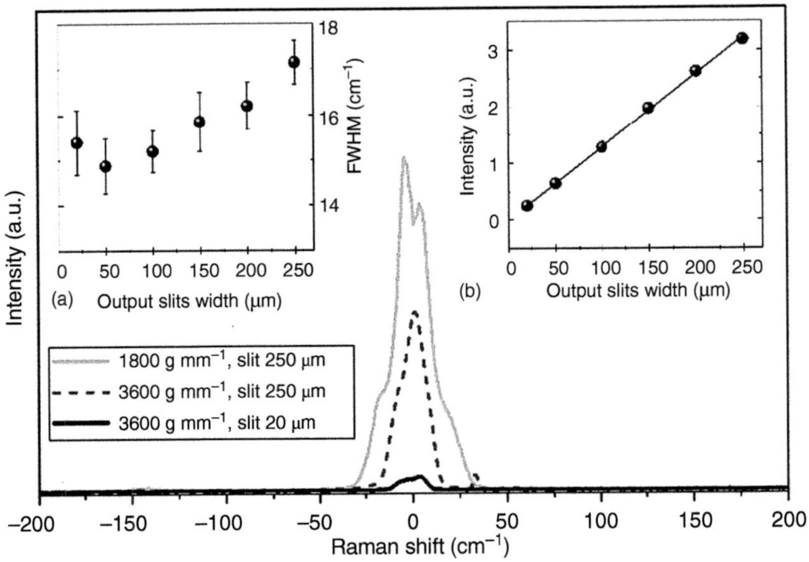

Figure 7.10 Image on the CCD of the elastic peak for 250 nm of excitation wavelength recorded by varying the selected grating and the output slits of the monochromator; insets (a) full width at half maximum (FWHM) of the image peak and (b) peak intensity as a function of the width of the monochromator output slits.

gratings and the output slits of the monochromator. For the best resolution grating option, 3600 grooves mm^{-1}, the full width at a half maximum (FWHM) and the intensity of the peak are plotted against entrance slit widths in Figure 7.10 insets (a) and (b), respectively.

The results reported in Figure 7.10 suggest that closing the entrance slit of monochromator to less than 250 μm does not improve significantly the narrow linewidth of the exciting peak, but it results in a strong reduction of the peak intensity.

The UV emission at the selected wavelength is collimated by a lens and leads to the Raman analyzer system (Figure 7.8). Once focused on the sample with a size of a few mm, the collection of the backscattered beam occurs through the plano-convex lens and mirrors. This macro-configuration is particularly suitable to analyze bulk solid and liquid samples in optical quartz cuvettes. The geometrical characteristics of the Raman setup allow planning different samples environments and analysis design, i.e. thermalization of samples in a temperature range of −70 –200 °C, change of pressure, etc. Recently, an integrated experimental set-up has been also developed and tested on IUVS for in situ and *operando* SR-UVRR spectroscopy, as displayed in Figure 7.11.

The type of samples holder designed for SR-UVRR experiments depends on the phase (liquid, solid, or gas) and on the amount of sample available. Various methods help to prevent or limiting the sample's photodecomposition due to prolonged UV exposition during measurements. Depending on the kind of sample, continuous spinning or flow systems [25] for liquids or rotation of the sample holder for highly absorbing materials can be used [26].

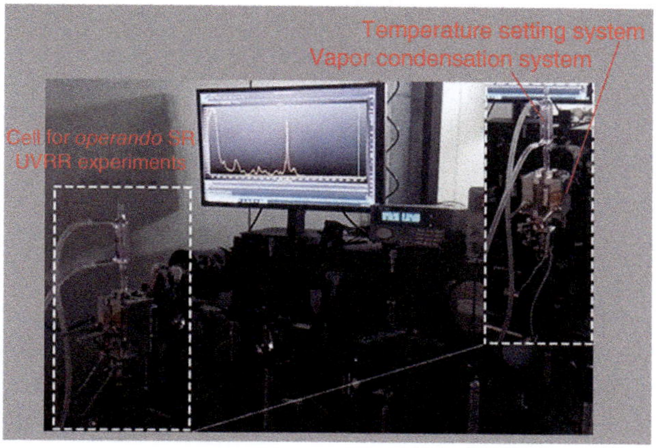

Figure 7.11 Example of experimental setup developed on IUVS@Elettra for in situ and *operando* SR-UVRR experiments. Source: Courtesy of S. Bordiga and A. Damin, Department of Chemistry, NIS Centre and INSTM Reference Centre University of Turin, Italy.

A three-stage spectrometer (TriVista 557, Princeton Instruments, USA) analyzes the Raman signal in a single configuration, each stage equipped with two flat holographic gratings (1800 and 3600 groves mm^{-1}) optimized for UV radiation. The UVRR layout is also equipped with a 266 nm laser source for *offline* complementary UVRR measurements (Figure 7.8). With this UV source, the TriVista spectrometer usually runs in triple configuration to accomplish high spectral resolution and high linear dispersion experiments (additive mode) or to get high stray light rejection (subtractive mode), avoiding definite deep UV bandpass filters. Finally, the scattered photons are detected by a Peltier-cooled UV-enhanced CCD camera. Polarized parallel (VV) and depolarized orthogonal (HV) UV Raman spectra can be collected by inserting Fresnel rhomb retarders (half-Wave retardance with broader wavelength range) and polarizers in the optical path.

7.3 SR-UVRR Characterization of Biopolymers

Polymers of natural origin have been attracting attention recently as constituents of new smart materials. Biopolymers are composed of monomeric units from natural renewable sources covalently linked to form larger molecules. These biodegradable compounds can be derived from living organisms, such as plants and animals, or be chemically synthesized from biological precursors. Examples of biopolymers include polynucleotides (deoxyribonucleic acid – DNA and ribonucleic acid – RNA), polypeptides (proteins), and polysaccharides (cellulose, lignin, glycogen, and starch).

Biopolymers are currently employed in food-related applications as coatings, packaging materials, and encapsulation matrices, and also in regenerative medicine, pharmaceutics, tissue engineering, and drug delivery. For example, biopolymers have turned out to be promising materials for the development of biosensors thanks to functional groups on their chains that impart excellent interfacial properties toward target molecules [27]. Many industrial sectors have been experiencing the improvement of functionalized biocompatible and biodegradable materials based on bio-macromolecules [28]. Proteins and especially their aggregates, such as amyloid fibrils to form hydrogels or aerogels, have been exploited for food preservation [29] and for the removal of environmental pollutants in water [30]. Even polysaccharides like gellan gum, cellulose, and lignin are amply used thanks to their ability to form networks through different molecular mechanisms, providing materials with tailored macroscopic properties [31–33].

Among biopolymers, deoxyribonucleic acid DNA has acquired a crucial role in new smart materials applications owing to selective interactions between nitrogenous bases. The low cytotoxicity and programmability arising from base-pairing make DNA the perfect building block for developing hydrogels with intriguing properties, interfacing biology with material science [34–37].

In recent years, the ability of ionic liquids (ILs) to induce gelation of biopolymers giving rise to an interesting class of polymeric materials, namely "ionogels," has been reported. These gels are physically or chemically cross-linked networks that can retain large amounts of water [38, 39]. These works report that DNA, dissolved at relative low concentration (from 0.1% to 1% w/v) in an aqueous solution of imidazolium-based ILs, exhibits a transition to a gel phase under thermal treatment, leading to the formation of a rigid ionogel [40, 41]. ILs' capacity to induce DNA gelation even at low concentration (i.e. 85% lower than for DNA hydrogel) suggests advancing applications of these biomaterials to several technological fields.

Figure 7.12 shows the steps for the synthesis of a DNA ionogel starting from double-strand ds-DNA (DNA sodium salt from salmon testes, sDNA, CAS number 438545-06-3, ∽2000 base-pairs, MW = 1.3106 Da) 1% w/v dissolved in an aqueous solution of 1-butyl-3-methylimidazolium chloride [BMIM]Cl at 1% w/v. The DNA-IL solution is first heated to 90 °C to provoke the separation of the DNA strands and then cooled at room temperature. During this process, hydrogen bonds (HB) form between the nucleotides and nitrogenous groups of imidazolium component of IL [41] keeping the DNA strands separate and leading to the physically cross-linked ionogel. The stability of the ionogel network is ensured also by hydrophobic bonds between pairs of single strands of DNA promoted by the overlapping of the tails of imidazolium cations of IL to avoid contact with water. Interestingly, the ionogel obtained through the procedure described above is thermo-reversible.

The tuning of several parameters such as the chemical structure of ILs, the type and length of the sequences of DNA oligomers, and the relative concentration of the gelling solutions' components allows designing ionogels differing in their local structure and functionality [40, 41]. The multistep self-assembly process described in Figure 7.12 can lead to various network building blocks, such as single polymer chains entangling, supramolecular coiled-coil structures, or colloidal networks.

A detailed description at the molecular level of these structural network building blocks is crucial to understand the close relationship between the gel architecture

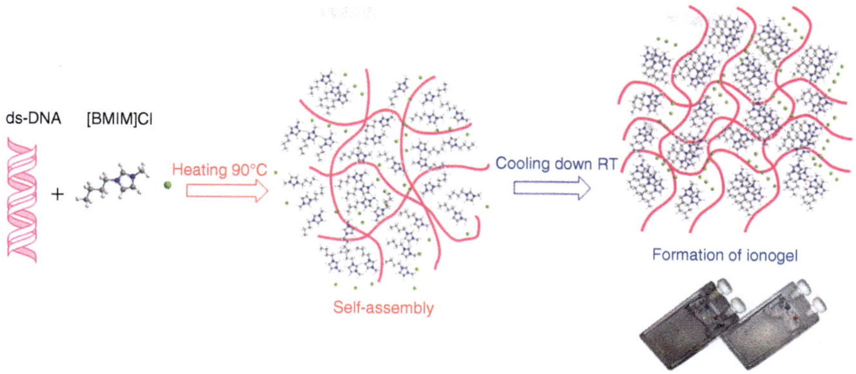

Figure 7.12 Steps of the formation of sDNA/[BMIM]Cl 1% w/v ionogel as described in the text.

and its macroscopic properties and how these functionalities can be opportunely modified.

SR-UVRR technique has demonstrated to be a valuable tool for detecting the effect of ILs on the conformational changes in the structure of DNA [42]. Previous studies performed on DNA by UVRR spectroscopy suggest the potential of the technique to give insight into the molecular analysis of this complex biopolymer [13, 43–45]. The excitation wavelengths in the UV range mainly intensify the vibrational signals arising from the nucleic acid bases [46, 47], while the bands associated to sugar and phosphate groups of DNA and any other interfering signal of the solvent and IL are nearly negligible. The choice of the exciting wavelength is fundamental to disentangle the spectral components associated with the different nucleobases in the complex Raman spectra of DNA, thanks to the selective enhancement of specific chromophores [13, 45].

Figure 7.13 displays the SR-UVRR spectra collected on sDNA/[BMIM]Cl ionogel using 266 and 250 nm as excitation wavelengths. In the same graphs, we report the UVRR spectra of equimolar solutions of the pure deoxynucleotides 2′-deoxyadenosine 5′- triphosphate sodium salt solution (dATP), 2′-deoxycytidine 5′- triphosphate disodium salt (dCTP), 2′-Deoxyguanosine 5′- triphosphate trisodium salt solution (dGTP), and thymidine 5′- triphosphate sodium salt (dTTP) for better identifying the contribution of each nitrogenous base in the corresponding 250 and 266 nm-excited spectra of ionogel. Since the energy wavelength at 250 nm is very close to the electronic transition on the imidazole ring of guanine residue dG, this excitation satisfies the best resonance condition for the nucleobase, as evident by inspection of Figure 7.13b [13]. Likewise, resonance Raman bands assigned to the adenine (dA) residue stand out with the 266 nm excitation wavelength (see Figure 7.13a). The 266 nm wavelength also enhances several signals associated with vibrational modes of thymine (dT).

George J. Thomas Jr. and Thomas Spiro previously reported detailed assignments of the normal modes composition for the DNA UVRR spectral features at different excitation wavelengths [13, 46]. Further, the same authors demonstrated that

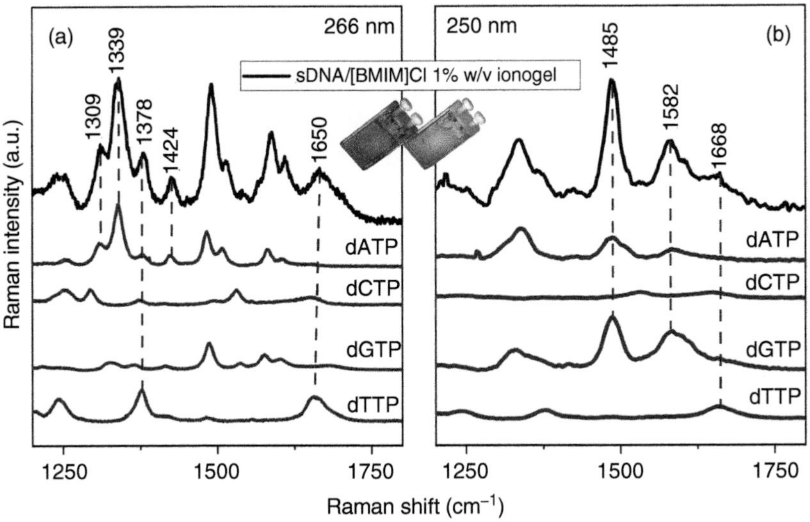

Figure 7.13 SR-UVRR spectra of sDNA/[BMIM]Cl 1% w/v ionogel obtained using 266 (a) and 250 nm (b) as excitation wavelengths. In each panel, the UVRR spectra of deoxynucleotides dNTP collected with the same excitation wavelength are reported for comparison. Source: Based on Fodor et al. [13].

Raman bands assigned to purine and pyrimidine vibrations in the 1200–1800 cm^{-1} spectral range of the UVRR spectrum of DNA are particularly sensitive to (i) the base stacking/unstacking interactions and (ii) the alterations in the hydrogen-bonding occurring at the ring sites of nucleobases.

Table 7.3 summarizes the assignments of the main spectral features observed in the UVRR spectra of sDNA/[BMIM]Cl 1% w/v ionogel displayed in Figure 7.13.

The analysis of the multiwavelengths excited UVRR spectra of DNA ionogels allows us to gain insight into the network building blocks focusing on the local structures involving guanine, adenine, and thymine regions of the DNA's sequence.

Figure 7.14a displays the temperature evolution of the 266 nm-excited UVRR spectra of sDNA/[BMIM]Cl ionogel. It is noteworthy that, as the temperature increases from 300 to 360 K, the signals labeled as dA-I ~ 1339 cm^{-1}, dT-I ~ 1378 cm^{-1}, and dT-II ~ 1650 cm^{-1} undergo strengthening of their intensity and shift of the frequency position. These changes relate to specific structural rearrangements promoted by the rise in temperature that involves adenine dA and thymine DNA residues.

The Raman band dA-I ∽1339 cm^{-1} in the spectra of ionogel in Figure 7.14a is assigned to the coupled stretching vibrations of N7=C8 and C5—N7 bonds of adenine [13, 43, 48]. This mode is a highly distinctive feature of the adenine nucleoside conformers [50] and the hydrogen-bonding state at the acceptor site N7 of adenine [51, 52]. In particular, the intensity of the Raman signal dA-I ∽1339 cm^{-1} is sensitive to the stacking/unstacking interactions of the nitrogen bases through the so-called hypochromic/hyperchromic effect related to modifications of the Raman resonance cross section. The intensity recovery of dA-I band is attributed to the attenuation exerted by stacked bases on their neighbors [46, 50, 53, 54]. The

Table 7.3 Assignment of the prominent Raman bands of DNA observed in the UVRR spectra of sDNA/[BMIM]Cl 1% w/v ionogel obtained using 250 and 266 nm as excitation wavelengths.

Excitation at 266 nm	Excitation at 250 nm	Assignment [13, 46]
Adenine (dA)	Adenine (dA)	
1309 cm^{-1}		ν(C8N9) + ν(C2N3) + δ(C8H) + δ(C2H)
1339 cm^{-1}	1339 cm^{-1}	ν(C8N7) + ν(C5N7)
1424 cm^{-1}		ν(C4N9) + δ(C8H)
1485 cm^{-1}		ν(N9C8) + δ(C2H) + δ(C8H9)
1582 cm^{-1}		ν(C5C4) + ν(C4N3)
Thymine (dT)	Thymine (dT)	
1378 cm^{-1}		ν ring + δ(C6H)
1650 cm^{-1}		ν(C4O) + ν(C5C4)
	1668 cm^{-1}	ν(C4O)
Guanine (dG)	Guanine (dG)	ν(C8N7) + ν(N9-C8) + δ(C8H)
	1485 cm^{-1}	ν(C5C4) + ν(C4N3) + ν(N7C5)

ν: stretching, δ: bending.

disruption of the vertical base–base stacking interactions leads to an increment of the resonance Raman of the modes associated with the nucleotides' aromatic ring. On the contrary, when the base–base stacking is strengthened, an intensity reduction of the same signals occurs.

Figure 7.14b shows the temperature – relative intensity dependence of the mode dA-I at ∽1339 cm^{-1} for the sDNA/[BMIM]Cl ionogel. The slight recovery of Raman hyperchromicity of the dA-I band observed at about 320 K reveals the unstacking of dA bases promoted by temperature. This behavior is consistent with localized structural changes involving the adenine residues of DNA and is affected by the inclusion of the imidazolium cations of IL in the ionogel network. If the intensity of the signal dA-I is related to the delocalization of the electronic cloud of the nitrogenous bases rings, its frequency strongly depends on solvation. The temperature-dependent frequency shifts of the bands dT-II and dA-I shown in Figure 7.14c reflect the perturbation of the hydrogen bond states of these base residues. The red-shift of the mode dA-I suggests the progressive weakening of H-bonding at the acceptor sites N7 of adenine upon the increment of temperature [50, 51] consistent with the structural rearrangement localized on dA tracts discussed above. The changes in the frequency of dA-I band could be related to conformational rearrangements of dA residues from C2'-*endo* to C3'-*endo* caused by temperature [43, 50, 54]. The Raman band dT-II ~ 1650 cm^{-1} assigned to a vibrational mode mainly localized on the C4=O group of thymine is particularly responsive to the solvation environment [13, 49, 51]. The red-shift observed for dT-II mode (Figure 7.14c) can be explained with a strengthening of H-bonding on the

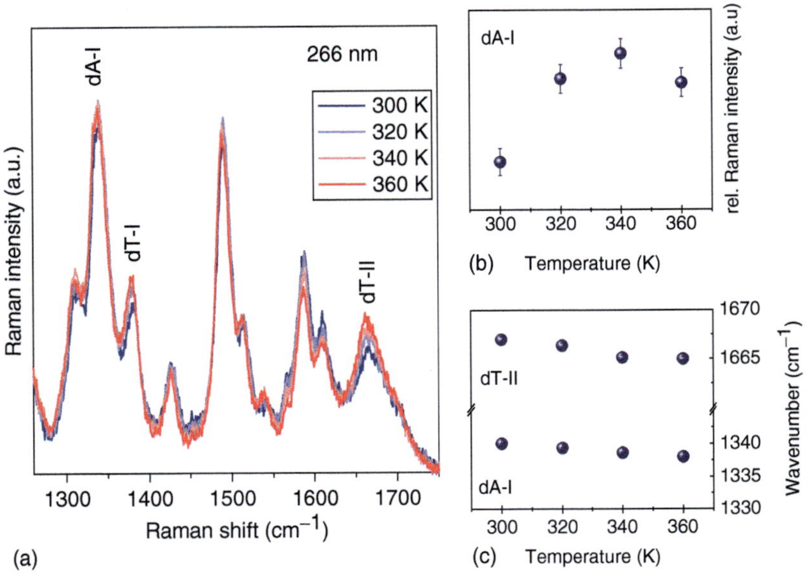

Figure 7.14 (a) 266 nm-excited UVRR spectra of sDNA/[BMIM]Cl 1% w/v ionogel as a function of temperature; evolution of (b) the relative UVRR intensity of dA-I at ~1339 cm^{-1} and (c) the central wavenumber position for the bands dA-I and dT-II as a function of temperature. Source: Refs. [13, 43, 48, 49].

thymine site [49]. This confirms a perturbation induced by the IL on the hydration shell of the C4=O group of thymine with the formation of slightly stronger HB at high temperatures. Hence, the small shift toward lower frequencies observed for the dT-I mode (see Figure 7.14a) echoes the weakening of hydrogen bonding at the thymine N3H donor site [50].

Figure 7.15a displays the temperature evolution of UVRR spectra of sDNA/[BMIM]Cl ionogel obtained with 250 nm as excitation wavelength and the reduction of the resonance Raman intensity of the mode dG-I ~ 1485 cm^{-1}.

Similar to the Raman band dA-I, the signal dG-I ~ 1485 cm^{-1} is ascribable to vibrational modes involving N7=C8 and C8—N9 ring stretching coupled with C8—H in-plane deformation of purine group of guanine residue [13, 46, 55, 58]. This mode exhibits marked Raman hypochromism due to stacked or ordered nucleic base structures [55–57]. Figure 7.15b describes the temperature dependence of the relative Raman intensity estimated for the dG-I band pointing out a sharp decrease between about 300 and 330 K. This trend shows a structural transformation involving the guanine pairs with a progressive reinforcement of the stacking interactions in this range, while there are no significant changes for temperatures above 330 K (see Figure 7.15b).

Interestingly, the opposite temperature behavior observed for dA-II and dG-I signals (Figures 7.14b and 7.15b, respectively) hints at dissimilar stacking arrangements for the guanine and adenine residues during the formation of the ionogel network. We can suppose the establishment of a preferential interaction of the imidazole ring of IL with the purine ring of the guanine residue that entails a

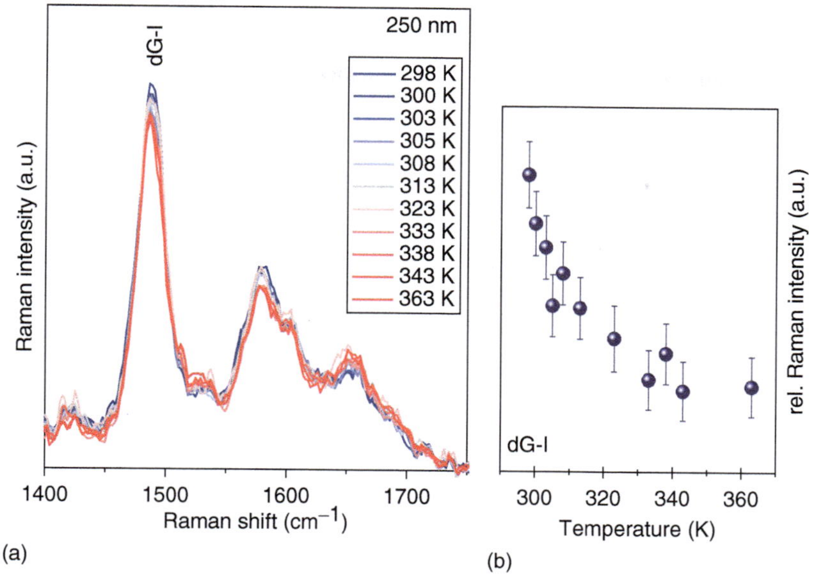

Figure 7.15 (a) 250 nm-excited UVRR spectra of sDNA/[BMIM]Cl 1% w/v ionogel as a function of temperature; (b) temperature-dependent relative UVRR intensity of dG-I band at ~1485 cm^{-1} estimated for the sample of ionogel. Source: Refs. [55–57].

"zipper" mechanism between the two aromatic rings. This is confirmed by the strong delocalization of the electronic charge on the purine ring of dG reflected by the marked Raman hypochromic effect exhibited by the dG-I signal as a function of temperature. Consistently, the analysis of the 266 nm-excited UVRR spectra discussed above proves that the dominant interactions between imidazolium cations and DNA first involve guanine and then adenine.

7.4 UV Resonance Raman Studies on Polymeric Hydrogels

Polymeric hydrogels are three-dimensional networks that can absorb and retain a large amount of water or aqueous solutions while maintaining their elastic texture [59–61]. Their matrix with interstitial spaces is created through covalent or physical cross-links among polymer chains [62, 63]. The degree of the polymer's entanglement, which depends on its formulation and the proportions of components, strongly determines its swelling behavior. On the base of the synthesis process, polymeric hydrogels can be classified as homopolymers, copolymers, semi-interpenetrating networks, and interpenetrating networks [64].

Hydrogels find widespread applications from agriculture and wastewater treatment to pharmaceutical and biomedical purposes. In particular, their biocompatibility and biodegradability, and mechanical similarity with natural tissues make them very intriguing materials for drug delivery [65–68], tissue engineering, and regenerative medicine [69–72].

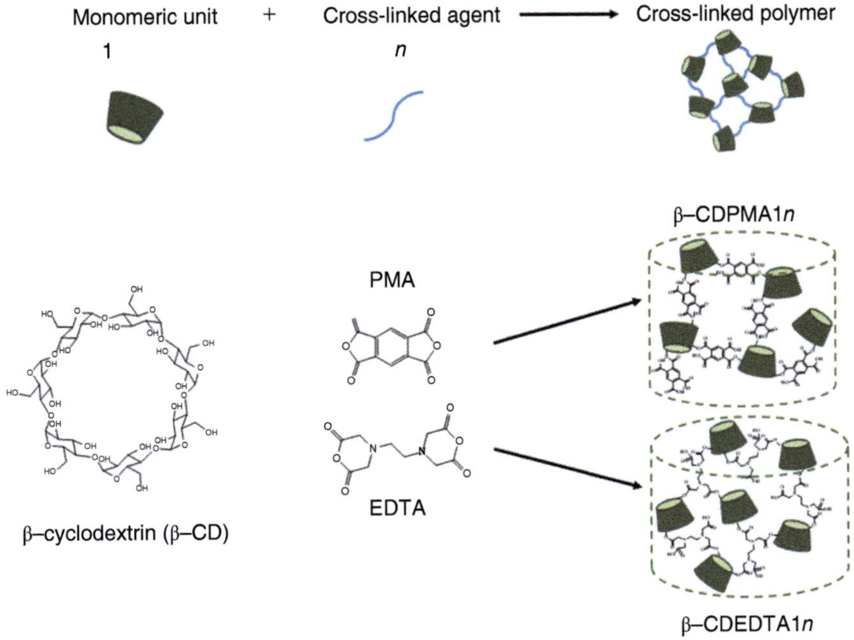

Figure 7.16 Scheme of synthesis of cyclodextrin nanosponges cross-linked polymers through the use of different cross-linkers, i.e. pyromellitic anhydride (PMA) and ethylenediaminetetraacetic acid dianhydride (EDTA). Source: Based on Venuti et al. [75].

Here we report the case study of natural and biodegradable cyclodextrin-based hydrogels, namely cyclodextrin nanosponges (NS) [73–75]. NS are cross-linked polymers synthesized by a condensation reaction between the OH groups of the glucose units of β-cyclodextrins (β-CD) and suitable polyfunctional cross-linker agents (CL) [76], such as pyromellitic anhydride (PMA) and ethylenediaminetetraacetic acid dianhydride (EDTA), as seen in Figure 7.16.

Compared to other smart hydrogels, NS permit to tune the properties of the final polymer network by modifying the synthesis parameters, i.e. the chemical structure of CL and the relative amount of CL with respect to the monomer β-CD (n = molar ratio between the cross-linking agent and CD), see Figure 7.16 [75]. This provides versatile materials for multiple applications, especially in biomedical science [77]. Besides their potential employment as nanocarriers of bioactive agents, NS are good model systems for investigations of water–water and water–polymer interactions in polymeric hydrogel phases [78, 79].

7.4.1 Water Confinement in Polysaccharide Hydrogels

The NS polymer matrix has proved to retain large fractions of water within its structure, affording homogeneous gel phases [78]. The investigation, at the molecular level, of water in NS hydrogels can give information on hydrogen bond water dynamics' dependence from (i) the confinement effects in the hydrogel's nano-spaces

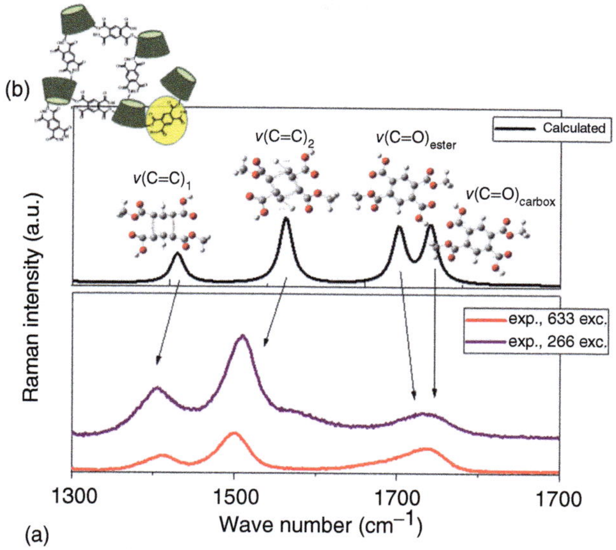

Figure 7.17 (a) Experimental Raman spectra of the β-CDPMA14 polymeric hydrogel in the wavenumber region 1500–1800 cm^{-1} collected using 633 and 266 nm as excitation wavelengths; (b) Raman activity calculated on the model of the 1,2,4,5-tetracarboxybenzoic acid dimethyl ester that mimics the ester bridges of β-CDPMA nanosponges; the sketches reported in panel (b) visualize the vibrational modes associated to the corresponding calculated frequencies. Source: Adapted from Rossi et al. [86].

[80–82] and (ii) the presence of hydrophobic/hydrophilic groups at the polymer–solvent interface [83–85]. Both these factors play an important role in determining the water swelling and gelation phenomena of the NS polymers.

UVRR spectroscopy is a valid tool to retrieve information on the dynamics of the two main components of polymeric hydrogels, i.e. water and polymer matrix [86–90]. It is possible, in fact, to implement the analysis of the spectral parameters of specific vibrational features in the UVRR spectra of hydrogels for probing the molecular connectivity of water confined in polymeric NS gels [86, 91, 92]. Figure 7.17a,b compares the experimental Raman and UVRR spectra collected for the β-CDPMA14 polymer hydrogel with the Raman activity computed on the molecular model of 1,2,4,5-tetracarboxybenzoic acid dimethyl ester [93]. The latter molecule mimics the molecular environment of the cross-linking agent PMA after the condensation with the OH groups of β-CD to form the ester bridges of the NS network. The calculated vibrational profile of 1,2,4,5-tetracarboxybenzoic acid dimethyl ester well reproduces the experimental Raman spectrum of NS in the wavenumber range 1500–1800 cm^{-1} where there are no interfering vibrational bands from monomeric units of cyclodextrins [94, 95].

Based on the frequencies and relative intensities of the experimental and computed peaks in Figure 7.17a,b, the experimental broad band at ~1720 cm^{-1} in the Raman spectrum of NS hydrogel is associated with the stretching vibration modes of the carbonyl groups of the PMA. This asymmetric band can be split into two different components, $\nu(C{=}O)_{ester}$, and $\nu(C{=}O)_{carbox}$, assigned to the stretching vibrations of the C=O groups belonging to the ester and the carboxylic moieties of PMA, respectively (see sketches in Figure 7.17b) [93, 96–99]. The other two signals in the experimental Raman spectrum of NS at ~1553 and 1604 cm^{-1} $\nu(C{=}C)_1$ and $\nu(C{=}C)_2$ are ascribable to C=C bonds stretching of the aromatic ring of PMA [93]. The vibrational signal $\nu(C{=}C)_1 \sim 1553\,cm^{-1}$ relates to a ring-breathing mode, while the vibration $\nu(C{=}C)_2 \sim 1604\,cm^{-1}$ involves the stretching of the aromatic ring of PMA together with the bending of the C—H groups. By looking at Figure 7.17a, a significant enhancement of the Raman intensity is observed for the NS bands $\nu(C{=}C)_1$ and $\nu(C{=}C)_2$ in the spectra excited using a 266 nm wavelength compared to visible light thanks to the resonance with the aromatic π–π* transitions of the ring of PMA residue in NS. This occurrence, along with the significant lessening of fluorescence background achieved by exciting Raman spectra under UV radiation, allows detecting even slight modifications of the Raman vibrations associated with the hydrophobic aromatic moiety of the NS network.

Figure 7.18a displays the isotropic UV Raman spectra of β-CDPMA14 hydrogels at different hydration levels h, where h is defined as the weight ratio of H_2O/NS. The isotropic Raman intensity has been obtained according to the relation: $I_{iso} = I_{VV} - \frac{4}{3} I_{HV}$, where I_{VV} and I_{HV} are the intensities of the light inelastically diffused from the sample polarized parallel (VV) and perpendicular (VH) to the incident radiation. As demonstrated in previous studies [86, 87, 91], the isotropic Raman spectrum of the NS hydrogel is a sensitive marker of structural rearrangements and interactions involving the molecular groups of the polymer matrix. In particular, modifications of Raman signals displayed in Figure 7.18a give information about the molecular perturbation occurring on specific chemical groups of the polymer skeleton when surrounded by water molecules.

The spectra in Figure 7.18a point out the striking enhancement of the intensity of the mode $\nu(C{=}C)_1$, as a consequence of the higher water content in NS hydrogels, while no significant spectral modifications can be detected for the Raman bands assigned to $\nu(C{=}C)_1$, $\nu(C{=}O)_{ester}$, and $\nu(C{=}O)_{carbox}$ upon the increase of h. These findings are also illustrated by the different spectral intensities I_{diff} (Figure 7.18b) obtained as: $\boldsymbol{I(h)_{diff} = I(h) - I(h=4)}$, where $\boldsymbol{I(h)}$ is the Raman spectrum of NS polymer hydrated at hydration level h and $\boldsymbol{I(h=4)}$ is the curve collected on the NS hydrated at $h = 4$.

The hydration dependence observed for the mode $\nu(C{=}C)_1$ and $\nu(C{=}C)_2$ is displayed in Figure 7.19. The intensity of the Raman signals $\nu(C{=}C)_1$ and $\nu(C{=}C)_2$ has been accurately estimated through a fitting procedure of the experimental UVRR profiles [89], as shown in Figure 7.19a.

The plot in Figure 7.19b demonstrates that for $4 < h < 5$ the band of $\nu(C{=}C)_1$ heightens as a consequence of the hydration level in the hydrogel. This parameter can be used as a spectroscopic descriptor of the perturbation induced by water molecules on the vibrational dynamics of the CH groups of PMA in the NS polymer network [86]. The setting up of hydrogen bonds between the activated CH groups of the aromatic ring of PMA and the water molecules confined in the hydrogel pores and surrounding the functional groups of NS explains the sudden recovery of the UV Raman intensity observed for $\nu(C{=}C)_1$. The remarkable sensitivity to hydration

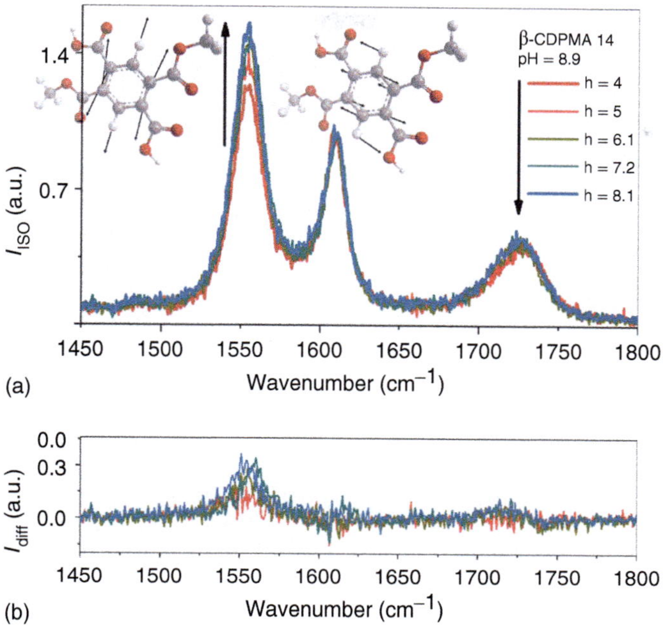

Figure 7.18 (a) Hydration dependence of isotropic UV Raman profiles for β-CDPMA14 hydrogel. The sketches reported in the graph illustrate the vibrational modes associated with the Raman signals $\nu(C{=}C)_1$ and $n(C{=}C)_2$. (b) Difference spectral intensities I_{diff} obtained as described in the text at various hydration values h. Source: Rossi et al. [87].

exhibited by the mode $\nu(C{=}C)_1$ compared to $\nu(C{=}C)_2$ supports the directional character of the C—H...O—H interactions, as expected from a genuine hydrogen bond. For $h > 5$, the data in Figure 7.19b show that the ratio between the intensity of the two modes $\nu(C{=}C)_1$ and $\nu(C{=}C)_2$ becomes almost hydration independent, suggesting that the dynamic perturbation induced on C—H groups of PMA by water tends to vanish at higher hydration levels. It is noteworthy that, as observed in the spectra of Figure 7.18, the intensity of the Raman signal associated with C=O stretching vibrations of carbonyl groups of NS is not influenced by the water content. This finding confirms that only interactions between the CH groups located in the hydrophobic part of the polymer (the PMA connector) and the surrounding water molecules prevail during the solvation of NS hydrogels.

Figure 7.20 compares the isotropic UV Raman profiles of different types of β-CDPMA1n hydrogels at the same hydration level $h = 5$.

As discussed in previous studies [93, 96, 97, 100], the molar ratio n affects the degree of crosslinking in NS polymers and modulates their elastic properties. In particular, it has been found that $n = 6$ is a critical value at which an ideal balance between activated carboxylic groups and free hydroxyl functions onto the cyclodextrin units leads to the highest interconnectivity between CD cavitands in the NS network. The spectra in Figure 7.20 point out a significant modification in the intensity of the $\nu(C{=}C)_1$ mode for PMA-NSs depending on the cross-linking of the polymer. Indeed, the inset of Figure 7.20 shows the relationship between n and

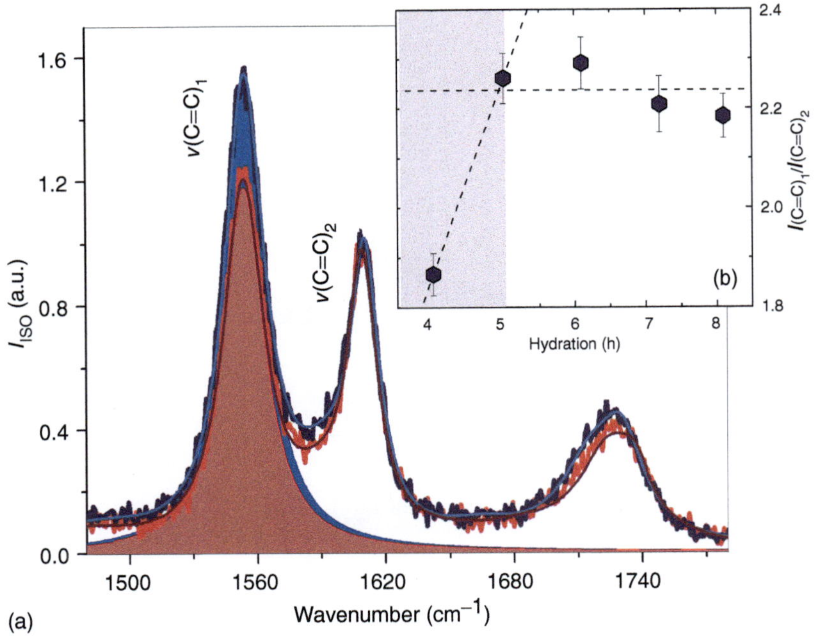

Figure 7.19 (a) Experimental isotropic Raman spectra of NS hydrogels hydrated at $h = 4$ (orange line) and $h = 8$ (blue line), reported together with the best fits (wine and cyan lines) obtained for the two spectra. (b) Hydration dependence of the ratio between the area of $\nu(C{=}C)_1$ and $\nu(C{=}C)_2$ modes. Dotted lines are guide to the eye to highlight the h-evolution. Source: Rossi et al. [87].

$\nu(C{=}C)_1/\nu(C{=}C)_2$ intensity ratio, with a maximum for $n = 6$. We can infer that the perturbation induced by water molecules on the vibrational motion of hydrophobic CH groups of NS is determined by the cross-linking degree of the gel and by the molar ratio. A more cross-linked polymer network undergoes broad perturbations owing to collisions between engaged water molecules and vibrating aromatic rings of NS. Nevertheless, according to previous investigations, NS polymers with $n = 6$ retain much less water than those synthesized using different starting conditions [78, 79, 97, 98, 101]. The reduced water-holding capacity of β-CDPMA16 can be explained, at the molecular level, by an interplay between higher rigidity and reduced dimensions of the pores of the NS polymeric network [78] and the strong water repulsion operated by the more hydrophobic groups of these systems at $n = 6$.

7.4.2 Phase Transition in Thermo-Sensitive Polysaccharide Hydrogels

Polymeric hydrogels exhibit the coexistence of solid-like and liquid-like phases at different experimental conditions [102, 103]. Several factors account for this complex behavior, such as the unique properties of the water solvent, the structural rearrangement of specific chemical groups of the polymer, water diffusion, and matrix–solvent interactions. Since these phenomena occur over broad time and space scales, an

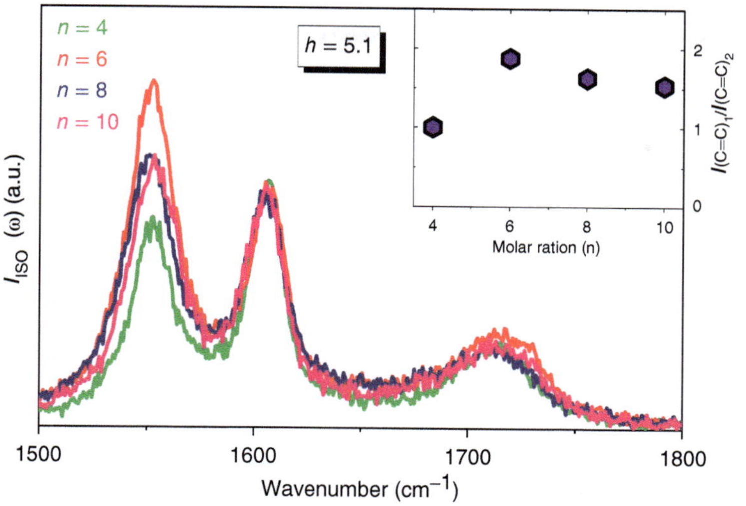

Figure 7.20 Isotropic UV Raman profiles of the β-CDPMA1*n* hydrogel (*n* = 4, 6, 8, and 10) at the same hydration level *h* = 5 in the spectral range 1500–1800 cm^{-1}. Inset: *n*-dependence of the ratio between the area of $\nu(C=C)_1$ and $\nu(C=C)_2$ modes. Source: Adapted from Rossi et al. [86].

exhaustive description of the gelling process is a challenging task that requires specific physical analytical methods for exploring hydrogels' structure and dynamics.

When water comes into contact with NS polymers, it gradually penetrates the interspaces of the polymer matrix. Then, the noncovalent (physical) aggregation of NS domains follows, giving rise to the swelling process of NS polymeric hydrogels. Experimental observations [78, 79, 101] have evidenced that NS hydrogels undergo the change of their physical state, from a highly viscous gel to a liquid phase, as a result of modifications in the system's hydration or temperature [98, 104, 105].

The UV Raman spectra shown in Figure 7.21 reveal once again that the UV Raman signals are very sensitive probes of the dynamical rearrangement of the intermolecular H-bond interactions occurring in NS hydrogels in response to changes in temperature. As already noted [86], UVRR spectra excited with a light wavelength close to the π–π* electronic transitions of the aromatic groups of PMA, show a moderate enhancement of the intensity of the Raman peaks assigned to vibrations involving the aromatic moieties compared to those associated with carbonyl groups.

The temperature effects on the spectral characteristics of the mode $\nu(C=C)_1$ are very visible in the UV isotropic Raman spectra of the β-CDPMA14 hydrogel (Figure 7.21a) that reflect the pure vibrational component of the total Raman signal.

To emphasize the temperature evolution of isotropic profiles, Figure 7.21b reports the curves $\boldsymbol{I(T)_{diff} = I(T) - I(T = 297\ K)}$, where $\boldsymbol{I(T)}$ represents the isotropic Raman spectrum acquired at the temperature T and $\boldsymbol{I(T = 297\ K)}$ is the curve collected at $T = 297$ K. The difference intensities evidence the rising of the spectral components $\nu(C=C)_1$ and $\nu(C=C)_2$ upon increasing temperature. More interestingly, we observe a progressive shift toward lower wavenumbers and an increase in intensity of the mode $\nu(C=C)_1$ as a function of temperature.

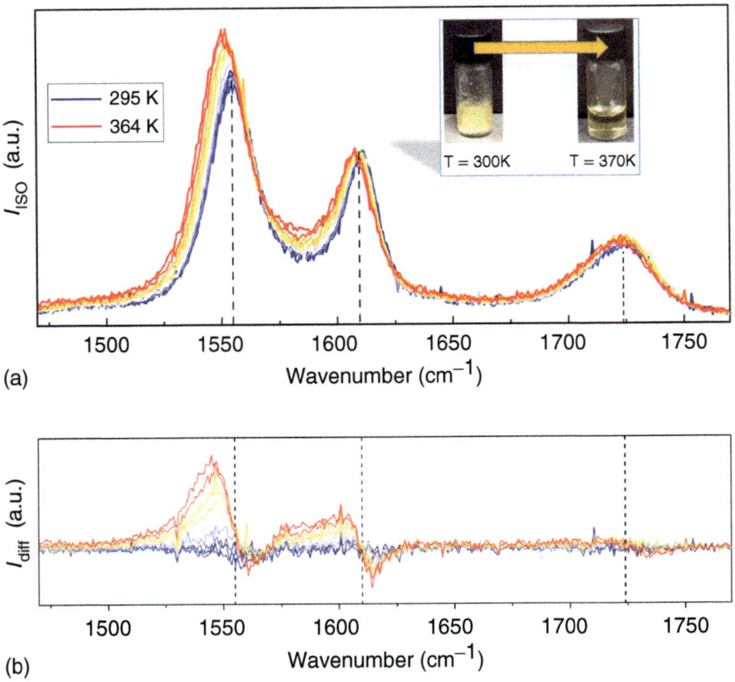

Figure 7.21 (a) Temperature evolution of isotropic Raman profiles for β-CDPMA14 nanosponge hydrated at $h = 4$ and pH = 8.9; the images display the modifications of the rigid gel into a liquid solution owing to the increment of temperature. (b) Different spectral intensities I_{diff} obtained as described in the text at various temperatures T. Source: Rossi et al. [91].

This red-shift may be related to the strengthening of HB interactions between the CH groups of PMA residues and the water molecules engaged in the polymer matrix of NS. This behavior is counterintuitive: indeed, as the temperature increases we might foresee the disruption, rather than the establishment, of C—H· · ·H—O hydrogen bonds. However, a growth of thermal motion is known to induce a destructuring effect on bulk water [86] freeing a consistent quota of water molecules available for water–polymer hydrogen bonds. This is consistent with the observed reinforcement of the dynamical perturbation on CH groups of PMA moieties provoked by surrounding water, upon the increment of temperature.

We can describe the temperature modifications observed in Figure 7.21 through the Kubo–Anderson analysis of the isotropic UV Raman spectra, already successfully applied to water–organic mixtures [106–108] and NS hydrogels [86]. This type of analysis accounts for the spectral shape modifications observed in the UV isotropic profiles of NS hydrogels under different temperature, hydration, or pH experimental conditions. The isotropic Raman spectra have been fitted using a combination of Kubo–Anderson functions, following the steps described in detail by Rossi et al. [86, 91]. For Raman peaks approximated by a Lorentzian function, the corresponding dephasing time τ_{deph} has been computed using the relation: $\tau_{deph} = {}^{1}/_{\pi\Gamma}$, where Γ is the estimated Lorentzian linewidth [106]. The physical quantity τ_{deph} defines the time during which a molecular vibration loses the initial phase relation of its vibrational amplitude. As suggested also by other authors [109, 110], in liquid systems, the inverse of τ_{deph} can be seen as a reasonable estimation of the collision rate of the solvent's molecules on the vibrating chemical groups. The temperature dependence of the dephasing time τ_{deph} associated with the Raman mode $\nu(C{=}C)_1$ is shown in Figure 7.22 for β-CDPMA14 hydrogels at different pH.

The data reported in the Figure 7.22 evidence that τ_{deph} remains nearly temperature-independent up to a value T^* that is influenced by pH. For $T > T^*$, we can observe a linear decrease of the dephasing time. This trend confirms the hypothesis made above of a sudden advance of the collision rate between water molecules and the hydrophobic CH groups on PMA moieties in NS hydrogels for temperatures greater than the triggering temperature T^*. The thermo-responsiveness of NS hydrogels can be interpreted, from a molecular point of view, as a progressive reinforcement of the dynamical perturbation of water around the more hydrophobic groups of the NS polymer, resulting in a stronger solvation of the system. This finding also appears to be of particular practical interest in view of the possible applications of NS as smart thermo-responsive systems. Figure 7.22 points out that the values estimated for T^* are significantly lowered upon increasing the pH of the gel, ranging from 337 to 310 K. Then, C—H groups act as hydrogen bond donors thanks to a higher pH of the surrounding medium. The accuracy of the estimation of T^* can be further confirmed by rescaling the T scale of the τ_{deph} data in terms of T/T^* as shown in the inset of Figure 7.22. Within the experimental error, the plots of τ_{deph} associated with different samples of NS hydrogels are perfectly overlapped for $T/T^* \geq 1$ collapsing into a single master curve. The master curve evidences that the

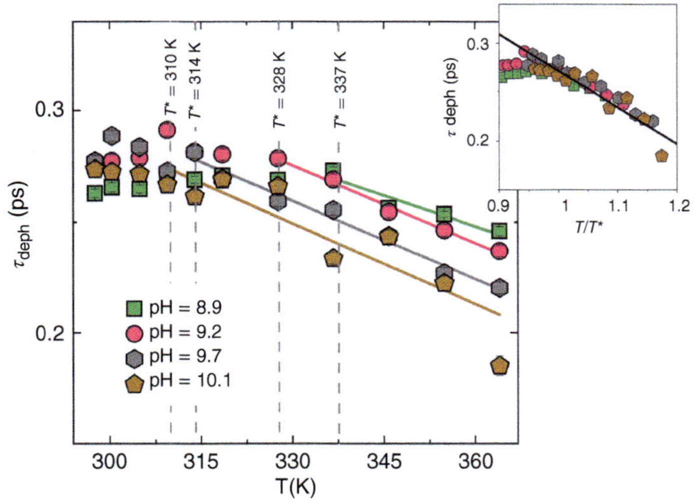

Figure 7.22 Temperature evolution of the dephasing time τ_{deph} associated with the $\nu(C{=}C)_1$ Raman mode for β-CDPMA14 hydrated at $h = 4$ and at different pH. Lines are linear fit of the experimental data. Inset: dephasing time for different samples of hydrogels scaled at the estimated values of T^* (see text for details). Source: Rossi et al. [91].

rate of decreasing observed for τ_{deph} is essentially the same for the investigated gel samples. This result is consistent with the thermo-response mechanisms exhibited by all NS, triggered at different critical temperatures T^* by pH variations. From a molecular point of view, the basic pH value of the surrounding aqueous medium enhances the slight acidic behavior of the CH groups on the aromatic rings of PMA, exposing these hydrophobic sites to the solvent molecules. Therefore, in this interpretation of the phenomena, a rise of pH promotes the accessibility of the polar water solvent by the hydrophobic sites of the NS polymer network and an activation of the hydrogel response at lower temperatures.

Figure 7.23 shows the temperature dependence of the frequency position of UVRR peaks related to aromatic ring vibrations, $\nu(C{=}C)_1$ and $\nu(C{=}C)_2$, and to carbonyl stretching modes, $\nu(C{=}C)_{ester}$, and $\nu(C{=}C)_{carbox}$. The data have been obtained for β-CDPMA13 polymeric hydrogels at different pH values. The plots in Figure 7.23 confirm the red-shift on increasing the temperature for all UV Raman modes of NS polymer in the investigated spectral window 1400–1800 cm^{-1}, as qualitatively noted in Figure 7.21.

The frequency red-shifts reported in Figure 7.23 can be interpreted as the vibrational signature of the progressive phase modification of NS polymer, which appears completely solvated at high temperatures. It is noteworthy that the shift for the mode $\nu(C{=}C)_1$ extends on increasing pH, whereas for $\nu(C{=}C)_2$ and for the stretching modes associated with the C=O groups is almost pH-independent. This finding further supports the conclusion that the UV Raman signal $\nu(C{=}C)_1$ is particularly sensitive to the molecular rearrangements of the polymer structure in response to pH changes, as discussed above. This sensitivity is linked to interactions with water molecules that strongly influence this mode [86].

By inspecting Figure 7.23, the effect of pH on the wavenumber of the mode $\nu(C{=}C)_1$ appears to be more evident in the high-temperature region (above ~330 K). This suggests that the force constant of this ring-breathing mode tends to decrease more rapidly as a function of T and pH (see the top panel of Figure 7.23). As explained before, the basic pH conditions promote the phase change in NS polymeric hydrogels and a different combination for water–polymer interactions involving PMA units. The structural rearrangement of the polymeric network is markedly caused by pH, which favors the sensitivity of the NS polymer to temperature variations (see Figure 7.22).

7.4.3 Water and Polymer Dynamics in pH-Responsive Polysaccharide Hydrogels

The transition between different molecular states in response to external stimuli, such as temperature or pH, may be highly desired in polymeric hydrogels to address specific technological applications [60, 111]. For this reason, great effort has been devoted to investigate the diffusion mechanism of molecules in pH-responsive NS hydrogels as a function of changes in acid or basic environmental conditions [112–116]. Despite the wide literature concerning the development of formulations of

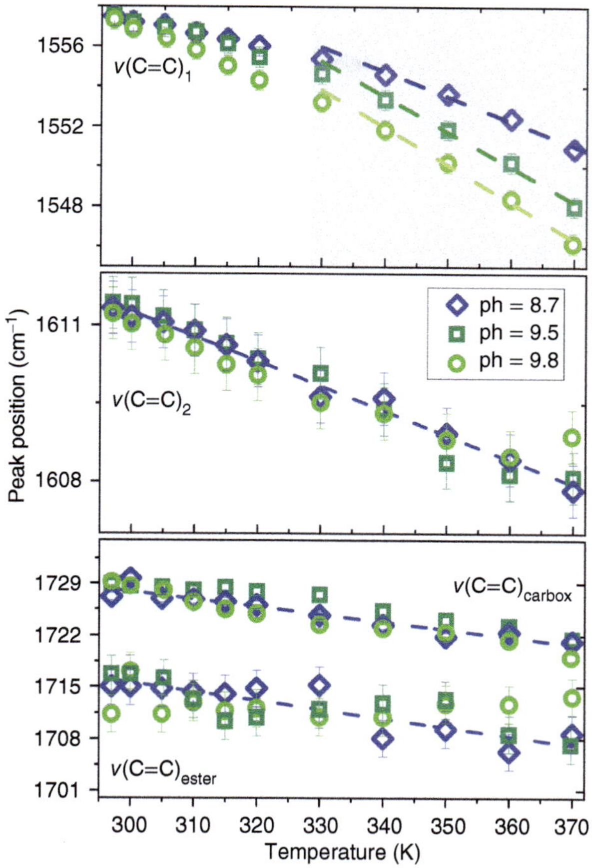

Figure 7.23 Temperature evolution of the wavenumber position of the modes $\nu(C{=}C)_1$, $\nu(C{=}C)_2$, $\nu(C{=}C)_{ester}$, and $\nu(C{=}C)_{carbox}$ (from the top to the bottom) obtained for β-CDPMA13 hydrogels at different pH values. Solid lines represent linear fits of the experimental data. The rectangular box in the top panel indicates the *T* range where a linear behavior is expected. Source: Refs. [105, 106].

pH-responsive polymeric hydrogels, a deep knowledge of the complex molecular features leading to the superior properties exhibited by these materials is still lacking.

UV Raman spectroscopy can be efficiently applied for investigating water–polymer interactions in NS polymeric hydrogels by analyzing the high-wavenumber region (2800–3800 cm^{-1}) of the spectra that includes the CH stretching signals of polymer network and the OH stretching band of water [117]. This latter Raman signal has turned out to be a sensitive probe for the H-bond organization of water in hydrogel systems [118–121].

Figure 7.24 displays the effect of pH on the spectral distribution of the β-CDEDTA18 polymer at a fixed hydration level (h = 4). For the sake of comparison, the spectra have been normalized to the maximum intensity of the band at ≈3400 cm^{-1}. The CH signals of EDTA-NS polymer can be clearly recognized between 2800 and 3100 cm^{-1}. Since the characteristic Raman band associated

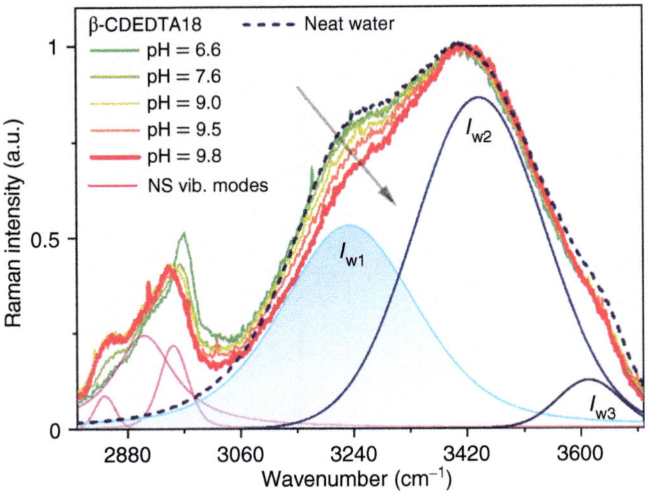

Figure 7.24 pH evolution of UV Raman spectra collected on β-CDEDTA18 gel swollen at fixed value of hydration $h = 4$. The arrows indicate the increase of pH measured in the gel phase; the UVRR spectrum of neat water is reported on the same graph for comparison. An example of best-fitting result obtained for the experimental profile corresponding to the β-CDEDTA18 gel at pH = 9.8 (red line) is reported in the same panel with the single components assigned to the vibrational modes of NS and to the OH stretching components of water. Source: Rossi et al. [117].

to the water solvent dominates the wavenumber region 3200–3600 cm^{-1}, we can ignore the spectral contributions arising from the polymer's OH groups. The UVRR spectrum of water evidences the presence of the three distinct components located at $\omega_1 \approx 3200$, $\omega_2 \approx 3450$, and $\omega_3 \approx 3600$ cm^{-1}. A common interpretation [118, 122–124] assigns the lowest wavenumber component to collective in-phase vibrations due to water molecules locally arranged in ordered *ice-like* tetrahedral configurations [122, 123]. The intermediate contribution is usually associated to water structures where H-bonds are partially elongated and/or distorted [122] and phase correlations among nearest oscillators vanish, whereas the shoulder at higher wavenumbers is assigned to OH groups not involved (as H-donors) in H-bonds.

As shown in Figure 7.24, the shape of OH stretching band in the spectrum of the β-CDEDTA18 polymeric gels greatly differs from that of pure solvent. Thus, the intermolecular structure of water in the compartmentalized spaces of the polymer network is not similar to bulk water, as already observed for other gel systems [118–121]. At the same time, by increasing pH, we can notice significant spectral changes of CH stretching signals assigned to the polymer network of NS (see Figure 7.24). This indicates modifications on the intermolecular interactions experienced by the CH groups of swollen NS polymer network that are considerably dependent on the pH.

The qualitative differences in UVRR profiles of NS hydrogels can be quantified by a suitable curve-fitting procedure that allows to analyze each spectral parameter (intensity and frequency positions) of the Raman signals [125–128]. An example of fitting results obtained for β-CDEDTA18 hydrogel at pH = 9.8 is presented in

Figure 7.24. Following a well-assessed procedure reported in literature [125–127], the OH distribution has been reproduced as the sum of three distinct components: a mixed Gaussian–Lorentzian form for the ice-like contribution ($\omega_1 \approx 3200\,\text{cm}^{-1}$) and two Gaussian profiles for the contributions arising from water molecules involved in distorted configurations and H-bond-free OH groups ($\omega_2 \approx 3450$ and $\omega_3 \approx 3600\,\text{cm}^{-1}$), respectively. In addition, we have used three Voight functions for reproducing the CH stretching signals of NS polymer. Starting from the results of fitting procedure, we have evaluated the quantity $C(h, \text{pH})$ as the relative intensity of the ω_1 component $I_1(h, \text{pH})$ with respect to the total one, $I_{\text{OHtot}} = I_{\omega 1} + I_{\omega 2} + I_{\omega 3}$: $C(h, \text{pH}) = \frac{I_1(h,pH)}{I_{\text{OHtot}}(h,pH)}$, where h and pH are the hydration level and the pH of the NS polymeric hydrogel.

The quantity $C(h, \text{pH})$ is directly proportional to the fraction of OH groups of water involved in ordered ice-like configurations [123, 126–129].

As displayed in Figure 7.25a, for low hydration values (h = 1–1.5), the value of parameter C is significantly lower compared to that of the pure solvent. This outcome evidences that, at these hydration levels, the tetrahedral ordering of a large fraction of water present into the NS matrix is significantly affected by the interactions with the polymer groups of NS.

The data in Figure 7.25a show that C increases together with the hydration degree up to $h = 4 \div 5$, reaching a plateau corresponding to the pure solvent. For instance, the addition of water leads to a progressive swelling of the NS polymer matrix with a corresponding increase of bulk water within the system up to its maximum at $h = 4 \div 5$.

The shift of the frequency positon of the NS mode at ~2970 cm^{-1} just up to $h = 5$ (inset of Figure 7.25a) with minor changes for higher hydration levels supports the conclusion that the main structural modifications on the polymer network occur in the hydration range h = 1–5. Interestingly, the inset in Figure 7.25b displays that the features of the CH stretching signals of NS change upon the increment of pH, with variations similar to those induced by hydration. The progressive red-shift of the CH band at 2970 cm^{-1} can be interpreted as a characteristic of the modification on the overall structure of the NS polymer network promoted by pH. Consistently, in Figure 7.25b, we can observe noticeable reduction of the quantity C(pH), especially for high pH values. This can be mainly ascribed to the larger area of the polymer surface exposed to solvent, as a consequence of pH-induced structural changes and/or the increase of ionic sites in the polymer network such as $-COO^-$ groups of the cross-linker [130]. The basic conditions also disrupt the tetrahedral organization of water inside the polymer matrix, as proved by the depletion of the ice-like component (3200 cm^{-1}) of the OH band (see spectra reported in Figure 7.24).

7.5 Conclusions

Deep-UVRR spectroscopy enables to characterize at molecular level different organic and biological samples, including polymers. The resonance enhancement not only reduces the amount of sample required, thanks to the significant increment

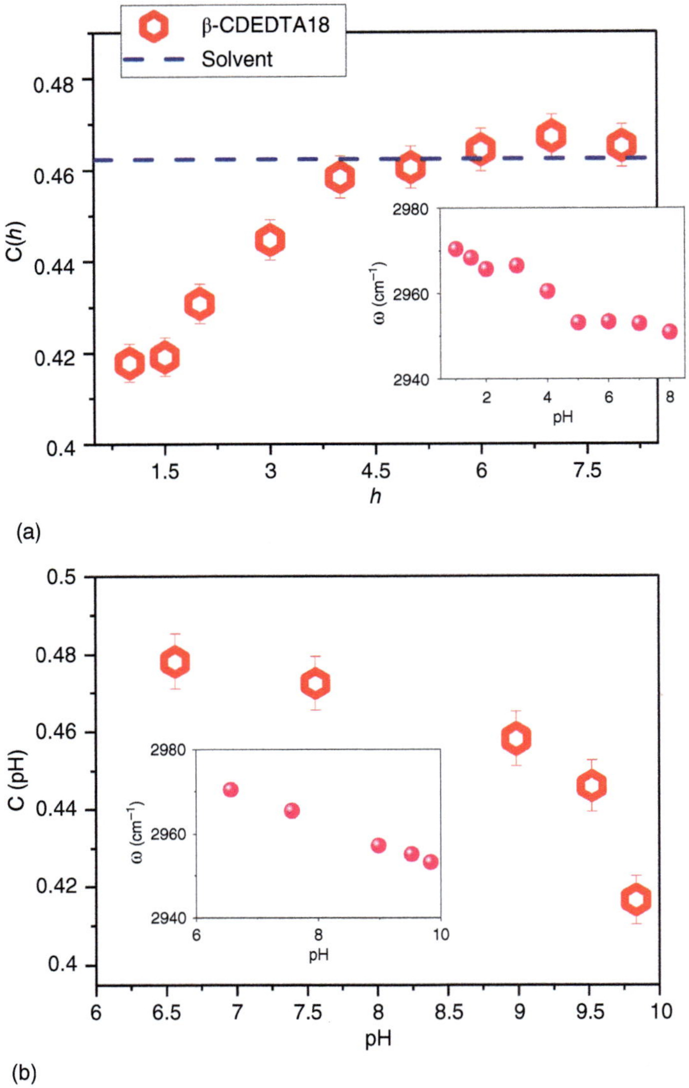

Figure 7.25 (a) Hydration dependence of parameter C for β-CDEDTA18 gel swollen at fixed pH ≈ 9. The dashed line represents the value of C obtained for the solvent used for hydrating NS. Inset: *h*-dependence of the frequency position of the most intense mode at about 2970 cm^{-1} of NS polymer. (b) pH-dependence of C for β-CDEDTA18 gel swollen at fixed $h = 4$. Inset: pH-evolution of the frequency position of the most intense mode at about 2970 cm^{-1} of NS. Source: Rossi et al. [117].

of the detection limit, but also allows probing specific parts of molecules in a very selective way. The SR-UVRR instrument developed at BL10.2-IUVS beamline of the Elettra SR facility takes advantage from the wide and tunable emission in the UV range of the synchrotron source. This offers the unique opportunity to fix on the photon energy to exactly match the electronic absorption transitions of the sample, thus enhancing the Raman signal coming from a set of chromophoric segments in the molecules. SR-UVRR technique can be successfully exerted to provide qualitative and quantitative information on complex macromolecules such as natural and synthetic polymers in different conformations and phase state, including description of polymer–water interactions in polymeric hydrogels. Finally, the possibility to push synchrotron-based UVRR spectroscopy toward shorter wavelengths (namely down to 120–130 nm) would allow exploiting electronic transitions of polymers, such as the ones responsible of valence to Rydberg states. The upcoming extension of SR-UVRR to the far-ultraviolet region is expected to provide a promising analytical technique for a further development in polymer science.

Acknowledgment

Mariagrazia Tortora acknowledges the InCIMa4 project, funded by the European Regional Development Fund and Interreg V-A Italy Austria 2014–2020.

References

1 Lin-Vien, D., Colthup, N.B., Fateley, W.G., and Grasselli, J.G. (1991). *The Handbook of Infrared and Raman Characteristic Frequencies of Organic Molecules*. London: Academic Press.

2 Evans, J.C. (1954). Infrared spectrum and thermodynamic functions of formamide. *J. Chem. Phys.* 22 (7): 1228–1234.

3 Nguyen, T.T. (1986). Infrared spectroscopic study of the formamide-Na-montmorillonite complex. Conversion of s-triazine to formamide. *Clay Clay Miner.* 34 (5): 521–528.

4 Park, S.K., Min, K.C., Lee, C.K. et al. (2009). Intermolecular hydrogen bonding and vibrational analysis of *N,N*-dimethylformamide hexamer cluster. *Bull. Korean Chem. Soc.* 30 (11): 2595–2602.

5 Asher, S.A. (1993). UV resonance Raman spectroscopy for analytical, physical, and biophysical chemistry. Part 1. *Anal. Chem.* 65 (2): 59A–66A.

6 Hildebrandt, P., Tsuboi, M., and Spiro, T.G. (1990). Ultraviolet resonance Raman spectroscopy of formamide: evidence for n-. pi.* interferences and intermolecular vibronic coupling. *J. Phys. Chem.* 94 (6): 2274–2279.

7 Asher, S.A. (1988). UV resonance Raman studies of molecular structure and dynamics: applications in physical and biophysical chemistry. *Annu. Rev. Phys. Chem.* 39 (1): 537–588.

8 Yeung, E.S., Heiling, M., and Small, G.J. (1975). Pre-resonance Raman intensities. *Spectroch. Acta M-A* 31 (12): 1921–1931.

9 Asher, S.A. (1993). UV resonance Raman spectroscopy for analytical, physical, and biophysical chemistry. Part 2. *Anal. Chem.* 65 (4): 201A–210A.

10 Hu, X. and Spiro, T.G. (1997). Tyrosine and tryptophan structure markers in hemoglobin ultraviolet resonance Raman spectra: mode assignments via subunit-specific isotope labeling of recombinant protein. *Biochemistry-US* 36 (50): 15701–15712.

11 Demchenko, A.P. (1986). *Ultraviolet Spectroscopy of Proteins*. Heidelberg: Springer-Verlag.

12 Oladepo, S.A., Xiong, K., Hong, Z., and Asher, S.A. (2011). Elucidating peptide and protein structure and dynamics: UV resonance Raman spectroscopy. *J. Phys. Chem. Lett.* 2 (4): 334–344.

13 Fodor, S.P., Rava, R.P., Hays, T.R., and Spiro, T.G. (1985). Ultraviolet resonance Raman spectroscopy of the nucleotides with 266-, 240-, 218-, and 200-nm pulsed laser excitation. *J. Am. Chem. Soc.* 107 (6): 1520–1529.

14 Asher, S.A. and Johnson, C.R. (1984). Raman spectroscopy of a coal liquid shows that fluorescence interference is minimized with ultraviolet excitation. *Science* 225 (4659): 311–313.

15 Hashimoto, S., Ikeda, T., and Takeuchi, H. (1993). Utilization of a prism monochromator as a sharp-cut bandpass filter in ultraviolet Raman spectroscopy. *Appl. Spectrosc.* 47 (8): 1283–1285.

16 Lednev, I.K., Ermolenkov, V.V., and Ming Xu, W.H. (2005). Deep-UV Raman spectrometer tunable between 193 and 205 nm for structural characterization of proteins. *Anal. Bioanal. Chem.* 381: 431–437.

17 Shutov, A.D., Petrov, G.V., Wang, D.-W. et al. (2019). Highly efficient tunable picosecond deep ultraviolet laser system for Raman spectroscopy. *Opt. Lett.* 44 (23): 5760–5763.

18 Balakrishnan, G., Hu, Y., Bronsted, S.N., and Spiro, G.T. (2005). Tunable kHz deep ultraviolet (193–210 nm) laser for Raman applications. *Appl. Spectrosc.* 59 (6): 776–781.

19 Russell, M.P., Vohnfk, S., and Thomas, G.J. Jr., (1995). Design and performance of an ultraviolet resonance Raman spectrometer for proteins and nucleic acids biophysical. *Biophys. J.* 68: 1607–1612.

20 Sterzi, A., Schneider, U., Sambalova, O. et al. (2019). Tunable deep-UV Raman spectroscopy reveals nitrate photolysis. *Proc. SPIE* 11086:1108605-1,9.

21 D'Amico, F., Saito, M., Bencivenga, F. et al. (2013). UV resonant Raman scattering facility at Elettra. *Nucl. Instrum. Methods Phys. Res. Sect. A* 703: 33–37.

22 Tanaka, T. and Kitamura, H. (1995). Figure-8 undulator as an insertion device with linear polarization and low on-axis power density. *Nucl. Instrum. Methods Phys. Res. Sect. A Accel. Spectrom. Detect. Assoc. Equip.* 364 (2): 368–373.

23 Diviacco, B., Bracco, R., Millo, D., and Zangrando, D. (2002). Design of a figure-8 undulator for ELETTRA. In: *Proceedings of the Eighth European Particle Accelerator Conference*, 2610–2612. Paris: European Physical Society.

24 Di Gennaro, R., Gee, B., Guigli, J. et al. (1988). A water-cooled mirror system for synchrotron radiation. *Nucl. Instrum. Methods Phys. Res. Sect. A Accel. Spectrom. Detect. Assoc. Equip.* 266 (1–3): 498–506.

25 Hiramatsu, H. and Saito, T. (2014). Vertical flow apparatus for enhancement and efficient collection of Raman signal. *J. Raman Spectrosc.* 45 (2): 208–210.

26 Kiefer, W. and Bernstein, H.J. (1971). Rotating Raman sample technique for colored crystal powders; resonance Raman effect in solid $KMnO_4$. *Appl. Spectrosc.* 25 (6): 609–613.

27 Dias, A.D., Kingsley, D.M., and Corr, D.T. (2004). Recent advances in bioprinting and applications for biosensing. *Biosensors* 4 (2): 111–136.

28 Mezzenga, R., Schurtenberger, P., Burbidge, A., and Michel, M. (2005). Understanding foods as soft materials. *Nat. Mater.* 4 (10): 729–740.

29 Cao, Y. and Mezzenga, R. (2019). Food protein amyloid fibrils: origin, structure, formation, characterization, applications and health implications. *Adv. Colloid Interf. Sci.* 269: 334–356.

30 Peydayesh, M., Suter, M.K., Bolisetty, S. et al. (2020). Amyloid fibrils aerogel for sustainable removal of organic contaminants from water. *Adv. Mater.* 32 (12): 1907932.

31 Diener, M., Adamcik, J., Bergfreund, J. et al. (2020). Rigid, fibrillar quaternary structures induced by divalent ions in a carboxylated linear polysaccharide. *ACS Macro Lett.* 9 (1): 115–121.

32 Tondi, G., Link, M., Kolbitsch, C. et al. (2016). Lignin-based foams: production process and characterization. *Bioresources* 11 (2): 2972–2986.

33 Tondi, G. and Schnabel, T. (2020). Bio-based polymers for engineered green materials. *Polymers* 12 (4): 775.

34 Lattuada, E., Leo, M., Caprara, D. et al. (2020). DNA-GEL, novel nanomaterial for biomedical applications and delivery of bioactive molecules. *Front. Pharmacol.* 11: 1345.

35 Bomboi, F., Romano, F., Leo, M. et al. (2016). Re-entrant DNA gels. *Nat. Commun.* 7: 13191.

36 Shahbazi, M.-A., Bauleth-Ramos, T., and Santos, H.A. (2018). DNA hydrogel assemblies: bridging synthesis principles to biomedical applications. *Adv. Therap.* 1: 1800042.

37 Moran, M.C., Miguel, M.G., and Lindman, B. (2007). DNA gel particles: particle preparation and release characteristics. *Langmuir* 23: 6478–6481.

38 Buchtová, N., Guyomard-Lack, A., and Le Bideau, J. (2014). Biopolymer based nanocomposite ionogels: high performance, sustainable and solid electrolytes. *Green Chem.* 16: 1149–1152.

39 Cerclier, C.V., Zanotti, J.-M., and Le Bideau, J. (2015). Ionogel based on biopolymer–silica interpenetrated networks: dynamics of confined ionic liquid with lithium salt. *Phys. Chem. Chem. Phys.* 17: 29707–29713.

40 Pandey, P.K., Rawat, K., Aswal, V.K. et al. (2017). DNA ionogel: structure and self-assembly. *Phys. Chem. Chem. Phys.* 19: 804–812.

41 Pandey, P.K., Rawat, K., Aswal, V.K. et al. (2018). Imidazolium based ionic liquid induced DNA gelation at remarkably low concentration. *Colloid Surf. A* 538: 184–191.

42 Bottari, C., Mancini, I., Mele, A. et al. (2019). Conformational stability of DNA in hydrated ionic liquid by synchrotron-based UV resonance Raman. *Proc. SPIE* 11086: 110860Q-1,8.

43 Benevides, J.M., Overman, S.A., and Thomas, G.J. Jr., (2005). Raman, polarized Raman and ultraviolet resonance Raman spectroscopy of nucleic acids and their complexes. *J. Raman Spectrosc.* 36: 279–299.

44 Perno, J.R., Grygon, C.A., and Spiro, T.G. (1989). Ultraviolet Raman excitation profiles for the nucleotides and for the nucleic acid duplexes poly(rA)-poly(rU) and poly(dG-dC). *J. Phys. Chem.* 93: 5672–5678.

45 Fodor, S.P.A. and Spiro, T.G. (1986). Ultraviolet resonance Raman spectroscopy of DNA with 200-266-nm laser excitation. *J. Am. Chem. Soc.* 108: 3198–3205.

46 Wen, Z.Q. and Thomas, G.J. Jr., (1998). UV resonance Raman spectroscopy of DNA and protein constituents of viruses: assignments and cross sections for excitations at 257, 244, 238, and 229 nm. *Biopolymers* 45 (5): 247–256.

47 Thomas, G.J. Jr., and Benevides, J.M. (1985). An A-helix structure for poly(dA–dT)· · ·poly(dA–dT). *Biopolymers* 24: 1101–1105.

48 Fujimoto, N., Toyama, A., and Takeuchi, H. (1998). Effects of hydrogen bonding on the UV resonance Raman bands of the adenine ring and its C8-deuterated analog. *J. Mol. Struct.* 447: 61–69.

49 Mukerji, I. and Williams, A.P. (2002). UV resonance Raman and circular dichroism studies of a DNA duplex containing an A(3)T(3) tract: evidence for a premelting transition and three-centered H-bonds. *Biochemistry* 41: 69–77.

50 Movileanu, L., Benevides, J.M., and Thomas, G.J. Jr., (1999). Temperature dependence of the Raman Spectrum of DNA. Part I–Raman signatures of premelting and melting transitions of poly(dA–dT)·poly(dA–dT). *J. Raman Spectrosc.* 30: 637–649.

51 Toyama, A., Takeuchi, H., and Harada, I. (1991). Ultraviolet resonance Raman spectra of adenine, uracil and thymine derivatives in several solvents. Correlation between band frequencies and hydrogen-bonding states of the nucleic acid bases. *J. Mol. Struct.* 242: 87–98.

52 Chan, S., Austin, R.H., Mukerji, I., and Spiro, T.G. (1997). Temperature-dependent ultraviolet resonance Raman spectroscopy of the premelting state of dA* dT DNA. *Biophys. J.* 72: 1512–1520.

53 Chinsky, L. and Turpin, P.Y. (1978). Ultraviolet resonance Raman study of DNA and of its interaction with actinomycin D. *Nucleic Acids Res.* 5 (8): 2969–2978.

54 Tomlinson, B. and Peticolas, W.L. (1970). Conformational dependence of Raman scattering intensities in polyadenylic acid. *J. Chem. Phys.* 52: 2154.

55 Duguid, J.G., Bloomfield, V.A., Benevides, J.M., and Thomas, G.J. Jr., (1995). Raman spectroscopy of DNA-metal complexes. I. The thermal denaturation of DNA in the presence of Sr' , Ba2, Mg2^{+}, Ca2^{+}, Mn2^{+}, C02^{+}, Ni2^{+}, and Cd2^{+}. *Biophys. J.* 69: 2623–2641.

56 Turpin, P.Y., Chinsky, L., Laigle, A., and Jollès, B. (1989). DNA structure studies by resonance Raman spectroscopy. *J. Mol. Struct.* 214: 43–70.

57 Erfurth, S.C. and Peticolas, W.I. (1975). Melting and premelting phenomenon in DNA by laser Raman scattering. *Biopolymers* 14: 247–264.

58 Duguid, J.D., Bloomfield, V.A., Benevides, J.M., and Thomas, G.J. Jr., (1996). DNA melting investigated by differential scanning calorimetry and Raman spectroscopy. *Biophys. J.* 71: 3350–3360.

59 Peppas, N.A., Slaughter, B.V., and Kanzelberger, M.A. (2012). Hydrogels. In: *Polymer Science: A Comprehensive Reference*, vol. 9 (eds. K. Matyjaszewski and M. Möller), 385–395. Amsterdam, The Netherlands: Elsevier.

60 Peppas, N.A., Hilt, J.Z., Khademhosseini, A., and Langer, R. (2006). Hydrogels in biology and medicine: from molecular principles to bionanotechnology. *Adv. Mater.* 18: 1345–1360.

61 Schacht, E.H. (2004). Polymer chemistry and hydrogel systems. *J. Phys. Conf. Ser.* 3: 22–28.

62 Hennink, W.E. and van Nostrum, C.F. (2002). Novel crosslinking methods to design hydrogels. *Adv. Drug Deliv. Rev.* 54: 13–36.

63 Flory, P.J. (1953). *Principles of Polymer Chemistry*. Ithaca, NY: Cornell University Press.

64 Rossi, D., Kajiwara, K., Osada, Y., and Yamauchi, A. (1991). *Polymer Gels*. New York: Plenum.

65 Shoichet, M.S. (2010). Polymer scaffolds for biomaterials applications. *Macromolecules* 43 (2): 581–591.

66 Perale, G., Rossi, F., Santoro, M. et al. (2011). Drug release from hydrogel: a new understanding of transport phenomena. *J. Biomed. Nanotechnol.* 7 (3): 476–481.

67 Santoro, M., Marchetti, P., Rossi, F. et al. (2011). Smart approach to evaluate drug diffusivity in injectable agar – Carbomer hydrogels for drug delivery. *J. Phys. Chem. B* 115 (11): 2503–2510.

68 Hoffman, A.S. (2022). Hydrogels for biomedical applications. *Adv. Drug Deliv. Rev.* 43: 3–12.

69 Lau, T.T. and Wang, D.A. (2013). Bioresponsive hydrogel scaffolding systems for 3D constructions in tissue engineering and regenerative medicine. *Nanomedicine* 8 (4): 655–228.

70 Tibbitt, M.W. and Anseth, K.S. (2009). Hydrogels as extracellular matrix mimics for 3D cell culture. *Biotechnol. Bioeng.* 103 (4): 655–663.

71 Greiner, A.M., Jackel, M., Scheiwe, A.C. et al. (2014). Multifunctional polymer scaffolds with adjustable pore size and chemoattractant gradients for studying cell matrix invasion. *Biomaterials* 35: 611–619.

72 Yue, K., Trujillo-de Santiago, G., Alvarez, M.M. et al. (2015). Synthesis, properties, and biomedical applications of gelatin methacryloyl (GelMA) hydrogels. *Biomaterials* 73: 254–271.

73 Sherje, A.P., Dravyakar, B.R., Kadam, D., and Jadhav, M. (2017). Cyclodextrin-based nanosponges: a critical review. *Carbohydr. Polym.* 173: 37–49.

74 Caldera, F., Tannous, M., Cavalli, R. et al. (2017). Evolution of cyclodextrin nanosponges. *Int. J. Pharm.* 531 (2): 470–479.

75 Venuti, V., Rossi, B., Mele, A. et al. (2017). Tuning structural parameters for the optimization of drug delivery performance of cyclodextrin-based nanosponges. *Expert Opin. Drug Deliv.* 14 (3): 331–340.

76 Trotta, F., Zanetti, M., and Cavalli, R. (2012). Cyclodextrin-based nanosponges as drug carriers Beilstein. *J. Org. Chem.* 8: 2091–2099.

77 Lembo, D., Trotta, F., and Cavalli, R. (2018). Cyclodextrin-based nanosponges as vehicles for antiviral drugs: challenges and perspectives. *Nanomedicine* 13 (5): 477–480.

78 Crupi, V., Majolino, D., Mele, A. et al. (2014). Direct evidence of gel-sol transition in cyclodextrin-based hydrogel as revealed by FTIR-ATR spectroscopy. *Soft Matter* 10: 2320–2326.

79 Crupi, V., Fontana, A., Majolino, D. et al. (2014). Hydrogen-bond dynamics of water confined in cyclodextrin nanosponges hydrogel. *J. Incl. Phenom. Macrocycl. Chem.* 80: 69–75.

80 Bellissent-Funel, M.C. (2003). Status of experiments probing the dynamics of water in confinement. *Eur. Phys. J. E: Soft Matter Biol. Phys.* 12: 83–89.

81 Tsukahara, T., Hibara, A., Ikeda, Y., and Kitamori, T. (2007). NMR study of water molecules confined in extended nanospaces. *Angew. Chem. Int. Ed.* 46: 1180–1183.

82 Le Caer, S., Pin, S., Esnouf, S. et al. (2011). A trapped water network in nanoporous material: the role of interfaces. *Phys. Chem. Chem. Phys.* 13: 17658–17666.

83 Laage, D. and Hynes, J. (2010). A molecular jump mechanism of water reorientation. *Science* 311: 832–835.

84 Nicodemus, R.A., Ramasesha, K., Roberts, S.T., and Tokmakoff, A. (2010). Hydrogen bond rearrangements in water probed with temperature-dependent 2D IR. *J. Phys. Chem. Lett.* 1: 1068–1072.

85 Bagchi, B. (2005). Water dynamics in the hydration layer around proteins and micelles. *Chem. Rev.* 105: 3197–3219.

86 Rossi, B., Venuti, V., D'Amico, F. et al. (2015). Water and polymer dynamics in a model polysaccharide hydrogel: the role of hydrophobic/hydrophilic balance. *Phys. Chem. Chem. Phys.* 17: 963–971.

87 Rossi, B., Venuti, V., Mele, A. et al. (2015). Probing the molecular connectivity of water confined in polymer hydrogels. *J. Chem. Phys.* 142 (1): 014901, 1-13.

88 Venuti, V., Rossi, B., D'Amico, F. et al. (2015). Combining Raman and infrared spectroscopy as a powerful tool for the structural elucidation of cyclodextrin-based hydrogels. *Phys. Chem. Chem. Phys.* 17: 10274–10282.

89 Rossi, B., Venuti, V., Mele, A. et al. (2016). Vibrational signatures of the dependence of water dynamics upon confinement in nanoporous polymer network. *Phys. Chem. Chem. Phys.* 18: 12252–12259.

90 Rossi, B., D'Amico, F., and Masciovecchio, C. (2019). *Vibrational spectroscopic methods for nanosponges*. In: *Nanosponges* (eds. F. Trotta and A. Mele). Wiley-VCH.

91 Rossi, B., Venuti, V., D'Amico, F. et al. (2015). Toward an understanding of the thermosensitive behaviour of pH-responsive hydrogels based on cyclodextrins. *Soft Matter* 11: 5862–5871.

92 Rossi, B., Venuti, V., D'Amico, F. et al. (2016). Guest-matrix interactions affect the solvation of cyclodextrin-based polymeric hydrogels: a UV Raman scattering study. *Soft Matter* 12: 8861–8868.

93 Castiglione, F., Crupi, V., Majolino, D. et al. (2012). Effect of cross-linking properties on the vibrational dynamics of cyclodextrins-based polymers: an experimental–numerical study. *J. Phys. Chem. B* 116 (27): 7952–7958.

94 Crupi, V., Majolino, D., Paciaroni, A. et al. (2010). The effect of hydrogen bond on the vibrational dynamics of genistein free and complexed with β-cyclodextrins. *J. Raman Spectrosc.* 41 (7): 764–770.

95 Crupi, V., Majolino, D., Venuti, V. et al. (2010). Temperature effect on the vibrational dynamics of cyclodextrin inclusion complexes: investigation by FTIR-ATR spectroscopy and numerical simulation. *J. Phys. Chem. A* 114 (25): 6811–6817.

96 Castiglione, F., Crupi, V., Majolino, D. et al. (2012). Nside new materials: an experimental numerical approach for the structural elucidation of nanoporous cross-linked polymers. *J. Phys. Chem. B* 116 (43): 13133–13140.

97 Crupi, V., Fontana, A., Giarola, M. et al. (2013). Connection between the vibrational dynamics and the cross-linking properties in cyclodextrins-based polymers. *J. Raman Spectrosc.* 44 (10): 1457–1462.

98 Crupi, V., Majolino, D., Mele, A. et al. (2013). Modelling the interplay between covalent and physical interactions in cyclodextrin-based hydrogel: effect of water confinement. *Soft Matter* 9: 6457–6464.

99 Castiglione, F., Crupi, V., Majolino, D. et al. (2013). Vibrational spectroscopy investigation of swelling phenomena in cyclodextrin nanosponges. *J. Raman Spectrosc.* 44 (10): 1463–1469.

100 Crupi, V., Fontana, A., Giarola, M. et al. (2014). Vibrational density of states and elastic properties of cross-linked polymers: combining inelastic light and neutron scattering. *J. Phys. Chem. B* 118 (2): 624–633.

101 Castiglione, F., Crupi, V., Majolino, D. et al. (2014). Gel-sol evolution of cyclodextrin-based nanosponges: role of the macrocycle size. *J. Incl. Phenom. Macrocycl. Chem.* 80: 77–83.

102 Ahmed, E.M., Aggor, F.S., Awad Ahmed, M., and El-Aref Ahmed, T. (2013). An innovative method for preparation of nanometal hydroxide superabsorbent hydrogel. *Carbohydr. Polym.* 91: 693–698.

103 Rees, D.A. (1969). Structure, conformation, and mechanism in the formation of polysaccharide gels and networks. *Adv. Carbohydr. Chem. Biochem.* 24: 267–332.

104 Rossi, B., Paciaroni, A., Venuti, V. et al. (2017). SANS investigation of water adsorption in tunable cyclodextrin-based polymeric hydrogels. *Phys. Chem. Chem. Phys.* 19: 6022–6029.

105 Bottari, C., Comez, L., Corezzi, S. et al. (2017). Correlation between collective and molecular dynamics in pH-responsive cyclodextrin-based hydrogels. *Phys. Chem. Chem. Phys.* 19: 22555–22563.

106 D'Amico, F., Bencivenga, F., Gessini, G. et al. (2012). Investigation of acetic acid hydration shell formation through Raman spectra line-shape analysis. *J. Phys. Chem. B* 116: 13219–13227.

107 D'Amico, F., Bencivenga, F., Camisasca, G. et al. (2013). Thermodynamic hydration shell behavior of glycine. *J. Chem. Phys.* 139: 015101.

108 D'Amico, F., Bencivenga, F., Gessini, A., and Masciovecchio, C. (2010). Temperature dependence of hydrogen-bond dynamics in acetic acid–water solutions. *J. Phys. Chem. B* 114: 10628–10633.

109 Rothschild, W.G. (1976). Motional characteristics of large molecules from their Raman and infrared band contours: vibrational dephasing. *J. Chem. Phys.* 65: 455.

110 Mariani, L., Morresi, A., Cataliotti, R.S., and Giorgini, M.G. (1996). Application of the Kubo–Anderson band shape equation to vibrational relaxation studies in the frequency domain and to an improved determination of spectral second moments from experimental data. *J. Chem. Phys.* 104: 914.

111 Willner, I. (2017). Stimuli-controlled hydrogels and their applications. *Acc. Chem. Res.* 50 (4): 657–658.

112 Reis, A.V., Guilherme, M.R., Cavalcanti, O.A. et al. (2006). Synthesis and characterization of pH-responsive hydrogels based on chemically modified Arabic gum polysaccharide. *Polymer* 47: 2023–2029.

113 Gupta, P., Vermani, K., and Garg, S. (2002). Hydrogels: from controlled release to pH-responsive drug delivery. *Drug Discov. Today* 7 (10): 569–579.

114 Hoare, T.R. and Kohane, D.S. (2008). Hydrogels in drug delivery: progress and challenges. *Polymer* 49: 1993–2007.

115 Chen, G.H. and Hoffman, A.S. (1995). Graft copolymers that exhibit temperature-induced phase transitions over a wide range of pH. *Nature* 373: 49–52.

116 Longo, G.S. and Szleifer, I. (2016). Adsorption and protonation of peptides and proteins in pH responsive gels. *J. Phys. D. Appl. Phys.* 49: 323001.

117 Rossi, B., Bottari, C., Comez, L. et al. (2018). Structural and molecular response in cyclodextrin-based pH-sensitive hydrogels by the joint use of Brillouin, UV Raman and Small Angles Neutron Scattering techniques. *J. Mol. Liq.* 271: 738–746.

118 Maeda, Y. and Kitano, H. (1995). The structure of water in polymer systems as revealed by Raman spectroscopy. *Spectrochim. Acta A* 51: 2433–2446.

119 Pastorczak, M., Kozanecki, M., and Ulanski, J. (2009). Water–polymer interactions in PVME hydrogels – Raman spectroscopy studies. *Polymer* 50: 4535–4542.

120 Joachimiak, A., Halamus, T., Wojciechowski, P., and Ulanski, J. (2005). Structure of hydrogels based on lyotropic phases of cellulose derivative as studied by Raman spectroscopy. *J. Macromol. Chem. Phys.* 206 (1): 59–65.

121 Ratajska-Gadomska, B. and Gadomski, W. (2004). Water structure in nanopores of agarose gel by Raman spectroscopy. *J. Chem. Phys.* 121: 12583–12588.

122 Walrafen, G.E. (1967). Raman spectral studies of the effects of temperature on water structure. *J. Chem. Phys.* 47: 114–126.

123 Green, J.L., Lacey, A.R., and Sceats, M.G. (1986). Spectroscopic evidence for spatial correlations of hydrogen bonds in liquid water. *J. Phys. Chem.* 90: 3958–3964.

124 Eaves, J.D., Loparo, J.J., Fecko, C.J. et al. (2005). Hydrogen bonds in liquid water are broken only fleetingly. *Proc. Natl. Acad. Sci. U.S.A.* 102: 13019–13022.

125 Bottari, C., Comez, L., Paolantoni, M. et al. (2018). Hydration properties and water structure in aqueous solutions of native and modified cyclodextrins by UV Raman and Brillouin scattering. *J. Raman Spectrosc.* 49, 6: 1076–1085.

126 Sassi, P., Paolantoni, M., Cataliotti, R.S. et al. (2004). Water/alcohol mixtures: a spectroscopic study of the water-saturated 1-octanol solution. *J. Phys. Chem. B* 108: 19557–19565.

127 Di Michele, A., Freda, M., Onori, G. et al. (2006). Modulation of hydrophobic effect by cosolutes. *J. Phys. Chem. B* 110: 21077–21085.

128 Marinov, V.S., Nickolov, Z.S., and Matsuura, H. (2001). Raman spectroscopic study of water structure in aqueous nonionic surfactant solutions. *J. Phys. Chem. B* 105: 9953–9959.

129 Paolantoni, M., Faginas Lago, N., Albertí, M., and Laganà, A. (2009). Tetrahedral ordering in water: Raman profiles and their temperature dependence. *J. Phys. Chem. A* 113: 15100–15105.

130 Ferro, M., Castiglione, F., Punta, C. et al. (2014). Anomalous diffusion of ibuprofen in cyclodextrin nanosponges hydrogels: an HRMAS NMR study. *Beilstein J. Org. Chem.* 10: 2715–2723.

8

Sum Frequency Generation Spectroscopy for Understanding the Polymer Dynamics at Buried Interfaces

Daisuke Kawaguchi[1,2] *and Keiji Tanaka*[1,2,3]

[1] *Kyushu University, Department of Applied Chemistry, 744 Motooka Nishi-ku, Fukuoka 819-0395, Japan*
[2] *Kyushu University, Center for Polymer Interface and Molecular Adhesion Science, 744 Motooka Nishi-ku, Fukuoka 819-0395, Japan*
[3] *International Institute for Carbon-Neutral Energy Research (WPI-I2CNER), Kyushu University, 744 Motooka Nishi-ku, Fukuoka 819-0395, Japan*

8.1 Introduction

Sum frequency generation (SFG) vibrational spectroscopy has been used to study the molecular structure at surfaces and interfaces of condensed matters [1–7]. SFG is a phenomenon in which photons with two different frequencies of ω_1 and ω_2 are incident on a substance to generate a photon with a frequency of $(\omega_1 + \omega_2)$. An intriguing feature of SFG is that resonating ω_2 (or ω_1) with a molecular vibrational, vibrational spectroscopy becomes possible. Also, the centrosymmetry of the system must be broken so that SFG signals are obtained. This condition can be generally satisfied at interfaces. Thus, using visible (Vis) and tunable infrared (IR) beams as excitation source, SFG enables us to gain direct access to the molecular characterization at the interfaces. Notably, the depth resolution of SFG is much better than that of other techniques such as attenuated-total-reflectance Fourier-transform IR spectroscopy [8] and surface-enhanced Raman spectroscopy [9], and is less than 1 nm which has been obtained compared with our own full atomistic molecular dynamics simulation. Needless to say, SFG can be also applied to an interface buried in a material as long as the excited beams can reach there. Another advantage of SFG vibrational spectroscopy is the ability to gain deep understanding of the orientation of functional groups on the basis of polarization combinations of input and output beams. So far, SFG measurements have been applied to interfaces of various polymers such as polystyrene (PS) [10–13], poly(methyl methacrylate) (PMMA) [14–19], poly(vinyl alcohol) (PVA) [20–22], poly(vinyl chloride) (PVC) [23], poly(2-vinylpyridine) (P2VP) [24], polyacrylonitrile (PAN) [25], polybutadiene (PB) [26], poly(dimethyl siloxane) (PDMS) [27–31], polyethylene (PE) [32, 33], polypropylene (PP) [32, 34], poly(4-methylpentene-1) (P4MP1) [35], poly(lactide) (PLA) [36–38], poly(ethylene glycol) (PEG) [39–41], polyurethane (PU) [42], and

Spectroscopic Techniques for Polymer Characterization: Methods, Instrumentation, Applications, First Edition. Edited by Yukihiro Ozaki and Harumi Sato.

poly(ethylene terephthalate) (PET) [43–45]. In most cases, the static local conformation of polymers has been discussed based on SFG signals in C–H, O–H, carbonyl, and cyano regions.

The dynamics of polymers at buried interfaces is of pivotal importance from the viewpoint of technological applications as well as fundamental science. The thermal molecular motion of polymer chains near a buried interface has been hitherto studied using other interfacial sensitive spectroscopy [46–51]. For example, evanescent-wave excitation fluorescence lifetime measurements revealed that glass transition temperature (T_g) of PS was higher in the interfacial region with a silicon substrate than in the bulk [48], meaning that the chain mobility was suppressed due to the presence of the solid wall even without the specific interaction. Besides, dielectric relaxation spectroscopy, nuclear magnetic resonance, and quasi-elastic neutron scattering for polymer composite materials provided the information about the thermal molecular motion of polymers near the buried interfaces [52–58]. However, the analytical depth for those techniques could be roughly estimated to a few tens of nanometer. Taking into account the difference in the energy state between the interface and bulk, the dynamics of chains, or segments, which directly contact the interface, should be studied to obtain a better understanding of the polymer behavior at the interface. Such could be realized by SFG spectroscopy at various temperatures. In this chapter, we discuss the dynamics of local conformation of various polymers at buried interfaces after the brief introduction of the SFG principle.

8.2 Principle

The SFG intensity (I^{SFG}) is proportional to the square of the absolute value of the effective sum frequency susceptibility tensor ($\chi_{\mathrm{eff}}^{(2)}$), which is itself related to the second-order nonlinear susceptibility tensor $\chi^{(2)}$ in the laboratory coordinate system, as well as the intensities of the two input beams ($I_{\mathrm{i}}(\omega_{\mathrm{Vis}})$ and $I_{\mathrm{i}}(\omega_{\mathrm{IR}})$). I^{SFG} in the reflective direction can be formulated as [59]

$$I^{\mathrm{SFG}} = \frac{8\pi^3\omega^2\sec^2\theta_{\mathrm{SF}}}{c_0^3 n_{\mathrm{i}}(\omega_{\mathrm{SF}})n_{\mathrm{i}}(\omega_{\mathrm{Vis}})n_{\mathrm{i}}(\omega_{\mathrm{IR}})}|\chi_{\mathrm{eff}}^{(2)}|^2 I_{\mathrm{i}}(\omega_{\mathrm{Vis}})I_{\mathrm{i}}(\omega_{\mathrm{IR}})AT \tag{8.1}$$

where $n_{\mathrm{i}}(\omega)$ is the refractive index of medium i at ω and c_0 is the speed of light in the vacuum. In Eq. (8.1), A and T are the beam overlap cross-section at the interface and the pulse width, respectively. In the case of an azimuthally isotropic interface, there are only four independent nonvanishing components of $\chi^{(2)}$. With the laboratory coordinates chosen such that z is along the interface normal and x is in the incidence plane, they are $\chi_{xxz} = \chi_{yyz}$, $\chi_{xzx} = \chi_{yzy}$, $\chi_{zxx} = \chi_{zyy}$, and χ_{zzz}. In the following, SFG spectra shown were collected with the ssp and ppp polarization combinations. The $\chi_{\mathrm{eff}}^{(2)}$ for the two polarization combinations can be expressed as follows [59];

$$\chi_{\mathrm{eff,ssp}}^{(2)} = L_{yy}(\omega_{\mathrm{SF}})L_{yy}(\omega_{\mathrm{Vis}})L_{zz}(\omega_{\mathrm{IR}})\sin\theta_{\mathrm{IR}}\chi_{yyz} \tag{8.2a}$$

$$\begin{aligned}\chi_{\mathrm{eff,ppp}}^{(2)} = &-L_{xx}(\omega_{\mathrm{SF}})L_{xx}(\omega_{\mathrm{Vis}})L_{zz}(\omega_{\mathrm{IR}})\cos\theta_{\mathrm{SF}}\cos\theta_{\mathrm{Vis}}\sin\theta_{\mathrm{IR}}\chi_{xxz}\\ &- L_{xx}(\omega_{\mathrm{SF}})L_{zz}(\omega_{\mathrm{Vis}})L_{xx}(\omega_{\mathrm{IR}})\cos\theta_{\mathrm{SF}}\sin\theta_{\mathrm{Vis}}\cos\theta_{\mathrm{IR}}\chi_{xzx}\end{aligned}$$

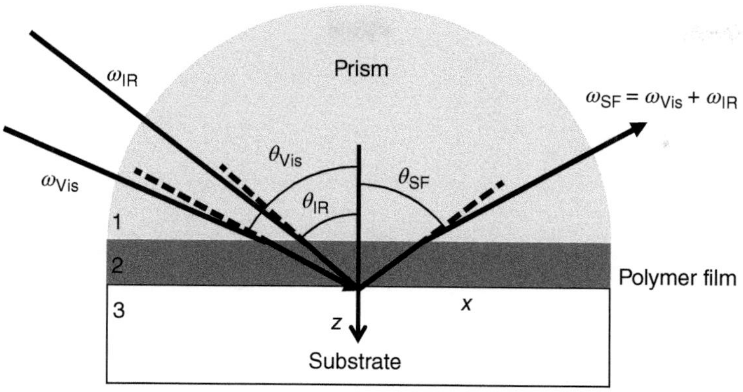

Figure 8.1 A schematic representation of the optical geometry used in SFG spectroscopy experiments. *Source:* Modified from Noguchi et al. [60].

$$+ L_{zz}(\omega_{SF})L_{xx}(\omega_{Vis})L_{xx}(\omega_{IR}) \sin\theta_{SF} \cos\theta_{Vis} \cos\theta_{IR} \chi_{zxx}$$
$$+ L_{zz}(\omega_{SF})L_{zz}(\omega_{Vis})L_{zz}(\omega_{IR}) \sin\theta_{SF} \sin\theta_{Vis} \sin\theta_{IR} \chi_{zzz} \tag{8.2b}$$

where $L_{ii}(i = x, y, z)$ are the Fresnel coefficients and θ is the angle between the surface normal and the beam. Figure 8.1 shows a schematic representation of the optical geometry used in SFG spectroscopy. Here, a polymer thin film prepared on a prism and contacting with a solid substrate is taken as an example. To calculate the Fresnel factor for polymer/substrate interface, the following equations are used [60],

$$L_{xx}(\omega) = \frac{2n_1(\omega)\cos\theta_1}{n_2(\omega)\cos\theta_1 + n_1(\omega)\cos\theta_2} \times \frac{2n_2(\omega)\cos\theta_2}{n_3(\omega)\cos\theta_2 + n_2(\omega)\cos\theta_3} \times \frac{\cos\theta_3}{\cos\theta_1} \tag{8.3a}$$

$$L_{yy}(\omega) = \frac{2n_1(\omega)\cos\theta_1}{n_1(\omega)\cos\theta_1 + n_2(\omega)\cos\theta_2} \times \frac{2n_2(\omega)\cos\theta_2}{n_2(\omega)\cos\theta_2 + n_3(\omega)\cos\theta_3} \tag{8.3b}$$

$$L_{zz}(\omega) = \frac{2n_1(\omega)\cos\theta_1}{n_2(\omega)\cos\theta_1 + n_1(\omega)\cos\theta_2} \times \frac{2n_2(\omega)\cos\theta_2}{n_3(\omega)\cos\theta_2 + n_2(\omega)\cos\theta_3} \times \frac{n_1}{n_3} \times \left(\frac{n_3(\omega)}{n'(\omega)}\right)^2 \tag{8.3c}$$

where n' is the refractive index of the interfacial layer of polymer with the substrate. In Eqs. (8.3a)–(8.3c), θ_i is the incidence angle in the medium i. When the IR frequency is near vibrational resonance, $\chi_{eff}^{(2)}$ can be formulated as [5]

$$\chi_{eff}^{(2)} = \chi_{NR}^{(2)} + \sum_q \frac{A_q}{\omega_{IR} - \omega_q + i\Gamma_q} \tag{8.4}$$

where $\chi_{NR}^{(2)}$ arises from the nonresonant background contribution. Here, A_q, and Γ_q are the strength and damping coefficient of the qth vibrational mode, respectively. Each peak on an SFG spectrum was fitted by Eq. (8.4) using A_q, ω_q, and Γ_q as fitting parameters to find the peak intensity, I^{SFG} if necessary.

8.3 Examples

8.3.1 Nonsolvent Interface

8.3.1.1 Polystyrene

Many polymers have found a wide range of applications as functional materials in the presence of a liquid like some used as solid electrolytes and separator films for cells, liquid filtration membranes, biochips for tailor-made diagnosis, contact lenses, etc. [61–63]. In these cases, the liquid serves as a nonsolvent for the polymer for the obvious reason not to dissolve the polymer. For a better understanding about how the design and construction of these highly functionalized materials could be improved, the aggregation states and physical properties of such polymers at the "nonsolvent" interfaces should be systematically studied. Here, we first show the density profiles of a PS film in water, methanol, and hexane are examined by neutron reflectivity (NR) to confirm whether PS in the outermost region in the film is changed upon contact with the liquids [62]. Then, the local conformation of PS at the liquid interfaces is determined by SFG vibrational spectroscopy [62]. Our results indicate that the PS surface can be reorganized at a molecular level upon contact with a liquid even if the liquid does not form strong favorable interaction with PS.

As a sample, perdeuterated PS (dPS) with a number-average molecular weight (M_n) of 317 k and a molecular weight distribution (M_w/M_n), where M_w is a weight-average molecular weight, of 1.05 was purchased from Polymer Source Inc. dPS was here used instead of the conventional (protonated) PS to confer the chemical contrast at the interfaces with liquids. The bulk T_g (T_g^b) of the dPS found by differential scanning calorimetry (DSC6220, SII Nanotechnology Inc.) was 376 K. A dPS film with a thickness of approximately 60 nm in a dried state was prepared from a toluene solution on a quartz block by a spin-coating method. The film was annealed under vacuum at 393 K for 24 hours. Water, methanol, and hexane, which are typical nonsolvents for PS, were used. Prior to the measurements, the dPS film was aged in these respective liquids for two hours, which was apparently enough to cause a structural change in the film. The density profiles of the dPS film in contact with the liquids were examined by NR measurements using horizontal-type neutron reflectometer SOFIA at a materials and life science experimental facility, J-PARC. All measurements were made using a single dPS film. Once the measurement in a liquid was completed, the film was taken away from the setup. Then, it was well dried and again annealed under vacuum for 12 hours at 393 K. The liquids were used in the following order: water, methanol, and hexane. The reflectivity was also calculated on the basis of the scattering length density (b/V) profile along the depth direction. The (b/V) values of SiO_2, dPS, water, methanol, and hexane used for the calculations were 3.48×10^{-4}, 6.22×10^{-4}, -5.61×10^{-5}, -3.74×10^{-5}, and -5.75×10^{-5} nm^{-2}, respectively.

As a sample for SFG, PS with M_n of 62 k and M_w/M_n of 1.08 was used. Since the initiator fragments and end groups of PS were observed in SFG spectra, PS was synthesized by anionic polymerization using potassium naphthalenide and deuterated methanol as an initiator and a terminator, respectively. The T_g^b of the PS by

DSC was 374 K. PS films were spin-coated onto quartz prisms from a toluene solution and were annealed under vacuum at 393 K for 24 hours. The thickness of the films determined by ellipsometry was approximately 200 nm. As nonsolvents, deuterium oxide (D_2O), deuterated methanol (d_4-methanol), and deuterated hexane (d_{14}-hexane) were used instead of water, methanol, and hexane to avoid peak overlapping of the vibrational mode of PS with those of the liquids. The SFG spectra were collected at a Vis wavelength of 532 nm with tunable IR beams traveling through the prism and polymer film and overlapping at the air and liquid interfaces on the polymer films.

Panel (a) of Figure 8.2 shows the scattering vector, q [$= (4\pi/\lambda)\sin\theta$], dependence of NR for a dPS film in contact with air, water, methanol, and hexane. For clarity, each data set for the dPS film in liquids was intentionally offset by a decade. The open symbols and solid curves denote experimental data and calculated reflectivities based on the model (b/V) profiles shown in panel (b) of Figure 8.2, respectively. Only

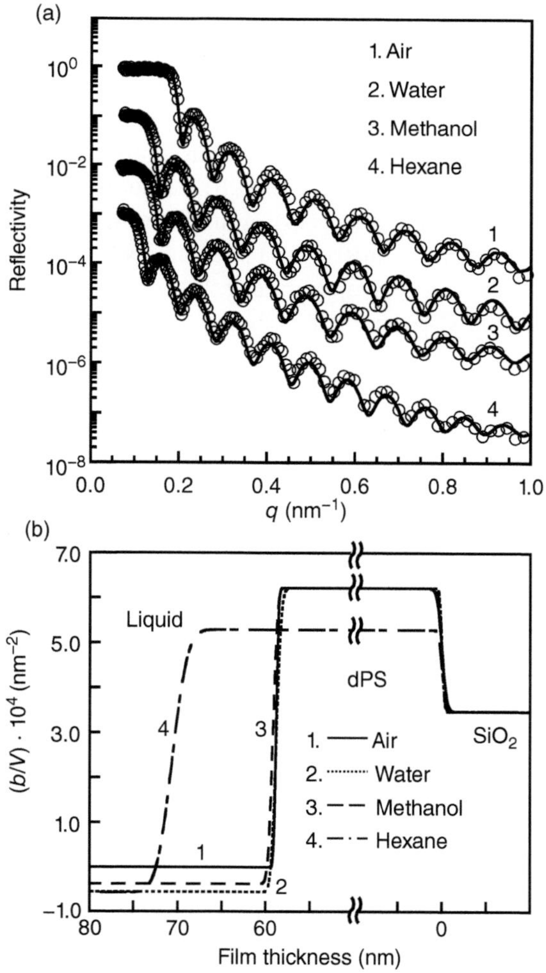

Figure 8.2 (a) Neutron reflectivity for a dPS film in air, water, methanol, and hexane. Open symbols depict experimental data, and solid lines represent the reflectivity calculated on the basis of the scattering length density (b/V) profiles shown in (b). For clarity, the values in panel (a) are offset from one another (on the y axis, reflectivity) by a factor of 10. *Source:* Horinouchi et al. [62].

in the case of the measurement in air neutron beams were guided into the sample from the air side. The parameters used for the fitting were t and σ representing the total thickness of the film and Gaussian roughness to express the interfacial width, respectively. Comparing the calculated NR curve with the experimental one, it was judged whether the (b/V) profile well reflected the density profile of the dPS film along the direction normal to the interface.

The (b/V) profiles of the dPS film in water looked similar to that of the dPS film in methanol, suggesting that the aggregation states of the PS film are comparable in both liquids. The interface of the film with water was also in good accordance with the data by Seo and Satija [64]. On the other hand, in hexane, the (b/V) value in the internal region of the dPS film was much lower than those in the two previous cases. This result makes it clear that hexane molecules penetrated deeply into the film. This is the reason that the film became thicker. That is, the dPS film was significantly swollen by hexane. The interfacial width between dPS and hexane was much broader than the dPS interfaces with the other two liquids. This interfacial broadening induced by nonsolvents was also observed in the system of PMMA/alcohols [65].

We then discuss the local conformation of PS at the liquid interfaces using SFG spectroscopy, which has the best depth resolution among available techniques [2, 5]. Figure 8.3a shows the SFG spectra for PS at air, D_2O, d_4-methanol, and d_{14}-hexane interfaces with the ssp polarization combination. On the basis of our previous study using deuterated PS (d_8-PS) with a *sec*-butyl end group and d_8-PS with an *n*-butyl end group [61], the SFG peaks around 2847 and 2921 cm^{-1} are assignable to symmetric and antisymmetric C–H stretching vibrations of methylene (CH_2s and CH_2as) groups. The peak appearing at 2902 cm^{-1}, which has not been assigned, seems to be from C–H stretching vibrations of methyne (CH) groups. In addition, typical

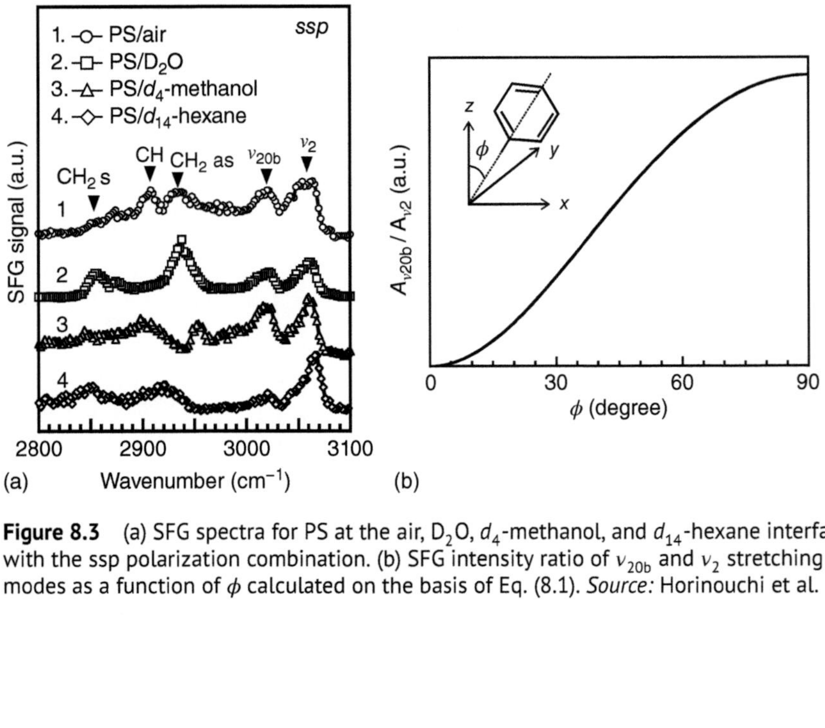

Figure 8.3 (a) SFG spectra for PS at the air, D_2O, d_4-methanol, and d_{14}-hexane interfaces with the ssp polarization combination. (b) SFG intensity ratio of ν_{20b} and ν_2 stretching modes as a function of ϕ calculated on the basis of Eq. (8.1). *Source:* Horinouchi et al. [62].

peaks originating from phenyl groups were observed in the wavenumber region of 3000–3100 cm^{-1} [10, 11]. Recently, Patterson's group has revisited the assignment of SFG peaks originating from phenyl rings and reported that the vibrational modes except for ν_{20b}, ν_2, and ν_{7a} would not appear in the SFG spectra [66, 67]. In our spectra, the peaks around 3020 and 3060 cm^{-1}, which are assignable to the ν_{20b} and ν_2 modes, were clearly observed. Although the backbone of PS is composed only of hydrophobic groups, the SFG spectrum changed its shape upon contact with D_2O. This spectrum change was also observed in the case of d_4-methanol. This implies that the local conformation of PS at the outermost region in the film changed to a more stable conformation in contact with other phases (such a liquid) as long as the chain mobility permits.

The tilt angle of phenyl rings relative to the interface normal (ϕ) can be discussed on the basis of the intensity ratio of the ν_{20b} to ν_2 modes, the most dominant peaks in our spectra [66–69].

$$\left|\frac{A_{\nu 20b}}{A_{\nu 2}}\right| = \left|\left(\frac{\beta_{caa,\ \nu 20b}}{\beta_{aac,\ \nu 2}}\right)\left(\frac{2(\langle\cos 3\phi\rangle - \langle\cos\phi\rangle)}{(7+2r)\langle\cos\phi\rangle + (1-2r)\langle\cos 3\phi\rangle}\right)\right| \tag{8.5}$$

Here, $A_{\nu 2}$ and $A_{\nu 20b}$ represent the amplitudes of the peaks for the ν_2 and ν_{20b} modes, respectively. $\beta_{aac,\nu 2}$ and $\beta_{caa,\nu 20b}$ are their respective hyperpolarizabilities. The r value, 1.13, is given by the β_{ccc}/β_{aac} ratio for the ν_2 mode [11]. Unfortunately, there is no information regarding the $\beta_{aac,\nu 2}$ and $\beta_{caa,\nu 20b}$ values. However, the tilt angle of phenyl rings can be somehow compared among the different interfaces by observing the trend in the peak amplitudes.

Panel (b) of Figure 8.3 shows the SFG intensity ratio of the ν_2 to ν_{20b} modes as a function of ϕ calculated on the basis of Eq. (8.1), provided that the ratio of $\beta_{caa,\nu 20b}$ to $\beta_{aac,\nu 2}$ is assumed to be 1.0. Figure 8.3b makes it clear that the intensity ratio of ν_{20b} to ν_2 increases with increasing ϕ. The $A_{\nu 20b}/A_{\nu 2}$ values from the SFG spectra at the air, D_2O, d_4-methanol, and d_{14}-hexane interfaces are 0.92, 0.69, 0.82, and 0.43, respectively. On the basis of these values, it seems most likely that the phenyl rings are the most tilted at the air interface and that the extent follows the order of d_4-methanol, D_2O, and d_{14}-hexane interfaces. Invoking the fact that the change in ϕ is mainly enthalpy-driven, the interaction between the phenyl rings and the liquid should be crucial. Taking into account that the surface free energies of air, water, methanol, and hexane are 0, 72.8, 22.3, and 18.4 mJ m^{-2} [70], respectively, it is plausible that the tilt angle of phenyl rings is correlated to the interfacial energy except in the case of the hexane interface. To conclude, the PS interface with other nonsolvents should be made. For the moment, the reason that the hexane case is ruled out of the correlation has yet to be established. However, since the interface was diffuse and the inside of the film was swollen, as shown in panel (b) of Figure 8.2, the situation may not be simply compared to others.

The surface free energy of a phenyl ring is larger than those of methylene and methyne groups of the main chain [71]. Hence, it seems reasonable that the orientation of phenyl rings toward a liquid with a higher surface free energy is effective at minimizing the interfacial free energy. Also, the interfacial orientation of the phenyl rings would be better for the interaction of the π electrons with hydroxyl groups of

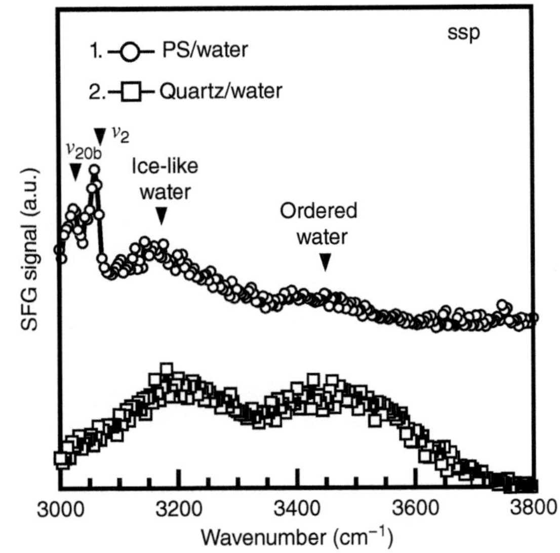

Figure 8.4 SFG spectra for water on PS and quartz in the O–H vibrational region with the ssp polarization combination. *Source:* Horinouchi et al. [62].

water and methanol via weak hydrogen bonding [72]. To discuss and establish the relationship of hydrogen bonding to our observations, SFG measurements for water and methanol on PS films were carried out in the O–H vibrational region.

Figure 8.4 shows the SFG spectrum for water on the PS film with the ssp polarization combination. For comparison, the result for water on the quartz substrate, which was pretreated with a piranha solution, is also presented in Figure 8.4. In the spectrum acquired for the PS film, a clear peak and a shoulder were observed at around 3150 and 3450 cm^{-1}, respectively. The former can be assigned to the O–H vibrational mode of water molecules, which interact with one another via strong hydrogen bonding. Although the name of this peak is controversial [73], here we call this ice-like water as commonly used [6, 74, 75]. The latter is assignable to the O–H vibration of ordered water molecules perhaps induced by an attractive interaction with π electrons of the phenyl rings of PS. Such a peak at around 3400–3600 cm^{-1} was also observed on PMMA [76], poly(2-hydroxyethyl acrylate) [77], and PVA [60]. These polymers possess functional groups, which can attractively interact with water. In the case of water on the hydrophilic quartz, this can be clearly seen. Intense peaks corresponding to the ice-like and ordered water were observed around 3150 and 3450 cm^{-1}. In particular, the latter is probably due to the O–H vibration of water hydrogen bonded to the quartz surface. Taking into account the above discussion, it is plausible that the π electrons of the PS phenyl rings interact with water molecules at the interface.

Panel (a) of Figure 8.5 shows SFG spectra for methanol on PS and quartz in the O–H vibrational region with the ssp polarization combination. In the case of methanol on the PS film, no other significant peak except for the peaks due to the phenyl groups (ν_{20b} and ν_2) were observed. This indicates that hydroxyl groups of methanol could be hardly aligned at the interface. Given that the hydroxyl groups of methanol interact with the π electrons of PS phenyl rings, it seems that the interaction is not sufficiently strong enough to orient the hydroxyl groups at

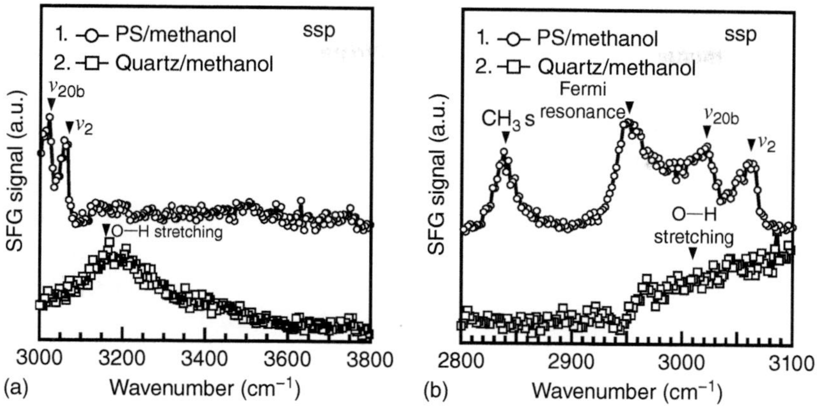

Figure 8.5 SFG spectra for PS and quartz at the methanol interface in the (a) O–H and (b) C–H vibrational regions with the ssp polarization combination. *Source:* Horinouchi et al. [62].

the PS interface. On the other hand, for methanol on the quartz, in which the strong interaction existed, a clear peak appeared at around 3150 cm^{-1}. This means that the hydroxyl groups of methanol face the quartz surface with some extent of orientation. To discuss the orientation of methanol at the PS interface, SFG measurements were also performed in the C–H vibrational region.

Panel (b) of Figure 8.5 shows the SFG spectrum acquired at the methanol/PS interface in the C–H vibrational region with the ssp polarization combination. For comparison, the data at the methanol/quartz interface are also presented. Four peaks were observed for the methanol/PS interface. Two peaks observed were assigned to the ν_{20b} and ν_2 modes of PS phenyl rings, as already discussed. Since the peaks around 2840 and 2951 cm^{-1} were not observed using d_4-methanol, it is clear that they originated from methanol at the PS interface. The former and latter are due to the C–H symmetric stretching vibration of methyl groups [78] and the Fermi resonance of the vibration with the overtone of the methyl-bending mode [78]. SFG with the ssp polarization combination enables us to gain access to the dipole, or the functional group, along the perpendicular direction. Hence, the peak at around 2840 cm^{-1} revealed that the methyl groups of methanol were oriented along the direction normal to the interface.

On the contrary, for the methanol/quartz interface, there were no clear peaks derived from the methyl groups of methanol. Although the baseline of the spectrum gradually increased with increasing wavenumber, it corresponds to the left shoulder of the peak arising from hydroxyl groups. This indicates that the methyl groups of methanol can hardly align on the quartz surface. Thus, it can be claimed that the chemical nature of a solid surface changes the aggregation states of a liquid at the interface. That is, the kind of liquid in contact with a polymer alters the aggregation states of the polymer at the interface, and at the same time, the kind of solid in contact with a liquid alters the aggregation states of the liquid at the interface.

To what extent the phenyl rings of PS oriented at each liquid interface is discussed here. Our experimental design to realizing this is to pursue the signal intensity as a function of time after introducing a liquid onto the PS surface. Figure 8.6 shows the time dependence of the SFG signal intensities at 3060 cm^{-1} before and after contact with (a) D_2O, (b) d_4-methanol, and (c) d_{14}-hexane. For comparison, each data set was normalized by the average value of the signal intensities at the air interface. In all cases, the signal intensities drastically decreased right after introducing a liquid onto the film surface. Here, a decrease in signal intensity could signify a decrease in the number of oriented phenyl rings; however, full consideration must be given to the optical nature of the interface, which requires us to discuss the Fresnel factors [60, 79].

Even if the number of phenyl rings oriented remains constant, the signal intensity changes at each interface. Thus, it is necessary to predict to what extent the signal intensity changes at each liquid interface. The signal intensity with the ssp polarization combination is proportional to the square of the absolute value of the effective sum frequency susceptibility tensor of the surface ($\chi_{\text{eff,ssp}}{}^{(2)}$) given by Eq. (8.2a). Also, χ_{xxz} is the second-order nonlinear susceptibility tensor in the laboratory coordinate system, which depends on the vibrational mode of the functional group [79, 80]. Since the SFG signal is detected here at a given wavenumber, the χ_{xxz} value can be assumed to be constant. Thus, the calculation of only L_{ii} is needed using Eqs. (8.3a)–(8.3c). Here, subscripts 1, 2, and 3 correspond to layers 1 (quartz), 2 (PS), and 3 (a liquid), respectively. The refractive index of the PS/liquid interfacial layer is taken as n'. The n values of quartz, PS, air, D_2O, d_4-methanol, and d_{14}-hexane used in the calculation were 1.46, 1.59, 1, 1.33, 1.32, and 1.37, respectively [79, 81–83]. Also, the values of n', which are unknown, were simply given by averaging the refractive indices between PS and a liquid.

The ordinate of Figure 8.6 was normalized by the signal intensity for the ν_2 mode at the air interface, and thus, it was unity before introducing a liquid onto the film surface. On the other hand, the normalized signal intensities at D_2O, d_4-methanol, and d_{14}-hexane interfaces were calculated to be 3.8, 3.7, and 2.3, respectively, as drawn by the dashed line in Figure 8.6. The intensity after the liquid introduction was much lower than the corresponding value calculated on the assumption that the number of phenyl rings oriented at each liquid interface is the same as that at the air interface. This makes it clear that the orientation of phenyl rings became more random upon contact with the liquid. Also, the decrement depended strongly on which kind of liquid was used. After D_2O contact, the intensity became smaller than the calculated value by 92%. This decrement is larger than those for the cases of d_4-methanol (80%) and d_{14}-hexane (78%). Thus, it is clear that the aggregation states of PS in the outermost region in the film were most altered by contact with water among the three liquids employed. This may be again explained in terms of the interfacial energy. Either way, the interaction between the phenyl ring and liquid molecule should be studied more using other nonsolvents with the aid of first-principles calculations.

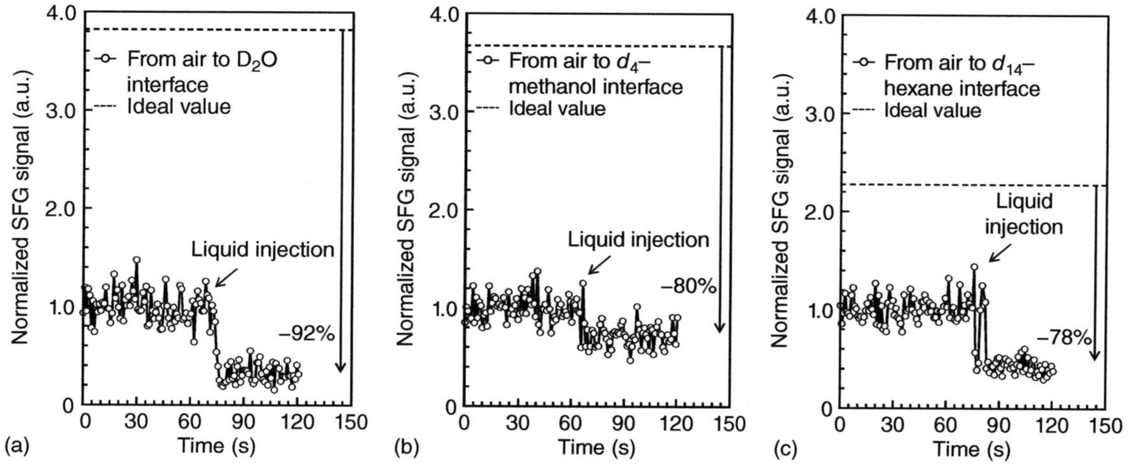

Figure 8.6 Time dependence of the intensity of the SFG signal at 3060 cm^{-1} before and after contact with (a) D_2O, (b) d_4-methanol, and (c) d_{14}-hexane. For the purpose of direct comparison, each data point was normalized by the average value of the signal intensity at the air interface in the time range from 0 to 60 seconds. The dashed line is ideal value estimated by Eqs. (8.2a), (8.3a)–(8.3c). *Source:* Horinouchi et al. [62].

8.3.2 Solid Interface

The interface between polymers and inorganic materials is of pivotal importance because of both technological importance and inherent scientific interest. An example of the former is a nanocomposite, which is a class of materials composed of polymers mixed with inorganic fillers [84]. Nanocomposites have been now used for biomaterials, sensors, power sources, etc. in addition to the popular and traditional usage as a structural material [85–87]. In general, the performance of a nanocomposite strongly relies upon the interfacial structure and physical properties of the composite. This means that understanding interfaces is the first benchmark to achieve in constructing highly functionalized nanocomposites.

We here try to provide a better understanding of the effect of a neutral solid substrate on the aggregation states of polymer chains. To do so, a typical noninteractive pair is here chosen. First, the local conformation of PS chains at the quartz interface is examined by SFG spectroscopy. We present that the interfacial conformation is strongly affected by how the film is prepared. The effect of annealing on the interfacial conformation of PS is studied [61]. Our results imply that polymer chains, even at a neutral substrate, are only slightly relaxed under the usual conditions. That is, there exists a dead layer in terms of chain mobility at the substrate. Then, we turned to rubbery polymers such as polyisoprene (PI) [88] and poly(styrene-*co*-butadiene) styrene-butadiene rubber (SBR) [89]. Using PI, we show our hypothesis that the mobility of chains at the solid interface is extremely slow in comparison with that in the bulk is universal, meaning that the concept is no longer a hypothesis but a fact. Finally, using SBR, we demonstrate how the interfacial glass transition behavior of chains can be discussed.

8.3.2.1 Polystyrene

As a sample PS with M_n of 57 k and M_w/M_n, of 1.07 was used. The T_g^b was estimated by DSC to be 376 K. Hydrophobic substrates were prepared by immersing quartz prisms and windows into a 0.15 wt% deuterated octadecyltrichlorosilane (*d*-OTS) toluene solution for two hours. The contact angle of a water droplet on the prepared substrate was 104°. Hydrophilic substrates were prepared by depositing SiO_x with a thickness of 20 nm on quartz or CaF_2 prisms and windows and following with a plasma treatment for five minutes. Two films prepared on the prism and the wafer were then bound together and annealed at 393 K in vacuo for 24 hours to prepare a sandwiched film. Thus, we could extract the information only from the polymer/solid interface under this sandwich film geometry.

Figure 8.7a shows the SFG spectrum of PS film spin-coated on the *d*-OTS substrate with the ssp polarization combination. Four clear peaks were here observed. The peaks at 2904 and 2970 cm^{-1} were assigned to the C–H stretching vibration of methyne (CH) groups and the antisymmetric vibration of methyl groups (CH_3as), respectively. Symmetric and antisymmetric C–H stretching vibrations of methylene groups (CH_2s and CH_2as) in the main chain part were not observed. Taking into account that the methyl groups exist only in the initiator fragment, *sec*-butyl group, at the chain end portion, the peak at 2970 cm^{-1} implies the orientation of the chain

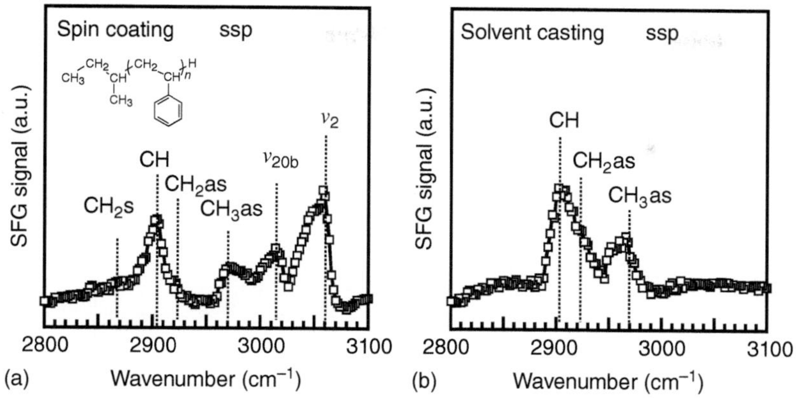

Figure 8.7 SFG spectra for PS (a) spin-coated and (b) solvent-cast films on *d*-OTS substrate with the ssp polarization combination. *Source:* Inutsuka et al. [63].

end groups rather than the main chain at the hydrophobic substrate. In addition, peaks originated from phenyl groups were observed around 3025 and 3060 cm^{-1}. Based on previous reports [10, 11, 62, 67, 90] in addition to the former section, these are assignable to the ν_{20b} and ν_2 modes, respectively. This makes it clear that the phenyl groups were also oriented at the *d*-OTS interface in the spin-coated PS film. On the other hand, the peaks originating from phenyl groups were not observed in the spectrum for the solvent-cast film (Figure 8.7b). This means that the phenyl groups were not oriented at the solid interface. The peaks at 2904 and 2970 cm^{-1}, which were assigned to the C–H stretching vibration of methyne groups and the antisymmetric vibration of methyl groups, were observed here [61]. This indicates that only *sec*-butyl end groups are oriented at the hydrophobic interface in the solvent-cast PS film.

Figure 8.8a shows the SFG spectra for a spin-coated PS film on hydrophilic quartz substrates with the ssp polarization combination. Peaks observed were similar to those for the spin-coated PS film on the hydrophobic substrate, meaning that the local conformation of PS in a spin-coated film on the quartz substrate is very similar to that on the hydrophobic *d*-OTS substrate, except for the finding of the interfacial orientation of the chain end portions in the spin-coated film on the *d*-OTS substrate. This can be explained in terms of a higher affinity between the chain ends and the hydrophobic substrate surface. Such localized end groups possibly alter the interfacial aggregation states [91].

The aggregation states and dynamics of polymer chains in films are strongly affected by how they were prepared. Thus, one usually anneals the polymer films to delete the preparation history to produce conditions where the polymer chains can be fully relaxed. We here examined how polymer chains at the interface relax with increasing temperature. Figure 8.8a shows the SFG spectra with the ssp polarization combination for a spin-coated film annealed at 393 K for 24 hours, and at 423 and 453 K for 3 hours. Recalling that the T_g^b of the sample by DSC was 375 K, the chains in the bulk should be fully relaxed at all these temperatures. Nevertheless, a signature of the interfacial orientation of phenyl rings, or the main chain part

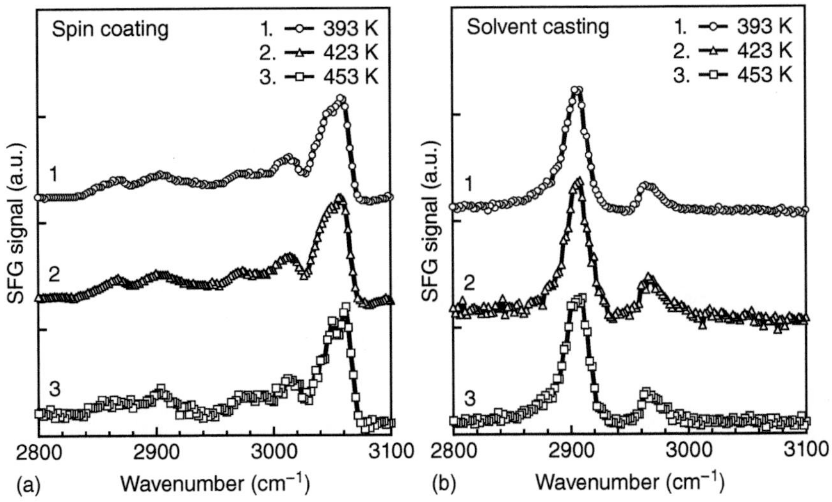

Figure 8.8 SFG spectra for PS (a) spin-coated and (b) solvent-cast films on SiO_x substrate annealed at 393 K for 24 hours and at 423 and 453 K for 3 hours with the ssp polarization combination. *Source:* Tsuruta et al. [61].

of the molecule, was clearly observed in the wavenumber region from 3000 to 3100 cm^{-1} even at these temperatures. This is in contradiction to our expectation that the spectra of the spin-coated film after sufficient annealing at a temperature higher than the T_g^b would produce conformation similar to those found in the case of solvent-casting. A possible, but unlikely, explanation for this finding is the interfacial orientation of phenyl rings is energetically preferable to the random orientation of phenyl rings at the interface. Otherwise, we can only surmise that the polymer chains at the substrate interface were not relaxed even at temperatures well in excess of the T_g^b.

An interfacial orientation of chains was not observed for the film prepared by the solvent-casting method. If the interfacial orientation of phenyl rings is energetically preferable, the SFG spectra for the solvent-cast film would be transformed to those for the spin-coated one after sufficient annealing. Figure 8.8b shows the SFG spectra with the ssp polarization combination for a solvent-cast film annealed at 393 K for 24 hours and at 423 and 453 K for 3 hours. The shape of the SFG spectra was insensitive to the annealing temperature, meaning that the randomly oriented chain conformation was not transformed to a conformation where the phenyl rings are oriented along the direction normal to the interface. Therefore, it seems reasonable to conclude that chain dynamics at the substrate interface are very much slower than those in the bulk. This conclusion is in excellent accord with our previous result [48].

8.3.2.2 Polyisoprene

PI with M_n of 91 K and M_w/M_n of 1.08, which contained 71% *cis*-1,4- and 22% *trans*-1,4-, 7% 3,4-linkages, was synthesized in toluene by living anionic polymerization using *n*-butyllithium and methanol as an initiator and a terminator, respectively. The T_g^b value of PI was 211 K determined by DSC. Quartz prisms and windows were used as substrates. Films of PI were prepared from a toluene solution onto the substrates by either a spin-coating or a solvent-casting method. The spinning rate was changed from 2000 to 5000 rpm in 1000 rpm steps. To prepare the films with a sandwiched geometry, two films on the prism and window prepared under the same condition were attached together and then dried at room temperature under vacuum for 24 hours.

SFG measurements were conducted with visible and tunable IR beams traveling through a prism and overlapping at the PI/quartz interface. To discuss the effect of centrifugal force (F) induced during the spin-coating process on the local conformation of chains at the substrate interface, SFG signals were collected at a position from the spinning center (r_{spin}) of 4 mm for the PI films prepared with various spinning rates and at various r_{spin}s for the PI film prepared with the spinning rate of 5000 rpm.

Figure 8.9a shows SFG spectra for PI films prepared by a spin-coating method with various spinning rates and a solvent-casting one. The data were obtained at a given position, r_{spin} of 4 mm. Figure 8.9b shows SFG spectra for a PI 5000 rpm-spin-coated film at r_{spin}s of 2, 4, 6, and 8 mm. Here, the ssp (SFG/s; visible/s; and IR/p) polarization combination, which collects the information along the direction normal to the interface, was employed. Open circles are experimental data, and blue curves denote the best-fit ones to each component obtained using Eqs. (8.1) and (8.4). A red curve

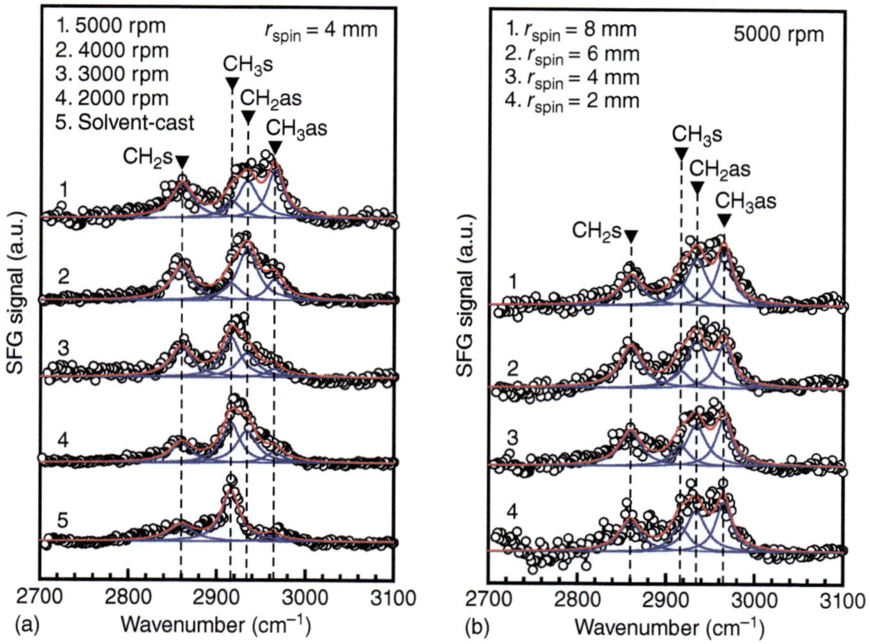

Figure 8.9 (a) SFG spectra with the ssp polarization combination for PI films prepared by a spin-coating method with various spinning rates and a solvent-casting one. The data were acquired at r_{spin} of 4 mm for all films. (b) SFG ssp spectra for a PI film prepared with the spinning rate of 5000 rpm. The data were collected at r_{spin} of 2, 4, 6, and 8 mm. Open circles are experimental data. Blue and red curves are the best fit to each component and the sum of them, respectively. *Source:* Sugimoto et al. [88].

is the sum of blue curves for a film. Peaks observed at 2860 and 2935 cm^{-1} were assigned to the symmetric and antisymmetric C-H stretching vibrations of methylene groups (CH_2s and CH_2as), respectively [92, 93]. Peaks at 2915 and 2965 cm^{-1} were assignable to those of methyl groups (CH_3s and CH_3as) [92, 93]. The shape of a spectrum, namely the local conformation of PI chains at the substrate interface, varied, depending on how the film was prepared.

To discuss the local conformation of PI at the quartz interface, we focus on the orientation of CH_3 groups, which are pendant groups of PI. The tilt angle (ϕ) of CH_3 groups with the direction normal to the interface was examined on the basis of the method proposed by Hirose et al. [69]. Since CH_3as and CH_3s modes have the C_{3v} symmetry, the intensity ratio of the CH_3as and CH_3s peaks I_{as}/I_s can be expressed as [79],

$$\left|\frac{I_{as}}{I_s}\right| = \left|\frac{\chi_{\text{eff,CH3as}}}{\chi_{\text{eff,CH3s}}}\right|^2 = \left|\frac{\chi_{yyz\text{,CH3as}}}{\chi_{yyz\text{,CH3s}}}\right|^2 \approx \left|\frac{\beta_{caa}}{\beta_{ccc}}\frac{2[\langle\cos\phi\rangle - \langle\cos^3\phi\rangle]}{\langle\cos\phi\rangle(1+r) - \langle\cos^3\phi\rangle(1-r)}\right|^2 \tag{8.6}$$

where r is given by β_{aac}/β_{ccc}, where β is the symmetric stretch hyperpolarizability tensor. In this study, the r and β_{caa}/β_{ccc} values were assumed to be 3.4 and 1.0, respectively [94, 95]. The angle fluctuation $f(\phi)$ of CH_3 groups was also modeled by a Gaussian distribution,

$$\langle\cos^n\phi\rangle = \int_0^{\pi} \cos^n\phi f(\phi)\sin\phi d\phi \tag{8.7}$$

$$f(\phi) = N\exp[-(\phi - \phi_0)^2/2\sigma^2] \tag{8.8}$$

where ϕ_0, σ, and N are the average tilt angle, the angular distribution, and a normalization constant, respectively.

Figure 8.10a shows the I_{as}/I_s value as a function of ϕ_0 with various σ values. Red broken lines indicate the experimental I_{as}/I_s values for the spin-coated film with a spinning rate of 5000 rpm and the solvent-cast film, respectively, which were obtained from the curve-fitting results in Figure 8.9. Given that there is no angular distribution ($\sigma = 0$), the ϕ_0 value for the spin-coated film is around 80°. This means that CH_3 groups were oriented along the direction parallel to the interface. On the other hand, in the case of the solvent-cast film, ϕ_0 is estimated to be around 30°, indicating that CH_3 groups were relatively oriented along the direction normal to the interface.

Taking the film preparation condition into account, it can be understood that the difference in the local conformation of chains is induced by the centrifugal force (F) applied to them during the spinning process. To confirm this hypothesis, the effect of F on ϕ_0 is examined. Here, F applied to a single chain upon the spinning process can be given by,

$$F = mr_{\text{spin}}v^2 \tag{8.9}$$

where m and v are an average weight of a single chain and the angular velocity during spinning, respectively. In the case of the solvent-cast film, F can be regarded as zero.

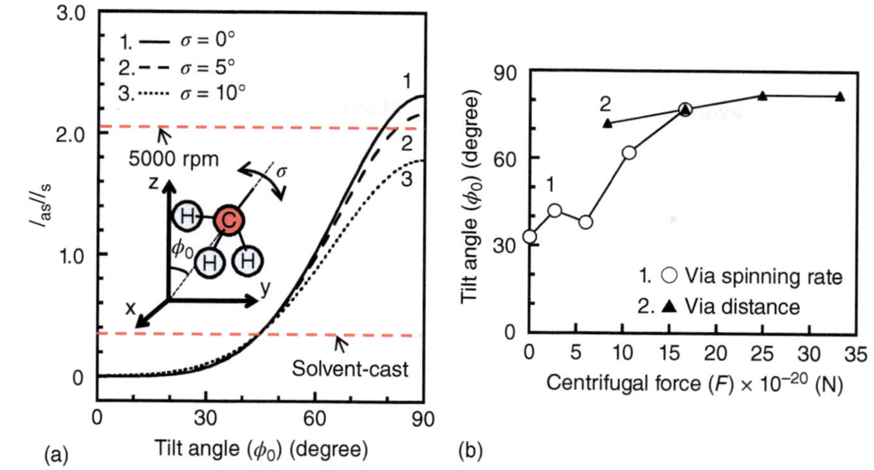

Figure 8.10 (a) SFG intensity ratio of the C–H antisymmetric to symmetric stretching modes for CH_3 groups (I_{as}/I_s) of PI as functions of average tilt angle ϕ_0 and angular distribution σ. The inset is a schematic illustration for a CH_3 group with ϕ_0. (b) Centrifugal force dependence of tilt angle for CH_3 groups at the interface. Open circles and filled triangles denote tilt angles for CH_3 groups derived from panels (a) and (b) of Figure 8.9, respectively. *Source:* Sugimoto et al. [88].

Figure 8.10b shows ϕ_0 as a function of F under the assumption of zero angular distribution. The open circles and filled triangles denote the ϕ_0 values obtained via the spinning rate (Figure 8.9a) and the distance from the spinning center (Figure 8.9b), respectively. It is apparent that ϕ_0 increased with increasing F, meaning that the parallel orientation of CH_3 groups to the interface was somehow induced by F. However, the data set via changing r_{spin} looks to be not simply superimposed on the data via changing ω, especially at a lower F region. This implies that the interfacial orientation of chains cannot be simply explained only in terms of the centrifugal force applied to chains. Recent reports showed that the evaporation rate of the solvent during the spin-coating process also affected the orientation of chains in a film [96–98]. Thus, the discrepancy of the ϕ_0 value at a lower F region may indicate that the interfacial orientation of chains is controlled by not only the centrifugal force but also the solvent evaporation rate during the solidification process.

Despite the T_g^b being much lower than room temperature, PI chains retained their orientation at the quartz interface. This means that the local orientation of PI at the interface could not be easily relaxed even at a temperature higher than T_g^b. To address a temperature at which the local orientation disappeared, the SFG peak intensity upon heating was examined by in situ measurements. Figure 8.11a shows the relationship between temperature and peak intensity of CH_3as at 2965 cm^{-1} for the PI film prepared at a spinning rate of 5000 rpm. The heating rate was fixed at 1 K min^{-1}. The peak intensity gradually decreased with increasing temperature up to 335 K, and then abruptly dropped to be nearly equal to 0. Postulating that the peak intensity is proportional to the population of CH_3 groups oriented at the interface, it can be

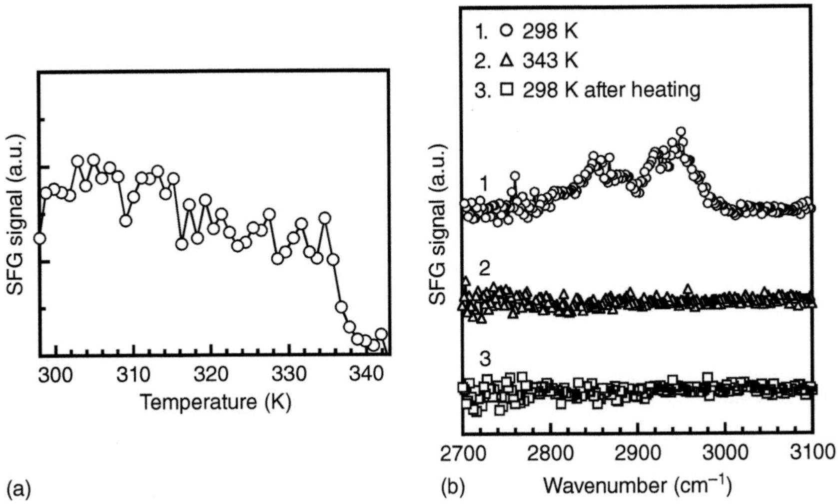

Figure 8.11 (a) Temperature dependence of SFG peak intensity at 2965 cm^{-1} for a PI spin-coated film. (b) SFG spectra for the PI spin-coated films measured at 298 K and 343 K and again at 298 K after heating at 343 K for one hour, respectively. *Source:* Sugimoto et al. [88].

claimed that chains at the interface relaxed at around 335 K. This characteristic temperature is higher than the T_g^b by approximately 120 K.

The local conformation above the characteristic temperature at the interface was further examined by in situ SFG measurements. Figure 8.11b shows SFG spectra for the PI spin-coated film measured at different temperatures. Although the SFG signals were obtained for an as-prepared sample measured at 298 K, all SFG signals disappeared at 343 K, which was higher than the characteristic temperature. This means that the functional groups were randomly oriented at the interface at 343 K. Since the interaction between the hydrophilic quartz substrate and hydrophobic PI segments is weak, this is quite reasonable. When the film was cooled down from 343 K to 298 K, the signals were still absent, as shown in the bottom curve in Figure 8.11b. Therefore, it seems reasonable to conclude that the orientation of chains at the substrate interface is observed when the segments are kinetically trapped during the film solidification processes.

8.3.2.3 Poly(styrene-*co*-butadiene) Rubber [89]

SBR with M_n of 177 K, M_w/M_n of 2.4 and a (PS: 1,4-polybutadiene (1,4-PB): 1,2-polybutadiene (1,2-PB)) weight ratio of 20 : 65 : 15 was supplied by Asahi Kasei Corporation. Dynamic mechanical analysis revealed that SBR had a single α-relaxation peak located around 213 K, and the T_g value obtained by DSC was 204 K. This number was higher and lower than the T_g values of 1,4-PB (174 K [99]) and PS (373 K) homopolymers, respectively. This means that PS and PB segments in this sample were in a miscible state. Films of SBR were prepared from toluene solutions with concentrations of 8.0 and 1.4 wt% onto quartz substrates by spin-coating at 3000 rpm and solvent-casting, respectively, resulting in the

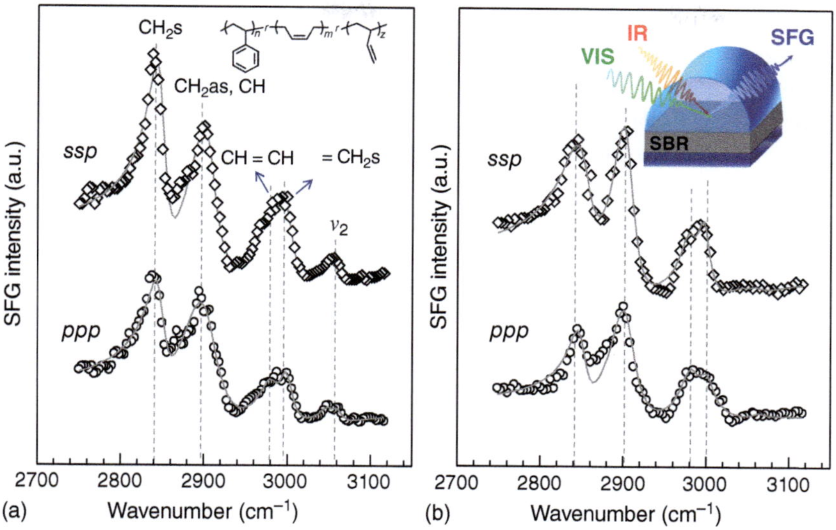

Figure 8.12 SFG spectra collected from (a) spin-coated and (b) solvent-cast SBR films sandwiched between a quartz prism and substrate. The inset of panel (b) shows the film geometry for detecting SFG signals from the SBR/quartz interface. *Source:* Zuo et al. [89].

thicknesses of ca. 800 nm. Two films prepared on a quartz prism and a quartz window were then attached together in a face-to-face geometry and annealed at 303 K under vacuum for 24 hours so that an SBR film sandwiched between quartz substrates was obtained, as shown in the inset of Figure 8.12a [61, 63]. To evaluate the conformational relaxation, the films were annealed at elevated temperatures under vacuum for one hour and quickly cooled, and then the SFG measurements were conducted at room temperature.

Panels (a) and (b) of Figure 8.12 show the SFG spectra from the SBR/quartz interfaces of the spin-coated and solvent-cast films, respectively. An intense peak appearing at 2840 cm^{-1} was attributed to the symmetric C–H stretching vibration of methylene groups (CH_2s), which were included both in PS and PB units. A peak at 2900 cm^{-1} could be originated from the antisymmetric C–H stretching vibration of methylene groups (CH_2as) in PB units [100] and/or the C–H stretching vibration of methyne groups (CH) in PS units [61, 63]. A broad peak observed in the wavenumber range from 2950 to 3015 cm^{-1} might be composed of two contributions from the C-H stretching of methyne groups in CH=CH (2979 cm^{-1}) from 1,4-PB units and symmetric C-H stretching of vinyl methylene groups (=CH_2s) (2999 cm^{-1}) in 1,2-PB units [26]. In addition, a peak observed at 3059 cm^{-1} was assignable to ν_2 vibration mode of phenyl groups [67, 101–103]. The appearance of these clear peaks in the SFG spectra indicates that the local conformation of SBR chains was ordered at the quartz interfaces both for spin-coated and solvent-cast films.

To discuss the local conformation of polymer chains at the interface, the orientation of side chain groups (i.e. phenyl groups in PS units and vinyl CH_2 groups in 1,2-PB units (=CH_2)) were focused on herein. In the case of the spin-coated film, only symmetric A1 vibrational mode [67] such as ν_2 (3059 cm^{-1}) was observed. That is, no B1 symmetric peaks were discerned. Such a spectral feature indicates that phenyl rings were oriented at the interface [67, 103]. On the other hand, no signals were detected in the wavenumber region of 3020–3100 cm^{-1} for the solvent-cast SBR film, meaning that phenyl rings in the solvent-cast film were randomized at the quartz interface. The quantitative analysis of Figure 8.12 revealed that the tilt angles of —CH_2= at the interface were estimated to be about 24° and 51° with respect to the surface normal for the spin-coated and solvent-cast films, respectively.

In short, the local conformation of SBR chains at the quartz interface is schematically summarized in Scheme 8.1. Overall, SBR chains are aligned on the quartz surface for both spin-coated and solvent-cast films. However, the extent of such an interfacial ordering was dependent on how the film was prepared. The side chain groups of SBR (i.e. phenyl groups and vinyl groups for 1,2-PB) were more likely to perpendicularly orient at the substrate interface for the spin-coated film, yet were randomly or less oriented at the interface for the solvent-cast one. The striking interfacial orientation for side chain groups of SBR observed for the spin-coated film can be interpreted as the preferential orientation of the main chain part along the direction parallel to the interface due to the centrifugal force applied to it during the spin-coating process, as evidenced by our previous works [61, 63, 104].

Figure 8.13a shows the SFG spectra with the ssp polarization combination for the spin-coated SBR film annealed at various temperatures. The intensity of all SFG peaks gradually decreased with increasing temperature, indicating the loss of the conformational ordering for SBR chains at the quartz interface. A similar trend was also found for the solvent-cast film, as shown in Figure 8.13b. These results imply that SBR chains at the quartz interface were thermally relaxed accompanied by the

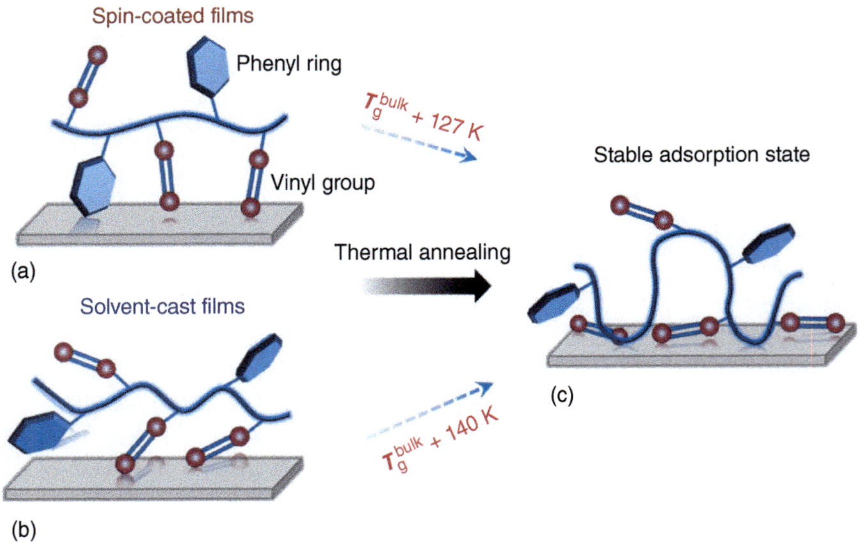

Scheme 8.1 Plausible local conformations of SBR chains at the quartz interface for (a) spin-coated and (b) solvent-cast films, respectively, and (c) the corresponding conformational relaxation induced by thermal annealing. *Source:* Zuo et al. [89].

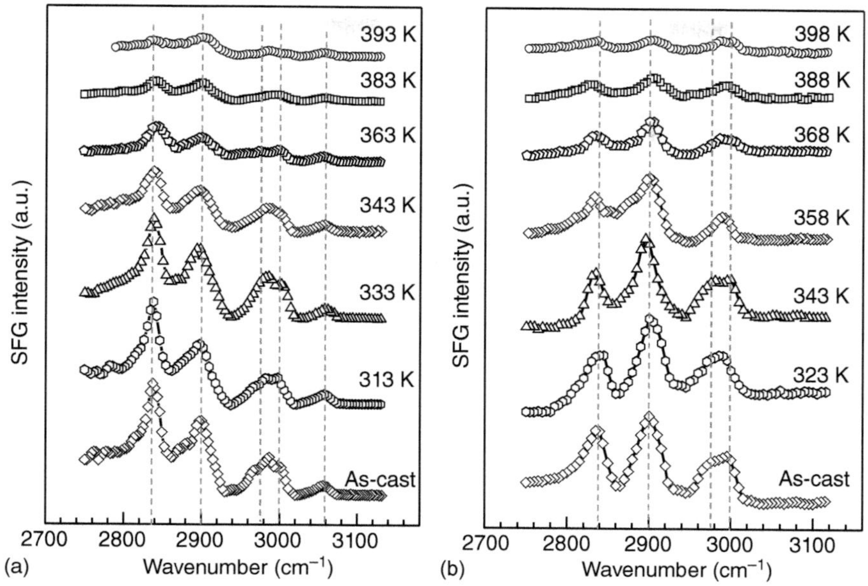

Figure 8.13 SFG ssp spectra for (a) spin-coated and (b) solvent-cast SBR films after the annealing for one hour at various temperatures under vacuum. *Source:* Zuo et al. [89].

loss of the ordering. That is, the relaxation behavior likely corresponds to the stabilizing process of chains at the interface. Taking into account that chains at the interface significantly lost their conformational entropy owing to the ordering structure, it seems reasonable that they changed the local conformation to be more random once the temperature went beyond the critical temperature (T_c), as shown in Scheme 8.1. Such can be realized by the arching up of some segments, which were weakly attached to the substrate surface [13, 104, 105]. In addition, others might get in closer touch with the substrate surface, resulting in an enthalpic gain [13, 104, 105]. These factors make it possible that short loops with high-density anchoring points (i.e. a matured flattened layer [13, 88, 104–106]) would be formed to minimize the total free energy. That is, these adsorbed chains at the interface should be extremely stable in terms of thermodynamics.

The above-mentioned interfacial relaxation induced by the thermal annealing has been observed for PI with a quartz substrate [88] but not for PS [61] and PMMA [79]. This difference can be explained in terms of T_g^b. Even if the interfacial T_g is much higher than the corresponding bulk value, the relaxation for a polymer having a lower T_g^b such as PI and SBR is detectable within the measurable temperature range and the experimentally accessible time scale. In contrast, the interfacial relaxation for a glassy polymer such as PS and PMMA cannot be observed due to the experimental temperature limit.

The peak intensity was detected as a function of temperature upon the continuous heating/cooling process. Figure 8.14a shows the temperature dependence of the intensity for the peak at 2900 cm^{-1}, which is composed of CH_2 as in PB units and/or CH in PS units, for the spin-coated film. The intensity decreased upon the heating,

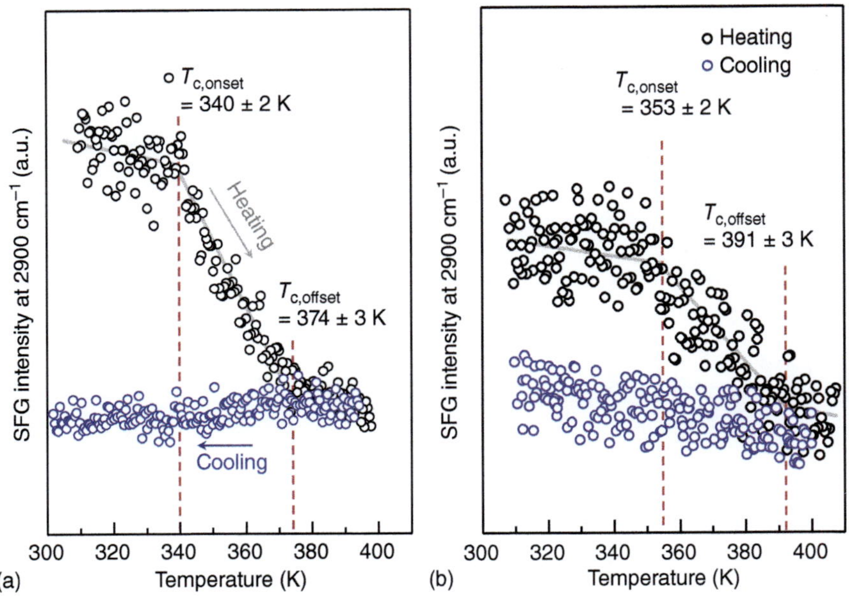

Figure 8.14 The temperature dependence of SFG intensity at 2900 cm^{-1} (with ssp) for (a) spin-coated and (b) solvent-cast SBR films upon heating and cooling processes. The heating and cooling rate was 2 K min^{-1}. *Source:* Zuo et al. [89].

and the slope of the plot was changed at 340 ± 2 K, which was herein defined as $T_{c,onset}$. It remained unchanged once the temperature went beyond approximately 374 ± 3 K, $T_{c,offset}$, implying that chains at the interface almost lost the ordering. Actually, in this temperature region, the SFG peak was almost indiscernible. Conversely, the peak intensity was not changed upon the cooling process, as shown by blue symbols in Figure 8.14a. This means that the local conformation at the interface was irreversibly changed to an energetically favorable one by the thermal annealing. T_c should be also determined on the basis of other SFG signals such as the one at 3059 cm^{-1}, which was originated from phenyl groups in PS units. The $T_{c,onset}$ and $T_{c,offset}$ so obtained for the spin-coated film were very close to those at 2900 cm^{-1} (not shown). The consistency in the $T_{c,onset}$ and $T_{c,offset}$ values determined by the SFG signals at two different wavenumbers of 2900 and 3059 cm^{-1} makes it clear that the conformational relaxation at the interface occurred both with PB and PS units. Thus, it seems reasonable to claim that the interfacial relaxation here observed possesses a relatively large length scale such as the segmental motion.

A similar temperature dependence of SFG signals was observed for the solvent-cast film, as shown in Figure 8.14b. However, the $T_{c,onset}$ value determined on the basis of the peak at 2900 cm^{-1} was shifted to 353 ± 2 K for the solvent-cast film, being higher than that of the spin-coated film by 13 K. This indicates that the chain relaxation at the interface is dependent on the film-preparation history via local conformation.

To discuss the interfacial relaxation dynamics, the time evolution of the SFG peak intensity at 2900 cm^{-1} at a given temperature was examined. Figure 8.15a provides

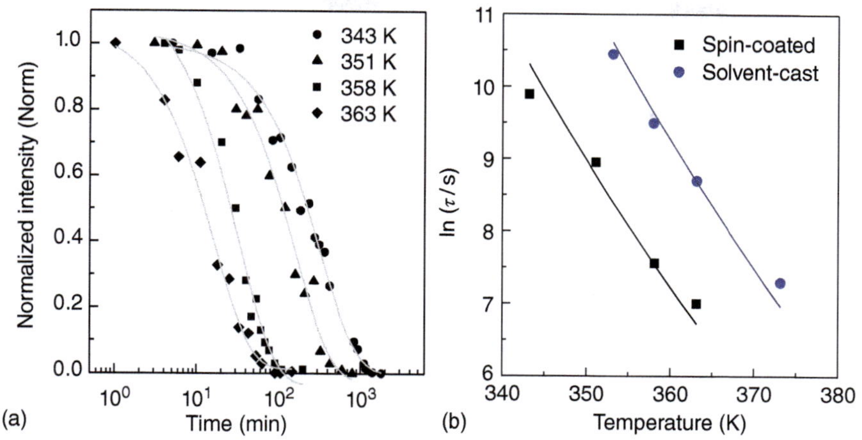

Figure 8.15 (a) Time dependence of the SFG intensity at 2900 cm^{-1} for the spin-coated SBR film. Solid curves were the best fit by the KWW equation. (b) Temperature dependence of relaxation time (τ) extracted from panel (a). Solid curves were the best fit by the VFT equation. *Source:* Zuo et al. [89].

the time dependence of normalized intensity (I_{Norm}) for the spin-coated film. I_{Norm} was simply obtained by the following Eq. (8.10),

$$I_{\text{Norm}} = (I_0 - I_t)/(I_0 - I_\infty) \tag{8.10}$$

where I_0, I_∞, and I_t denote the SFG intensities before and after the annealing for an infinite time, and at a given time t, respectively. The I_{Norm} decreased with increasing time, corresponding to the relaxation dynamics at the interface. The Kohlrausch–Williams–Watts (KWW) equation [107], which has been often used to describe the segmental relaxation for bulk polymers, was adopted to express the time vs. I_{Norm} plots.

$$1 - I_{\text{Norm}} = \text{A} \cdot \exp \cdot (-t/\tau)^{1-\gamma} \tag{8.11}$$

where τ, γ, and A are characteristic relaxation time, exponent related to the intermolecular cooperativity ($1 > \gamma > 0$) and a pre-exponential factor, respectively. The data shown could be well reproduced by the KWW equation, as shown in Figure 8.15a.

The γ values here estimated were 0.25–0.35, and were definitely lower than the reported bulk values for PB (0.55) [108] and PS (0.64) [109]. According to Ngai's coupling model [110, 111], the intermolecular coupling is believed to be the origin for the slowing down of the local segmental motion, and the γ value, or coupling parameter, is a measure of the extent of such a coupling. The reduced γ value observed here means that the intermolecular cooperativity, possibly for the segmental level, at the substrate interface is less than that in the bulk. Using broadband dielectric relaxation spectroscopy in conjunction with DSC, Klonos et al. [112] reported similarly reduced cooperativity and slowing down of dynamics for polydimethylsiloxane adsorbed onto a silica surface. A possible explanation for the reduced cooperativity is that the presence of the substrate becomes dominant over the intermolecular interaction because of an interaction between segments and substrate surface. As a result, segments may not easily cooperate, or couple, with each other in the interfacial layer.

Figure 8.15b shows τ, which was extracted on the basis of KWW fitting, both for spin-coated and solvent-cast films as a function of temperature. The τ value was found to be larger for the solvent-cast film than that for the spin-coated one, indicating slower interfacial dynamics for the solvent-cast film. Applying the Vogel–Fulcher–Tammann (VFT) equation to the data in Figure 8.15b, the temperatures at $\tau = 100$ seconds was estimated to be 372 and 389 K for the spin-coated and solvent-cast films, respectively. Interestingly, these temperatures are in good accordance with $T_{c,offset}$, rather than $T_{c,onset}$, obtained from Figure 8.14.

It has been accepted that the sample-preparation history is related to the residual stress within the film due to the nonequilibrium initial conformation of chains [61, 88, 113–116]. Such an applied internal stress reduces the T_g of the polymer and accelerates the segmental mobility [117–120]. In the case of the solvent-cast process, it takes time for chains to be solidified because of the slow evaporation of solvent molecules. Thus, chains are less oriented in the solvent-cast film than in the spin-coated one, and thereby possess a lesser nonequilibrity. This is more striking at the interface, as shown in the above and our previous publications [61, 88, 104]. Therefore, the spin-coated SBR film with a larger internal stress could make $T_{c,}$ or T_g^{inter}, higher.

8.4 Conclusions

In this chapter, we demonstrate how SFG vibrational spectroscopy, which is one of the most interfacial sensitive techniques, can be applied to study the aggregation states and dynamics of polymers at buried interfaces. The local conformation of PS at nonsolvent interfaces and solid interfaces was discussed. NR revealed that while water and methanol molecules could hardly penetrate into the PS thin films, hexane deeply penetrated into the film, resulting in a diffuse interfacial layer with hexane. Interestingly, these solvents altered the local conformation of PS at their interfaces though the extent was dependent on the interfacial energy except hexane. The conformational change induced by the nonsolvents occurred at the time scale of a few seconds. At solid interfaces, PS phenyl groups in a spin-coated film oriented relatively along the direction normal to the interface because of the centrifugal force generated during the film solidification process. On the other hand, this was not the case for a solvent-cast PS film, which had been solidified over time. These were common for both hydrophobic and hydrophilic substrates. This is because the interaction between PS and the surface is weak. Also, the conformational change of PS at the solid interface was much slower than what we understood on the basis of the polymer dynamics, which had been established for bulk polymers. To gain insight into the polymer dynamics at the solid interface, rubbery polymers such as PI and SBR were used. Even for these polymers having the T_g^b much lower than room temperature, chains contacting with the substrate surface could not be relaxed at room temperature. Applying KWW and VFT equations to the time dependence of the SFG signals at higher temperature, the interfacial T_g was extracted. The value so obtained was extremely higher than the T_g^b and was dependent on how the sample was prepared. This is in excellent accordance with our recent knowledge of polymer dynamics in the confined systems such as interfaces as well as ultrathin states based on a wide variety of experiments. As mentioned above, SFG provides detailed information about the molecular pictures at buried interfaces. Thus, it will provide fundamental knowledge on the design of interfacial polymer engineering such as adhesive and nanocomposite materials.

Acknowledgements

We thank Prof. Norifumi L. Yamada, Dr. Manabu Inutsuka, and Prof. Biao Zuo for their help and fruitful discussion.

References

1 Zhu, X.D., Suhr, H., and Shen, Y.R. (1987). Surface vibrational spectroscopy by infrared-visible sum frequency generation. *Phys. Rev. B* 35 (6): 3047–3050.
2 Shen, Y.R. (1989). Surface-properties probed by 2nd-harmonic and sum-frequency generation. *Nature* 337 (6207): 519–525.
3 Miranda, P.B. and Shen, Y.R. (1999). Liquid interfaces: a study by sum-frequency vibrational spectroscopy. *J. Phys. Chem. B* 103 (17): 3292–3307.
4 Wei, X., Zhuang, X.W., Hong, S.C. et al. (1999). Sum-frequency vibrational spectroscopic study of a rubbed polymer surface. *Phys. Rev. Lett.* 82 (21): 4256–4259.
5 Chen, Z., Shen, Y.R., and Somorjai, G.A. (2002). Studies of polymer surfaces by sum frequency generation vibrational spectroscopy. *Annu. Rev. Phys. Chem.* 53: 437–465.
6 Richmond, G.L. (2002). Molecular bonding and interactions at aqueous surfaces as probed by vibrational sum frequency spectroscopy. *Chem. Rev.* 102 (8): 2693–2724.
7 Chen, Z. (2007). Understanding surfaces and buried interfaces of polymer materials at the molecular level using sum frequency generation vibrational spectroscopy. *Polym. Int.* 56 (5): 577–587.
8 Wilks, P.A. (1962). Infrared analysis by attenuated total reflectance. *Spectrochim. Acta* 18 (6): 884–884.
9 Chen, Y.J., Chen, W.P., and Burstein, E. (1976). Surface-electromagnetic-wave-enhanced raman scattering by overlayers on metals. *Phys. Rev. Lett.* 36 (20): 1207–1210.
10 Gautam, K.S., Schwab, A.D., Dhinojwala, A. et al. (2000). Molecular structure of polystyrene at air/polymer and solid/polymer interfaces. *Phys. Rev. Lett.* 85 (18): 3854–3857.

11 Briggman, K.A., Stephenson, J.C., Wallace, W.E., and Richter, L.J. (2001). Absolute molecular orientational distribution of the polystyrene surface. *J. Phys. Chem. B* 105 (14): 2785–2791.

12 Tatek, Y.B. and Tsige, M. (2011). Structural properties of atactic polystyrene adsorbed onto solid surfaces. *J. Chem. Phys.* 135 (17): 174708.

13 Shimomura, S., Inutsuka, M., Tajima, K. et al. (2016). Stabilization of polystyrene thin films by introduction of a functional end group. *Polym. J.* 48 (9): 949–953.

14 Wang, J., Chen, C.Y., Buck, S.M., and Chen, Z. (2001). Molecular chemical structure on poly(methyl methacrylate) (PMMA) surface studied by sum frequency generation (SFG) vibrational spectroscopy. *J. Phys. Chem. B* 105 (48): 12118–12125.

15 Wang, J., Woodcock, S.E., Buck, S.M. et al. (2001). Different surface-restructuring behaviors of poly(methacrylate)s detected by SFG in water. *J. Am. Chem. Soc.* 123 (38): 9470–9471.

16 Miyamae, T. and Nozoye, H. (2003). Morphology and chemical structure of poly(methyl methacrylate) surfaces and interfaces: restructuring behavior induced by the deposition of SiO_2. *Surf. Sci.* 532: 1045–1050.

17 Rao, A., Rangwalla, H., Varshney, V., and Dhinojwala, A. (2004). Structure of poly(methyl methacrylate) chains adsorbed on sapphire probed using infrared–visible sum frequency generation spectroscopy. *Langmuir* 20 (17): 7183–7188.

18 Kweskin, S.J., Komvopoulos, K., and Somorjai, G.A. (2005). Molecular restructuring at poly(*n*-butyl methacrylate) and poly(methyl methacrylate) surfaces due to compression by a sapphire prism studied by infrared-visible sum frequency generation vibrational spectroscopy. *Langmuir* 21 (8): 3647–3652.

19 Wang, X., Ni, H., Xue, D. et al. (2008). Solvent effect on the film formation and the stability of the surface properties of poly(methyl methacrylate) end-capped with fluorinated units. *J. Colloid Interface Sci.* 321 (2): 373–383.

20 Ohe, C., Kamijo, H., Arai, M. et al. (2008). Sum frequency generation spectroscopic study on photoinduced isomerization of poly(vinyl alcohol) containing azobenzene side chain at the air-water interface. *J. Phys. Chem. C* 112 (1): 172–181.

21 Zhang, W., Zhang, Z.N., and Wang, X.P. (2009). Investigation on surface molecular conformations and pervaporation performance of the poly(vinyl alcohol) (PVA) membrane. *J. Colloid Interface Sci.* 333 (1): 346–353.

22 Moll, C.J., Meister, K., Kirschner, J., and Bakker, H.J. (2018). Surface structure of solutions of poly(vinyl alcohol) in water. *J. Phys. Chem. B* 122 (47): 10722–10727.

23 Zhang, X.X., Li, Y.X., Hankett, J.M., and Chen, Z. (2015). The molecular interfacial structure and plasticizer migration behavior of "green" plasticized poly(vinyl chloride). *Phys. Chem. Chem. Phys.* 17 (6): 4472–4482.

24 Voylov, D.N., Holt, A.P., Doughty, B. et al. (2017). Unraveling the molecular weight dependence of interfacial interactions in poly(2-vinylpyridine)/silica nanocomposites. *ACS Macro Lett.* 6 (2): 68–72.

25 Lachat, V., Varshney, V., Dhinojwala, A., and Yeganch, M.S. (2009). Molecular origin of solvent resistance of polyacrylonitrile. *Macromolecules* 42 (18): 7103–7107.

26 Fang, Y., Li, B., Yu, J. et al. (2013). Probing surface and interfacial molecular structures of a rubbery adhesion promoter using sum frequency generation vibrational spectroscopy. *Surf. Sci.* 615: 26–32.

27 Chen, C.Y., Wang, J., and Chen, Z. (2004). Surface restructuring behavior of various types of poly(dimethylsiloxane) in water detected by SFG. *Langmuir* 20 (23): 10186–10193.

28 Harp, G.P. and Dhinojwala, A. (2005). Direct probe of interfacial structure during mechanical contact between two polymer films using infrared visible sum frequency generation spectroscopy. *J. Adhes.* 81 (3–4): 371–379.

29 Ye, H.K., Gu, Z.Y., and Gracias, D.H. (2006). Kinetics of ultraviolet and plasma surface modification of poly(dimethylsiloxane) probed by sum frequency vibrational spectroscopy. *Langmuir* 22 (4): 1863–1868.

30 Kim, C., Gurau, M.C., Cremer, P.S., and Yu, H. (2008). Chain conformation of poly(dimethyl siloxane) at the air/water interface by sum frequency generation. *Langmuir* 24 (18): 10155–10160.

31 Zhang, C. and Chen, Z. (2013). Probing molecular structures of poly(dimethylsiloxane) at buried interfaces in situ. *J. Phys. Chem. C* 117 (8): 3903–3914.

32 Zhang, D., Shen, Y.R., and Somorjai, G.A. (1997). Studies of surface structures and compositions of polyethylene and polypropylene by IR plus visible sum frequency vibrational spectroscopy. *Chem. Phys. Lett.* 281 (4–6): 394–400.

33 Gracias, D.H., Zhang, D., Shen, Y.R., and Somorjai, G.A. (1998). Surface chemistry-mechanical property relationship of low density polyethylene: an IR plus visible sum frequency generation spectroscopy and atomic force microscopy study. *Tribol. Lett.* 4 (3–4): 231–235.

34 Gracias, D.H., Zhang, D., Lianos, L. et al. (1999). A study of the glass transition of polypropylene surfaces by sum-frequency vibrational spectroscopy and scanning force microscopy. *Chem. Phys.* 245 (1–3): 277–284.

35 Yamamoto, K., Kawaguchi, D., Sasahara, K. et al. (2019). Aggregation states of poly(4-methylpentene-1) at a solid interface. *Polym. J.* 51 (2): 247–255.

36 Johnson, C.M., Sugiharto, A.B., and Roke, S. (2007). Surface and bulk structure of poly-(lactic acid) films studied by vibrational sum frequency generation spectroscopy. *Chem. Phys. Lett.* 449 (1–3): 191–195.

37 de Aguiar, H.B., de Beer, A.G.F., and Roke, S. (2013). The presence of ultralow densities of nanocrystallites in amorphous poly(lactic acid) microspheres. *J. Phys. Chem. B* 117 (29): 8906–8910.

38 Okuno, M. and Ishibashi, T. (2018). Bulk-or-interface assignment of heterodyne-detected chiral vibrational sum frequency generation signal by its polarization dependence. *J. Chem. Phys.* 149 (24): 244703.

39 Chen, C.Y., Even, M.A., Wang, J., and Chen, Z. (2002). Sum frequency generation vibrational spectroscopy studies on molecular conformation of liquid

polymers poly(ethylene glycol) and poly(propylene glycol) at different interfaces. *Macromolecules* 35 (24): 9130–9135.
40 Kim, J., Opdahl, A., Chou, K.C., and Somorjai, G.A. (2003). Hydrophobic-interaction-induced alignment of polymers at the solid/liquid interface studied by infrared-visible sum frequency generation. *Langmuir* 19 (23): 9551–9553.
41 Even, M.A., Chen, C.Y., Wang, J., and Chen, Z. (2006). Chemical structures of liquid poly(ethylene glycol)s with different end groups at buried polymer interfaces. *Macromolecules* 39 (26): 9396–9401.
42 Clarke, M.L., Wang, J., and Chen, Z. (2003). Sum frequency generation studies on the surface structures on plasticized and unplasticized polyurethane in air and water. *Anal. Chem.* 75 (14): 3275–3280.
43 Miyamae, T. and Nozoye, H. (2001). Surface characterization and photochemical behavior of poly(ethylene terephthalate) and TiO_2/poly(ethylene terephthalate) interface by using sum-frequency generation. *J. Photochem. Photobiol. A* 145 (1–2): 93–99.
44 Loch, C.L., Ahn, D., Chen, C.Y. et al. (2004). Sum frequency generation studies at poly(ethylene terephthalate)/silane interfaces: hydrogen bond formation and molecular conformation determination. *Langmuir* 20 (13): 5467–5473.
45 Zhang, C., Myers, J.N., and Chen, Z. (2014). Molecular behavior at buried epoxy/poly(ethylene terephthalate) interface. *Langmuir* 30 (42): 12541–12550.
46 Priestley, R.D., Ellison, C.J., Broadbelt, L.J., and Torkelson, J.M. (2005). Structural relaxation of polymer glasses at surfaces, interfaces and in between. *Science* 309 (5733): 456–459.
47 Tanaka, K., Tsuchimura, Y., Akabori, K. et al. (2006). Time- and space-resolved fluorescence study on interfacial mobility of polymers. *Appl. Phys. Lett.* 89 (6): 061916.
48 Tanaka, K., Tateishi, Y., Okada, Y. et al. (2009). Interfacial mobility of polymers on inorganic solids. *J. Phys. Chem. B* 113 (14): 4571–4577.
49 Nguyen, H.K., Konomi, A., Sugimoto, S. et al. (2018). Rotational dynamics of a probe in rubbery polymers characterized by time-resolved fluorescence anisotropy measurement. *Macromol. Chem. Phys.* 219 (3): 1700329.
50 Nguyen, H.K., Sugimoto, S., Konomi, A. et al. (2019). Dynamics gradient of polymer chains near a solid interface. *ACS Macro Lett.* 8 (8): 1006–1011.
51 Burroughs, M.J., Napolitano, S., Cangialosi, D., and Priestley, R.D. (2016). Direct measurement of glass transition temperature in exposed and buried adsorbed polymer nanolayers. *Macromolecules* 49 (12): 4647–4655.
52 Schonhals, A., Goering, H., Schick, C. et al. (2003). Glassy dynamics of polymers confined to nanoporous glasses revealed by relaxational and scattering experiments. *Eur. Polym. J. E* 12 (1): 173–178.
53 Elmahdy, M.M., Chrissopoulou, K., Afratis, A. et al. (2006). Effect of confinement on polymer segmental motion and ion mobility in peo/layered silicate nanocomposites. *Macromolecules* 39 (16): 5170–5173.
54 Bogoslovov, R.B., Roland, C.M., Ellis, A.R. et al. (2008). Effect of silica nanoparticles on the local segmental dynamics in poly(vinyl acetate). *Macromolecules* 41 (4): 1289–1296.

55 Robertson, C.G. and Roland, C.M. (2008). Glass transition and interfacial segmental dynamics in polymer-particle composites. *Rubber Chem. Technol.* 81 (3): 506–522.

56 Inoue, R., Nakamura, M., Matsui, K. et al. (2013). Distribution of glass transition temperature in multilayered poly(methyl methacrylate) thin film supported on a Si substrate as studied by neutron reflectivity. *Phys. Rev. E* 88 (3): 032601.

57 Panagopoulou, A. and Napolitano, S. (2017). Irreversible adsorption governs the equilibration of thin polymer films. *Phys. Rev. Lett.* 119 (9): 097801.

58 Oh, S.M., Abbasi, M., Shin, T.J. et al. (2019). Initial solvent-driven nonequilibrium effect on structure, properties, and dynamics of polymer nanocomposites. *Phys. Rev. Lett.* 123 (16): 167801.

59 Zhuang, X., Miranda, P.B., Kim, D., and Shen, Y.R. (1999). Mapping molecular orientation and conformation at interfaces by surface nonlinear optics. *Phys. Rev. B* 59 (19): 12632–12640.

60 Noguchi, H., Hiroshi, M., Tominaga, T. et al. (2008). Interfacial water structure at polymer gel/quartz interfaces investigated by sum frequency generation spectroscopy. *Phys. Chem. Chem. Phys.* 10 (32): 4987–4993.

61 Tsuruta, H., Fujii, Y., Kai, N. et al. (2012). Local conformation and relaxation of polystyrene at substrate interface. *Macromolecules* 45 (11): 4643–4649.

62 Horinouchi, A., Yamada, N.L., and Tanaka, K. (2014). Aggregation states of polystyrene at nonsolvent interfaces. *Langmuir* 30 (22): 6565–6570.

63 Inutsuka, M., Horinouchi, A., and Tanaka, K. (2015). Aggregation states of polymers at hydrophobic and hydrophilic solid interfaces. *ACS Macro Lett.* 4 (10): 1174–1178.

64 Seo, Y.-S. and Satija, S. (2006). No intrinsic depletion layer on a polystyrene thin film at a water interface. *Langmuir* 22 (17): 7113–7116.

65 Atarashi, H., Morita, H., Yamazaki, D. et al. (2010). Swelling structure of thin poly(methyl methacrylate) films in various alkyl length alcohols. *J. Phys. Chem. Lett.* 1 (5): 881–885.

66 Calchera, A.R., Curtis, A.D., and Patterson, J.E. (2012). Plasma treatment of polystyrene thin films affects more than the surface. *ACS Appl. Mater. Interfaces* 4 (7): 3493–3499.

67 Curtis, A.D., Calchera, A.R., Asplund, M.C., and Patterson, J.E. (2013). Observation of sub-surface phenyl rings in polystyrene with vibrationally resonant sum-frequency generation. *Vib. Spectrosc.* 68: 71–81.

68 Hirose, C., Akamatsu, N., and Domen, K. (1992). Formulas for the analysis of the surface SFG spectrum and transformation coefficients of cartesian SFG tensor components. *Appl. Spectrosc.* 46 (6): 1051–1072.

69 Hirose, C., Akamatsu, N., and Domen, K. (1992). Formulas for the analysis of surface sum-frequency generation spectrum by CH stretching modes of methyl and methylene groups. *J. Chem. Phys.* 96 (2): 997–1004.

70 Wu, S. (1982). *Polymer Interface and Adhesion*. Taylor & Francis.

71 Zisman, W.A. (1964). Relation of the equilibrium contact angle to liquid and solid constitution. In: *Contact Angle, Wettability, and Adhesion* (ed. F.M. Fowkes), 1–51. American Chemical Society.

72 van der Beek, G.P., Stuart, M.A.C., Fleer, G.J., and Hofman, J.E. (1991). Segmental adsorption energies for polymers on silica and alumina. *Macromolecules* 24 (25): 6600–6611.

73 Nihonyanagi, S., Ishiyama, T., Lee, T.-K. et al. (2011). Unified molecular view of the air/water interface based on experimental and theoretical $\chi(2)$ spectra of an isotopically diluted water surface. *J. Am. Chem. Soc.* 133 (42): 16875–16880.

74 Du, Q., Freysz, E., and Shen, Y.R. (1994). Surface vibrational spectroscopic studies of hydrogen bonding and hydrophobicity. *Science* 264 (5160): 826–828.

75 Kim, J. and Cremer, P.S. (2000). IR–visible SFG investigations of interfacial water structure upon polyelectrolyte adsorption at the solid/liquid interface. *J. Am. Chem. Soc.* 122 (49): 12371–12372.

76 Ursell, P.G., Spalton, D.J., Pande, M.V. et al. (1998). Relationship between intraocular lens biomaterials and posterior capsule opacification. *J. Cataract Refract. Surg.* 24 (3): 352–360.

77 Morita, S., Tanaka, M., and Ozaki, Y. (2007). Time-resolved in situ ATR–IR observations of the process of sorption of water into a poly(2-methoxyethyl acrylate) film. *Langmuir* 23 (7): 3750–3761.

78 Stanners, C.D., Du, Q., Chin, R.P. et al. (1995). Polar ordering at the liquid–vapor interface of *n*-alcohols (C1–C8). *Chem. Phys. Lett.* 232 (4): 407–413.

79 Tateishi, Y., Kai, N., Noguchi, H. et al. (2010). Local conformation of poly(methyl methacrylate) at nitrogen and water interfaces. *Polym. Chem.* 1 (3): 303–311.

80 Hunt, J.H., Guyot-Sionnest, P., and Shen, Y.R. (1987). Observation of C–H stretch vibrations of monolayers of molecules optical sum-frequency generation. *Chem. Phys. Lett.* 133 (3): 189–192.

81 Fink, Y., Ripin, D.J., Fan, S. et al. (1999). Guiding optical light in air using an all-dielectric structure. *J. Light. Technol.* 17 (11): 2039–2041.

82 Hell, S., Reiner, G., Cremer, C., and Stelzer, E.H.K. (1993). Aberrations in confocal fluorescence microscopy induced by mismatches in refractive index. *J. Microsc.* 169 (3): 391–405.

83 Muenchausen, R.E., Jacobsohn, L.G., Bennett, B.L. et al. (2006). A novel method for extracting oscillator strength of select rare-earth ion optical transitions in nanostructured dielectric materials. *Solid State Commun.* 139 (10): 497–500.

84 Zou, H., Wu, S.S., and Shen, J. (2008). Polymer/silica nanocomposites: preparation, characterization, properties, and applications. *Chem. Rev.* 108 (9): 3893–3957.

85 Pushparaj, V.L., Shaijumon, M.M., Kumar, A. et al. (2007). Flexible energy storage devices based on nanocomposite paper. *Proc. Natl. Acad. Sci. U.S.A.* 104 (34): 13574–13577.

86 Rhim, J.W., Park, H.M., and Ha, C.S. (2013). Bio-nanocomposites for food packaging applications. *Prog. Polym. Sci.* 38 (10–11): 1629–1652.

87 Kaushik, A., Kumar, R., Arya, S.K. et al. (2015). Organic-inorganic hybrid nanocomposite-based gas sensors for environmental monitoring. *Chem. Rev.* 115 (11): 4571–4606.

88 Sugimoto, S., Inutsuka, M., Kawaguchi, D., and Tanaka, K. (2018). Reorientation kinetics of local conformation of polyisoprene at substrate interface. *ACS Macro Lett.* 7 (1): 85–89.
89 Zuo, B., Inutsuka, M., Kawaguch, D. et al. (2018). Conformational relaxation of poly(styrene-co-butadiene) chains at substrate interface in spin-coated and solvent-cast films. *Macromolecules* 51 (6): 2180–2186.
90 Li, B.L., Lu, X.L., Ma, Y.H. et al. (2015). Method to probe glass transition temperatures of polymer thin films. *ACS Macro Lett.* 4 (5): 548–551.
91 Zhang, L. and Torkelson, J.M. (2015). Influence of initiator fragments as chain ends on the Tg-confinement effect and dewetting of thin films of ultralow molecular weight polymer. *Polymer* 65: 105–114.
92 Binder, J.L. (1963). The infrared spectra and structures of polyisoprenes. *J. Polym. Sci. Part A* 1 (1): 37–46.
93 Cornell, S.W. and Koenig, J.L. (1969). Raman spectra of polyisoprene rubbers. *Macromolecules* 2 (5): 546–549.
94 Lu, R., Gan, W., Wu, B.-h. et al. (2005). C—H stretching vibrations of methyl, methylene and methine groups at the vapor/alcohol (n = 1–8) interfaces. *J. Phys. Chem. B* 109 (29): 14118–14129.
95 Ye, S., Morita, S., Li, G. et al. (2003). Structural changes in poly(2-methoxyethyl acrylate) thin films induced by absorption of bisphenol A. An infrared and sum frequency generation (SFG) study. *Macromolecules* 36 (15): 5694–5703.
96 Zheng, K., Sun, F., Tian, X. et al. (2015). Tuning the interfacial thermal conductance between polystyrene and sapphire by controlling the interfacial adhesion. *ACS Appl. Mater. Interfaces* 7 (42): 23644–23649.
97 Na, J.Y., Kang, B., Sin, D.H. et al. (2015). Understanding solidification of polythiophene thin films during spin-coating: effects of spin-coating time and processing additives. *Sci. Rep.* 5 (1): 13288.
98 Shioya, N., Shimoaka, T., Eda, K., and Hasegawa, T. (2017). Controlling mechanism of molecular orientation of poly(3-alkylthiophene) in a thin film revealed by using pMAIRS. *Macromolecules* 50 (13): 5090–5097.
99 Robertson, C.G. and Roland, C.M. (2000). Breadth of the α-relaxation function in 1,4-polybutadiene. *Macromolecules* 33 (4): 1262–1267.
100 Hsu, S.L., Moore, W.H., and Krimm, S. (1975). A vibrational analysis of crystalline trans-1,4-polybutadiene. *J. Appl. Phys.* 46 (10): 4185–4193.
101 Duffy, D.C., Davies, P.B., and Bain, C.D. (1995). Surface vibrational spectroscopy of organic counterions bound to a surfactant monolayer. *J. Phys. Chem.* 99 (41): 15241–15246.
102 Zhang, D., Dougal, S.M., and Yeganeh, M.S. (2000). Effects of UV irradiation and plasma treatment on a polystyrene surface studied by IR—visible sum frequency generation spectroscopy. *Langmuir* 16 (10): 4528–4532.
103 Myers, J.N., Zhang, C., Chen, C., and Chen, Z. (2014). Influence of casting solvent on phenyl ordering at the surface of spin cast polymer thin films. *J. Colloid Interface Sci.* 423: 60–66.
104 Sen, M., Jiang, N., Cheung, J. et al. (2016). Flattening process of polymer chains irreversibly adsorbed on a solid. *ACS Macro Lett.* 5 (4): 504–508.

105 Jiang, N.S., Shang, J., Di, X.Y. et al. (2014). Formation mechanism of high-density, flattened polymer nanolayers adsorbed on planar solids. *Macromolecules* 47 (8): 2682–2689.

106 Gin, P., Jiang, N., Liang, C. et al. (2012). Revealed architectures of adsorbed polymer chains at solid-polymer melt interfaces. *Phys. Rev. Lett.* 109: 1–5.

107 Williams, G. (1970). Non-symmetrical dielectric relaxation behaviour arising from a simple empirical decay function. *Trans. Faraday Soc.* 66: 80–85.

108 Frick, B. (1989). *Study of the Glass Transition of Polybutadiene by Neutron Scattering*, 164–171. Darmstadt: Steinkopff.

109 Lindsey, C.P., Patterson, G.D., and Stevens, J.R. (1979). Photon correlation spectroscopy of polystyrene near the glass–rubber relaxation. *J. Polym. Sci.: Polym. Phys. Ed.* 17 (9): 1547–1555.

110 Ngai, K.L., Rizos, A.K., and Plazek, D.J. (1998). Reduction of the glass temperature of thin freely standing polymer films caused by the decrease of the coupling parameter in the coupling model. *J. Non-Cryst. Solids* 235-237: 435–443.

111 Ngai, K.L. and Rendell, R.W. (1991). From conformational transitions in a polymer chain to segmental relaxation in a bulk polymer. *J. Non-Cryst. Solids* 131-133: 942–948.

112 Klonos, P., Kulyk, K., Borysenko, M.V. et al. (2016). Effects of molecular weight below the entanglement threshold on interfacial nanoparticles/polymer dynamics. *Macromolecules* 49 (24): 9457–9473.

113 Tian, H., Yang, Y., Ding, J. et al. (2014). Surface dynamics of poly(methyl methacrylate) films affected by the concentration of casting solutions. *Soft Matt.* 10 (33): 6347–6356.

114 Chandran, S., Handa, R., Kchaou, M. et al. (2017). Time allowed for equilibration quantifies the preparation induced nonequilibrium behavior of polymer films. *ACS Macro Lett.* 6 (11): 1296–1300.

115 Chowdhury, M., Sheng, X., Ziebert, F. et al. (2016). Intrinsic stresses in thin glassy polymer films revealed by crack formation. *Macromolecules* 49 (23): 9060–9067.

116 Chowdhury, M., Al Akhrass, S., Ziebert, F., and Reiter, G. (2017). Relaxing nonequilibrated polymers in thin films at temperatures slightly above the glass transition. *J. Polym. Sci., Part B: Polym. Phys.* 55 (6): 515–523.

117 Lee, H.-N., Paeng, K., Swallen, S.F., and Ediger, M.D. (2009). Direct measurement of molecular mobility in actively deformed polymer glasses. *Science* 323 (5911): 231–234.

118 Reiter, G., Hamieh, M., Damman, P. et al. (2005). Residual stresses in thin polymer films cause rupture and dominate early stages of dewetting. *Nat. Mater.* 4 (10): 754–758.

119 Bending, B., Christison, K., Ricci, J., and Ediger, M.D. (2014). Measurement of segmental mobility during constant strain rate deformation of a poly(methyl methacrylate) glass. *Macromolecules* 47 (2): 800–806.

120 Eyring, H. (1936). Viscosity, plasticity, and diffusion as examples of absolute reaction rates. *J. Chem. Phys.* 4 (4): 283–291.

9

Application of Two-Dimensional Correlation Spectroscopy (2D-COS) in Polymer Studies

Yeonju Park[1], *Isao Noda*[2], *and Young Mee Jung*[1,3]

[1] *Kangwon National University, Kangwon Radiation Convergence Research Support Center, Chuncheon 24341, South Korea*
[2] *University of Delaware, Department of Materials Science and Engineering, Newark, DE 19716, USA*
[3] *Kangwon National University, Chemistry, Institute for Molecular Science and Fusion Technology, Chuncheon 24341, South Korea*

9.1 Introduction

To understand the various properties of polymers, many analytical techniques, such as thermal analysis, rheology, microscopy, scattering, and various spectroscopic techniques, are routinely applied. It is often very important to understand the correlation between the physical properties of the polymer and structural information provided by each obtained datum. Two-dimensional correlation spectroscopy (2D-COS) is a powerful and effective analytical method for better understanding the correlation between properties and the molecular structure of polymers.

2D-COS was firstly proposed by Noda in 1986 to investigate the time-dependent IR dichroism during the dynamic stretching of polymer films [1]. The technique was later extended as the generalized 2D-COS for a wide range of applications in all waveforms, perturbations, and analytical probes [2]. Generalized 2D-COS provides useful information, such as detection of intra- or inter-molecular interactions, resolution of hidden or overlapped bands, determination of sequential order of the events occurring in the system under the perturbation, and many others, which are hardly detected in conventional analytical techniques. The most popularly studied samples by 2D-COS are polymers [3–120] and the most actively used probe in polymer studies is IR spectroscopy [7, 10, 13, 17, 19, 23, 24, 26, 28, 29, 32, 34–36, 41, 43–46, 49, 54, 56, 59, 67–69, 71, 73, 74, 79, 80, 82–84, 87, 90, 93–95, 98, 102, 104, 106, 110, 112, 116–120].

Some examples of the application of 2D-COS in the characterization of biodegradable and thermo-responsive polymers are highlighted in this chapter. The newly developed methods of 2D-COS for polymer studies and its various applications are also discussed.

Spectroscopic Techniques for Polymer Characterization: Methods, Instrumentation, Applications, First Edition. Edited by Yukihiro Ozaki and Harumi Sato.

9.2 Theory

9.2.1 Background

The essential theory of the generalized 2D-COS is briefly discussed first. More details have been described in the review articles, books, and book chapters [2, 3, 25, 70, 92, 121–134].

Figure 9.1 shows the schematic description for obtaining generalized 2D correlation spectra. In conventional spectroscopic measurements, some type of electromagnetic probes, such as IR, Raman, nuclear magnetic resonance (NMR), X-ray, and so on, is applied to the system of interest, and then the measured spectra are analyzed to obtain the detailed information. In the 2D-COS, an additional external perturbation, such as thermal, chemical, electrical, and magnetic, is applied to the system. A set of spectral intensity variations induced by the perturbation, i.e. dynamic spectra, are observed in a systematic manner and converted to 2D correlation spectra through simple cross-correlation analysis.

9.2.2 Properties of 2D-COS

As shown in Figure 9.1, synchronous and asynchronous 2D correlation spectra, which provide complementary information, are directly calculated from the dynamic spectra.

The synchronous 2D correlation spectrum shows that the spectral intensity changes for the two different spectral variables, ν_1 and ν_2, occur or exist simultaneously. In synchronous 2D correlation spectrum, the autopeaks appeared on the diagonal line represent overall spectral intensity changes that responded under external perturbation. The cross-peaks appearing in the off-diagonal line show coincidental or simultaneous variations of spectral intensities observed at two different spectral

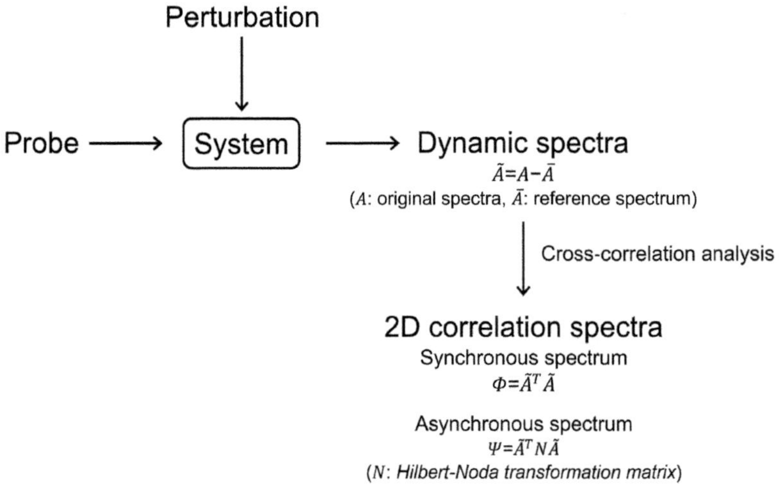

Figure 9.1 The scheme of the generalized 2D-COS.

variables. The positive cross-peak means that two spectral intensities are increasing or decreasing together, while the negative one represents the one that is increasing and the other is decreasing. Therefore, it is possible to detect intra- or inter-molecular interactions from correlations between two different spectral variables.

The asynchronous 2D correlation spectrum shows dissimilar spectral intensity changes of two different spectral variables. In the asynchronous 2D correlation spectrum, only cross-peaks appear, which occur only if two spectral intensities at ν_1 and ν_2 change out of phase with each other. Therefore, it can clearly resolve the overlapped bands corresponding to the spectral signals from different origins.

From the analysis of 2D correlation spectra, one can determine the sequential order of the events observed under the external perturbation. If the signs of the cross-peaks in synchronous and asynchronous spectra are the same, the intensity changes at ν_1 occur before those at ν_2. If the signs of the cross-peaks in synchronous and asynchronous spectra are different, the intensity changes at ν_1 occur after those at ν_2.

9.3 Applications of 2D-COS in Polymer Studies

Various polymers, such as biodegradable polymers [18, 20, 21, 29, 36, 49, 53, 54, 63, 66, 90, 117], thermo-responsive polymers [9, 12, 19, 22, 25, 28, 60, 61, 74, 76, 97, 98, 105, 115, 120], block copolymers [21, 32, 36–38, 53, 61, 66, 71, 85, 90, 103, 115, 117], water-soluble polymers [7, 13, 39, 48, 50, 54, 59, 78, 84, 87, 89, 102, 109], and others [5, 6, 8, 10, 11, 16, 18, 24, 30, 40–47, 52, 56–58, 64, 65, 67, 69, 70, 73, 75, 80, 81, 85–87, 92–95, 100, 101, 106, 107, 110–112, 114], have been studied by using 2D-COS to understand its interesting properties related to structure. For studies of polymer, various probes, such as Raman [32, 46, 49, 53, 101], IR [7, 10, 13, 17, 19, 23, 24, 26, 28, 29, 32, 34–36, 41, 43–46, 49, 54, 56, 59, 67–69, 71, 73, 74, 79, 80, 82–84, 87, 90, 93–95, 98, 102, 104, 106, 110, 112, 116–119], near-infrared (NIR) [6, 31, 44, 81, 100, 104], fluorescence [42, 108], X-ray [16, 48, 87, 90], terahertz (THz) [16, 87, 101], and so on [48, 52], are used in 2D-COS. Among them, IR spectroscopy is the most actively used in 2D-OCS to investigate the thermal behavior of polymers.

Here, some 2D-COS applications in biodegradable and thermo-responsive polymers are highlighted. 2D-COS combined with various analytical techniques, such as 2D gradient mapping, principal component analysis (PCA)-based 2D-COS (PCA 2D-COS), PCA 2D-COS combined with eigenvector manipulation transformation (EMT) technique, hetero-spectral correlation, projection 2D-COS, and smooth factor analysis (SFA) for polymer studies, are also discussed.

9.3.1 Applications of Conventional 2D-COS

9.3.1.1 Biodegradable Polymers

Chen et al. [117] explored the effect of poly(ethylene glycol) (PEG, Mn = 400, 1500, and 3400) having different chain length on the miscibility and thermal properties of biodegradable poly[(*R*)-3-hydroxybutyrate-*co*-(*R*)-3-hydroxyhexanoate] (PHBHx)

(Hx = 6.9 mol%)/PEG blend films using IR spectroscopy and 2D-COS. From differential scanning calorimetry (DSC) result, which showed a single glass transition temperature (T_g) and single melting temperature (T_m) in PHBHx/PEG 400 blend, they reported the lower-molecular-weight PEG (Mn = 400) is fully mixed with PHBHx.

To elucidate the variations of conformation and local molecular environment of PHBHx/PEG blends in response to temperature change, they measured the temperature-dependent infrared reflection–absorption spectroscopy (IRRAS) spectra of PHBHx/PEG blend films during the melting process. Figure 9.2 shows the temperature-dependent IRRAS spectra of PHBHx/PEG blends. As shown in Figure 9.2, the intensities of bands at 1724, 1295, and 1136 cm^{-1} decrease with increasing temperature, while those of bands at 1751 and 1195 cm^{-1} increase. In all spectra of neat PHBHx and its blends, band intensity changes show similar trends. There are no bands of PEG, in the C=O stretching region. Thus, they focused on this region to investigate the influence of PEG on the thermal property of PHBHx.

To more deeply investigate the effect of PEG on the thermal behavior of PHBHx/PEG blend with the addition of different PEG chain lengths, 2D-COS was applied to the temperature-dependent IRRAS spectra of a series of PHBHx/PEG blends as shown in Figure 9.3. A cross-peak at (1744, 1754 cm^{-1}) in the asynchronous 2D correlation spectrum (see Figure 9.3b) is detected only in PHBHx/PEG 400 blend, which indicates that two different amorphous states apparently exist in PHBHx/PEG 400 blend. It suggests that PEG 1500 and PEG 3400 have much less effect on the crystalline structure and thermal behavior of PHBHx because PHBHx and these two PEGs are only partially miscible. They concluded that the possible origin of the amorphous band at 1744 cm^{-1} may be observed only in the 70/30 PHBHx/PEG 400 blend, and this amorphous band probably corresponds to the amorphous mixture of PHBHx and PEG 400. In the related study [18] on morphology, molecular orientation, and crystal thermal stability of PHBHx film with increasing PEG 400 content, the cross-peak at (1744, 1754 cm^{-1}) was observed in the asynchronous 2D IR correlation spectra. From these observations, they confirmed that the amorphous mixture band of PHBHx and PEG at 1744 cm^{-1} occurs in the molten state of the PHBHx/PEG blend during the heating process.

9.3.1.2 Thermo-Responsive Polymers

Park et al. [28, 120] investigated the gelation process and phase transition of thermo-responsive polymers using PCA and 2D-COS. Poly(*n*-isopropyl acrylamide) (PNiPAAm) hydrogel is a well-known thermo-responsive polymer having a lower critical solution temperature (LCST) near body temperature. In this study, the gelation mechanism of PNiPAAm hydrogels below the LCST (22 °C) and above the LCST (38 °C) is explored for the first time at the molecular level in situ [28].

Figure 9.4 shows the in situ FTIR spectra of the gelation process below and above the LCST of PNiPAAm. Similar trends of spectral intensity changes in the in situ FTIR spectra measured at two different temperatures were observed, but the time required for completing the gelation reaction below and above the LCST was different.

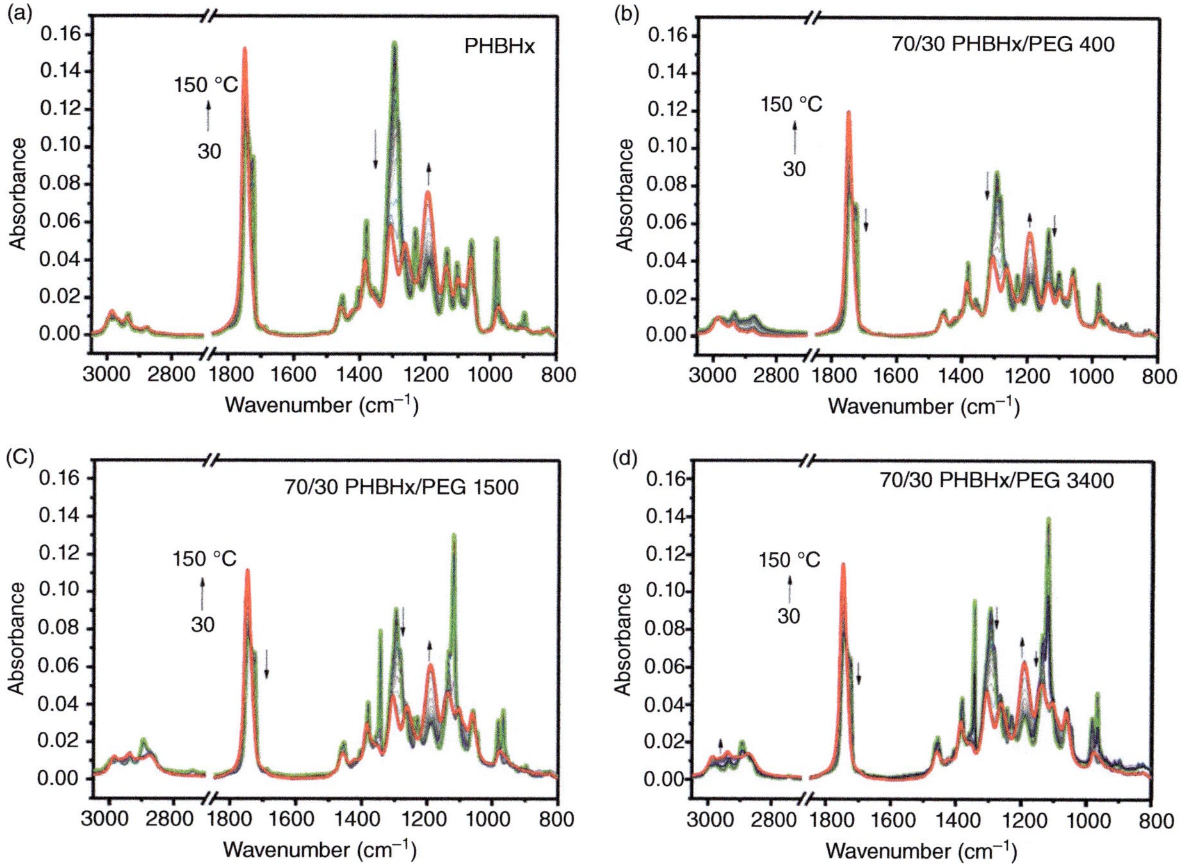

Figure 9.2 Temperature-dependent IRRAS spectra of spin-coated films of PHBHx (a), 70/30 PHBHx/PEG 400 blend (b), 70/30 PHBHx/PEG 1500 blend (c), and 70/30 PHBHx/PEG 3400 blend (d) during the heating from 30 to 150 °C at an interval of 5 °C. *Source:* Chen et al. [117].

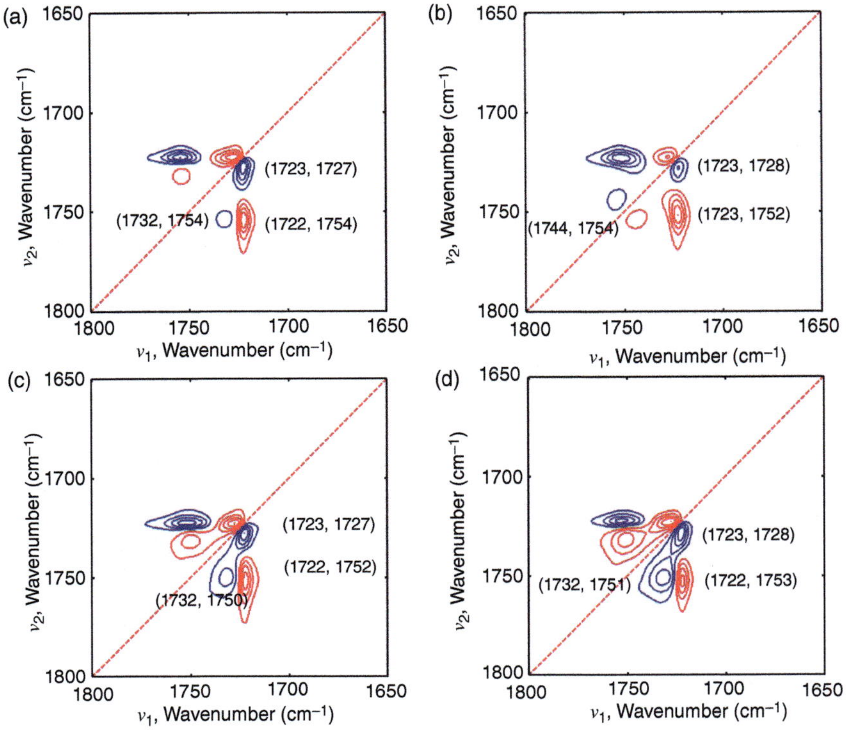

Figure 9.3 Asynchronous 2D correlation spectra of PHBHx (a), 70/30 PHBHx/PEG 400 blend (b), 70/30 PHBHx/PEG 1500 blend (c), and 70/30 PHBHx/PEG 3400 blend (d) in the C=O stretching vibration region generated from their corresponding temperature-dependent IRRAS spectra. The red and blue lines represent positive and negative cross-peaks, respectively. *Source:* Chen et al. [117].

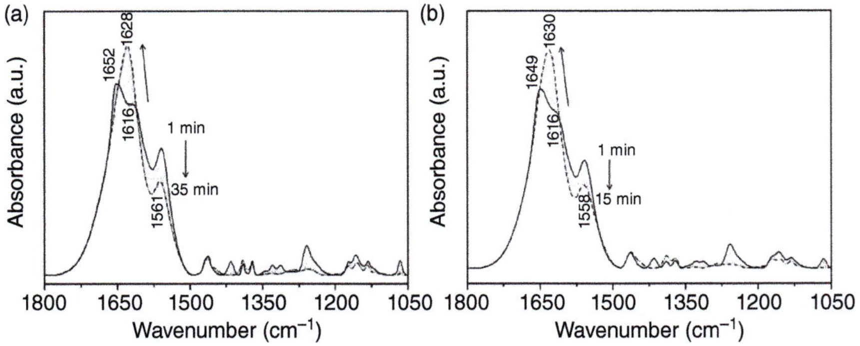

Figure 9.4 FTIR spectra of the NiPAAm gelation process measured at 22 (a) and 38 °C (b). Spectra are collected at every one minute. The solid and dashed lines represent the spectrum measured at 1 and 35 minutes (a) and 1 and 15 minutes (b), respectively, after initiation of the gelation reaction. The dotted lines represent the time evolution of the spectra in between the first and last spectra, and the arrows indicate the trend for the band intensity change with the upward arrow and the downward arrow showing the intensity increase and decrease with time, respectively. *Source:* Park et al. [28].

To better understand and determine the gelation mechanism, 2D-COS was applied to the time evolution of the in situ FTIR spectra of this system. They also performed PCA suggesting that the gelation process occurs in two stages of the reactions process. Based on PCA results, they applied 2D-COS to the selected at specific time spans for two-stage reactions process: before and after 20 and 8 minutes for in situ FTIR spectra below and above LCST, respectively. As shown in Figure 9.5, the 2D correlation spectra are completely different from each other, even though the conventional in situ FTIR spectra are quite similar in Figure 9.4. From the analysis of 2D correlation spectra, they determined the sequence of the spectral changes for the gelation process at two different temperatures, which clearly revealed two different gelation mechanisms below and above the LCST. It is the first report to demonstrate the detailed mechanisms of the gelation process of NiPAAm below and above the LCST.

Park et al. [120] also reported the phase transition of linear PNiPAAm during the heating and cooling processes using IR spectroscopy and 2D-COS. The linear PNiPAAm hydrogel has the LCST near 34 °C. Above the LCST, the hydrophilic, transparent, and swollen hydrogel is changed to the hydrophobic, turbid, and shrunken hydrogel, which is the reversible-phase transition. To elucidate the phase transition mechanism of PNiPAAm during the heating and cooling processes, they analyzed the temperature-dependent FTIR spectra using PCA and 2D-COS.

The temperature-dependent FTIR spectra of linear PNiPAAm gel during the heating and cooling processes are displayed in Figure 9.6. The spectral intensities of bands for the C–H stretching and C=O stretching, including N–H bending modes, increase during the heating process, while they decrease during the cooling process. PCA results show that the intensity changes of major components participating in the phase transition during the heating process are different from those during the cooling process. This observation suggests that the mechanism of transition from the coil to globule during the heating process is different from that during the cooling process. In other words, the phase transition of linear PNiPAAm gel is not completely a reversible process at the molecular level, unlike in its macroscopic appearance change.

For better understanding of the phase transition mechanisms of linear PNiPAAm gel with changing the temperature, they performed 2D-COS. Figure 9.7 shows the synchronous and asynchronous hetero-region 2D correlation spectra of linear PNiPAAm gel during the heating process for the 3020 to2850 and 1800 to 1580 cm^{-1} regions. As shown in Figure 9.7a,c, the synchronous 2D correlation spectra of linear PNiPAAm gel during the heating and cooling processes appear to be seemingly similar, because the trend of spectral intensity changes in C–H stretching and C=O stretching, including N–H bending modes (see Figure 9.6), during the heating and cooling processes is essentially the same. Interestingly, a band at 1681 cm^{-1} corresponding to free C=O stretching mode is negatively correlated with C–H stretching modes, which indicates that the intensity change of free C = O stretching mode is in the opposite direction to C–H stretching modes.

Asynchronous 2D-COS analysis, in turn, provides a much more intriguing insight. As shown in Figure 9.7b,d, the asynchronous 2D correlation spectra of

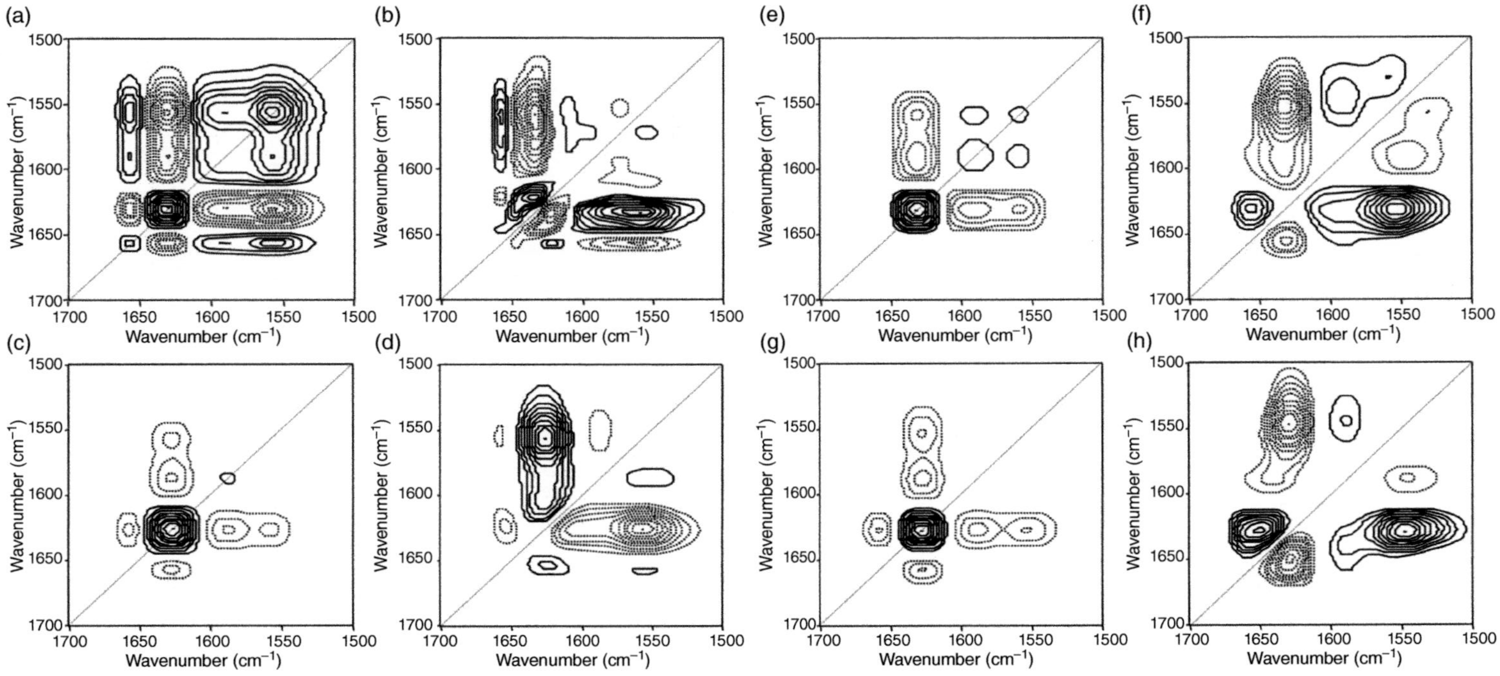

Figure 9.5 Synchronous (a, c, e, g) and asynchronous (b, d, f, h) 2D correlation spectra in the first-stage (a,b,e,f) and second-stage (c,d,g,h) reaction processes of the NiPAAm gelation process at 22 °C (a–d) and 38 °C (e–h). The solid and dotted lines represent positive and negative cross-peaks, respectively. *Source:* Park et al. [28].

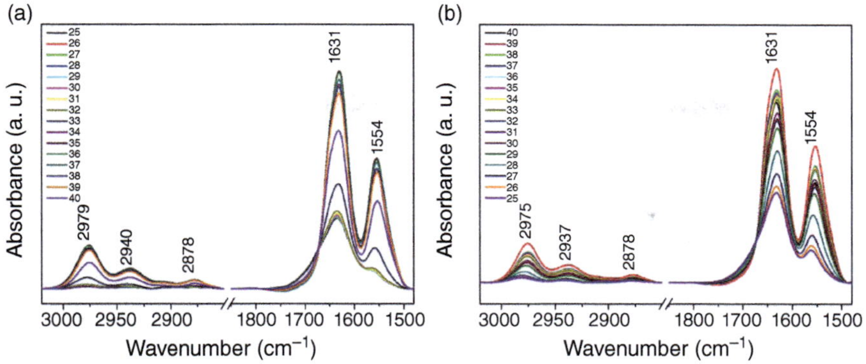

Figure 9.6 Temperature-dependent FTIR spectra of linear PNiPAAm gel during the heating (a) and cooling (b) processes between 25 and 40 °C. *Source:* Park et al. [120].

linear PNiPAAm gel during the heating and cooling processes are completely different. This result clearly suggests that the phase transition mechanism of linear PNiPAAm gel during the heating process is completely different from that during the cooling process. During the heating process, the intensity of the main chain in linear PNiPAAm gel changes after that of side chains. While, during the cooling process, the intensity of C–H stretching modes changes before that of the C=O stretching coupled N–H bending modes.

9.3.2 2D Hetero-Spectral Correlation Analysis

One of the powerful advantages of 2D-COS is the hetero-correlation analysis. Three kinds of 2D hetero-correlation analysis, such as hetero-spectral correlation [81, 86, 90, 135], hetero-sample correlation [136, 137], and hetero-perturbation (or hybrid) correlation [138, 139], are possible. 2D hetero-spectral correlation analysis is most actively applied in various fields, which can compare two different spectral data sets obtained by different analytical techniques but under the same external perturbation. 2D hetero-spectral correlation analysis can be applied to not only correlation between closely related spectroscopic techniques, such as IR and Raman, but also correlation between completely different types of spectroscopic measurements. It provides much more interesting information than conventional 2D-COS. It also sheds light on the correlation between two different spectral changes detected from different types of spectroscopic techniques, which is difficult to observe from a simple analysis of each spectral data alone.

Choi et al. [90] investigated the thermal behavior of a spin-coated film of PHBHx (Hx = 12.0 mol%) copolymers using IR spectroscopy, X-ray photoelectron spectroscopy (XPS), and 2D-COS. Especially the successful application of 2D hetero-spectral XPS/IR correlation spectroscopy to polymer study, which is the correlation between completely different types of spectroscopy, was firstly reported. Useful information for the local atomic environment of the material surface is detected by XPS. Its chemical state and the surface composition of constituent elements can be determined. However, XPS spectra of organic materials are often

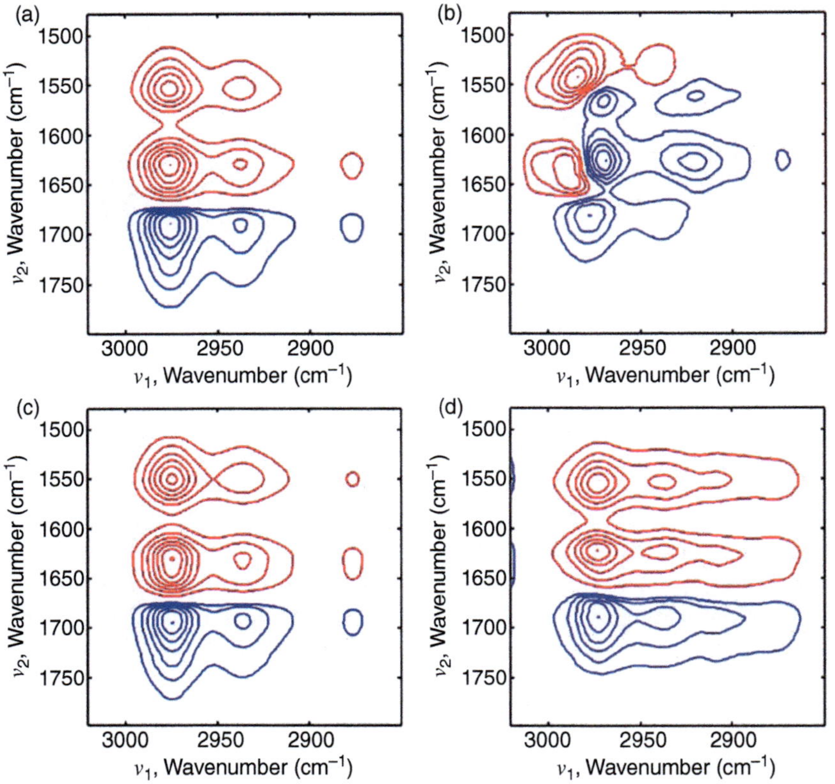

Figure 9.7 Synchronous (a, c) and asynchronous (b, d) hetero-region 2D correlation spectra of linear PNiPAAm gel during the heating (a, b) and cooling (c, d) processes for the 3020 to 2850 and 1800 to 1580 cm^{-1} regions. The red and blue lines represent positive and negative cross-peaks, respectively. *Source:* Park et al. [120].

complicated due to the charging effect. In this study, they applied 2D-COS to the XPS spectra to investigate the details in the structural changes of PHBHx copolymers. The 2D XPS correlation spectra provide enhanced resolution and more information than peak deconvolution or curve fitting techniques.

Furthermore, they also performed the 2D hetero-spectral XPS/IR correlation spectroscopy, to ascertain the band assignment of two resolved binding energies assumed the amorphous (289.3 eV) and crystalline (288.3 eV) C=O in 2D XPS correlation spectra. The assignment of IR spectra for PHBHx copolymers, especially the C=O stretching mode, is establishes well that bands at around 1751 and 1720 cm^{-1} are assigned to amorphous and crystalline C=O components, respectively. 2D hetero-spectral XPS/IR correlation spectra of PHBHx (Hx = 12.0 mol%) copolymer during the heating process are displayed in Figure 9.8. In the synchronous 2D hetero-spectral XPS/IR correlation spectra, the IR band at 1751 cm^{-1} for amorphous C=O stretching mode is positively correlated with the XPS band at 289.3 eV, but negatively correlated with the XPS band at 288.3 eV. While the IR band at 1726 cm^{-1} for crystalline C=O stretching mode is positively correlated with the XPS band at

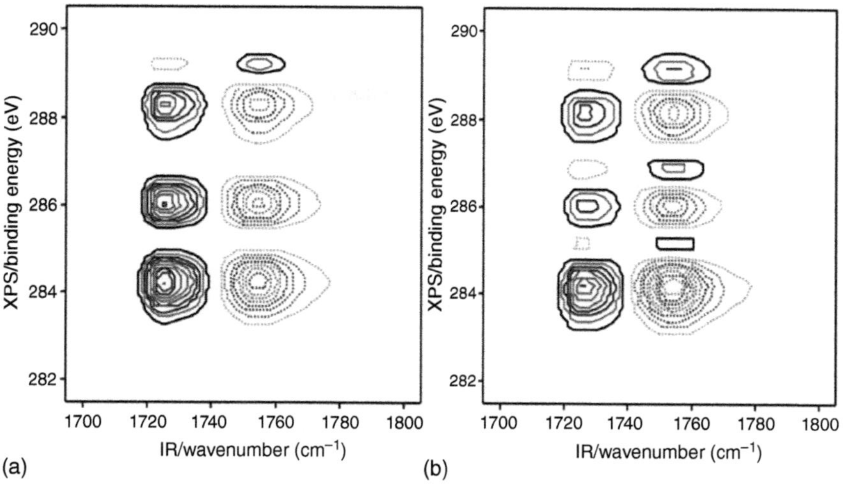

Figure 9.8 Synchronous (a) and asynchronous (b) 2D hetero-spectral XPS/IR correlation spectra of PHBHx (Hx = 12.0 mol%). The solid and dashed lines represent positive and negative cross-peaks, respectively. *Source:* Choi et al. [90].

288.3 eV, it is negatively correlated with the XPS band at 289.3 eV. These observations certainly confirm that the XPS bands at 289.3 and 288.3 eV, respectively, correspond to the amorphous and crystalline C=O components. From the analysis of 2D hetero-spectral XPS/IR correlation spectra, the sequence of the spectral intensity changes shows that intensity changes detected by the IR spectroscopy always occurred before those by the process involving structural changes detected at a quite different level of microscopic scales. Such intriguing subtle differentiation of the probe sensitivity is hardly obtained from a simple analysis of XPS or IR spectra alone.

9.3.3 Two-Dimensional (2D) Gradient-Mapping Method

Determination of a transition temperature is one of the important issues in polymer studies to understand the thermal property of the polymer. Jung et al. [140] have proposed a 2D gradient map as a very simple, useful, and direct presentation method for the determination of the transition temperature of polymers. In the 2D gradient-mapping method, the first derivatives of the absorbance with respect to temperature are displayed on a single map. In other words, the first derivatives form, $dA(\nu, T)/dT$, is presented with wavenumber (ν) and temperature (T) on a 2D map.

Ji et al. [14] also determined the transition temperature of the spin-coated film of PHBHx (Hx = 7.2 mol%) from the 2D gradient mapping method. The 2D gradient map of the temperature-dependent IRRAS spectra of PHBHx (Hx = 7.2 mol%) is shown in Figure 9.9. The highest value of $dA(\nu, T)/dT$ is observed near 140 °C for all bands. Thus, the transition temperature of a spin-coated film of PHBHx (Hx = 7.2 mol%) copolymer determined by a 2D map is about 140 °C. The transition temperature determined by this method is very close to the melting temperature of the corresponding bulk polymer [141, 142]. Based on the results of the 2D gradient

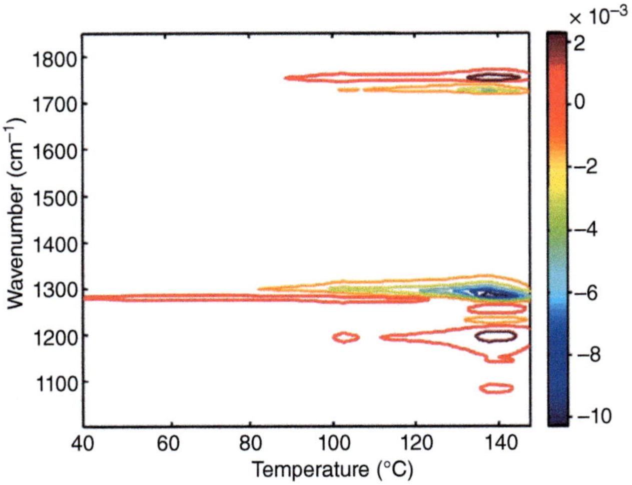

Figure 9.9 2D gradient map of *dA(ν, T)/dT* as a function of *ν* and *T*. Here, *ν* is the wavenumber and *T* is the temperature. *Source:* Ji et al. [14].

mapping method, they applied 2D-COS to two different temperature ranges of the spectral data below and above the transition temperature (140 °C). In this study, 2D correlation spectra show that the bands at 1298 and 1280 cm^{-1} are assigned to less-ordered and well-ordered crystalline components, respectively. Results of 2D-COS confirm that the intensity of the band at 1298 cm^{-1} changes much more significantly below the transition temperature, while that of 1280 cm^{-1} changes more greatly above the transition temperature.

Kim et al. [118] demonstrated the thermal behavior of spin-coated films of PHBHx/PEG blends having different PEG wt% using 2D-COS. They also used a 2D gradient mapping method to determine the transition temperatures of PHBHx/PEG blend films. It shows that the transition temperature of PHBHx/PEG blends decreased from 137.5 to 132.5 °C, with increasing PEG content of PHBHx/PEG blends, which is consistent with the DSC result. This result suggests that the transition temperature obtained from the 2D gradient map reflects the melting point of PHBHx/PEG blends as compared to the DSC results.

9.3.4 Chemometric Techniques Combined with 2D-COS

The successful combination of chemometric techniques and 2D-COS can provide more useful information to interpret very subtle spectral changes of various systems, which is hardly explained in conventional 2D-COS [143–148]. Jung et al. introduced the effective combination of 2D-COS and PCA for the first time, which is named PCA 2D-COS. In this PCA 2D-COS technique, PCA is used as an essential and integral part of the subsequent 2D-COS analysis [143, 146]. They also subsequently proposed a new and interesting concept of EMT for PCA 2D-COS [145, 148].

PCA 2D-COS and EMT (eigenvalue manipulating transformation) techniques are briefly described.

Let the spectral data be expressed by n-by-m data matrix $\mathbf{A}$.

The **A** matrix can be represented in terms of the PCA form as follows:

$$\mathbf{A} = \mathbf{W}\mathbf{V}^{\mathrm{T}} + \mathbf{E} \tag{9.1}$$

Here, **W** is n-by-r orthogonal score matrix, **V** is m-by-r orthogonal loading vector matrix, and **E** is the residual matrix often associated with pure noise. The number of PCA factors r is usually selected to be much less than the total number of spectra (n) for the analysis.

As a result, a new PCA model for the data matrix **A*** can be reconstructed only using the selected score and loading matrices **W** and **V** as follows:

$$\mathbf{A}^{*} = \mathbf{W}\mathbf{V}^{\mathrm{T}} \tag{9.2}$$

This reconstructed data set represented in the new matrix **A*** is applied to the 2D-COS instead of the raw data **A**, which can often reduce the effect of noise in 2D correlation spectra without any substantial loss of information of the system.

Ji et al. [14] demonstrated the thermal behavior of PHBHx (Hx = 7.2 mol%) during the heating process using PCA 2D-COS. The temperature-dependent IR spectra were reconstructed with loading vectors and scores using only three major principal components (PCs) decomposed by PCA. Because three PCs have all essential information (over 99.9%) to demonstrate the temperature-dependent IR spectra of PHBHx, the reconstructed data, which no longer contains noise contributions, were used to 2D-COS instead of the original data.

The EMT reconstruction of the data matrix provides the enhancement of the spectral selectivity for 2D-COS, particularly for the synchronous 2D correlation spectrum. EMT-reconstructed data are obtained by altering the eigenvalues associated with the spectral data set.

The PCA-reconstructed data matrix **A*** is also presented as a singular value decomposition (SVD) form.

$$\mathbf{A}^{*} = \mathbf{U}\mathbf{S}\mathbf{V}^{\mathrm{T}} \tag{9.3}$$

Here, **U** and **V** are orthonormal matrices. The PCA score matrix **W** is now represented in the form of $\mathbf{W} = \mathbf{US}$ and can be obtained directly from $\mathbf{W} = \mathbf{A}^{*}\mathbf{V}$. A diagonal matrix **S** is given by $\mathbf{S} = \mathbf{L}^{1/2}$, and **L** given by $\mathbf{W}^{\mathrm{T}}\mathbf{W}$ is a diagonal matrix where each diagonal element is associated with the eigenvalue of a PC. For the EMT operation, the new transformed data matrix **A**** is obtained by replacing eigenvalues of **A*** as follows:

$$\mathbf{A}^{**} = \mathbf{U}\mathbf{S}^{**}\mathbf{V}^{\mathrm{T}} \tag{9.4}$$

S** is given by varying the corresponding singular values, i.e. positive square roots of eigenvalues, in **S** by changing them to the power parameter m.

$$\mathbf{S}^{**} = \mathbf{S}^{\mathbf{m}} \tag{9.5}$$

If the $\mathbf{m} = 1$, **A**** corresponds to the normal PCA-reconstructed data matrix, because it is given by $\mathbf{A}^{*} = \mathbf{U}\mathbf{S}\mathbf{V}^{\mathrm{T}} = \mathbf{W}\mathbf{V}^{\mathrm{T}}$. If the $\mathbf{m} > 1$, a larger eigenvalue becomes more prominent resulting in a gradual noise reduction. If the $\mathbf{m} < 1$, a smaller eigenvalue becomes more prominent and finer details from contributions of minor components become amplified resulting in spectral selectivity enhancement.

Ji et al. [79] applied PCA 2D-COS combined with EMT technique to the temperature-dependent IR spectra of PHBHx (Hx = 3.8, 7.2, 10.0 mol%) spin-coated films. The 2D correlation spectra obtained using the PCA-reconstructed data are displayed (Figure 9.10). The intensity changes of the bands at 1136, 1282, and 1753 cm^{-1} for PHBHx (Hx = 10.0 mol%) copolymer appear much more significantly than those for PHBHx (Hx = 3.8, 7.2 mol%) copolymers. There is also a new band at 1265 cm^{-1} in PHBHx (Hx = 10.0 mol%) copolymer compared to PHBHx (Hx = 3.8, 7.2 mol%) copolymers. Here, the bands at 1136, 1265, 1282, and 1753 cm^{-1} are related to crystalline C–O–C stretching mode, amorphous C–O stretching mode, well-ordered crystalline C–O stretching mode, and less ordered secondary crystalline C=O stretching mode, respectively. This result indicates that amount of Hx mol% affects intensity variations of these bands.

To more clearly detect the subtle difference of thermal behavior of PHBHx (Hx = 3.8, 7.2, 10.0 mol%) films, the EMT technique was applied to the original data by decreasing the power parameter. Here, the power parameter m was 1/4, as shown in Figure 9.11. The intensity variation of a band at 1282 cm^{-1} is obviously observed in not only PHBHx (Hx = 10.0 mol%) copolymer but also in PHBHx (Hx = 3.8, 7.2 mol%) copolymers. The relative intensities of two bands at 1282 and 1295 cm^{-1} (due to less-ordered crystals) are also changed with increasing Hx molar contents in PHBHx copolymers. Based on these results, they demonstrated that the intensity change includes some contribution from the amorphous component that is not readily detected in conventional spectral analysis. Minor contribution component, such an amorphous, presented in PHBHx (Hx = 3.8, 7.2, 10.0 mol%) films during the heating process can be further remarkably elucidated by the EMT technique.

9.3.5 Smooth Factor Analysis

Noise is a common problem in spectral analysis. Particularly it becomes serious in the 2D-COS analysis because its application to the noisy spectral data results in unwanted artifacts in 2D correlation spectra. The intrinsic strength of 2D-COS, which can effectively accentuate even subtle differences in spectral intensity changes, becomes the unfortunate source of noise interference. For this reason, the spectral data are often pretreated using various smoothing methods, such as moving average or Savitzky–Golay filters. If smoothing is overly performed to raw spectral data, it can distort the features and lose their real information, causing to give misreading 2D-COS results. The use of the conventional smoothing method needs caution and training.

PCA reduces the dimensionality of a data set by expressing the pertinent information content parsimoniously using only a limited number of factors. Therefore, it is a denoising method for a set of spectral data having small and uncorrected noise components. Jung et al. [146, 147] used PCA 2D-COS for removing noise from the original spectra without any smoothing treatment. However, PCA is not always an effective denoising technique if the level of noise is high.

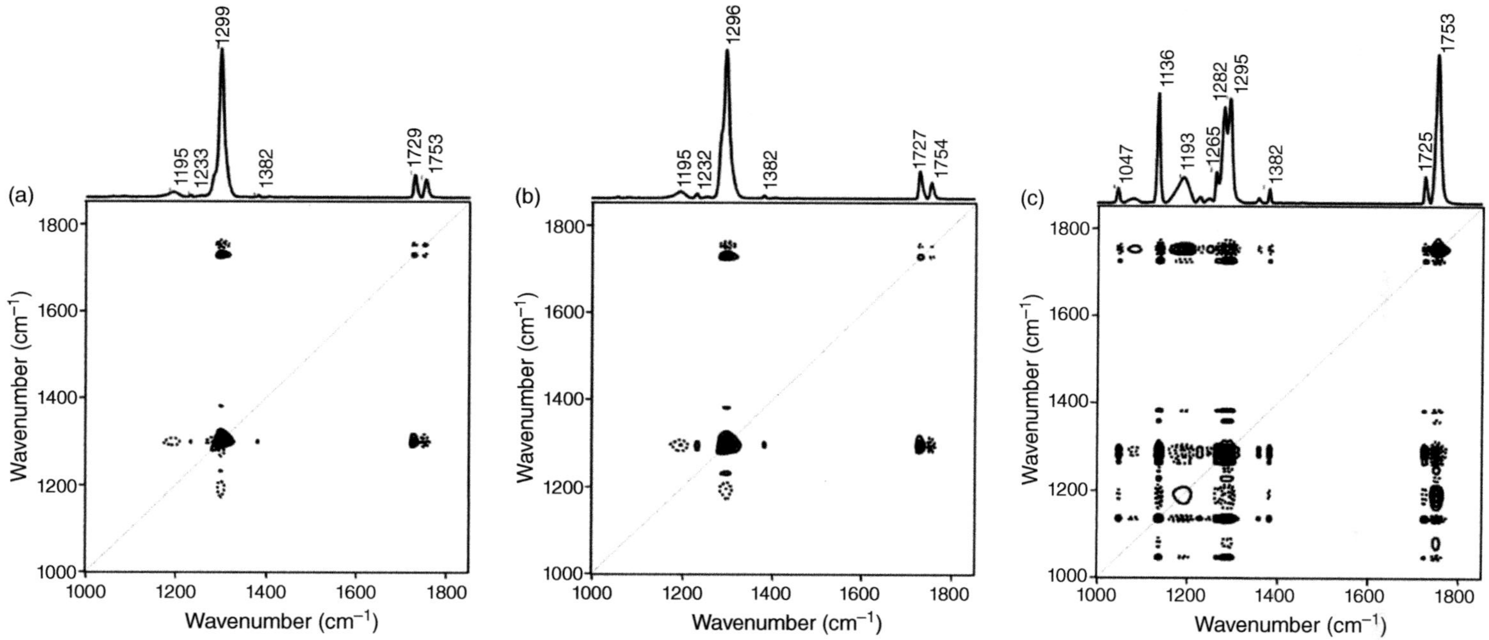

Figure 9.10 Synchronous 2D correlation spectra of spin-coated films of PHBHx (Hx = 3.8 (a), 7.2 (b), and 10.0 (c) mol%) copolymers obtained using the PCA-reconstructed data from loading vectors and scores of PC1, PC2, and PC3. The solid and dashed lines represent positive and negative cross-peaks, respectively. *Source:* Ji et al. [79].

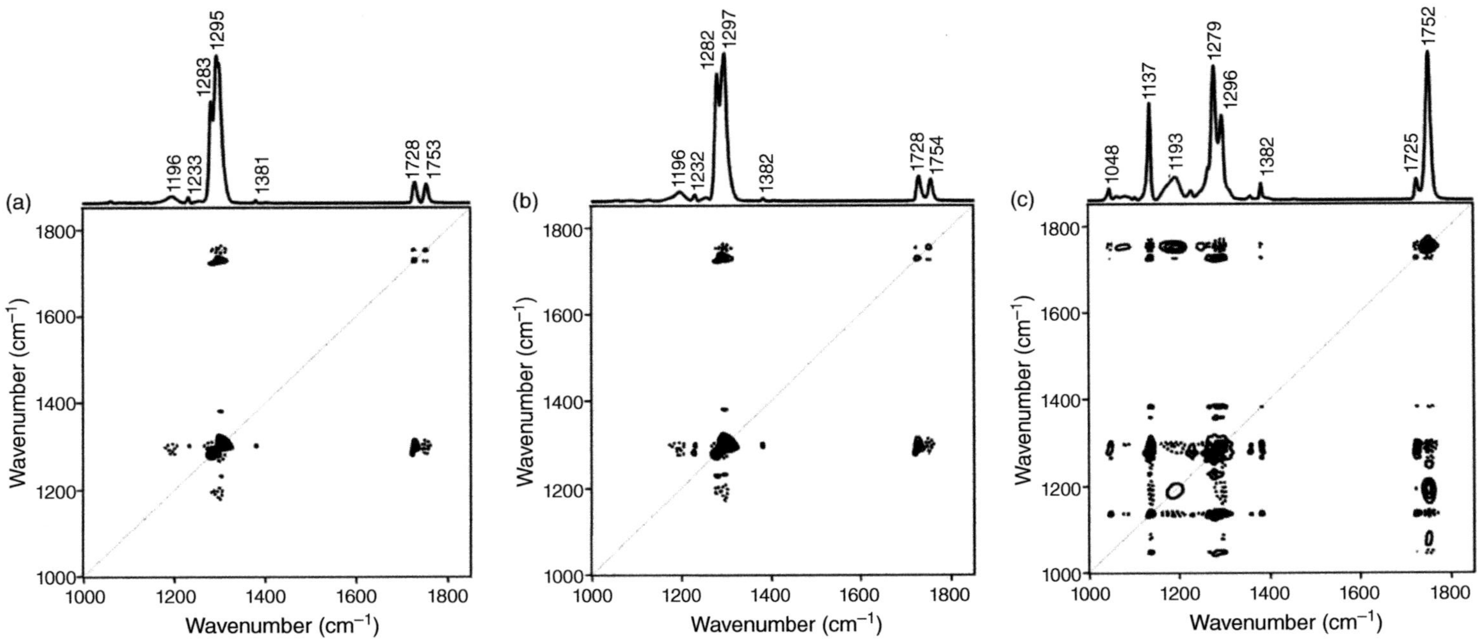

Figure 9.11 Synchronous PCA 2D correlation spectra of spin-coated films of PHBHx (Hx = 3.8 (a), 7.2 (b), and 10.0 (c) mol%) copolymers constructed from obtained from the EMT-reconstructed data by varying the value of power parameter m, at m = 1/4. The solid and dashed lines represent positive and negative cross-peaks, respectively. *Source:* Ji et al. [79].

Recently, Park et al. [20] introduced a very effective noise removal method in heavily noisy spectra, which is called SFA. SFA is based on the alternative nonlinear iterative partial least squares (NIPALS) algorithm.

Let a set of heavy noisy spectra $\mathbf{A}(i, j)$ consist of ith stimuli and jth spectral variables, then the reconstructed to noise removal spectra through modified NIPALS algorithm is as follows:

Step 1. Let A_k be the data set for the kth NIPALS stage. Start with $\mathbf{A}_1 = \mathbf{A}$.
Step 2. Obtain the loading and score vectors (**p** and **t**) by the standard PCA method, such as SVD or power method. Assume the kth normalized score vector $\mathbf{t}_k$, typically from a column of $\mathbf{A}_k$.
Step 3. Estimate the corresponding kth loading vector by $\mathbf{p}_k = \mathbf{t}_k{}^{\mathrm{T}}\mathbf{A}$. Superscript T represents the transpose of the vector of the matrix.
Step 4. Smooth $\mathbf{p}_k$ extensively by any of the standard techniques, e.g. Savitzky–Golay filter. If appropriate, a nonlinear robust filter may also be used. Importantly, minor distortion introduced by the oversmoothing is tolerated.
Step 5. Estimate the score vector $\mathbf{t}_k = \mathbf{A}\mathbf{p}_k$.
Step 6. Extensively smooth $\mathbf{t}_k$ first then normalize it.
Step 7. Go back to Step 2 and repeat until it converges (i.e. almost the same $\mathbf{t}_k$ and $\mathbf{p}_k$).
Step 8. Enter the score $\mathbf{t}_k$ to the kth column of U, the norm $||\mathbf{p}_k||$ to the kth diagonal element of S, and the normalized loading $\mathbf{p}_k/||\mathbf{p}_k||$ to the kth column of V.
Step 9. Calculate the new data matrix $\mathbf{A}_{k+1} = \mathbf{A}_k - \mathbf{t}_k\mathbf{p}_k{}^{\mathrm{T}}$.
Step 10. Increment the NIPALS stage from k to $k + 1$ and go back to Step 1.
Step 11. Stop when no more "meaningful" factors can be extracted: $\mathbf{A} = \mathbf{USV}^{\mathrm{T}} + \mathbf{A}_{k+1}$

Figure 9.12 shows the powerful advantage of the SFA method for denoising from noisy spectral data. As shown in Figure 9.12a,b, the temperature-dependent IR spectra of PHB thin film in C=O stretching region were measured at 4 cm^{-1} resolution and 512 scans and at a 0.5 cm^{-1} resolution and a single scan during the heating process for noise-free IR spectra and noisy IR spectra, respectively. In Figure 9.12, the asynchronous 2D correlation spectrum for noisy IR spectra is too noisy to analyze using 2D-COS. To remove this noise contribution, the SFA method was applied to noisy IR spectra (shown in Figure 9.12b). The first seven smoothing factors (SFs) of the 81-point and 3-point Savitzky–Golay double smoothing for loading vectors and scores, respectively, during NIPALS, were used in SFA-reconstructed spectra. The asynchronous 2D correlation spectrum (shown in Figure 9.12e) for SFA-treated noisy IR spectra is surprisingly the same as that (shown in Figure 9.12c) for the noisy free IR spectra without any distortion of features.

9.3.6 Projection 2D-COS

Projection 2D-COS was introduced by Noda [149] for simplifying and streamlining the complex 2D correlation spectra observed from too complicated systems such as polymers. The combination of the projection and null-space projection operations can attenuate or augment select features in 2D correlation spectra with highly

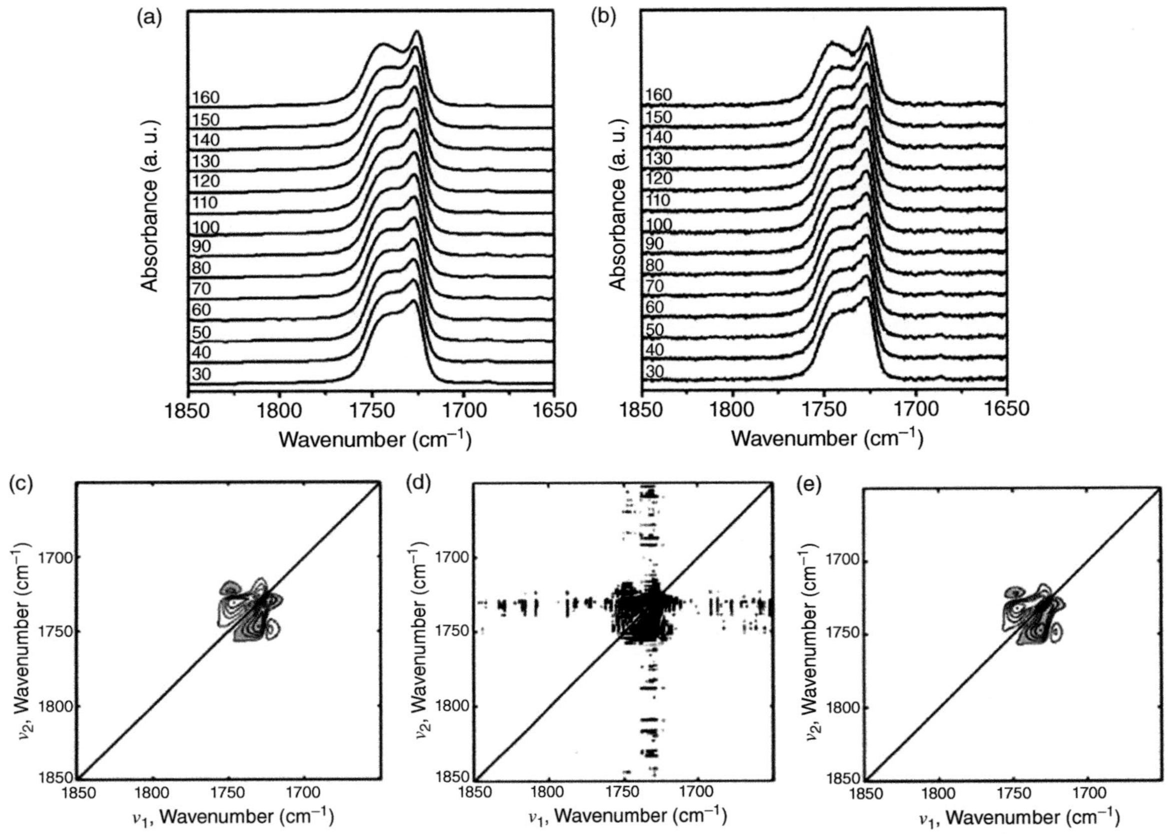

Figure 9.12 Temperature-dependent IR spectra of PHB thin film measured at 4 cm^{-1} resolution and 512 scans (a) and at a 0.5 cm^{-1} resolution and a single scan (b) during the heating process. Asynchronous 2D correlation spectra obtained from noise-free IR spectra (c), noisy IR spectra (d), and SFA-treated noisy IR spectra (e). Negative cross-peaks are indicated by shading. *Source:* Park et al. [20].

overlapped cross-peaks to more easily interpret 2D correlation spectra. The mathematical matrix projection is used for this method. For obtaining the projection 2D correlation spectra, the projection matrix $\mathbf{P_Y}$ of an arbitrary matrix $\mathbf{Y}$ is prepared, which is typically selected from a portion of the spectral data matrix $\mathbf{A}$.

$$\mathbf{P_Y} = \mathbf{Y}(\mathbf{Y}^T\mathbf{Y})^{-1}\mathbf{Y}^T \tag{9.6}$$

Here, the subscripts T and -1 indicate the transpose and inverse operation of the matrix, respectively. The projected data matrix $\mathbf{A_p}$ can be obtained by the simple multiplication of $\mathbf{P_Y}$ with $\mathbf{A}$.

$$\mathbf{A_P} = \mathbf{P_Y A} \tag{9.7}$$

The corresponding positively null-space projected data matrix $\mathbf{A_N}$ is produced as

$$\mathbf{A_N} = \mathbf{A} - \mathbf{A_P} \tag{9.8}$$

This null-space projection is possible to filter out the $\mathbf{A_p}$ contribution from $\mathbf{A}$.

Therefore, the above-projected matrix converts to the 2D correlation spectra as follows:

$$\mathbf{\Phi_P} = \mathbf{A_P^T A_P} \tag{9.9}$$

$$\mathbf{\Psi_P} = \mathbf{A_P^T N A_P} \tag{9.10}$$

Here, $\mathbf{N}$ is m-by-m Hilbert-Noda transformation matrix given by $N_{jk} = 1/\{\mathrm{p}(k-j)\}$.

The null-space projection 2D correlation spectra are given by

$$\mathbf{\Phi_N} = \mathbf{A_N^T A_N} \tag{9.11}$$

$$\mathbf{\Psi_N} = \mathbf{A_N^T N A_N} \tag{9.12}$$

More details of the background of projection 2D-COS are referred to in Noda (2010) [149].

Kim et al. [119] applied the projection 2D-COS to the temperature-dependent IR spectra of PHBHx/PEG blend film, to clearly identify the contribution of PEG in PHBHx/PEG blend system. All information in the conventional 2D correlation spectrum indicates the PHBHx contribution from not PEG in PHBHx/PEG blend film during the heating process as shown in Figure 9.13a. In this study, for projection 2D correlation spectra, they used the transformation data by the vector projection with a band at 1724 $\mathrm{cm^{-1}}$ assigned to crystalline C=O stretching bands of PHBHx. As shown in Figure 9.13b, the projection 2D correlation spectrum is the same as the conventional 2D correlation spectrum (see Figure 9.13a). This result suggests that all spectral intensity changes in the conventional 2D correlation spectrum are from the contribution of PHBHx in PHBHx/PEG blend film with increasing temperature. No contribution of PEG is observed in the conventional 2D correlation spectrum. To remove the PHBHx contribution in PHBHx/PEG blend film, the null-space projection 2D-COS was performed. The completely different null-space projection 2D correlation spectrum from conventional 2D-COS is shown in Figure 9.13c. Interestingly, the new bands at 1313, 1105, and 1065 cm^{-1} can be observed, which might be assigned to PEG. Therefore, null-space projection 2D-COS is a powerful method to easily detect the very subtle contribution of the component in the polymer system.

9.3.7 2D-COS for Hyperspectral Imaging

Hyperspectral imaging (HSI) is a useful method to visualize the phase transition, crystallinity, phase separation, etc. of polymers. However, it is not easy to interpret and elucidate HIS result as it is collected, because its results typically consist of big data. We show that 2D-COS is also a very useful tool for analyzing the HSI data.

Park et al. [66] applied 2D-COS to FTIR spectra extracted from chemical images of the C=O stretching region obtained at each temperature of PHBHx/PEG blend film during the melting process. In their previous study [21, 117], it was found that only PEG 400 was completely miscible with PHBHx, while PEG 1500 and PEG 3400 were partially mixed. Based on this observation, they investigated the possible mixing behavior and the existence of domains in PHBHx/PEG film with a higher-molecular-weight PEG (Mw = 3400) using chemical images and 2D-COS. The existence of domains for PHBHx/PEG 3400 (70/30 and 50/50) blend films was examined using chemical images of the C=O stretching region as shown in Figure 9.14. The variations of chemical images of PHBHx/PEG 3400 (70/30 and 50/50) with increasing temperature indicate that the PHBHx and PEG are not fully miscible.

For better understanding the thermal behavior of each domain, they analyzed the temperature-dependent FTIR spectra extracted at different positions in the chemical images. Interestingly, as shown in Figure 9.15a,b, the temperature-dependent FTIR spectra extracted at parts A and B of PHBHx/PEG = 70/30 blend film show a similar tendency of the spectral intensity changes of two bands at 1735 and 1722 cm^{-1}, which are assigned to amorphous and crystalline C=O stretching modes, respectively. This observation means that the distinction that appeared in the chemical images (see Figure 9.14a) is simply the difference of thickness variations on blend film. On the other hand, intensity changes are very different in parts A and B of PHBHx/PEG = 50/50 blend film (see Figure 9.15c,d), which indicates that PEG has an effect to change the kinetic and degree of PHBHx crystallization.

They applied 2D-COS to the temperature-dependent FTIR spectra extracted at parts A and B of PHBHx/PEG = 70/30 and 50/50 blend films, to further understand in detail the effect of PEG of PHBHx crystallization. As shown in Figure 9.16a–d, 2D correlation spectra in parts A and B are very similar, strongly supporting that differences in chemical images reflect only the thickness variation. However, the corresponding 2D correlation spectra for PHBHx/PEG 50/50 blend film (see Figure 9.16e–h) are clearly different from each other. In part A, the new amorphous band at 1748 cm^{-1} was observed, which means the existence of a partially mixed structure in the PHBHx/PEG = 50/50 blend film. The application of 2D-COS in HSI spectral data thus provides deeper insight for understanding the thermal behavior of PHBHx/PEG blend films.

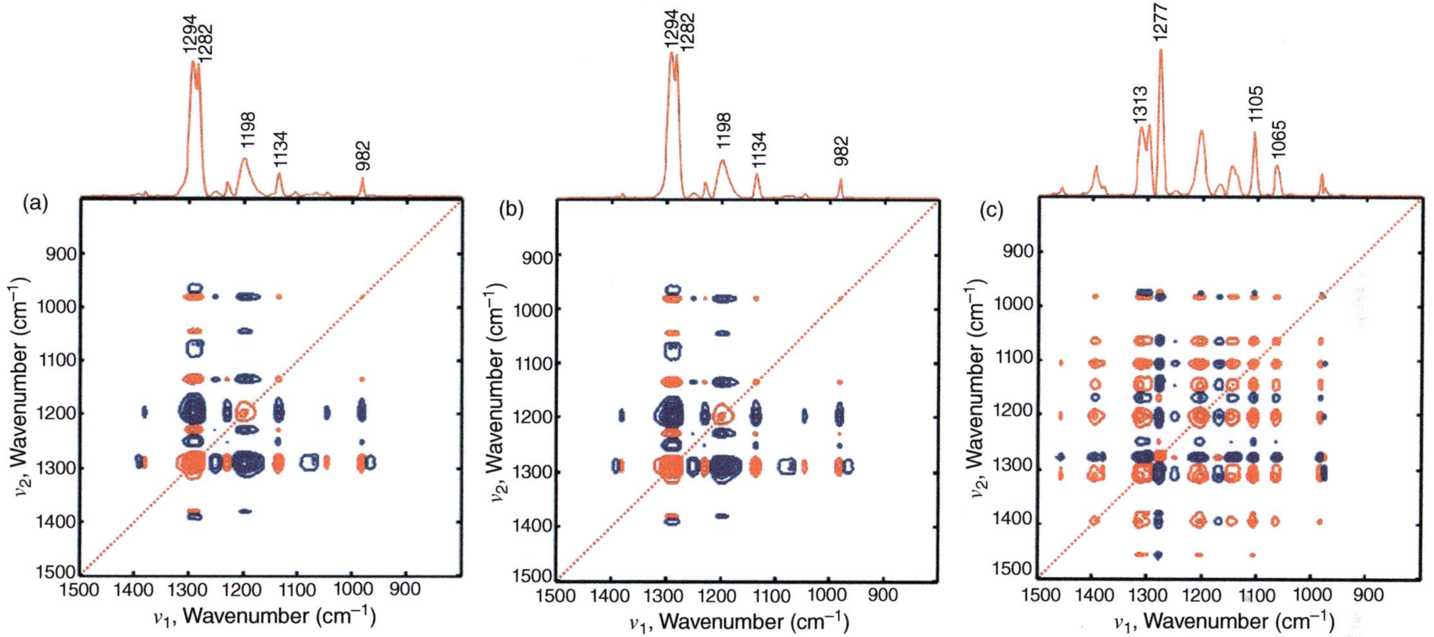

Figure 9.13 Synchronous (a), synchronous projection (b), and synchronous null-space projection (c) 2D correlation spectra, in the region of C–H deformation C–O–C stretching modes in 1000–1500 cm^{-1} for a spin-coated film of PHBHx/PEG blend. The red and blue lines represent positive and negative cross-peaks, respectively. *Source:* Kim et al. [119].

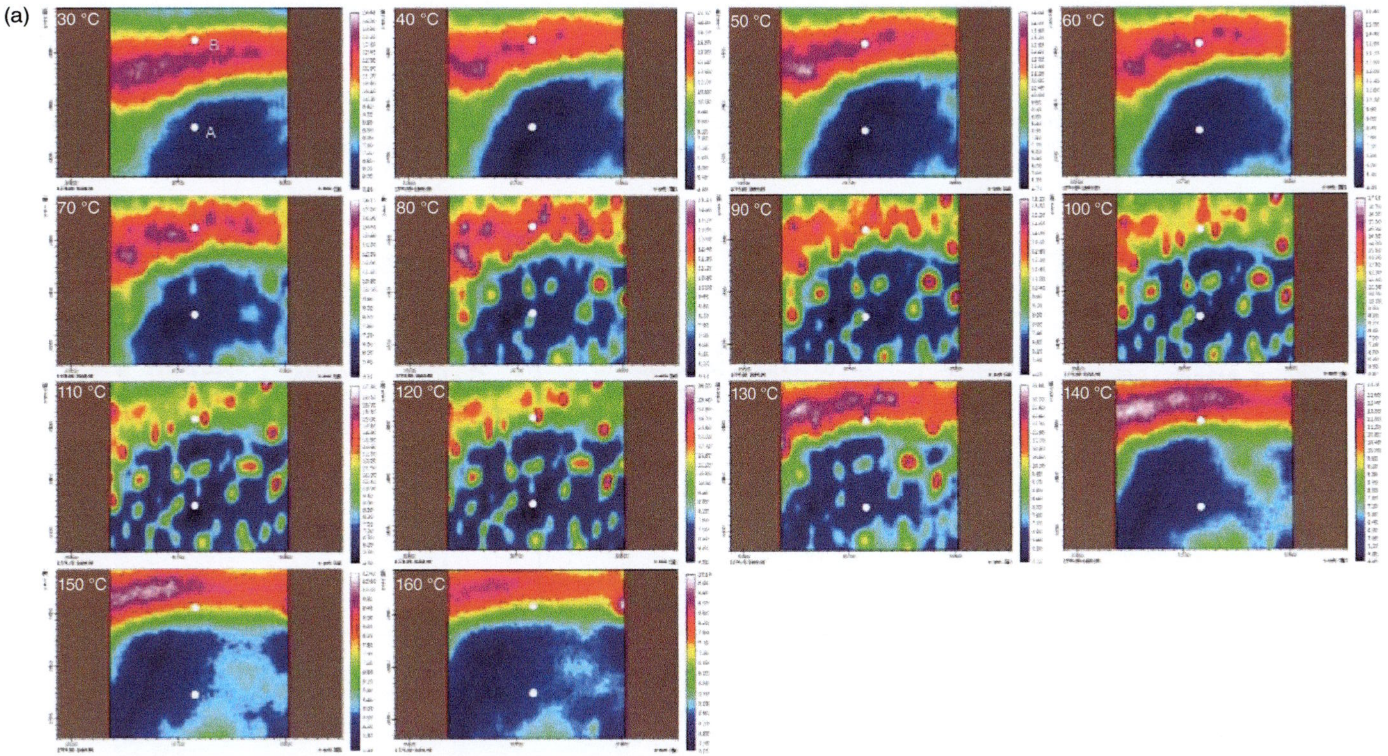

Figure 9.14 Chemical images of the C=O stretching region at each temperature of PHBHx/PEG = 70/30 (a) and 50/50 (b) blend films. *Source:* Park et al. [66].

(b)

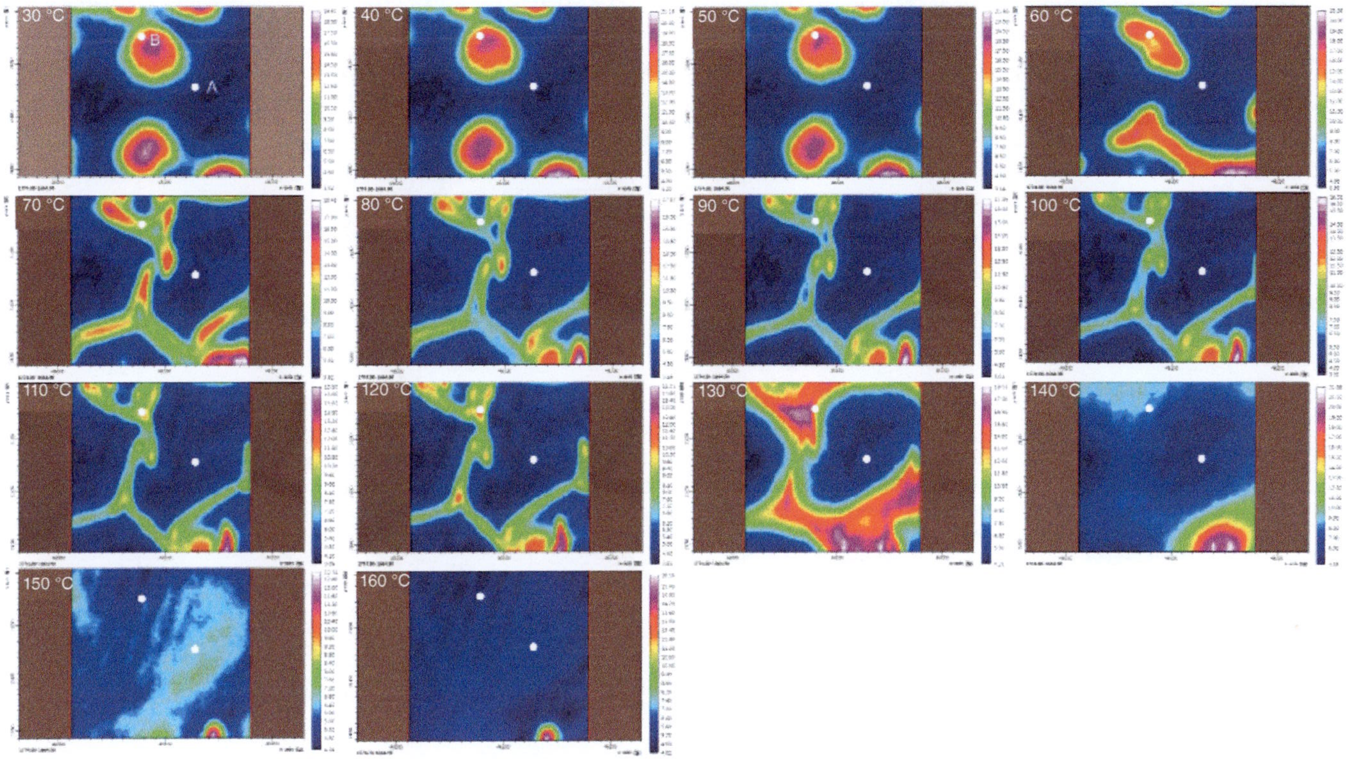

Figure 9.14 (*Continued*)

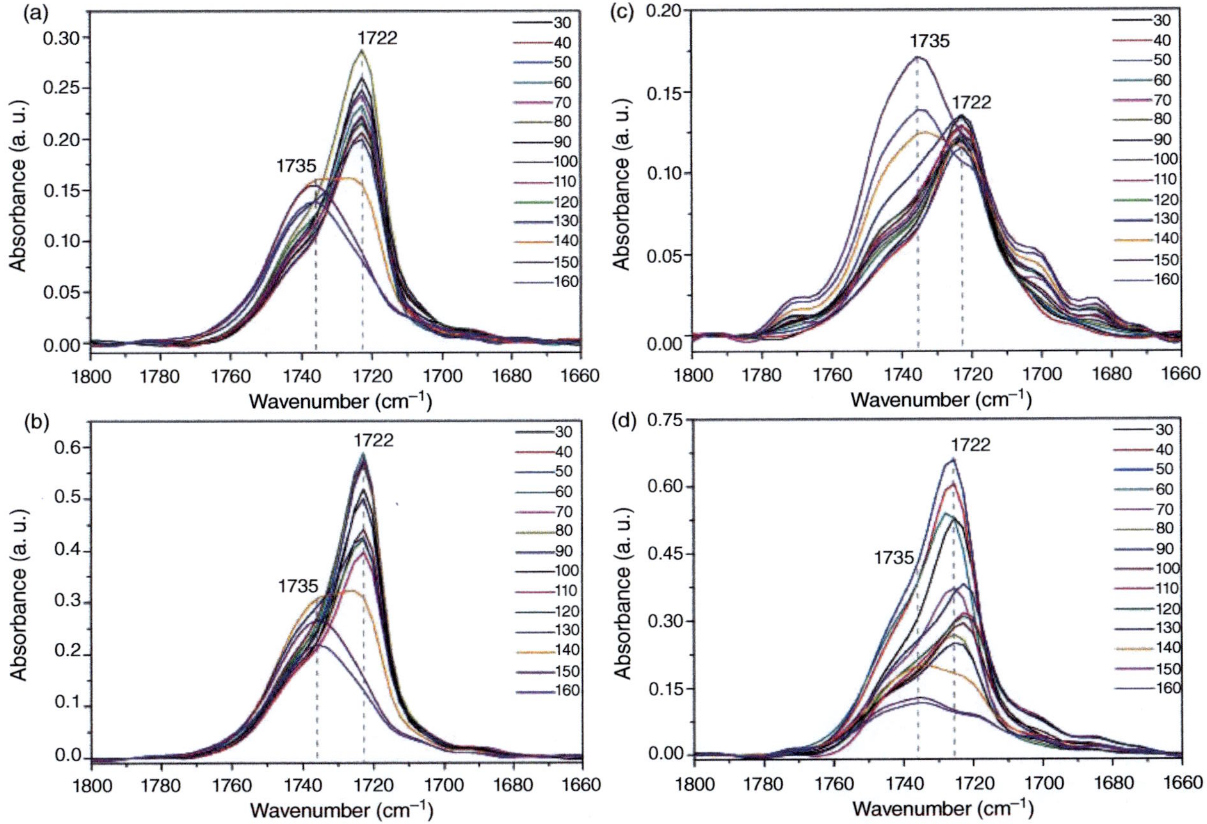

Figure 9.15 Temperature-dependent FTIR spectra extracted at parts A (a, c) and B (b, d) in Figures 9.14a,b, respectively. *Source:* Park et al. [66].

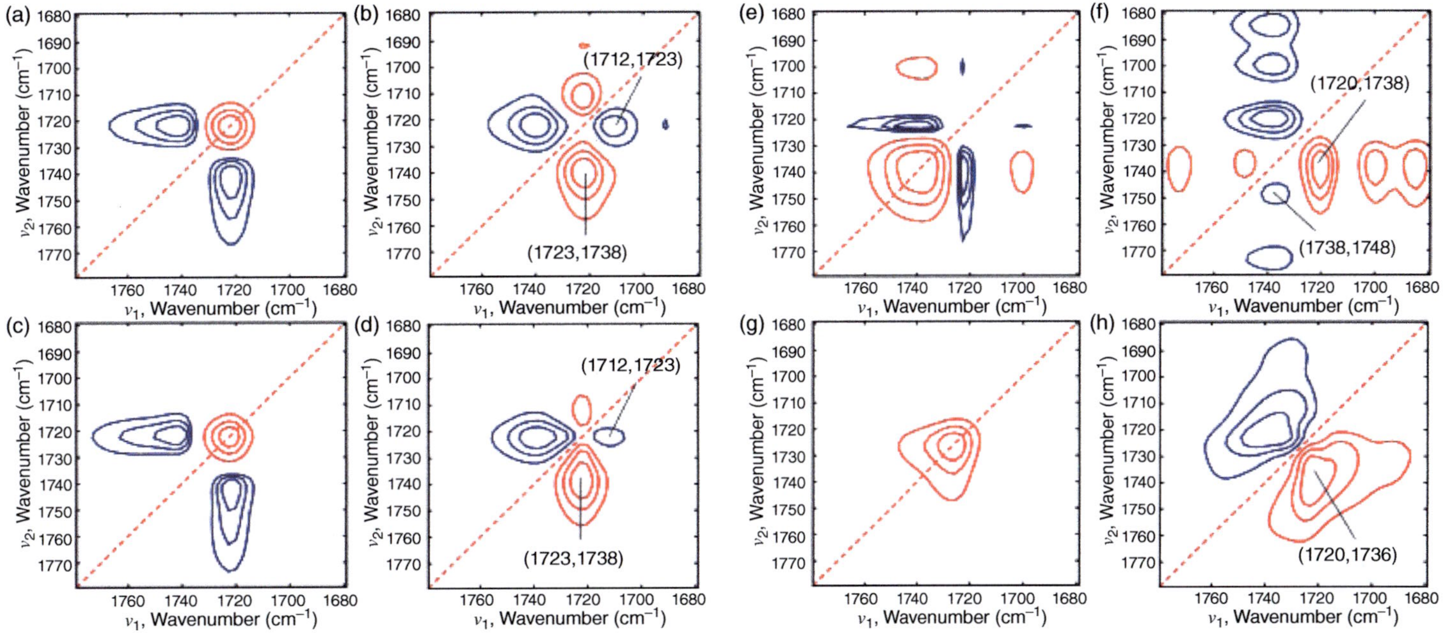

Figure 9.16 Synchronous (a, c, e, g) and asynchronous (b, d, f, h) 2D correlation spectra obtained from Figure 9.15. *Source:* Park et al. [66].

9.4 Conclusions

Versatile applications of 2D-COS to biodegradable and thermo-responsive polymers are presented with various techniques and some examples. 2D-COS is shown to provide the in-depth elucidation of the polymer system, which is often not readily detected in conventional spectral analysis. Newly developed methods for 2D-COS, such as 2D gradient mapping, PCA 2D-COS, EMT technique, hetero-spectral correlation, projection 2D-COS, and SFA and their successful applications to polymer studies, are also discussed. 2D-COS and related techniques would certainly open a new way for in-depth studies of various polymer systems.

References

1 Noda, I. (1986). Two-dimensional infrared (2D IR) spectroscopy of synthetic and biopolymers. *Bull. Am. Phys. Soc.* 31: 520.

2 Noda, I. (1993). Generalized two-dimensional correlation method applicable to infrared, Raman, and other types of spectroscopy. *Appl. Spectrosc.* 47: 1329–1336. https://doi.org/10.1366/0003702934067694.

3 Wei, H., Yu, H., Pan, H., and Gao, H. (2018). Application of UV-visible absorption spectroscopy combined with two-dimensional correlation for insight into DOM fractions from native halophyte soils in a larger Estuarine delta. *Environ. Sci. Pollut. Res.* 25: 14197–14205. https://doi.org/10.1007/s11356-018-1645-7.

4 Zhang, J., Hori, N., and Takemura, A. (2019). Optimization of agricultural wastes liquefaction process and preparing bio-based polyurethane foams by the obtained polyols. *Ind. Crop. Prod.* 138: 111455. https://doi.org/10.1016/j.indcrop.-2019.06.018.

5 Shi, Y., Hu, D., Xin, Z. et al. (2016). Conformation order of poly(L-lactic acid) chains during the melt crystallization process: infrared and two-dimensional infrared correlation spectroscopy study. *J. Mater. Sci.* 51: 4880–4887. https://doi.org/10.1007/s10853-016-9793-y.

6 Shinzawa, H., Kanematsu, W., and Noda, I. (2015). Tensile deformation of isotactic polypropylene (IPP) and IPP-nanocomposite studied by Rheo-optical near-infrared (NIR) spectroscopy. *Vib. Spectrosc.* 78: 34–38. https://doi.org/10.1016/j.vibspec.20-15.03.007.

7 Jia, L., Su, G., Yuan, Q. et al. (2018). Difference in the micro-dynamics mechanism between aromatic nylon and aliphatic nylon during water absorption: spectroscopic evidence. *Phys. Chem. Chem. Phys.* 20: 26764–26776. https://doi.org/10.1039/C8CP05432K.

8 Wang, Y.-M., Pan, M., Liang, X.-Y. et al. (2017). Electromagnetic wave absorption coating material with self-healing properties. *Macromol. Rapid Commun.* 38: 1700447. https://doi.org/10.1002/marc.201700447.

9 Li, T., Tang, H., and Wu, P. (2015). Structural investigation of thermo-responsive poly(2-isopropyl-2-oxazoline) hydrogel across the volume phase transition. *Soft Matter* 11: 1911–1918. https://doi.org/10.1039/C4SM02812K.

10 Lee, J.-E., Eom, Y., Shin, Y.-E. et al. (2019). Effect of interfacial interaction on the conformational variation of poly(vinylidene fluoride) (PVDF) chains in PVDF/graphene oxide (GO) nanocomposite fibers and corresponding mechanical properties. *ACS Appl. Mater. Interfaces* 11: 13665–13675. https://doi.org/10.1021/acsami.8b22586.

11 Suzuki, Y. and Furuya, H. (2018). Insights into the helix-sense inversion of poly(β-phenethyl-L-aspartate) by two-dimensional Raman correlation spectroscopy. *J. Pept. Sci.* 24: e3079. https://doi.org/10.1002/psc.3079.

12 Li, T., Tang, H., and Wu, P. (2015). Remarkable distinctions in the heat-induced phase transition processes of two poly(2-isopropyl-2-oxazoline)-based mixed aqueous solutions. *Soft Matter* 11: 3046–3055. https://doi.org/10.1039/C5SM0018-6B.

13 Pannico, M. and La Manna, P. (2019). Sorption of water vapor in poly(L-lactic acid): a time-resolved FTIR spectroscopy investigation. *Front. Chem.* 275 https://www.frontiersin.org/article/10.3389/fchem.2019.00275.

14 Ji, H., Kim, S.B., Noda, I., and Jung, Y.M. (2009). Details of thermal behavior of spin-coated film of biodegradable poly(3-hydroxybutyrate-*co*-3-hydroxyhexanoate) copolymer studied by principal component analysis-based two-dimensional (PCA2D) correlation spectroscopy. *Spectrochim. Acta A Mol. Biomol. Spectrosc.* 71: 1873–1876. https://doi.org/10.1016/j.saa.2008.07.007.

15 Bartelle, B.B., Barandov, A., and Jasanoff, A. (2016). Molecular FMRI. *J. Neurosci.* 36: 4139–4148. https://doi.org/10.1523/JNEUROSCI.4050-15.2016.

16 Nishimura, F., Hoshina, H., Ozaki, Y., and Sato, H. (2019). Isothermal crystallization of poly(glycolic acid) studied by terahertz and infrared spectroscopy and SAXS/WAXD simultaneous measurements. *Polym. J.* 51: 237–245. https://doi.org/10.1038/s41428-018-0150-7.

17 Sun, S. and Zhao, A. (2018). Effect of microstructure morphology on mechanical properties of quenching and partitioning steel. *Mater. Sci. Technol.* 34: 347–354. https://doi.org/10.1080/02670836.2017.1390901.

18 Chen, Y., Noda, I., and Jung, Y.M. (2017). Influence of polyethylene glycol on the phase transition of poly(butylene adipate) thin films: polymorphic control. *Spectrochim. Acta A Mol. Biomol. Spectrosc.* 184: 228–234. https://doi.org/10.1016/j.saa.2017.05.003.

19 Su, G., Jia, L., Zhang, X. et al. (2018). Exploration of the unusual two-step volume phase transition of the poly(*N*-vinylcaprolactam-*co*-hydroxyethyl methacrylate) hydrogel. *Phys. Chem. Chem. Phys.* 20: 23013–23024. https://doi.org/10.1039/C8CP02429D.

20 Park, Y., Noda, I., and Jung, Y.M. (2018). Smooth factor analysis (SFA) to effectively remove high levels of noise from spectral data sets. *Appl. Spectrosc.* 72: 765–775. https://doi.org/10.1177/0003702817752126.

21 Chen, Y., Park, Y., Yang, C. et al. (2017). Reorientation of the poly (3-hydroxybutyrate-*co*–3-Hydroxyhexanoate) crystal in thin film induced by polyethylene glycol. *Polymer* 120: 59–67. https://doi.org/10.1016/j.polymer.2017.0-5.050.

22 Shi, S., Yang, C., and Nie, M. (2017). Enhanced interfacial strength of natural fiber/polypropylene composite with mechanical-interlocking interface. *ACS Sustain. Chem. Eng.* 5: 10413–10420. https://doi.org/10.1021/acssuschemeng.7b02-448.

23 Yuan, Q., Zhou, T., Li, L. et al. (2015). Hydrogen bond breaking of TPU upon heating: understanding from the viewpoints of molecular movements and enthalpy. *RSC Adv.* 5: 31153–31165. https://doi.org/10.1039/C5RA03984C.

24 Lee, B.-M. and Hur, J. (2016). Adsorption behavior of extracellular polymeric substances on graphene materials explored by fluorescence spectroscopy and two-dimensional fourier transform infrared correlation spectroscopy. *Environ. Sci. Technol.* 50: 7364–7372. https://doi.org/10.1021/acs.est.6b01286.

25 Ma, L. and Wu, P. (2018). The role of unique spatial structure in the volume phase transition behavior of poly(*N*-isopropylacrylamide)-based interpenetrating polymer network microgels including a thermosensitive poly(ionic liquid). *Phys. Chem. Chem. Phys.* 20: 8077–8087. https://doi.org/10.1039/C8CP00340H.

26 Shinzawa, H. and Mizukado, J. (2017). Near-infrared (NIR) disrelation mapping analysis for poly(lactic) acid nanocomposite. *Spectrochim. Acta A Mol. Biomol. Spectrosc.* 181: 1–6. https://doi.org/https://doi.org/10.1016/j.saa.2017.03.026.

27 Chokki, J., Darracq, G., Poelt, P. et al. (2019). Investigation of poly(ethersulfone)/polyvinylpyrrolidone ultrafiltration membrane degradation by contact with sodium hypochlorite through FTIR mapping and two-dimensional correlation spectroscopy. *Polym. Degrad. Stab.* 161: 131–138. https://doi.org/10.1016/j.polymdegradstab.2019.01.017.

28 Park, Y., Hashimoto, C., Hashimoto, T. et al. (2013). Reaction-induced self-assembly of gel structure: a new insight into chemical gelation process of *N*-isopropylacrylamide as studied by two-dimensional infrared correlation spectroscopy. *Macromolecules* 46: 3587–3602. https://doi.org/10.1021/ma400457e.

29 Chon, Y.J., Koo, J.M., Park, Y.J. et al. (2017). Synthesis of a high-performance citric acid-based polyester elastomer by a hot-pressing technique. *Polymer* 125: 283–291. https://doi.org/10.1016/j.polymer.201-7.08.013.

30 Zhang, S., Yang, P., Bai, Y. et al. (2017). Polybenzoxazines: thermal responsiveness of hydrogen bonds and application as latent curing agents for thermosetting resins. *ACS Omega* 2: 1529–1534. https://doi.org/10.1021/acsomega.7b00075.

31 Shinzawa, H. and Mizukado, J. (2018). Rheo-optical near-infrared (NIR) spectroscopy study of partially miscible polymer blend of polymethyl methacrylate (PMMA) and polyethylene glycol (PEG). *Spectrochim. Acta A Mol. Biomol. Spectrosc.* 192: 236–243. https://doi.org/https://doi.org/10.1016/j.saa.2017.11.027.

32 Geitner, R., Kötteritzsch, J., Siegmann, M. et al. (2015). Two-dimensional Raman correlation spectroscopy reveals molecular structural changes during temperature-induced self-healing in polymers based on the Diels–Alder reaction. *Phys. Chem. Chem. Phys.* 17: 22587–22595. https://doi.org/10.1039/C5CP02151K.

33 Tasumi, M. (2014). *Introduction to Experimental Infrared Spectroscopy: Fundamentals and Practical Methods.* Wiley.

34 Park, Y., Noda, I., and Jung, Y.M. (2015). Two-dimensional correlation spectroscopy in polymer study. *Front. Chem.* 3 https://doi.org/10.3389/fchem.2015.00014.

35 Zhou, T., Zhou, T., and Zhang, A. (2015). Separation of the molecular motion from different components or phases using projection moving-window 2D correlation FTIR spectroscopy for multiphase and multicomponent polymers. *RSC Adv.* 5: 14832–14842. https://doi.org/10.1039/C4RA16373G.

36 Chen, Y., Noda, I., and Jung, Y.M. (2018). Revealing thermal behavior of poly(3-hydroxybutyrate-*co*-3-hydroxyhexanoate) and its polyethylene glycol blends thin films: effect of 3-hydroxyhexanoate comonomer content. *J. Mol. Struct.* 1162: 140–144. https://doi.org/10.1016/j.molstruc.2018.02.053.

37 Tan, X., Li, J., and Guo, S. (2018). Temperature-dependent order-to-order transition of polystyrene-block-poly(ethylene-*co*-butylene)-*block*-polystyrene triblock copolymer under multilayered confinement. *Macromolecules* 51: 2099–2109. https://doi.org/10.1021/acs.macromol.7b02651.

38 Wang, Q., Tang, H., and Wu, P. (2015). Aqueous solutions of poly(ethylene oxide)-poly(*N*-isopropylacrylamide): thermosensitive behavior and distinct multiple assembly processes. *Langmuir* 31: 6497–6506. https://doi.org/10.1021/acs.langmu-ir.5b00878.

39 de Nicola, A., Correa, A., Milano, G. et al. (2017). Local structure and dynamics of water absorbed in poly(ether imide): a hydrogen bonding anatomy. *J. Phys. Chem. B* 121: 3162–3176. https://doi.org/10.1021/acs.jpcb.7b00992.

40 Shinzawa, H., Uchimaru, T., Mizukado, J., and Kazarian, S.G. (2017). Non-equilibrium behavior of polyethylene glycol (PEG)/polypropylene glycol (PPG) mixture studied by fourier transform infrared (FTIR) spectroscopy. *Vib. Spectrosc.* 88: 49–55. https://doi.org/10.1016/j.vibspec.2016.11.001.

41 Zhou, Y., Tang, H., and Wu, P. (2017). Intra-molecular interactions dominating the dehydration of a poly(2-isopropyl-2-oxazoline)-based densely grafted polymer comb in aqueous solution and hysteretic liquid–liquid phase separation. *Phys. Chem. Chem. Phys.* 19: 6626–6635. https://doi.org/10.1039/C6CP08574A.

42 Wei, D., Li, M., Wang, X. et al. (2016). Extracellular polymeric substances for Zn (II) binding during its sorption process onto aerobic granular sludge. *J. Hazard. Mater.* 301: 407–415. https://doi.org/10.1016/j.jhazmat.2015.09.018.

43 Ma, Y. and Zhou, T. (2019). Preliminary study of experimental parameters for projection moving-window two-dimensional correlation FTIR spectroscopy. *J. Mol. Struct.* 1176: 777–790. https://doi.org/10.1016/j.molstruc.2018.09.022.

44 Watanabe, R., Shinzawa, H., Kunioka, M. et al. (2017). Reinforcement mechanism of functionalized polypropylene containing hydroxyl group nanocomposites studied by Rheo-optical near-infrared spectroscopy. *Eur. Polym. J.* 92: 86–96. https://doi.org/10.1016/j.eurpolymj.2017.0-4.032.

45 Al Lafi, A.G., Abboudi, M., and Aljoumaa, K. (2017). Natural sunlight ageing of control and sterilized poly(ethylene terephthalate): two-dimensional infrared correlation spectroscopic investigation. *J. Appl. Polym. Sci.* 134: 44736. https://doi.org/10.1002/app.44736.

46 Sutton, J.J., Nguyen, T.L., Woo, H.Y., and Gordon, K.C. (2019). Variable-temperature resonance Raman studies to probe Interchain ordering for semiconducting conjugated polymers with different chain curvature. *Chem. Asian J.* 14: 1175–1183. https://doi.org/10.1002/asia.201801667.

47 Chae, B., Seo, J.H., Moon, K.H. et al. (2018). Thermal induced structural changes of polyhydroxyamide by two-dimensional (2D) infrared correlation study. *J. Mol. Struct.* 1167: 169–173. https://doi.org/10.1016/j.molstruc.2018.04.070.

48 Feng, K., Hou, L., Tang, B., and Wu, P. (2015). Does thermal treatment merely make a H_2O-saturated Nafion membrane lose its absorbed water at high temperature? *Phys. Chem. Chem. Phys.* 17: 9106–9115. https://doi.org/10.1039/C5CP00203F.

49 Noda, I., Roy, A., Carriere, J. et al. (2017). Two-dimensional Raman correlation spectroscopy study of poly[(*R*)-3-ydroxybutyrate-*co*-(*R*)-3-hydroxyhexanoate] copolymers. *Appl. Spectrosc.* 71: 1427–1431. https://doi.org/10.1177/0003702817707219.

50 Al Lafi, A.G. and Hay, J.N. (2015). State of the water in crosslinked sulfonated poly(ether ether ketone). Two-dimensional differential scanning calorimetry correlation mapping. *Thermochim. Acta* 612: 63–69. https://doi.org/10.1016/j.tca.2015.05.008.

51 Li, K. and Kobayashi, T. (2016). Ultrasound response of aqueous poly(ionic liquid) solution. *Ultrason. Sonochem.* 30: 52–60. https://doi.org/10.1016/j.ultsonch.2015.10.021.

52 D'hooge, D.R., Van Steenberge, P.H.M., Derboven, P. et al. (2015). Model-based design of the polymer microstructure: bridging the gap between polymer chemistry and engineering. *Polym. Chem.* 6: 7081–7096. https://doi.org/10.1039/C5PY01069A.

53 Kumar, B.N.V., Guo, S., Bocklitz, T. et al. (2016). Demonstration of carbon catabolite repression in naphthalene degrading soil bacteria via Raman spectroscopy based stable isotope probing. *Anal. Chem.* 88: 7574–7582. https://doi.org/10.1021/acs.analchem.6b01046.

54 Feng, K., Hou, L., Schoener, C.A. et al. (2015). Exploring the drug migration process through ethyl cellulose-based films from infrared-spectral insights. *Eur. J. Pharm. Biopharm.* 93: 46–51. https://doi.org/10.1016/j.ejpb.2015.03.011.

55 Xu, J.-Z., Xu, L., Liang, Y.-Y. et al. (2015). Temperature dependence of molecular conformation in uniaxially deformed isotactic polypropylene investigated by combination of polarized FTIR spectroscopy and 2D correlation analysis. *J. Polym. Sci. B Polym. Phys.* 53: 673–684. https://doi.org/10.1002/polb.23683.

56 Xu, K., Zhang, F., Zhang, X. et al. (2015). Molecular insights into the damping mechanism of poly(vinyl acetate)/hindered phenol hybrids by a combination of experiment and molecular dynamics simulation. *RSC Adv.* 5: 4200–4209. https://doi.org/10.1039/C4RA06644H.

57 Al Lafi, A.G. and Ajji, Z. (2017). Radiation grafting of acrylic acid and *N*-vinyl imidazole onto polyethylene films for Lead-ion removal: a two-dimensional correlation infrared spectroscopy investigation. *J. Appl. Polym. Sci.* 134: 44781. https://doi.org/10.1002/app.44781.

58 Chen, P., Wu, H., Yang, L. et al. (2019). A 2D correlation infrared spectroscopic study on the temperature-induced molecular motion mechanism concerning self-formed composite structure of 3D printed PA6. *Polymer* 167: 48–53. https://doi.org/10.1016/j.polymer.2019.01.075.

59 Lv, J., Liu, Z., Zhang, J. et al. (2017). Bio-based episulfide composed of cardanol/cardol for anti-corrosion coating applications. *Polymer* 121: 286–296. https://doi.org/10.1016/j.polymer.2017.06.036.

60 Sun, S. and Wu, P. (2017). Spectral insights into microdynamics of thermoresponsive polymers from the perspective of two-dimensional correlation spectroscopy. *Chin. J. Polym. Sci.* 35: 700–712. https://doi.org/10.1007/s10118-017-1938-1.

61 Dai, Y. and Wu, P. (2017). Toward the two-step microdynamic phase transition mechanism of an oligo(ethylene glycol)methacrylate-based copolymer with a LCST-type poly(ionic liquid) block. *Phys. Chem. Chem. Phys.* 19: 18556–18564. https://doi.org/10.1039/C7CP02942J.

62 Park, Y., Noda, I., and Jung, Y.M. (2016). Novel developments and applications of two-dimensional correlation spectroscopy. *J. Mol. Struct.* 1124: 11–28. https://doi.org/10.1016/j.molstruc.2016.01.028.

63 Xue, B., Xie, L., and Zhang, J. (2017). Detailed molecular movements during poly(L-lactic acid) cold-crystallization investigated by FTIR spectroscopy combined with two-dimensional correlation analysis. *RSC Adv.* 7: 47017–47028. https://doi.org/10.1039/C7RA08921J.

64 Hu, D., Wang, G., Feng, J., and Lu, X. (2016). Exploring supramolecular self-assembly of a bisamide nucleating agent in polypropylene melt: the roles of hydrogen bond and molecular conformation. *Polymer* 93: 123–131. https://doi.org/10.1016/j.polymer.2016.04.033.

65 Wiedemair, V., Mayr, S., Wimmer, D.S. et al. (2017). Novel molecular spectroscopic multimethod approach for monitoring water absorption/desorption kinetics of CAD/CAM poly(methyl methacrylate) prosthodontics. *Appl. Spectrosc.* 71: 1600–1612. https://doi.org/10.1177/0003702816682742.

66 Park, Y., Jin, S., Park, Y. et al. (2019). Studies on chemical IR images of poly(hydroxybutyrate-*co*-hydroxyhexanoate)/poly(ethylene glycol) blends and two-dimensional correlation spectroscopy. *Polymers* 11: 507. https://doi.org/10.3390/polym11030507.

67 Wang, X.-J., Zhao, X.-Y., Li, Q.-G. et al. (2016). Artificial neural network modeling and mechanism study for relaxation of deformed rubber. *Ind. Eng. Chem. Res.* 55: 4059–4070. https://doi.org/10.1021/acs.iecr.6b00010.

68 Wang, H., Zhou, Y.W., and Cai, W.B. (2017). Recent applications of *in situ* ATR-IR spectroscopy in interfacial electrochemistry. *Curr. Opin. Electrochem.*: 73–79. https://doi.org/10.1016/j.coelec.2017.01.008.

69 Zimmerer, C., Ziegler, L., Heinrich, G., and Steiner, G. (2018). Time resolved characterization of the solid-state reaction between polycarbonate and primary amine. *Eur. Polym. J.* 98: 313–320. https://doi.org/10.1016/j.eurpolymj.2017.11.029.

70 Zhang, S., Ran, Q., Fu, Q., and Gu, Y. (2019). Thermal responsiveness of hydrogen bonding and dielectric property of polybenzoxazines with different mannich bridge structures. *Polymer* 175: 302–309. https://doi.org/10.1016/j.polymer.2019.05.058.

71 Wen, L., Zhang, J., Zhou, T., and Zhang, A. (2016). Hydrogen bonding in micro-phase separation of poly(polyamide 12-*block*-polytetrahydrofuran) alternating block copolymer: enthalpies and molecular movements. *Vib. Spectrosc.* 86: 160–172. https://doi.org/10.1016/j.vibspec.2016.07.001.

72 Cai, S., Zhao, H., Li, T. et al. (2017). Influence of molecular interplay on the HPAM/UR rheological properties in an aqueous solution. *RSC Adv.* 7: 37055–37064. https://doi.org/10.1039/C7RA05263D.

73 Fu, Z., Liu, B., Liu, Y. et al. (2018). Detailed cyclization pathways identification of polyacrylonitrile and poly(acrylonitrile-*co*-itaconic acid) by in situ FTIR and two-dimensional correlation analysis. *Ind. Eng. Chem. Res.* 57: 8348–8359. https://doi.org/10.1021/acs.iecr.8b01162.

74 Su, G., Zhou, T., Liu, X., and Zhang, Y. (2017). Two-step volume phase transition mechanism of poly(*N*-vinylcaprolactam) hydrogel online-tracked by two-dimensional correlation spectroscopy. *Phys. Chem. Chem. Phys.* 19: 27221–27232. https://doi.org/10.1039/C7CP04571A.

75 Jiang, H., Tovar-Carrillo, K., and Kobayashi, T. (2016). Ultrasound stimulated release of mimosa medicine from cellulose hydrogel matrix. *Ultrason. Sonochem.* 32: 398–406. https://doi.org/10.1016/j.ultsonch.2016.04.008.

76 Xia, M., Cheng, Y., Theato, P., and Zhu, M. (2015). Thermo-induced double phase transition behavior of physically cross-linked hydrogels based on oligo(ethylene glycol) methacrylates. *Macromol. Chem. Phys.* 216: 2230–2240. https://doi.org/10.1002/macp.201500242.

77 Gao, P.-F., Zhang, S., Li, H.-W. et al. (2015). A two-step binding process of Eu-containing polyoxometalates to bovine serum albumin. *Langmuir* 31: 10888–10896. https://doi.org/10.1021/acs.langmuir.5b02868.

78 Hou, L. and Wu, P. (2019). Exploring the hydrogen-bond structures in sodium alginate through two-dimensional correlation infrared spectroscopy. *Carbohydr. Polym.* 205: 420–426. https://doi.org/10.1016/j.carbpol.2018.10.091.

79 Ji, H., Hwang, H., Kim, S.B. et al. (2008). Characterization of spin-coated films of biodegradable poly(3-hydroxybutyrate-*co*-3-hydroxyhexanoate) copolymers by two-dimensional correlation spectroscopy. *J. Mol. Struct.* 883–884: 167–172. https://doi.org/10.1016/j.molstruc.2007.12.031.

80 Hou, L. and Wu, P. (2019). Two-dimensional correlation infrared spectroscopy of heat-induced esterification of cellulose with 1,2,3,4-butanetetracarboxylic acid in the presence of sodium hypophosphite. *Cellulose* 26: 2759–2769. https://doi.org/10.1007/s10570-019-02255-w.

81 Yamasaki, H. and Morita, S. (2018). Multivariate curve resolution using a combination of mid-infrared and near-infrared spectra for the analysis of isothermal epoxy curing reaction. *Spectrochim. Acta A Mol. Biomol. Spectrosc.* 197: 114–120. https://doi.org/https://doi.org/10.1016/j.saa.2017.11.043.

82 Park, Y., Jin, S., Noda, I., and Jung, Y.M. (2018). Recent progresses in two-dimensional correlation spectroscopy (2D-COS). *J. Mol. Struct.* 1168: 1–21. https://doi.org/10.1016/j.molstruc.2018.04.099.

83 Zhou, Z., Huang, G., Xiong, Y. et al. (2017). Unveiling the susceptibility of functional groups of poly(ether sulfone)/polyvinylpyrrolidone membranes to NaOCl: a two-dimensional correlation spectroscopic study. *Environ. Sci. Technol.* 51: 14342–14351. https://doi.org/10.1021/acs.est.7b03952.

84 Guo, W., Chen, J., Sun, S., and Zhou, Q. (2018). Investigation of water diffusion in hydrogel pore-filled membrane via 2D correlation time-dependent ATR–FTIR spectroscopy. *J. Mol. Struct.* 1171: 600–604. https://doi.org/10.1016/j.molstruc.2018.06.048.

85 Huang, Y., Liu, J., Zhang, A., and Zhou, T. (2019). Crystallization behavior of poly(tetramethylene oxide) influenced by the crystallization condition of poly(butylene succinate) in their copolymers. *J. Wuhan Univ. Technol. Mater. Sci. Ed.* 34: 496–506. https://doi.org/10.1007/s11595-019-2079-x.

86 Ramsay, M., Beutier, C., McGarvey, G.B., and Hore, D.K. (2019). Adsorption of heptane–toluene binary mixtures on a hydrophobic polymer surface. *J. Chem. Phys.* 150: 014702. https://doi.org/10.1063/1.5066555.

87 Hoshina, H., Iwasaki, Y., Katahira, E. et al. (2018). Structure and dynamics of bound water in poly(ethylene-vinylalcohol) copolymers studied by terahertz spectroscopy. *Polymer* 148: 49–60. https://doi.org/10.1016/j.polymer.2018.06.020.

88 Popescu, M.-C., Dogaru, B.-I., Sun, D. et al. (2019). Structural and sorption properties of bio-nanocomposite films based on κ-carrageenan and cellulose nanocrystals. *Int. J. Biol. Macromol.* 135: 462–471. https://doi.org/10.1016/j.ijbiomac.2019.05.194.

89 Zhang, D., Li, K., Li, Y. et al. (2018). Characteristics of water absorption in amine-cured epoxy networks: a molecular simulation and experimental study. *Soft Matter* 14: 8740–8749. https://doi.org/10.1039/C8SM01516C.

90 Choi, H.C., Ryu, S.R., Ji, H. et al. (2010). Two-dimensional heterospectral correlation analysis of X-ray photoelectron spectra and infrared spectra for spin-coated films of biodegradable poly(3-hydroxybutyrate-*co*-3-hydroxyhexanoate) copolymers. *J. Phys. Chem. B* 114: 10979–10985. https://doi.org/10.1021/jp103288x.

91 Jung, Y.M. and Noda, I. (2018). Two-dimensional correlation spectroscopy: new developments and applications. In: *Encyclopedia of Analytical Chemistry* (ed. R.A. Meyers), 1–27. https://doi.org/10.1002/9780470027318.a9453.pub2.

92 Yamane, S., Shinzawa, H., Ata, S. et al. (2018). A thermal oxidative degradation study of triallyl Isocyanurate crosslinking moiety in fluorinated rubber by

two-dimensional infrared correlation spectroscopy. *Vib. Spectrosc.* 98: 30–34. https://doi.org/10.1016/j.vibspec.2018.07.002.
93 Al Lafi, A.G. and Hay, J.N. (2019). 2D-COS-FTIR analysis of high molecular weight poly(*N*-vinyl carbazole) undergoing phase separation on purification and thermal annealing. *J. Mol. Struct.* 1175: 152–162. https://doi.org/10.1016/j.molstruc.2018.07.077.
94 Schuchardt, P., Unger, M., and Siesler, H.W. (2018). 2DCOS and PCMW2D analysis of FT-IR/ATR spectra measured at variable temperatures on-line to a polyurethane polymerization. *Spectrochim. Acta A Mol. Biomol. Spectrosc.* 188: 478–482. https://doi.org/https://doi.org/10.1016/j.saa.2017.07.044.
95 Xu, L., Teng, J., Li, L. et al. (2019). Hydrophobic graphene oxide as a promising barrier of water vapor for regenerated cellulose nanocomposite films. *ACS Omega* 4: 509–517. https://doi.org/10.1021/acsomega.8b02866.
96 Chen, G.Q. (2010). Plastics from bacteria. In: *Microbiology Monographs* (ed. G.G.-Q. Chen). Berlin Heidelberg: Springer https://doi.org/10.1007/978-3-642-03287-5.
97 Sun, W. and Wu, P. (2018). A molecular level study of the phase transition process of hydrogen-bonding UCST polymers. *Phys. Chem. Chem. Phys.* 20: 20849–20855. https://doi.org/10.1039/C8CP04147D.
98 Ma, L., Tang, H., and Wu, P. (2017). Volume phase transition mechanism of poly[di(ethylene glycol)ethyl ether acrylate]-based microgels involving a thermosensitive poly(ionic liquid). *Langmuir* 33: 12326–12335. https://doi.org/10.1021/acs.langmuir.7b02884.
99 Yan, W., Wang, H., and Jing, C. (2016). Adhesion of *Shewanella oneidensis* MR-1 to goethite: a two-dimensional correlation spectroscopic study. *Environ. Sci. Technol.* 50: 4343–4349. https://doi.org/10.1021/acs.est.6b00066.
100 Shinzawa, H. and Mizukado, J. (2018). Rheo-optical two-dimensional (2D) near-infrared (NIR) correlation spectroscopy for probing strain-induced molecular chain deformation of annealed and quenched nylon 6 films. *J. Mol. Struct.* 1158: 271–276. https://doi.org/https://doi.org/10.1016/j.molstruc.2018.01.025.
101 Bertoldo Menezes, D., Reyer, A., and Musso, M. (2018). Investigation of the brill transition in nylon 6,6 by Raman, THz-Raman, and two-dimensional correlation spectroscopy. *Spectrochim. Acta A Mol. Biomol. Spectrosc.* 190: 433–441. https://doi.org/10.1016/j.saa.2017.09.055.
102 Wang, H., Liu, Y., Zhang, J. et al. (2015). Effect of curing conversion on the water sorption, corrosion resistance and thermo-mechanical properties of epoxy resin. *RSC Adv.* 5: 11358–11370. https://doi.org/10.1039/C4RA13678K.
103 Zhou, Y. and Wu, P. (2018). Block length-dependent phase transition of poly(*N*-isopropylacrylamide)-*b*-poly(2-isopropyl-2-oxazoline) diblock copolymer in water. *Polymer* 153: 250–261. https://doi.org/https://doi.org/10.1016/j.polymer.201-8.08.027.
104 Nishida, M., Ogura, T., Shinzawa, H. et al. (2016). Tensile properties of polyhydroxyalkanoate/polycaprolactone blends studied by rheo-optical near-infrared (NIR) spectroscopy. *J. Mol. Struct.* 1124: 92–97. https://doi.org/10.1016/j.molstruc.2016.04.020.

105 Lei, Z. and Wu, P. (2018). Zwitterionic skins with a wide scope of customizable functionalities. *ACS Nano* 12: 12860–12868. https://doi.org/10.1021/acsnano.8b08-062.

106 Jiang, Q., Zhao, Y., Zhang, C. et al. (2016). Investigation on the overlapping bands of syndiotactic polystyrene by using 2D-IR spectroscopy. *J. Mol. Struct.* 1124: 98–102. https://doi.org/10.1016/j.molstruc.2016.03.103.

107 Feng, K., Tang, B., and Wu, P. (2016). A self-protection phenomenon in the nafion membrane when it breathes in methanol-saturated air. *Phys. Chem. Chem. Phys.* 18: 19440–19450. https://doi.org/10.1039/C6CP03176E.

108 Hou, L. and Wu, P. (2015). Comparison of LCST-transitions of homopolymer mixture, diblock and statistical copolymers of NIPAM and VCL in water. *Soft Matter* 11: 2771–2781. https://doi.org/10.1039/C5SM00026B.

109 Li, L., Zhou, Z.-H., Yang, B. et al. (2019). Robust cellulose nanocomposite films based on covalently cross-linked network with effective resistance to water permeability. *Carbohydr. Polym.* 211: 237–248. https://doi.org/10.1016/j.carbpol.2019.01.084.

110 Kim, S., Shinzawa, H., Chung, H., and Ozaki, Y. (2015). Measurement of polyethylene pellets near the glass transition temperature to enhance Raman spectral selectivity among samples and improve accuracy for density determination. *Analyst* 140: 1906–1912. https://doi.org/10.1039/C4AN02144D.

111 Nishikawa, Y., Itoh, H., and Noda, I. (2016). Analysis of molecular interactions in polymer systems by using a pulsed compression time-resolved FT-IR/2D-IR spectroscopy. Generation of ring down compression pulses. *Vib. Spectrosc.* 86: 149–159. https://doi.org/10.1016/j.vibspec.2016.07.006.

112 Ye, H.-M., Song, Y.-Y., Meng, X., and Zhou, Q. (2016). Fractionated crystallization, polymorphism and crystal transformation of poly(butylene adipate) confined in electrospun immiscible blend fibers with polystyrene. *RSC Adv.* 6: 55961–55969. https://doi.org/10.1039/C6RA09117B.

113 Al Lafi, A.G. (2018). Application of 2D correlation methods to the analysis of XPS spectra of ion irradiated poly(ether ether ketone). *J. Polym. Res.* 25: 105. https://doi.org/10.1007/s10965-018-1504-8.

114 Shinzawa, H. and Mizukado, J. (2016). Near-infrared (NIR) monitoring of nylon 6 during quenching studied by projection two-dimensional (2D) correlation spectroscopy. *J. Mol. Struct.* 1124: 188–191. https://doi.org/https://doi.org/10.1016/j.molstruc.2016.02.016.

115 Sun, W., An, Z., and Wu, P. (2017). Revealing the distinct thermal transition behavior between PEGA-based linear polymers and their disulfide cross-linked nanogels. *Phys. Chem. Chem. Phys.* 19: 25746–25753. https://doi.org/10.1039/C7CP05084D.

116 Shinzawa, H., Mizukado, J., and Kazarian, S.G. (2016). Fourier transform infrared (FT-IR) spectroscopic imaging analysis of partially miscible PMMA–PEG blends using two-dimensional disrelation mapping. *Appl. Spectrosc.* 71: 1189–1197. https://doi.org/10.1177/0003702816670917.

117 Chen, Y., Park, Y., Noda, I., and Jung, Y.M. (2016). Influence of polyethylene glycol (PEG) chain length on the thermal behavior of spin-coated thin films of biodegradable poly(3-hydroxybutyrate-*co*-3-hydroxyhexanoate)/PEG blends. *J. Mol. Struct.* 1124: 159–163. https://doi.org/10.1016/j.molstruc.2016.02.059.

118 Kim, M.-K., Ryu, S.-R., Noda, I., and Jung, Y.-M. (2011). 2D correlation analysis of spin-coated films of biodegradable P(HB-*co*-HHx)/PEG blends. *Bull. Kor. Chem. Soc.* 32: 4005–4010. https://doi.org/10.5012/bkcs.2011.32.11.4005.

119 Kim, M.K., Ryu, S.R., Noda, I., and Jung, Y.M. (2012). Projection 2D correlation analysis of spin-coated film of biodegradable P(HB-*co*-HHx)/PEG blend. *Vib. Spectrosc.* 60: 163–167. https://doi.org/10.1016/j.vibspec.2012.02.008.

120 Park, Y., Hashimoto, C., Ozaki, Y., and Jung, Y.M. (2016). Understanding the phase transition of linear poly(*N*-isopropylacrylamide) gel under the heating and cooling processes. *J. Mol. Struct.* 1124: 144–150. https://doi.org/10.1016/j.molstruc.2016.02.067.

121 Noda, I. (2000). Progress in 2D correlation spectroscopy. In: *AIP Conference Proceedings*, 3–17. https://doi.org/10.1063/1.1302843.

122 Noda, I., Dowrey, A.E., Marcoli, C. et al. (2000). Generalized two-dimensional correlation spectroscopy. *Appl. Spectrosc.* 54: 236A–248A. https://www.osapublishing.org/as/abstract.cfm?URI=as-54-7-236A.

123 Noda, I., Dowrey, A.E., and Marcott, C. (1993). Recent developments in two-dimensional infrared (2D IR) correlation spectroscopy. *Appl. Spectrosc.* 47: 1317–1323. http://as.osa.org/abstract.cfm?URI=as-47-9-1317.

124 Noda, I. (2006). General theory of two-dimensional (2D) analysis. In: *Handbook of Vibrational Spectroscopy* (ed. P.R. Griffiths), 2123–2134. Chichester, UK: Wiley https://doi.org/10.1002/0470027320.s4401.

125 Noda, I. (2009). Generalized two-dimensional correlation spectroscopy. In: *Frontiers of Molecular Spectroscopy* (ed. J. Laan), 367–381. Amsterdam: Elsevier https://doi.org/10.1016/B978-0-444-53175-9.00013-1.

126 Noda, I. (2008). Recent advancement in the field of two-dimensional correlation spectroscopy. *J. Mol. Struct.* 883–884: 2–26. https://doi.org/10.1016/j.molstruc.2007.11.038.

127 Noda, I. (2006). Progress in two-dimensional (2D) correlation spectroscopy. *J. Mol. Struct.* 799: 2–15. https://doi.org/10.1016/j.molstruc.2006.03.053.

128 Noda, I. (2018). Advances in two-dimensional correlation spectroscopy (2DCOS). In: *Frontiers and Advances in Molecular Spectroscopy* (ed. J. Laan), 47–75. Elsevier https://doi.org/10.1016/B978-0-12-811220-5.00002-2.

129 Noda, I. (2014). Frontiers of two-dimensional correlation spectroscopy. Part 2. Perturbation methods, fields of applications, and types of analytical probes. *J. Mol. Struct.* 1069: 23–49. https://doi.org/10.1016/j.molstruc.2014.01.016.

130 Noda, I. (2014). Frontiers of two-dimensional correlation spectroscopy. Part 1. New concepts and noteworthy developments. *J. Mol. Struct.* 1069: 3–22. https://doi.org/10.1016/j.molstruc.2014.01.025.

131 Noda, I. (2012). Close-up view on the inner workings of two-dimensional correlation spectroscopy. *Vib. Spectrosc.* 60: 146–153. https://doi.org/10.1016/j.vibspec.2012.01.006.

132 Noda, I. (2007). Two-dimensional correlation analysis useful for spectroscopy, chromatography, and other analytical measurements. *Anal. Sci.* 23: 139–146. https://doi.org/10.2116/analsci.23.139.

133 Noda, I. and Ozaki, Y. (2004). *Two-Dimensional Correlation Spectroscopy – Applications in Vibrational and Optical Spectroscopy*. Chichester, UK: Wiley https://doi.org/10.1002/0470012404.

134 Ozaki, Y. and Noda, I. (2006). Two-dimensional vibrational correlation spectroscopy in biomedical sciences. In: *Encyclopedia of Analytical Chemistry* (eds. R.A. Meyers and H.H. Mantsch), 1–18. Chichester, UK. https://doi.org/10.1002/9780470027318.a0116: Wiley.

135 Roy, S., Freiberg, S., Leblanc, C., and Hore, D.K. (2017). Surface structure of acrylate polymer adhesives. *Langmuir* 33: 1763–1768. https://doi.org/10.1021/acs.langmuir.6b03875.

136 Pi, F., Shinzawa, H., Czarnecki, M.A. et al. (2010). Self-assembling of oleic acid (cis-9-octadecenoic acid) and linoleic acid (cis-9, cis-12-octadecadienoic acid) in ethanol studied by time-dependent attenuated total reflectance (ATR) infrared (IR) and two-dimensional (2D) correlation spectroscopy. *J. Mol. Struct.* 974: 40–45. https://doi.org/10.1016/j.molstruc.2009.11.060.

137 Czarnik-Matusewicz, B., Kim, S.B., and Jung, Y.M. (2009). A study of urea-dependent denaturation of β-lactoglobulin by principal component analysis and two-dimensional correlation spectroscopy. *J. Phys. Chem. B* 113: 559–566. https://doi.org/10.1021/jp808396g.

138 Wu, Y., Meersman, F., and Ozaki, Y. (2006). A novel application of hybrid two-dimensional correlation infrared spectroscopy: exploration of the reversibility of the pressure- and temperature-induced phase separation of poly(*N*-isopropylacrylamide) and poly(*N*-isopropylmethacrylamide) in aqueous solution. *Macromolecules* 39: 1182–1188. https://doi.org/10.1021/ma0521755.

139 Wu, Y., Jiang, J.H., and Ozaki, Y. (2002). A new possibility of generalized two-dimensional correlation spectroscopy: hybrid two-dimensional correlation spectroscopy. *J. Phys. Chem. A* 106: 2422–2429. https://doi.org/10.1021/jp012140g.

140 Jung, Y.M., Shin, H.S., Czarnik-Matusewicz, B. et al. (2002). Characterization of transition temperatures of a Langmuir—Blodgett film of poly(tert-butyl methacrylate) by two-dimensional correlation spectroscopy and principal component analysis. *Appl. Spectrosc.* 56: 1568–1574. https://doi.org/10.1366/000370202321116039.

141 Noda, I., Green, P.R., Satkowski, M.M., and Schechtman, L.A. (2005). Preparation and properties of a novel class of polyhydroxyalkanoate copolymers. *Biomacromolecules* 6: 580–586. https://doi.org/10.1021/bm049472m.

142 Satkowski, M.M., Melik, D.H., Autran, J. et al. (2005). Physical and processing properties of polyhydroxyalkanoate (PHA) copolymers. In: *Biopolymers Online* (eds. Y. Doi and A. Steinbüchel). Wiley https://doi.org/10.1002/3527600035.bpol3b09.

143 Jung, Y.M. (2004). Principal component analysis based two-dimensional correlation spectroscopy for noise filtering effect. *Vib. Spectrosc.* 36: 267–270. https://doi.org/10.1016/j.vibspec.2003.11.019.

144 Jung, Y.M., Kim, H.J., Ryu, D.Y. et al. (2006). Application of principal component analysis-based two-dimensional correlation spectroscopy to characterization of order–disorder transition of polystyrene-*block*-poly(*n*-pentyl methacrylate) copolymer. *J. Mol. Struct.* 799: 96–101. https://doi.org/10.1016/j.molstruc.2006.03.021.

145 Jung, Y.M., Kim, S.B., and Noda, I. (2003). New approach to generalized two-dimensional correlation spectroscopy. III: Eigenvalue manipulation transformation (EMT) for spectral selectivity enhancement. *Appl. Spectrosc.* 57: 564–570. https://doi.org/10.1366/000370203321666605.

146 Jung, Y.M., Shin, H.S., Kim, S.B., and Noda, I. (2002). New approach to generalized two-dimensional correlation spectroscopy. 1: Combination of principal component analysis and two-dimensional correlation spectroscopy. *Appl. Spectrosc.* 56: 1562–1567. https://doi.org/10.1366/000370202321116020.

147 Jung, Y.M. (2003). Principal component analysis based two-dimensional (PCA-2D) correlation spectroscopy: PCA Denoising for 2D correlation spectroscopy. *Bull. Kor. Chem. Soc.* 24: 1345–1350. https://doi.org/10.5012/bkcs.2003.24.9.1345.

148 Jung, Y.M., Kim, S.B., and Noda, I. (2003). New approach to generalized two-dimensional correlation spectroscopy. IV: Eigenvalue manipulation transformation (EMT) for partial attenuation of dominant factors. *Appl. Spectrosc.* 57: 850–857. https://doi.org/10.1366/000370203322102951.

149 Noda, I. (2010). Projection two-dimensional correlation analysis. *J. Mol. Struct.* 974: 116–126. https://doi.org/10.1016/j.molstruc.2009.11.047.

10

Molecular Dynamics in Polymer Science

Mateusz Z. Brela, Marek Boczar, and Marek J. Wójcik

Jagiellonian University, Faculty of Chemistry, Gronostajowa 2, Krakow, 30-387, Poland

10.1 Introduction

In this chapter, we would like to introduce the current knowledge about usage of molecular dynamics (MD) for molecular modeling of polymers. Dynamic approach focuses on the analysis of the time course of the atom positions or/and electronic properties, and further the Fourier transform of the autocorrelation function of the considered descriptors, e.g. dipole moment operator involved in the infrared (IR) transitions [1–5]. One of the main advantages of MD is the possibility to introduce the appropriate external conditions [6–8]. Some of the studied phenomena like hydrogen bond require the deeper analysis of a harmonicity or delocalization effect, which has dynamic character [3, 9, 10].

Current trends in molecular modeling, as well as need of designing new structures and materials, show the great importance of understanding and providing correct description of interactions in molecular systems. The great theoretician John Pople in his Nobel lecture [11] has described the "features of theoretical models" as predictions: "if the model has been properly validated according to some such criteria, it may be applied to chemical problems to which the answer is unknown or in dispute." This elucidates the necessity of cooperation between experiment and theory. Recently, this concept has been realized by modern spectroscopy where comparison between experimental data and calculated results gives possibility to describe and characterize interactions in the considered systems. Fifteen years after People's lecture, other great chemists have been awarded "for the development of multiscale models for complex chemical system" by the Royal Swedish Academy of Sciences. Karplus, Levitt, and Warshel work increase abilities of computational methods for complex and large systems [12–15]. Recently, we observe the developments of new methods, such as coarse grained, that allow us to simulate macroscale systems [16–18]. It should be pointed out that model systems used in calculations very often have more than micrometer dimensions [9].

Spectroscopic Techniques for Polymer Characterization: Methods, Instrumentation, Applications, First Edition. Edited by Yukihiro Ozaki and Harumi Sato.

At the same time, experimental techniques have been significantly developed. Nowadays, the characterization of studied material is complex, accurate, and precise. Spectroscopy has matured. It has been applied in many fields, such as chemistry, biology, and molecular physics [19–37]. In addition to classical approaches of Raman scattering (RS) and IR absorption, new optical technologies gave rise to Raman optical activity (ROA), vibrational circular dichroism (VCD), terahertz-time domain (THz–TD), and multiphoton vibrational spectroscopy. However, interpretation of the spectroscopic data obtained from advanced and modern techniques is not easy and straightforward, especially for complicated systems. It should be pointed out that spectroscopic data are very sensitive to inter- and intramolecular interactions. Therefore, analysis of weak interactions is very difficult. Much more reliable quantum mechanical simulations are desirable, allowing one to interpret experimental observations more deeply and model molecular properties in a consistent way.

The twentieth century was the age of the characterization of materials, while twenty-first century is the age of modeling and designing new materials. Modeling desirable molecular properties is not an easy and straightforward process. The properties of considered materials depend on interactions between molecules or macromolecules as well as conformational liability that is essential for polymers chains. Interactions in model systems often involve van der Waals interactions or hydrogen bonds, which should be described correctly. For such cases, vibrational spectroscopy is a very powerful and informative method. This was demonstrated during studies of the IR spectra of hydrogen-bonded systems [1–9]. Quantitative reconstruction and interpretation of the complex spectra of hydrogen bonded systems require reasonable modeling and complete description of the interactions within and between hydrogen bonds and the molecular environment. Molecular dynamics give precise information on the nature of hydrogen bonds, which is useful for understanding the molecular properties.

Polymers have many definitions; commonly, they are defined as a material consisting of molecules or macromolecules, composed of many repeating units (mers). Polymers can be divided into two types: natural and synthetic. The DNA is one of the famous examples of natural polymers. The synthetic polymers surround us in everyday life. Due to their broad spectrum of properties, polymers play a pivotal role in the global industry. It should be stressed that some of the polymers applications are essential and ubiquitous. Polymers range from familiar synthetic plastics, such as polystyrene or nylon, to natural biopolymers such as proteins that are fundamental to biological structure [28–31]. However, the polymer properties often vary because of different interactions present in their structures. Theoretical chemistry gave us instruments to analyze these phenomena. In hydrogen-bonded polymers, there exist some properties of hydrogen bonds, which complicate their theoretical description and give rise to a number of macroscopic physical properties that are unique to such materials [32–34]. It should be stressed that hydrogen bonds are responsible for the remarkable properties of water, folding of proteins, and self-assembly of many advanced materials [35, 36].

The main problem for the computational investigation of hydrogen bond phenomenon is quantum effects of proton movement. The hydrogen is the lightest atom, which can easily tunnel. Taking into account all quantum effects, this process is very complicated. The dynamics of proton motion in hydrogen bonds is determined by complex interplay of vibrational interactions [37–40]. These interactions are responsible for the complicated structure of the IR and Raman spectra of hydrogen-bonded systems [38–40] and for the dynamics of proton tunneling [41, 42].

In this chapter, we present several dynamics methods for simulation of mechanical, physical, and chemical polymer properties. The selected methods are based on different approaches as well as approximations. Study of spectroscopic properties, such as UV–vis spectra, evidently requires to take into account electrons [43–45]. On other hand, the calculation of the Young module or share module for crystalline polymers can be done based on the atomic positions designated by classical molecular dynamics [46, 47]. Understanding some phenomena, such as membrane sculpting, requires simulations of complex systems which often consists of millions of atoms. For such cases, coarse-grained models that mimic the behavior of a considered system by breaking it down into simpler subcomponents are appropriate.

In this chapter, we focus on the presentation of selected applications. Molecular dynamics, based on the different approaches, have been used for studies of the vibrational spectra of different polymers, elastic properties as well as interphase interactions. We present several applications for synthetic polymers. First, we will present simulation of IR spectra of the nylon and polyhydroxybutyrate (PHB) [48, 49], discussed in literature from decades. In this case, we will present the hydrogen bond effects described by Born–Oppenheimer molecular dynamics (BOMD) methods assisted by *post*-MD analysis. Then, the elastic properties of bulk modules and share modules of nylon will be discussed in detail. Afterwards, the polymer interactions with solvent will be discussed in terms of hybrid molecular dynamics results. The last but not least-discussed example will be interphase interactions and examination of complex processes by coarse-grained methods.

The last part of this chapter contains short summary and indicates possible future directions of applications of various methods in static and dynamic approaches. We discuss also the perspectives of modeling and design of polymers by theoretical methods.

10.2 Historical and Theoretical Background

Theoretical calculations have been more and more popular as a support for experimental analyses. During past several years, density functional theory (DFT) has become a promising method in computational chemistry. This method, developed by Hohenberg, Kohn, and Sham [50, 51], gave possibility to perform relatively accurate calculations for big systems [7, 52–56]. Following this, expansion of theoretical methods has been enormous. Quantum chemical methods are able to predict many important properties, such as electronic structures [56–63], bonding characteristics [64–67], interaction mechanisms [64, 67], optics [68–75], magnetic [76–79], and thermal properties [80, 81]. DFT calculations of vibrational spectra, which have been primarily restricted to organic compounds and small polyatomic systems [82–86], were recently used for transition metal complexes [87–103]. Extensively complicated systems, such as metal halides, have been subject of growing interest, due to their scientific importance and various industrial applications [104–107]. At this point, we need to state that it is impossible to perform catalytic mechanism analysis without correct and accurate vibrational analysis. Thermodynamic functions are based on calculations of IR vibrations and intensities.

The main problem with the static methods is influence of environmental effects. Vibrational analysis performed for molecules in solvents is extremely complicated. The *explicit* solvent model in which solvents molecules are included in calculations is exceptionally expensive. Addition of new molecules results in complication of the analysis of spectra. However, the vibrational analysis is possible in the *implicit* solvent model (PCM, COSMO) [108–110]. Interpretation of the stationary points in the potential energy surface changed by some external potentials is still unclear.

Molecular dynamics using potentials, based on independent electronic structure calculations, has been well established as a powerful tool, especially for investigation of many-body condensed matter. The wide and broad description of molecular dynamics technique was presented in several monographs and reviews [7, 52, 53]. The main assumption of adiabatic molecular dynamics is to describe the movement of nuclei in terms of potential energy surfaces designed by electronic energy. The Born–Oppenheimer approach gives us possibility to treat the nuclei and electrons independently. It should be noted that quantization of nuclear motion should be done during MD simulation, which requires full quantum treatment. However, full quantum method, which includes quantum effects for all atoms, is still computationally too expensive to be a "standard" treatment.

The calculation of the IR spectra of hydrogen-bonded complexes by MD calculations is based on linear response theory, in which the spectral density is the Fourier transform of the autocorrelation function of the dipole moment operator involved in the IR transition. Recently, molecular dynamics was performed by using several approaches, which have been used to simulate IR spectra of hydrogen-bonded systems [49, 111].

One of the most popular approaches is the BOMD. This method is based on integration of the classical equation of motion on ab initio molecular potential surface [112–114]. Such approach provides an information about electronic structure from the first principles by using quantum chemistry methods, the same as in static calculations. Ab initio methods may be used to calculate potential energy surface on the fly. The potential surface calculated by ab initio methods is vital for systems with high possibility of breaking and formation of chemical bonds. Such systems are hydrogen-bonded structures, especially with strong hydrogen bonds. It should be stressed that the dynamics of proton motion in hydrogen bonds is determined by a complex interplay of vibrational interactions. These interactions are responsible for the complicated structure of the IR and Raman (R) spectra of hydrogen-bonded systems, and for the dynamics of proton tunneling. For many such systems, ab initio molecular dynamics has been used.

The cost of MD calculations is the cost of ab initio calculations multiplied by number of trajectory steps. The long trajectories are beneficial for spectroscopic investigations, especially in the low-frequency region, that represents slow motions. It should be pointed out that time step is an important parameter for MD simulation. The decreasing time step in numerically calculated trajectory leads to the more accurate simulations. Nevertheless, computational cost increases with number of steps. In practice, the limit of the maximum time steps is determined by the vibrational period of the mode with the highest energy.

The extension of BOMD methods is the Car–Parrinello molecular dynamics (CPMD) [7], which couples the electronic degrees of freedom with the classical coordinates system by assigning a fictitious mass to electrons. This approximation excludes minimization of the electron wavefunctions at every step in the trajectory. CPMD uses fictitious dynamics to keep the electrons close to the ground state, preventing the need for the costly self-consistent iterative minimization at each time step. To stay on the Born–Oppenheimer surface, the time step should be comparatively small. In consequence, the computational cost of simulation is relatively small. The costly self-consistent iterative minimization is done only at the first step of the MD simulation.

Simultaneously with improving molecular dynamics methods, based on the Born–Oppenheimer approach, the force field methodology has been developed. This computational method is based on the estimation of the forces acting between atoms within molecules and also between molecules. The force field (classical molecular dynamics) refers to the functional form and parameter sets used to calculate the potential energy of considered system. It should be stressed that functional form depends on the kind of force field: classical, polarizable, or reactive. However, in many cases, combination of classical force field with the quantum methods is required.

The hybrid methods have been recently developed to analyze bigger systems, such as proteins. The most common is the QM/MM method introduced by Warshel and Levit in 1976 [12–14]. The enormous increase of system sizes was a trigger to develop the new method that merge accurate and exact molecular dynamics based on BO approach with the fast and efficient molecular mechanics (MM). These methods are based on selection of two or more regions: first, treated by quantum mechanics; second, treated by classic molecular mechanics; and the others, treated by lower level of theory. The main advantage of such approaches is the possibility of treatment of chemical reactions in such big systems as proteins and biomolecules. However, an open question is always the selection of size of QM region. Recent study [14] shows that there is no advantage in investigation of big QM regions. We would like to point out that there is some possibility of combining QM and MM regions, which has been done by one of the most popular ONIOM models [15, 115, 116].

The hybrid molecular dynamics is more useful for the chemical process analyses than for spectroscopic investigations. However, this method is still expanding. The new QM/MM approaches has been developed recently and new investigations use these new methods. In last decade, the QM/MM method was successfully used to understand the selected vibrations of biomolecules.

The MD simulations produce trajectories with information about time dependence of some system descriptors. It should be stressed that even classical molecular dynamics gave us information about change of properties of atoms under given conditions. Post-molecular dynamics analysis gave a possibility to analyze the descriptors describing, e.g. electron densities, atomic charges, interaction energies, or nuclei characterization, such as proton potentials. One such example is quantization of nuclear motion by snapshot methodology developed by Mavri and Stare [117–119]. This technique allows us to explore the changes in electronic structure during MD simulations. An approach based on Wannier localization function has been especially popular [15]. Information about distribution of Wannier centers that represent pairs of electrons (in close-shell calculations) or one electron (in open-shell calculations) gives us possibility to discuss changes in interaction character and its direction along MD. The Wannier localization functions may be also used for assignment of dipole moments, which is important for crystal structures treated by periodic boundary conditions. This methodology is still pioneering and has wide perspectives.

10.3 Applications

Polymers have unique physicochemical properties, for example elasticity or hardness. Nowadays, polymers are commonly used in all branches of chemical industry. New polymers with desired properties are needed. Molecular modeling is one of the promising methods for designing new materials. Understanding of interactions in studied materials is crucial. In polymers, we may divide interactions into three kinds: interactions between polymer chains – weak interactions or hydrogen bonds, interactions between polymer chains and molecules above, and interactions between polymer chains and molecules from other phase.

Interactions between polymer chains influence mechanical properties, such as elastics modules. Hydrogen bonding is a ubiquitous interaction that is important in a diverse range of properties, including inter- and intramolecular interactions, solvation, and self-assembly of macromolecules. As an example, there has been some new investigations for weak hydrogen bonds between carbonyl methyl groups in polymer structures using molecular dynamics [49]. A variety of different experimental and theoretical evidences are characteristic for hydrogen bonding.

Analyses of spectroscopic features of macromolecules gave a possibility to verify theoretical models as well as characterize existing interactions. IR spectroscopy has been used several times for characterization of bacterial co-polyester, poly(3-hydroxybutyrate-*co*-3-hydroxyhexanoate) [120]. Temperature-dependent IR spectral variations have been extensively used for analyses of weak interactions as well as hydrogen bond studies.

Computational methods have been widely used for investigation of catalysis. Theoretical methods proved their power to understand the chemical processes in inorganic as well as enzymatic reactions. In polymer science, molecular dynamics is the key method for understanding interactions between solid polymer electrolytes (SPEs) and Li metal surface. In research concerning renewable energy as well as energy storage, one usually uses polymeric materials. However, this research is extremely complicated because of inter-phase processes that occur in these systems.

10.3.1 Vibrational Spectra of Hydrogen-Bonded Polymers

Crystal structures and physical properties of polymers have been investigated using various spectroscopic and X-ray diffraction methods, such as IR [121], near-infrared (NIR) [122], Raman [123, 124], terahertz [124, 125], far-ultraviolet (FUV) [126], and wide-angle X-ray diffraction (WAXD) [127]. Moreover, crystal structures, physical properties, and intermolecular interactions in polymers were analyzed by several theoretical methods [128–131].

The main goal of our studies of vibrational spectra of hydrogen-bonded polymers was the study of structure – properties relationships of hydrogen-bonded polymer systems, such as PHB. We aimed to combine spectroscopic data with computational studies to obtain better understanding of the relations between the structure of the biodegradable polymers and their properties.

Figure 10.1 presents the analysis of geometric parameters along ab initio molecular dynamics simulations. The BOMD simulations were performed for super cells of PHB. The analysis of the C=O stretching modes has resulted in reasonably good agreement with the experimental IR data after inclusion of thermal motions of the nuclei. Our investigation has shown that weak interactions have dynamic character. However, the surroundings of the C=O group do not change its wavenumber. Weak hydrogen bond interactions are present all the time during the simulations despite the methyl group rotations and polymer chain movements. The CH· · ·OC interactions are not stable along the MD simulations; both forming and breaking processes occur simultaneously at the same time.

Our results clearly show that the molecular modeling of polymer structures should take into account dynamics effects. It should be pointed out that considering only interaction energies between the polymer chains in some selected structures is not enough to discuss the strength of the interchains interactions. The big fluctuation of the number of hydrogen bond interactions along the BOMD simulation was observed and allowed to explain the C=O band positions [49].

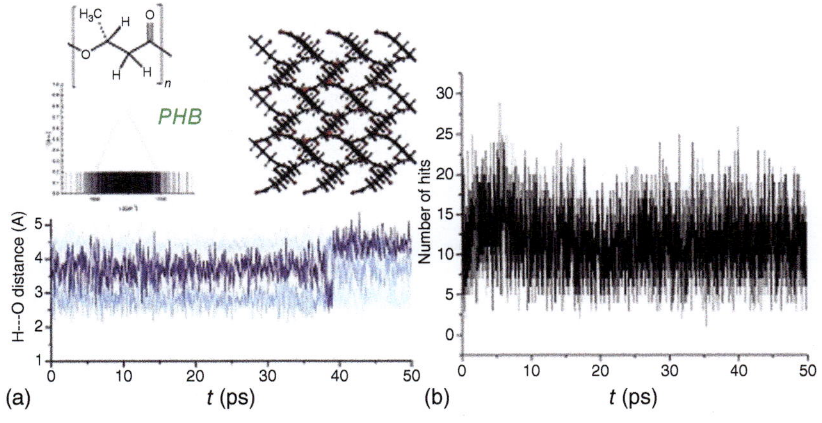

Figure 10.1 (a) Shows the structure of PHB and evolution of O–H distance along simulations. (b) Presents the number of O· · ·C interactions in a weak hydrogen bond observed from BOMD simulations; the applied cut-off donor–acceptor distance was 3.4 Å and donor – hydrogen – acceptor angle was 140°. Source: Brela et al. [49].

The detailed analyses of the IR and Raman spectra of PHB were performed by large-scale theoretical calculations using BOMD, to explain interactions between the polymer chains [131, 132]. Hydrogen bonds and weak interactions were the spotlight of this study.

In next step, multiscale molecular simulations (QM/MM) will be performed for model systems that will validate by previous analyses and spectroscopic data. The mechanisms of polymer degradation will be studied on the basis of the analyses of elementary reactions.

Since polymer degradation processes involve chemical reactions, it is essential to take into account the multiscale simulations to include the fully representative polymer models. Multiscale molecular simulation is a powerful method for treatment of enzymes reactions [133] and reactions in solutions [134]. The results obtained from these simulations will be used to describe the polymer structures and interchain interactions.

Our previous works [48, 49] show the importance of dynamics associated with intramolecular hydrogen bonds in hydrogen-bonded polymers. Two crystal forms of nylon 6 (α and γ forms) have been discussed in detail. Analysis of the different crystal structures with only slight difference in the high-frequency region of IR spectra prompted us to investigate strong interchain hydrogen bonds by ab initio molecular dynamics supported by quantization of motion methodology. BOMD simulations for supercells of each form were performed and four hydrogen bonds selected were considered by post-MD quantization of motions. The 1D proton potentials associated with the N–H stretching modes were calculated, and based on these results the contours of N–H motion were analyzed for each form. This approach gave us a possibility of comparing the strength of hydrogen bonds in the α and γ forms.

Our results [48] indicate that γ form has slightly stronger hydrogen bonds than α form. This result was explained by presence of number of interchain hydrogen bond interactions along MD simulations. In the γ form, the distribution of the O–H distances along simulations is shifted to shorter values, which indicates the presence of stronger hydrogen bonds. Red shift of the N–H band in γ form compared to the N–H band in α form was observed. This was the first study that analyzed hydrogen bond interactions in nylon polymorphs by ab initio molecular dynamics, giving evidences of nuclear quantum effects. It demonstrated the importance of quantum effects in the simulations of the IR spectra and analyses of interactions in hydrogen-bonded polymers.

10.3.2 Studies of Interactions Between Polymers and Water

Many applications of polymer materials require contact with solvents, especially water. Wide range of polymer applications require knowledge of its specific properties when polymer interacts with water molecules. The molecular dynamics may elucidate the nature of these interactions. Particularly, water uptake and swelling are the properties, which can be determined by MD methods. From theoretical point of view, molecular dynamics is a complementary method to static calculations. For example, determination of solvation energy by cluster method as well thermodynamic cycle require knowledge configuration of water molecules. The starting configuration typically is the structure obtained by molecular dynamics. The study of polyelectrolyte and anion exchange in membranes also requires knowledge of counterion position that usually is obtained by molecular dynamics calculations [135, 136]. For example, the difference between electrostatic interactions of BF_4^- and covalently interacted OH^- has been described by BOMD calculations.

Another application focuses on the structure and thermodynamic properties of a system of flexible polymer chains grafted to a flat substrate. In this work, macromolecules are described by a coarse-grained bead–spring model, and solvent molecules by pointlike particles, assuming Lennard-Jones-type interactions between pairs of monomers, solvent molecules, and solvent monomer. Authors of this work [137] obtained density profiles of solvent particles and monomers by performing several molecular dynamics calculations. Deep analysis of chain structural properties has been presented. The authors discussed possible phase transitions and made comparison with previous work using implicit solvent models. It should be pointed out that variation of the brush height and the interfacial width of the transition zone between pure solvent and the brush agrees very well with corresponding experiments [137].

It should be stressed that studying equilibrium properties of polymers in solution by atomistic simulations is a challenging task. Relaxation of systems as well as their equilibration requires long simulations because of a large number of degrees of freedom. Available computation time is often not sufficient to ensure representative sampling of the phase space. There are several solutions presented in the literature. One approach to tackle this problem is to create a simulation scenario, which is simple enough to enable adequate sampling of equilibrium states. Typically, researchers reduce the system to the essential parts of the physics of the polymer in solution and in the aftermath they create the molecular model. On the other hand, changing or mixing the level of theory is possible, as has been already discussed above.

The advantage of using the first approach has been presented by García et al. [138]. The authors presented and tested a scenario, which is designed for studying whether a given polymer will aggregate or dissolve in a given solvent. Two periodic polymer molecules were simulated in the solvent. The distance between the polymer chains lends itself as an order parameter so that advanced sampling techniques, such as umbrella sampling, can be applied easily. The authors studied several states corresponding to dissolved polymers and to aggregated polymers, please see Figure 10.2. This approach missed intramolecular collapse of a single chain, but it retained full atomistic detail regarding polymer–solvent and the intermolecular polymer–polymer interactions. The results show that the quantitative agreement of calculated results with the experimental equilibrium data.

One of the most advanced and promising approaches for dealing with solvation is application of the thermodynamics cycle concept. The main advantage of this methods is taking into account specific interactions, such as hydrogen bonds, as well as continuum character of solvent. This method is especially advantageous

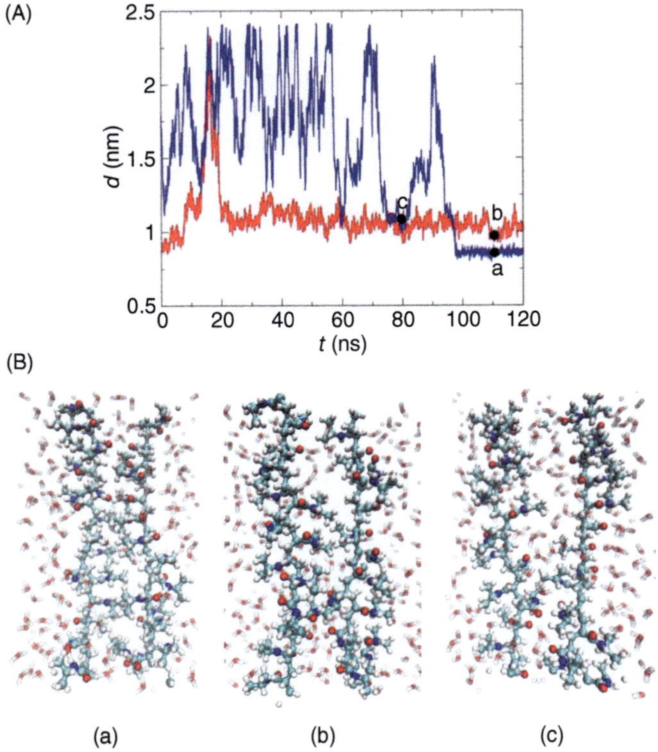

Figure 10.2 Simulation results for PNIPAM in water at 280 K obtained with the two-chain scenario ($\xi = 1.0$). (A) chain–chain distance as a function of simulation time at 280 K. The simulations were started from different configurations, dini = 0.9 nm (red line) and dini = 1.8 nm (blue line). (B) Snapshots of favorable conformations: (a) fully aggregated ($d = 0.87$ nm); (b) partially aggregated ($d = 0.96$ nm), and (c) water-mediated aggregation ($d = 1.12$ nm). Only water molecules in the layer of 1 nm around PNIPAM chains are depicted. Source: García et al. [138].

for polymers with functional groups, which may form hydrogen bonds with water molecules [139]. This situation may be even complicated. Interchain interactions hardly influence the properties of considered materials. The absorbed water breaks up some of hydrogen bonds initially present in polymer structure and forms new ones, bonded both with the network groups and with other water molecules. It should be stressed that water absorption is anisotropic. For this reason, the molecular dynamics is a promising method for analyses of flexibility and orientation of uptake water molecules. The calculation of the time autocorrelation function for various HB pairs may easily characterize hydrogen bonding between water and polymer functional groups [139].

10.3.3 Mechanical Properties of Polymers

One of the main applications of hydrogen-bonded polymers studied by theoretical and experimental techniques is supramolecular chemistry. It has been shown that hydrogen bonding plays a determining role in the transport of water and protonated species in such processes as crystallization and dehydration of sugars or the diffusion mechanisms of protons through fuel-cell electrolytes [140–142]. Nowadays, hydrogen bonds have been used as a versatile tool to achieve the structure with desired properties [143, 144]. It should be stressed that developing new methods and techniques to assess the formation of hydrogen bonds has unmeasurable impact in different fundamental and applied fields.

Polymers are pivotal structures for industries and material science. Responsible management of our planet resources requires paying attention to biodegradable polymers as environment-friendly materials [145]. Recent trends in biodegradable polymers show significant developments in terms of novel design strategies and engineering to provide advanced polymers with comparably good performance [146]. Therefore, it is highly desirable to understand and control complex polymer structures and physical and chemical properties of polymers. There are many uncertainties in terms of technology, possible usage, and cost of production. Therefore, there is a strong need to have a fresh perspective on design, properties, and functions of polymers with design strategies for future developments and applications of polymers.

Hydrogen bonds influence mechanical properties of almost all polymers. A thorough analysis of the impact of hydrogen bond interactions requires selection of studied polymers. At first, we studied polyamides. The preliminary studies on polymorphs of nylon 6 have already been published ([48] and Figure 10.3). Then, the analysis of the influence of the aromatic rings on the HB interactions shows unusual usage of polyethylene terephthalate and Kevlar polymer in, e.g. bulletproof jacket. Further, the presence of chloride and fluoride atoms in the polymer structure changes dramatically its elastic properties [147]. This group of polymers contains polyethylene, polyvinyl chloride as well as polytetrafluoroethylene (Teflon).

The mechanical properties can be estimated by the Birch–Murnaghan equation of state [148–152]. The approximation of the bulk modules may be done by stressing the system by static calculations as well as steered molecular dynamics (SMD). The second approach gave us possibility to deal with preferred conditions. The SMD methodology gives a possibility to directly test the molecular behavior when submitted to virtual traction along its longitudinal axis [153]. Based on this approach, it is possible to reproduce the experimental setting used to carry out the measurements. Then, the considered systems can be stressed by changing parameters of simulations, such as pressure or temperature. Examination of molecular reaction after external stress reveals microscopic characteristic and properties involved in its response. Analysis of underlying intramolecular interactions responsible for preserving the molecular structure can be done at atomic level.

10.3.4 Interphase Interactions

Another application presented in this chapter is the molecular dynamics study of interphase interactions. Polymer materials are regularly used as a base for reactions. On the other hand, biomedical applications require very specific mechanical and

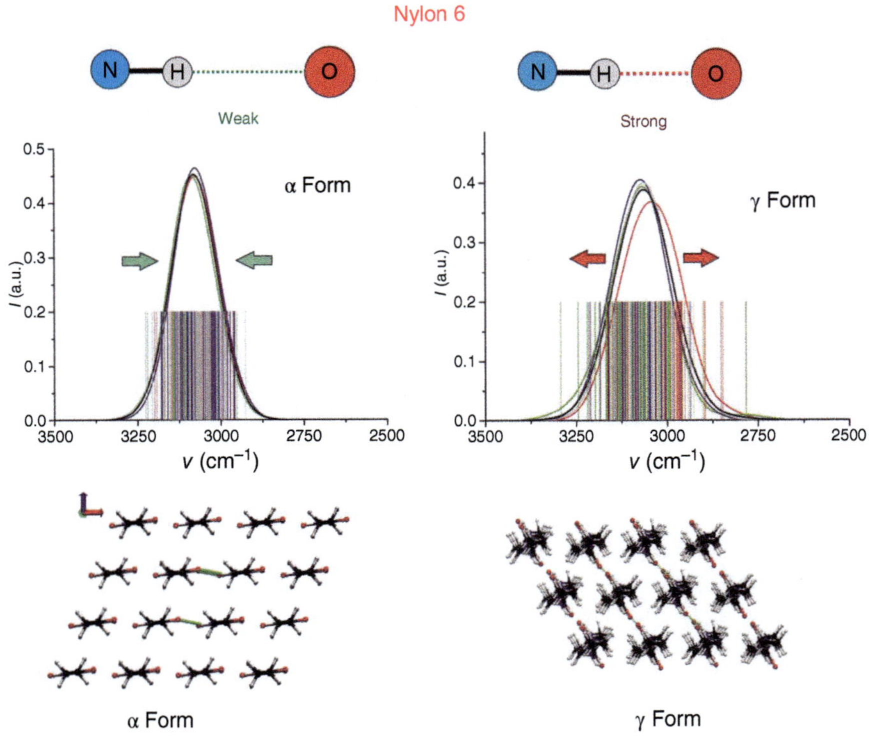

Figure 10.3 Hydrogen bonds differentiate polymer crystal forms. Solving the Schrödinger equation for proton potentials based on the snapshots extracted from ab initio MD provides a possible explanation to the red shift in the IR spectra of the N–H stretching band observed in the γ form compared with the α form of nylon 6. The HBs in the γ form are stronger than in the α form. This is mainly due to the dynamic effects. Source: Brela et al. [48].

chemical performance properties. Surgical sutures should absorb themselves easily. Potential bone replacements, e.g. polymerehydroxyapatite (HAP), should be durable and nonreactive. The HAP polymer interfacial molecular interactions have significant role in mechanical response in composite systems. Bhowmik et al. [154] used molecular dynamics simulations to evaluate the nature of these interfaces in polyacrylicacidehydroxyapatite composites.

These authors obtained parameters for monoclinic hydroxyapatite in CVFF (consistent valence force field) and indicated that potential sites for chelation and hydrogen bond formation between HAP and polyacrylic acid (PAAc) occur. Their theoretical results have been compared with experimental measurements. Binding energy for ex situ HAP composites is found to be lower as compared to that of in situ HAP. Figure 10.4 presents the model systems considered in the molecular modeling. Evaluation of the nature of interactions at mineral polymer interfaces is significant due to large impact of these interactions on overall mechanical behavior of the composite biomaterial. In composite systems, especially nanocomposite systems, interfaces may have significant role in mechanical properties [154–157].

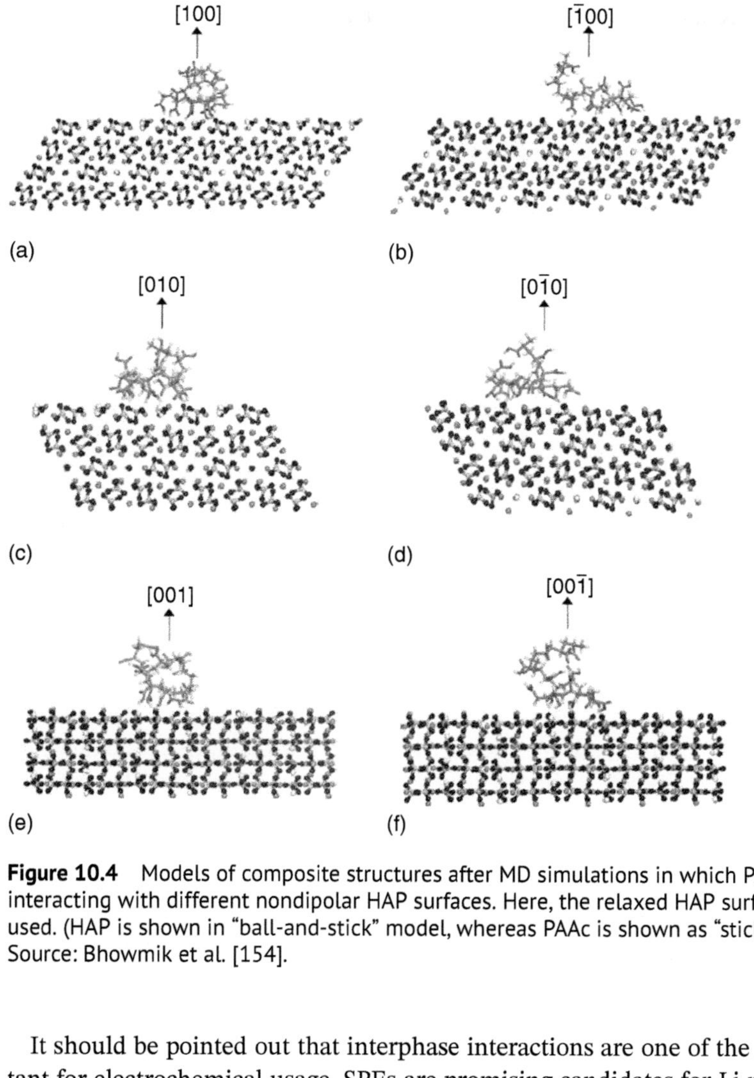

Figure 10.4 Models of composite structures after MD simulations in which PAAc is interacting with different nondipolar HAP surfaces. Here, the relaxed HAP surfaces were used. (HAP is shown in "ball-and-stick" model, whereas PAAc is shown as "stick" model). Source: Bhowmik et al. [154].

It should be pointed out that interphase interactions are one of the most important for electrochemical usage. SPEs are promising candidates for Li metal battery applications [158]. However, the interface between battery and polymers has not been studied deeply. The main goal in this research is to get insights into atomistic surface interactions and the early stages of solid electrolyte interphase formation between ionically conductive SPE host polymers and the Li metal electrodes. Molecular dynamics may gave us a lot of information about interactions as well as stability of studied systems [158].

10.4 Summary and Perspectives

This chapter presents new applications of ab initio molecular dynamics calculations as a powerful and competitive tool for simulations and analyses of polymer materials. Especially, this method is promising for understanding interactions present in polymers as well as the specific interactions between considered materials and environment. It should be stressed that weak interactions, as well as hydrogen-bonds, may be valuably analyzed by molecular dynamics methods. The main advantage of ab initio molecular dynamics is to perform calculations in predefined statistical ensemble, i.e. taking into account thermal effects. Of course, the quality of results depends on proper operations of thermostats.

Molecular dynamics simulations give also a possibility to include several additional effects, such as proton hopping or conformational variety. It was shown by Warshel and co-worker [14] that preorganization and electrostatic field of the neighborhood has a pivotal role for enzymatic reactions. Similar observations have been done for other macromolecules. Correct assignment of stationary points in multidimensional hyperspace is a difficult task for systems made up of thousands of atoms. There are too many local minima located close to each other. For this reason, the choice is molecular dynamics, which takes into account many states of the system. However, it should be pointed out that dynamic approaches are naturally more expensive than static.

The development of QM/MM and coarse-grained methods makes popular molecular dynamics. The possibility to choose adequate level of theory to considered systems and problems is a big challenge. In this chapter, we discussed only a snatch of applications. Recently, the number of studies that have given deeper interpretations based on MD calculations of complex systems, such as polymers or macromolecules, has dramatically increased. One of the examples is the analysis of the weak C—H···O=C hydrogen bonds in PHB [49]. This study has shown the dynamic changes of interactions between acetylic groups and methyl groups. The C—H···O=C interactions have been unstable along MD simulations, and the forming and the breaking processes occurred simultaneously. This effect is not influenced by methyl group rotations and polymer chains movements. It should be pointed out that consideration of the interaction energies between polymer chains in selected structures is not sufficient to discuss the strength of interchain interactions. Similar conclusions have been drawn from the analysis for two polymorphs of nylon 6 [48]. Spectroscopy techniques have been often used for validation of the postulated theoretical models.

The new models have been used in the several of the above-discussed applications in polymer science, studied by molecular dynamics. The variety of polymer structures and diversity of their usage provide a broad range of possible molecular dynamics applications. The present situation looks like a self-perpetuating circle. Complex and difficult problems require advanced and specific methods. New commercial computational techniques give possibility to deal with even more challenging problems that inspire developers to design new algorithms.

In our opinion, the design of new materials is impossible without understanding the structural features as well as interactions. Molecular dynamics is one of the most promising methods, especially for interphase studies.

Acknowledgment

This work was financially supported by National Science Centre Poland, grant 2016/21/B/ST4/02102.

References

1 Brela, M.Z., Boczar, M., Boda, Ł., and Wójcik, M.J. (2018). Molecular dynamics simulations of vibrational spectra of hydrogen-bonded systems. In: *Froniteirs of Quantum Chemistry* (eds. M.J. Wójcik, H. Nakatsuji, B. Kirtman and Y. Ozaki), 353–376. Springer.
2 Brela, M.Z., Boczar, M., Boda, Ł., and Wójcik, M.J. (2019). Dynamic and static quantum mechanical studies of vibrational spectra of hydrogen-bonded crystals. In: *Molecular Spectroscopy: A Quantum Chemistry Approach* (eds. Y. Ozaki, M.J. Wójcik and J. Popp), 327–352. Wiley-VCH.
3 Wójcik, M.J., Gług, M., Boczar, M., and Boda, Ł. (2014). *Chem. Phys. Lett.* 612: 162–166.
4 Gług, M., Brela, M.Z., Boczar, M. et al. (2017). *J. Phys. Chem. B* 121: 479–489.
5 Mavri, J., Pirc, G., and Stare, J. (2010). *J. Chem. Phys.* 132: 224506.
6 Cundari, T.R., Benson, M.T., Lutz, M.L., and Sommerer, S.O. (1996). Effective core potential approaches to the chemistry of the heavier elements. In: *Reviews in Computational Chemistry*, vol. 8 (eds. K.B. Lipkowitz and D.B. Boyd), 145–202. New York: VCH.
7 Marx, D. (1998). Proton transfer in ice. In: *Classical and Quantum Dynamics in Condensed Phase Simulations* Chapter 15 (eds. B.J. Berne, G. Ciccotti and D.F. Coker), 359–383. Singapore: World Scientific.
8 Tuckerman, M.E. and Hughes, A. (1998). Path integral molecular dynamics: a computational approach to quantum statistical mechanics. In: *Classicaland Quantum Dynamics in Condensed Phase Simulations*, Chapter 14 (eds. B.J. Berne, G. Ciccotti and D.F. Coker), 311. Singapore: World Scientific.
9 Durlak, P., Latajka, Z., and Berski, S.A. (2009). *J. Chem. Phys.* 131: 024308–024316.
10 Flakus, H.T., Hachula, B., and Garbacz, A. (2012). *J. Phys. Chem. A* 116: 11553–11567. Full Text.
11 Pople, J.A. (1998). Nobel lecture.
12 Warshel, A. (2014). Nobel lecture.
13 Warshel, A. and Levitt, M. (1976). *J. Mol. Biol.* 103: 227–249.
14 Jindal, G. and Warshel, A. (2016). *J. Phys. Chem. B* 120: 9913.
15 Vreven, T. and Morokuma, K. (2006). *Annual Reports in Computational Chemistry*, vol. 2, 35–51. Elsevier.
16 Ingólfsson, H.I., Lopez, C.A., Uusitalo, J.J. et al. (2014). *Wiley Interdiscip. Rev. Comput. Mol. Sci.* 4 (3): 225–248.

17 Kmiecik, S., Gront, D., Kolinski, M. et al. (2016). Coarse-grained protein models and their applications. *Chem. Rev.* 116: 7898–7936.

18 Dawid, A.E., Gront, D., and Kolinski, A. (2017). SURPASS low-resolution coarse-grained protein modeling. *J. Chem. Theory Comput.* 13: 5766–5779.

19 Durlak, P. and Latajka, Z. (2009). *Chem. Phys. Lett.* 477: 249–254.

20 Hadži, D. and Thompson, H.W. (1959). International union of pure and applied chemistry. *Hydrogen Bonding: Papers Presented at the Symposium on Hydrogen Bonding*, Ljubljana (29 July–3 August 1957). Pergamon Press: London.

21 Pimentel, G.C. and McClellan, A.L. (1960). *The Hydrogen Bond*. New York/San Francisco: W.H. Freeman/Reinhold Pub. Corp.

22 Schuster, P., Zundel, G., and Sandorfy, C. (1976). *The Hydrogen Bond: Recent Developments in Theory and Experiments*. Amsterdam/New York: North-Holland Pub. Co./American Elsevier Pub. Co.

23 Scheiner, S. (1997). *Hydrogen Bonding: A Theoretical Perspective*. New York/Oxford: Oxford University Press.

24 Jeffrey, G.A. (1997). *An Introduction to Hydrogen Bonding*. New York/Oxford: Oxford University Press.

25 Hadži, D. (1997). *Theoretical Treatments of Hydrogen Bonding*. Chichester: Wiley.

26 Grabowski, S. (2006). *Hydrogen Bonding New Insights*. Dordrecht: Springer.

27 Kobayashi, S. and Makino, A. (2009). *Chem. Rev.* 109: 5288–5353.

28 Wiltshire, J.T. and Qiao, G.G. (2006). *Macromolecules* 39: 9018–9027.

29 Mansky, P., Liu, Y., Huang, E. et al. (1997). *Science* 275: 1458–1460.

30 Liu, N.C. and Baker, W.E. (1992). *Adv. Polym. Tech.* 11: 249–262.

31 Van Krevelen, D.W. and TeNijenhuis, K. (2009). *Properties of Polymers: Their Correlation with Chemical Structure; Their Numerical Estimation and Prediction From Additive Group Contributions*. Elsevier.

32 Feldman, K.E., Kade, M.J., Meijer, E.W. et al. (2009). *Macromolecules* 42: 9072–9081.

33 Lange, R.F.M., Van Gurp, M., and Meijer, E.W. (1999). *J. Polym. Sci., Part A: Polym. Chem.* 37: 3657–3670.

34 Heymann, B. and Grubmüller, H. (1999). *Chem. Phys. Lett.* 307: 425–432.

35 Baukh, W.V., Huinink, H.P., Adan, O.C.G. et al. (2011). *Macromolecules* 44 (12): 4863–4871.

36 Flagg, L.Q., Bischak, C.G., Onorato, J.W. et al. (2019). *J. Am. Chem. Soc.* 141 (10): 4345–4354.

37 Wojcik, M.J. (1976). *Int. J. Quantum Chem.* 10: 747.

38 Ozaki, Y., Pliura, D.H., Carey, P.R., and Storer, A.C. (1982). *Biochemistry* 21: 3102.

39 Wu, Y., Murayama, K., and Ozaki, Y. (2001). *J. Phys. Chem. B* 105: 6251.

40 Ozaki, Y., Aoyagi, K., Iriyama, K. et al. (1989). *J. Phys. Chem.* 93: 3842.

41 Wojcik, M.J., Boczar, M., and Boda, L. (2007). *J. Chem. Phys.* 127: 08430.

42 Wojcik, M.J., Kwiendacz, J., Boczar, M. et al. (2010). *Chem. Phys.* 372: 72.

43 Katayama, N., Ozaki, Y., Seki, T. et al. (1994). *Langmuir* 10: 1898–1902.

44 Shigehiro, Y., Kazuo, I., and Kohei, T. (1998). *Chem. Lett.* 27 (1): 89–90.

45 Chen, S., Teng, C., Zhang, M. et al. (2016). *Adv. Mater.* 28: 5969–5974.

46 Brandell, D., Karo, J., Liivat, A., and Thomas, J.O. (2007). *J. Mol. Model.* 13: 1039–1046.

47 Sok, R.M., Berendsen, H.J.C., and van Gunsteren, W.F. (1992). *J. Chem. Phys.* 96 (6): 4699–4704.

48 Brela, M.Z., Wójcik, M.J., Boczar, M. et al. (2018). *Int. J. Quantum Chem.* 118: e25595.

49 Brela, M.Z., Boczar, M., Wójcik, M.J. et al. (2017). *Chem. Phys. Lett.* 678: 112.

50 Hohenberg, P. and Kohn, W. (1964). *Phys. Rev.* 136: B864–B871.

51 Kohn, W. and Sham, L.J. (1965). *Phys. Rev.* 140: A1133–A1138.

52 Tuckerman, M.E. and Hughes, A. (1998). *Classical and Quantum Dynamics in Condensed Phase Simulations*, vol. 14 (eds. B.J. Berne, G. Ciccotti and D.F. Coker), 311. Singapore: World Scientific.

53 Babitzki, G., Mathias, G., and Tavan, P. (2009). *J. Phys. Chem. B* 113: 10496–10508.

54 Roos, G., Geerlings, P., and Messens, J. (2009). *J. Phys. Chem. B* 113: 13465–13475.

55 Albaugh, A., Boateng, H.A., Bradshaw, R.T. et al. (2016). *J. Phys. Chem. B* 120: 9811–9983.

56 Goyal, P., Qian, H.-J., Irle, S. et al. (2014). *J. Phys. Chem. B* 118: 11007–11027.

57 Goszczycki, P., Stadnicka, K., Brela, M.Z. et al. (2017). *J. Mol. Struct.* 1146: 337–346.

58 Wang, J., Wang, G., and Zhao, J. (2002). *Phys. Rev. B* 66: 35418.

59 Saarikoski, H., Puska, M.J., and Nieminen, R.M. (2003). *Int. J. Quantum Chem.* 91: 490.

60 Zhao, X.G., Richardson, W.H., Chen, J.L. et al. (1997). *Inorg. Chem.* 36: 1198.

61 Erkoc, S. (2003). *Chem. Phys. Lett.* 369: 605.

62 Shia, Y. and Zhoua, Z. (2004). *J. Mol. Struct. THEOCHEM* 674: 113.

63 Mataa, R.A. and Cabral, B.J.C. (2004). *J. Mol. Struct. THEOCHEM* 673: 155.

64 Janoschek, R. (1998). *J. Mol. Struct. THEOCHEM* 423: 219.

65 Tsushima, S. and Yang, T. (2005). *Chem. Phys. Lett.* 401: 68.

66 McClenaghan, N.D., Hu, P., and Hardacre, C. (2000). *Surf. Sci.* 464: 223.

67 Ferullo, R.M. and Castellani, N.J. (2004). *J. Mol. Catal. A: Chem.* 221: 155.

68 Grimme, S. (2001). *Chem. Phys. Lett.* 339: 380.

69 Brewer, S.H. and Franzen, S. (2004). *Chem. Phys.* 300: 285.

70 Wang, Y.X., Zhang, W.L., Wang, C.L., and Zhang, P.L. (2002). *Opt. Commun.* 201: 79.

71 Kobko, N., Masunov, A., and Tretiak, S. (2004). *Chem. Phys. Lett.* 392: 444.

72 Liyanage, P.S., de Silva, R.M., and Nalin de Silva, K.M. (2003). *J. Mol. Struct. THEOCHEM* 639: 195.

73 Stephens, P.J., Devlin, F.J., Cheeseman, J.R. et al. (2000). *Tetrahedron: Asymmetry* 11: 2443.

74 Bruschi, M., Fantucci, P., Pizzotti, M., and Rovizzi, C. (2003). *J. Mol. Catal. A: Chem.* 204/205: 793.

75 Cai, M., Yin, Z., and Zhang, M. (2004). *Chem. Phys. Lett.* 388: 223.

76 Wilson, P.J. and Tozer, D.J. (2002). *J. Mol. Struct.* 602/603: 191.

77 Hua, Z., Weia, H., and Chen, Z. (2004). *J. Mol. Struct. THEOCHEM* 668: 235.

78 Eschriga, H. and Pickett, W.E. (2001). *Solid State Commun.* 118: 123.

79 Weia, H., Hua, Z., and Chen, Z. (2005). *J. Mol. Struct. THEOCHEM* 713: 145.

80 Tappan, B.C., Beal, R.W., and Brill, T.B. (2002). *Thermochim. Acta* 388: 227.

81 Zhanpeisova, N.U., Sugimotoa, A., Mizunoa, K. et al. (2002). *J. Mol. Struct. THEOCHEM* 592: 149.

82 Bridgemana, A.J. and Cavigliasso, G. (2002). *Chem. Phys.* 279: 143.

83 Ball, B., Zhou, X., and Liu, R. (1996). *Spectrochim. Acta, Part A* 52: 1803.

84 Fu, A., Dua, D., and Zhou, Z. (2003). *Spectrochim. Acta, Part A* 59: 245.

85 Fabian, J. and Herzog, K. (1998). *Vib. Spectrosc.* 16: 77.

86 Xiao, H., Xie, D., and Xue, Y. (2002). *Spectrochim. Acta, Part A* 58: 2227.

87 Diaz-Acosta, I., Baker, J., Hinton, J.F., and Pulay, P. (2003). *Spectrochim. Acta, Part A* 59: 363.

88 Altun, A., Golcuk, K., and Kumru, M. (2003). *Vib. Spectrosc.* 33: 63.

89 Acevedo-Chaveza, R., Roblesb, J., and Costasa, M.E. (2003). *J. Mol. Struct. THEOCHEM* 664/665: 91.

90 Bridgeman, A.J. (2003). *Chem. Phys.* 287: 55.

91 Jonas, V. and Thiel, W. (1995). *J. Chem. Phys.* 102: 8474.

92 Jonas, V. and Thiel, W. (1996). *J. Chem. Phys.* 105: 3636.

93 Jonas, V. and Thiel, W. (1998). *Organometallics* 17: 353.

94 Zhou, M. and Andrews, L. (2000). *J. Am. Chem. Soc.* 122: 1531.

95 Hu, Z. and Boyd, R.J. (2000). *J. Chem. Phys.* 113: 9393.

96 Wang, X., Zhou, M., and Andrews, L. (2000). *J. Phys. Chem. A* 104: 7964.

97 Zhou, M. and Andrews, L. (1999). *J. Chem. Phys.* 111: 4548.

98 Schroeder, D., Wesendrup, R., Hertwig, R.H. et al. (2000). *Organometallics* 19: 2608.

99 Bencze, E., Papai, I., Mink, J., and Goggin, P.L. (1999). *J. Organomet. Chem.* 584: 118.

100 Nikolov, G.S., Mikosch, H., and Bauer, G. (2000). *Theochemistry* 499: 35.

101 Bare, W.D., Citra, A., Chertihin, G.V., and Andrews, L. (1999). *J. Phys. Chem. A* 103: 5456.

102 Henson, N.J., Hay, P.J., and Redondo, A. (1999). *Inorg. Chem.* 38: 1618.

103 Schimmelpfennig, B., Wahlgren, U., Gropen, O., and Haaland, A. (2001). *J. Chem. Soc., Dalton Trans.* (10): 1616.

104 Binnewies, M. (1998). *Chem. unserer Zeit* 32: 15.

105 Ozaki, T., Jiang, J.Z., Murase, K. et al. (1998). *J. Alloys Compd.* 265: 125.

106 Hendricks, J.H., Aquino, M.I., Maslar, J.E., and Zachariah, M.R. (1998). *Chem. Mater.* 10: 2221.

107 Ottosson, M., Andersson, T., and Carlsson, J.O. (1989). *Appl. Phys. Lett.* 54: 2476.

108 Klamt, A. (1995). *J. Phys. Chem.* 99: 2224.

109 Klamt, A., Jonas, V., Bürger, T., and Lohrenz, J.C. (1998). *J. Phys. Chem. A* 102: 5074.

110 Abegg, P.W. and Ha, T.-K. (1974). *Mol. Phys.* 27: 763–767.

111 Brela, M., Stare, J., Pirc, G. et al. (2012). *J. Phys. Chem. B* 116: 4510–4518.
112 Marx, D. and Hutter, J. (2009). *Ab Initio Molecular Dynamics*. Cambridge: Cambridge University Press.
113 Krack, M. and Parrinello, M. (2004). Forschungszentrum Jülich, NIC Series 25, 29.
114 Born, M. and Oppenheimer, R. (1927). *Ann. Phys.* 84: 457–484.
115 Brown, M.E. and Calvin, W.M. (2000). *Science* 287: 107.
116 Iedema, M.J., Dresser, M.J., Doering, D.L. et al. (1998). *J. Phys. Chem. B* 102: 9203.
117 Pirc, G., Mavri, J., and Stare, J. (2012). *Vib. Spectrosc.* 58: 153–162.
118 Stare, J., Jezierska, A., Ambrožič, G. et al. (2004). *J. Am. Chem. Soc.* 126: 4437–4443.
119 Stare, J., Mavri, J., Grdadolnik, J. et al. (2011). *J. Phys. Chem. B* 115: 5999–6010.
120 Sato, H., Murakami, R., Padermshoke, A. et al. (2004). Infrared spectroscopy studies of CH center dot center dot center dot O hydrogen bondings and thermal behavior of biodegradable poly(hydroxyalkanoate). *Macromolecules* 37 (19): 7203–7213.
121 Krimm, S. (1960). Infrared spectra of high polymers. *Fortschr. Hochpolym. Forsch.* 2: 51–172.
122 Ozaki, Y. (2012). Near-infrared spectroscopy – its versatility in analytical chemistry. *Anal. Sci.* 28: 545.
123 Yamamoto, S., Furukawa, T., Bour, P., and Ozaki, Y. (2014). Solvated states of poly-L-alanine a-helix explored by Raman optical activity. *J. Phys. Chem. A* 118: 3655–3662.
124 Yamamoto, S., Morisawa, Y., Sato, H. et al. (2003). *J. Phys. Chem. B* 117: 2180–2187.
125 Hoshina, H., Morisawa, Y., Sato, H. et al. (2010). *Appl. Phys. Lett.* 100: 101904.
126 Morisawa, Y., Yasunaga, M., Sato, H. et al. (2014). *J. Phys. Chem. B* 118: 11855–11861.
127 Sato, H., Miyada, M., Yamamoto, S. et al. (2016). *RSC Adv.* 6: 16817–16823.
128 Li, Y. and Goddard, W.A. III, (2002). *Macromolecules* 35: 8440–8455.
129 Egorov, S.A. (2008). *J. Chem. Phys.* 129: 064901.
130 Bouř, P., Sopková, J., Bednárová, L. et al. (1997). *J. Comput. Chem.* 18: 646–659.
131 Yamamoto, S., Li, X., Ruud, K., and Bour, P. (2012). *J. Chem. Theory Comput.* 8: 977–985.
132 Naira, L.S. and Laurencina, C.T. (2007). Biodegradable polymers as biomaterials. *Prog. Polym. Sci.* 32: 762–798.
133 Kamerlin, S.C.L. and Warshel, A. (2011). *Phys. Chem. Chem. Phys.* 13: 10401.
134 Brela, M.Z., Prah, A., Boczar, M. et al. (2019). *Molecules* 24 (23): 4359.
135 Henkensmeier, D., Cho, H., Brela, M. et al. (2014). *Int. J. Hydrogen Energy* 39 (6): 2842–2853.
136 Schibli, E.M., Stewart, J.C., Wright, A.A. et al. (2020). *Macromolecules* 53 (12): 4908–4916.
137 Dimitrov, D.I. (2007). *J. Chem. Phys.* 127: 084905.

138 García, E.J., Bhandary, D., Horsch, M.T., and Hasse, H. (2018). *J. Mol. Liq.* 268: 294–302.

139 Mijović, J. and Zhang, H. (2004). *J. Phys. Chem. B* 108: 2557–2563.

140 Auffinger, P. and Westhof, E. (1997). *J. Mol. Biol.* 274: 54–63.

141 Johnson, E.R., Keinan, S., Mori-Sanchez, P. et al. (2010). *J. Am. Chem. Soc.* 132: 6498–6506.

142 Martinez-Felipe, A., Lu, Z., Henderson, P.A. et al. (2012). *Polymer* 53: 2604–2612.

143 Yao, M., Chen, X.S., Dong, L. et al. (2018). *Liq. Cryst.* 45: 931–941.

144 Jansze, S.M., Martinez-Felipe, A., Storey, J.M.D. et al. (2015). A twist-bend nematic phase driven by hydrogen bonding. *Angew. Chem. Int. Ed.* 127: 653–656.

145 Mitrus, M., Wojtowicz, A., and Moscicki, L. (2009). Biodegradable polymers and their practical utility. In: *Thermoplastic Starch* (eds. L. Janssen and L. Moscicki), 1–33. Weinheim: Wiley-VCH Verlag GmbH & Co. KGaA.

146 Balk, M., Behl, M., Wischke, C. et al. (2016). Recent advances in degradable lactide-based shape-memory polymers. *Adv. Drug Delivery Rev.* 107: 136–152.

147 Tanner, D., Fitzgerald, J.A., and Phillips, B.R. (1989). *Angew. Chem. Int. Ed.* 28: 649–654.

148 Katsura, T. and Tange, Y. (2019). *Minerals* 9: 745.

149 Murnaghan, F.D. (1937). Finite deformations of an elastic solid. *Am. J. Math.* 59: 235–260.

150 Birch, F. (1947). Finite elastic strain of cubic crystals. *Physiol. Rev.* 71: 809–824.

151 Birch, F. (1952). Elasticity and constitution of the earth's interior. *J. Geophys. Res.* 57: 227–286.

152 Bass, J.D. (1995). Elasticity of minerals, glasses, and melts. In: *Mineral Physics & Crystallography: A Handbook of Physical Constants*, vol. 2 (ed. T.J. Ahrens), 45–63.

153 Izrailev, S. et al. (1999). Steered molecular dynamics. In: *Computational Molecular Dynamics: Challenges, Methods, Ideas*, Lecture Notes in Computational Science and Engineering, vol. 4 (eds. P. Deuflhard, J. Hermans, B. Leimkuhler, et al.), 39–65. Berlin, Heidelberg: Springer.

154 Bhowmik, R., Katti, K.S., and Katti, D. (2007). *Polymer* 48: 664–674.

155 Katti, K.S. (2004). *Colloids Surf., B* 39: 133.

156 Katti, K.S., Turlapati, P.K., Verma, D. et al. (2006). *Am. J. Biochem. Biotechnol.* 2: 73.

157 Katti, D.R., Katti, K.S., and Shanmugasundaram, V. (2002). Role of clay-solvent inter and intraparticle. *Interactions on Swelling Characteristics of Montmorillonite Nano-Meso-Micro Scale Particulate Systems, Materials Research Society Symposium – Proceedings*, 704, 257–262.

158 Ebadi, M., Marchiori, C., Mindemark, J. et al. (2019). *J. Mater. Chem. A* 7: 8394.

11

Spectroscopic Analysis of Structural Transformations Associated with Poly(lactic acid)

Shaw L. Hsu[1] *and Xiaozhen Yang*[2]

[1] *University of Massachusetts (Amherst), Polymer Science and Engineering, Amherst, MA 01003, USA*
[2] *Institute of Chemistry, Chinese Academy of Sciences, Beijing 100080, China*

11.1 Introduction

The characterization of various forms of semicrystalline poly(lactic acid) (PLA) presents a number of noteworthy and interesting challenges. Among all of the characterization techniques for polymers, vibrational spectroscopy is one of the most valuable for the elucidation of polymeric molecular structures to comprehend their physical properties. Spectroscopy is extremely convenient for the characterization of PLA configuration, polymer conformation, segmental orientation, and intermolecular interactions. The principal advantage of using vibrational spectroscopy as a characterization tool is that it is selective. Both crystalline and amorphous polymer structures can be characterized. Conformational order ranging from the shortest chemical repeat to hundreds of angstroms can be measured. Although numerous techniques can address polymer crystals, few provide information on the disordered chains as in solution, melt, or the amorphous region of semicrystalline samples. In this section, the applicability of both infrared and Raman spectroscopy to PLA structural characterization will be presented. The experimental studies will be complemented by normal coordinate analysis for both ordered and disordered states. Assignments of vibrational transitions based on group theoretical analysis will also be validated using highly ordered and oriented PLA samples.

PLA is a biodegradable and compostable thermoplastic derived from renewable plant sources such as starch and sugar [1]. It has attracted a great deal of attention not only because it is an interesting polymer to study, but also because this research subject resonates with societal concern regarding polymer waste on land and in the ocean. In most commercial PLAs, some small percentage of D isomers are usually incorporated into a predominantly L isomer chain. Both isomers can be derived from saccharides through fermentation [1]. Depending on the D content, PLA usually exists as a semicrystalline polymer with different degrees of crystallinity. Among the many environment-friendly polymers, PLA is one of the most promising candidates for being commercialized and may be used in large-scale applications. Over

Spectroscopic Techniques for Polymer Characterization: Methods, Instrumentation, Applications, First Edition. Edited by Yukihiro Ozaki and Harumi Sato.

the last couple of decades, PLA has been a popular choice for use in high-value medical devices due to its bioabsorbable characteristics [2, 3]. In the last few years, new polymerization routes have been developed, allowing for the more economical production of polymers with high molecular weight, to be used in more commercial sectors [4]. PLA has an ideal combination of attractive physical properties, such as high modulus, good film-forming properties, good heat-seal characteristics, and barrier to flavor and aroma. Therefore, it can be used to manufacture consumer goods, especially food packaging for use at low temperatures [4–6]. The use of PLA has also been explored across the globe for large-scale commercial products such as field covers [7, 8]. Recently, PLA has also been processed in conjunction with other filler materials to form composites, which possess unique properties, including those based on nanoclays, cellulose, and glass fibers [9, 10]. All of these applications, both small-scale and large-scale, have similar physical requirements, such as dimensional stability and maintenance of mechanical strength as a function of temperature, especially in the presence of moisture [7, 11, 12].

The correlation of physical properties to morphological features in PLA is no different from other polymers. The degree of crystallinity and crystallite size in the sample are important parameters. For PLA to be developed into a commercially important polymer, its physical properties need to be improved. PLA is brittle because of the high degree of crystallinity generated extremely rapidly during processing. The rate at which the degree of crystallinity is achieved as a function of processing speed is crucial. Unfortunately, PLA also exhibits a low glass transition temperature, making it inherently unstable at elevated temperatures. Because of the large crystallites, the lack of optical transparency is also a concern. Overcoming these shortcomings will significantly increase the number of applications for such an environment-friendly polymer. It is also known that PLA exhibits a number of phases, each possessing some structural imperfections. The amount and type of these structural defects are also important considerations in maintaining product stability. Vibrational spectroscopy can play an important role in resolving these issues associated with PLA.

As an example of the use of PLA in medical devices, an image of bioresorbable polymeric scaffolds is shown in Figure 11.1. This polymeric candidate is usually promoted as superior relative to traditional metallic stents [13]. This scaffold must be hemocompatible and biocompatible, both of which are satisfied by PLA. It degrades hydrolytically during its lifetime in the body, yet it must maintain sufficient radial strength, resistance to fracture, and acute recoil to provide blood vessel support even when the body deforms [13]. Since these scaffolds are deployed in a compressed form and then expanded, the junctions (as seen in Figure 11.1) experience considerable bending strain at a high rate. Although PLA is known to be a hydrophobic polymer, 0.5% of water can infiltrate into a semicrystalline PLA [14]. These requirements for a relatively simple yet crucially important biomedical device necessitate a detailed understanding of the morphology of the PLA used and how it changes as a function of time, temperature, and in the presence of water/moisture.

Although PLA can be processed using standard equipment, its unique material properties must be taken into consideration to optimize the conversion of PLA to

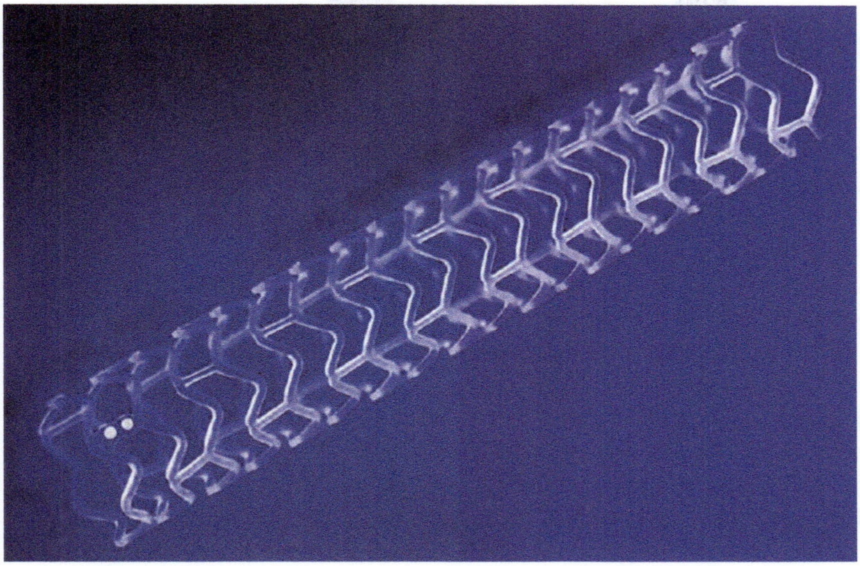

Figure 11.1 Schematic drawing of a bioresorbable polymeric scaffold. Source: Courtesy of Abbott Vascular, Santa Clara, CA, with permission.

molded parts, films, foams, and fibers. Similar to other products, the PLA used in bioresorbable polymeric scaffolds as shown in Figure 11.1 needs to be extruded into the proper diameters and then cut to perform properly [13]. The morphology of each part of the scaffold, such as struts and junctions, needs to be clarified. When used in large quantities in other applications, the most practical processing method involves film blowing. In that case, high melt viscosity must be maintained to achieve large elongation and expansion ratios [15–17]. In each of those applications, a clear understanding of the PLA's morphology must also be accomplished. As summarized in this review, vibrational spectroscopy has proven to be extremely valuable complementing other characterization techniques in elucidating the relationship between the morphological changes during processing and the subsequent physical properties achieved.

11.2 Spectroscopic Tools

As we show below, vibrational spectroscopy has contributed significantly to the clarification of PLA structures in solution, in melted and condensed states, and even during processing when PLA resin is deformed into fibers or films [11, 12, 18–21]. Vibrational spectroscopy, including infrared absorption and Raman scattering, is one of the oldest and most powerful techniques for analyzing polymer composition, segmental orientation, chain conformation, and intermolecular interactions. Due to significant advances in instrumentation and simulation software, most laboratories now possess instrumentation capable of producing excellent data. Fibers, films, solutions, and melts can be analyzed with great confidence. The sample under examination need not be static in nature. Time-dependent phenomena can be measured using time resolutions as short as microseconds. It is also possible to measure samples in the range of micrometers or smaller with dedicated accessories. Vibrational spectroscopy provides information complementary to other characterization techniques and has proven of enormous value to studies in polymer science and engineering.

Infrared absorption spectroscopy and Raman scattering have very different physical origins, each offering particular advantages for polymer characterization. Infrared active bands depend on the dipole change, whereas Raman scattering depends on changes in the polarizability tensor [22–24]. Different vibrational modes are thus active in Raman and infrared. The more symmetric the molecule, the greater the difference between Raman and infrared spectra. The two spectroscopic techniques are thus complementary. For both techniques, although more so for infrared, substantial literature exists regarding their utility for polymer characterization. Infrared spectroscopy is more appropriate for studies of chemical composition and side groups [25, 26]. Raman scattering is more effective for structural characterization, since intense Raman scattering is often associated with the significant polarizability changes of the symmetric all-carbon backbone of the polymer chain and can be used much more readily to analyze chain conformation [27].

Proper assignment of molecular vibrations is of fundamental importance, especially when morphological studies such as crystal orientation or amorphous deformation are conducted using vibrational spectroscopy. Even for ordered crystalline PLAs, a large number of vibrations (both infrared and Raman) have been difficult to explain [18]. For example, until recently no clear explanation was available for the rich features observed in the carbonyl stretch region (from 1700 to 1850 cm^{-1}). Multiple components are not expected since the relatively high frequency carbonyl stretching vibrations are localized (mainly C=O stretching motion). While helical symmetry was speculated as the cause for the multiple bands in this frequency region, no evidence was provided in previous studies to support this assignment. It was brought to our attention that Raman-active bands were mistakenly assigned incorrect optical species, including optically inactive ones, and that typographical errors were present in the character table presented [18]. Because of uncertain structures, understanding processing-induced structural changes in PLA, e.g. segmental orientation or crystallite size, has been difficult [5, 6, 18, 21]. Whether a spectral change is caused by a peak shift or by simultaneous intensity changes in two or more separate bands determines the accuracy of the molecular interpretation of spectral data. This can be determined only when proper normal mode assignments are established [18].

Since polarized incident radiation is readily available for both infrared absorption and Raman scattering, the polarization characteristics of the vibrational transitions play a crucial role in determining the optical activity of both crystalline and amorphous structures. Additionally, the polarization of vibrational spectroscopic transitions can provide more structural information than just segmental orientation. As we shall show below, the relative intensity of the bands of well-defined PLA samples can also provide information regarding the magnitude and specificity of intermolecular interactions in PLA [18, 21]. The optical activities of vibrational transitions can usually be predicted using the equilibrium geometric structure of the crystalline state. Since Raman scattering is a second-order transition, a scattering tensor is involved. In this case, even for samples of different degrees of disorder, the diagonal elements and the off-diagonal elements transform separately and are representative of the postulated partially ordered structures [23, 28]. Even for totally disordered structures, such as amorphous chains in solution or melt, these two types of tensor elements transform separately and their ratios can be extremely useful in the analysis of chain conformational distributions associated with amorphous structures [29, 30].

11.2.1 Vibrational Features of PLA Crystals

The crystalline structure of PLA has been investigated previously. The space group of the PLA α-crystal usually obtained has been unambiguously assigned as $P2_12_12_1$, but actual atomic coordinates, as well as the helical symmetry associated with PLA α-crystals, have been difficult to resolve [31, 32]. PLA chains exhibit a number of helical conformations. The initial structural studies were conducted by two separate groups [33–35]. These earlier studies have definitively established that it is possible for PLA to exist in at least four helical forms (2_1, 3_1, 4_1, and 5_1), each with different probabilities of occurring. Three skeletal bonds occur along the molecular chain in PLA: C—O(ester), O—C, and C—C. The C—O(ester) bond is always assumed to be trans due to conjugation with the C—O double bond. The O—C bond has two energy minima around −160° and −48°, and the C—C bond has two energy minima around 160° and −73°. Since 160° is close to 180°, the 160° angles are termed trans (t). In the same way, −48° and −73° are close to −60° and are described as gauche (g). Assuming the ester group to be planar, the remaining two rotational freedoms along the backbone allow the postulated PLA chains to transform among the various helices. It has also been established that PLA helical structure can transform from the usual dominant 3_1 helix to the more stable form of PLLA in the 10_3 helix as a function of time and at elevated temperatures [34]. The differences in the two most probable helices (10_3 and 3_1) seem to be relatively small. However, their actual atomic placements can be significantly different, as seen in Figure 11.2. Experimental and simulation analyses revealed a helix that is rather rigid, dominated by tgt conformation [20, 36]. The characteristic ratio calculated based on the conformational statistics is ~12, consistent with light scattering and theoretical results [20, 37].

Under most practical circumstances, only the PLA α-crystal is obtained. Therefore, most spectroscopic analysis is focused on the infrared or Raman spectra obtained using this structure. The $P2_12_12_1$ space group of the α-crystal has D_2 group symmetry, the character table of which is well known, having four symmetry species: A, B_1, B_2, and B_3 [18, 24]. Only the B_1, B_2, and B_3 species are infrared-active, with dipole derivative vector components well aligned to the crystallographic axes. All four symmetry species are Raman-active, where the A species is polarized and the other three are depolarized. Knowledge of these optical activities, combined with appropriate polarized spectroscopy, permits detailed analysis of experimental data.

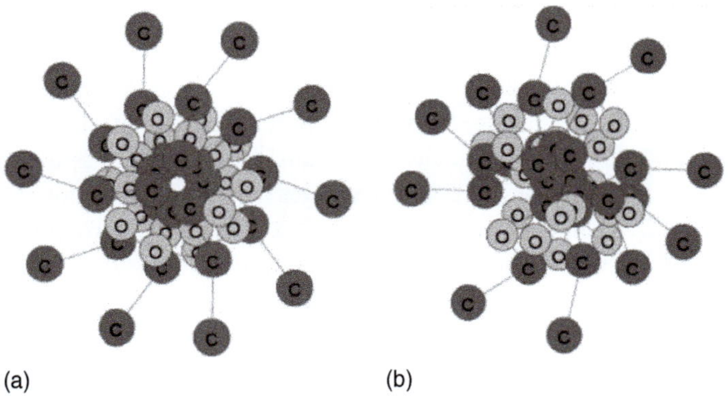

Figure 11.2 Two proposed forms of the PLA α-crystals are shown schematically: (a) Aleman/Puiggali/Lotz coordinates; (b) Sasaki/Asakura coordinates. Hydrogens are abbreviated to aid clarity. Source: Aou and Hsu [18].

Clearly, access to highly perfect crystalline samples is necessary. Analysis performed on oriented PLA films using the attenuated total reflectance (ATR) technique was not entirely satisfactory, because in ATR small shifts in frequency and relative intensity are hard to predict. Instead, extremely highly oriented electrospun fibers have become commonly available. These fibers form thin mats that can be oriented manually at 120 °C, then studied using transmission technique [18]. In addition, extremely large spherulitic samples can be prepared from a PLA solution of chloroform, as shown below in Figure 11.3. These samples are so large that optical microscopic images and X-ray diffractions of these crystals can clearly show the growth direction. Since transmission technique was utilized, it was easy to obtain infrared spectra at liquid nitrogen temperature using a custom-built cold cell. Intermolecular interactions strongly depend on temperature. These low-temperature spectra show the effects of these changes as crystal field splittings that can be easily observed, making band assignments much more reliable. The normal coordinate analysis computer program, developed at the University of Michigan, was used to compute normal frequencies of both 3_1 and 10_3 single-chain helices. The atomic coordinates used to calculate infrared and Raman features were transferred directly from a previous study, as were those used to calculate the vibrational spectrum of a

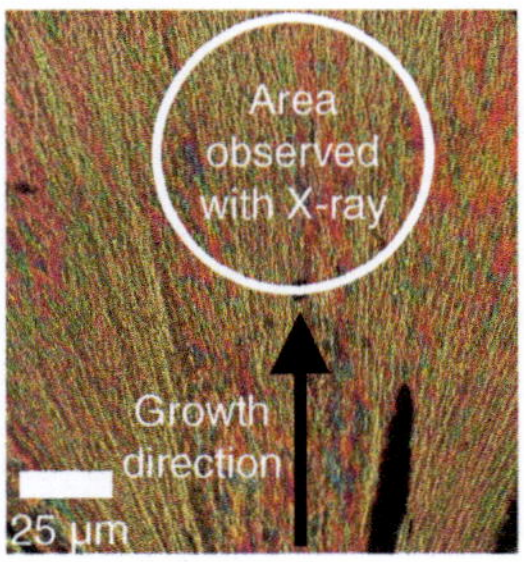

Figure 11.3 Spherulite observed under polarized optical microscopy. The approximate region probed by the X-ray is indicated by the circle. The center of the spherulite lies outside the photographed area. "Growth" indicates the radial growth direction. Source: Aou and Hsu [18].

helix. The force constants are the same as in previous studies, employing standard definitions of internal coordinates and symmetry coordinates. A character table for C10 (10_3 helix) was derived using the method prescribed by Wilson, and the character table for C3 (3_1 helix) can be taken directly from Wilson [24]. To simulate the infrared absorbance of the carbonyl bands, dipole derivatives of the carbonyl stretching vibrations were estimated by running an ab initio vibrational calculation on a model compound using Gaussian98 with appropriate basis set [18].

Both electrospun fibers and the large, thin PLA spherulite are in the α-crystalline form and are sufficiently large and thin to produce high-quality polarized infrared spectra as shown in Figure 11.4 [18]. When the symmetry groups are known, the optical activity of both infrared and Raman bands can be predicted and used to assign the experimental observations. In this way, previously unobserved bands have now been revealed. Even in the isolated localized carbonyl stretching region, multiple components are seen. This is especially true for the spectra obtained at low temperatures (Figure 11.5). High-quality Raman-active vibrations were also extremely informative when isotropic spectra became available (Figure 11.6). For example, infrared dichroic analysis of large two-dimensional PLA crystals can be conducted in detail [18]. By combining the two dichroic analyses, accurate band assignments for an α-crystal were achieved. The results of this analysis are only consistent with a 10_3 helical chain conformation, as opposed to the more simplistic 3_1 helix generally used throughout the literature. The orthorhombic structure, previously suggested by X-ray studies, was confirmed by vibrational analysis.

All experimental infrared and Raman data, particularly the lower-frequency conformation-sensitive data, strongly support that the 10_3 helix provides a better description of the stable crystalline structure of PLA than the more simplistic 3_1 helix. The numerous band splittings that emerge in highly crystalline PLA spectra obtained at low temperatures are attributed to crystal field splitting affected by the presence of intermolecular interactions, i.e. carbonyl dipole interactions in the case of band splittings in the 1700–1850 cm^{-1} frequency range. Poorly understood spectral features can then be understood. Examination of the symmetry and potential energy distribution of different vibrational bands indicated that either the overtone of the 873 cm^{-1} band or a combination of the 1368 and 412 cm^{-1} bands is the most likely higher-ordered vibration in Fermi resonance with the carbonyl stretching fundamental vibration, accounting for the multiple components observed in both infrared and Raman. This is true for both amorphous and crystalline PLA that exhibit splitting in the carbonyl stretching in the 1700 cm^{-1} region.

11.2.2 Analysis of Disordered PLA Chains

A careful analysis of the disordered PLA in solution, melt, or solid state is also very important. The inherent stiffness of the PLA chain dictates the kinetics and "routes" of the solidification process (formation of crystalline and glassy states), as well as structural changes during processing or simply due to aging. An area of great interest in polymer studies is the application of vibrational spectroscopy, particularly Raman, to the analysis of disordered chains. It is known that PLA chains have

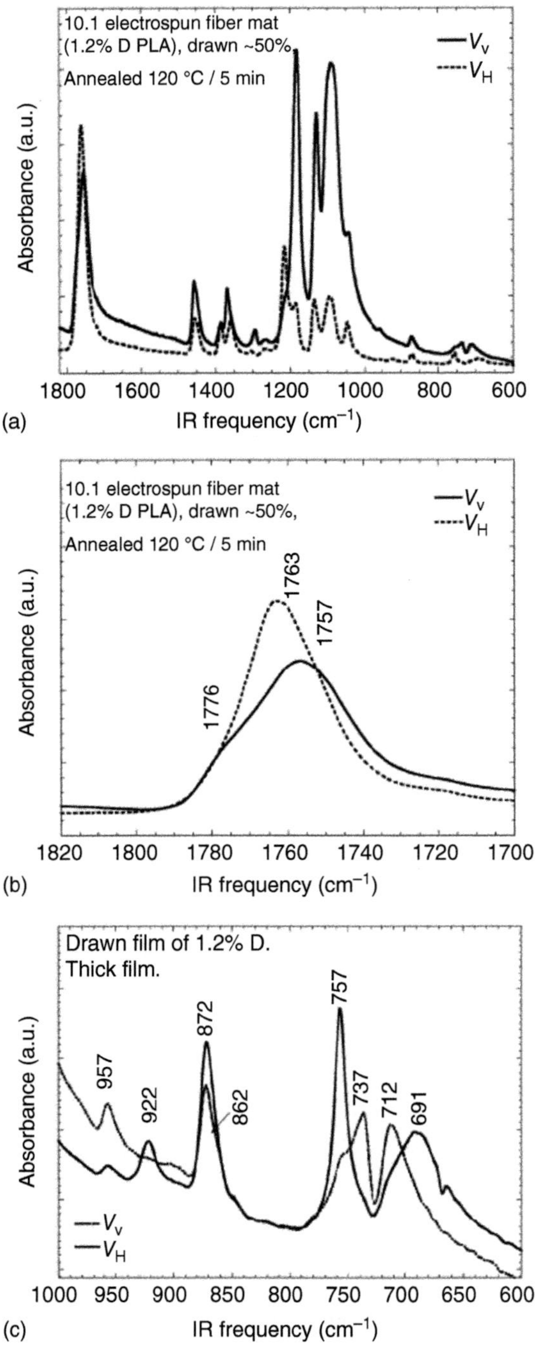

Figure 11.4 Polarized infrared spectra of PLA 1.2% D, electrospun into fiber mats from solution of DMF/chloroform mixture: (a) entire spectrum; (b) carbonyl stretch region; (c) backbone stretch region. V_V refers to the parallel spectrum and V_H to the perpendicular spectrum. Source: Aou and Hsu [18].

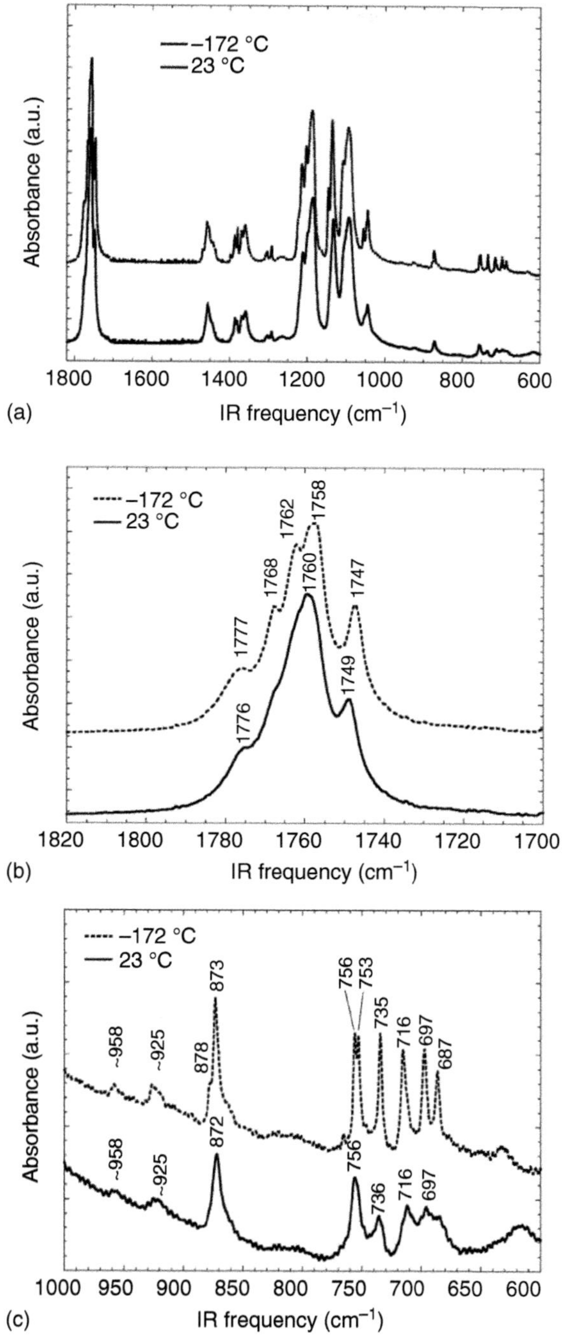

Figure 11.5 Cryogenic infrared spectra of PLA 1.2% D spherulite, prepared by melt-quenching and then crystallizing at 156 °C for 700 minutes: (a) entire spectrum; (b) carbonyl stretch region; (c) backbone stretch region. V_V refers to the parallel spectrum and V_H to the perpendicular spectrum. Source: Aou and Hsu [18].

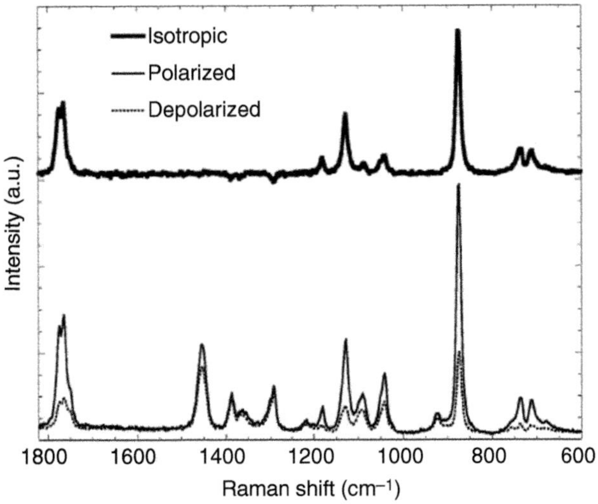

Figure 11.6 Polarized Raman spectra of 3D spherulite of PLA 1.2% D α-crystal. Source: Aou and Hsu [18].

conformational and configurational defects [11, 12, 38]. Disordered chains lack long-range order but may contain short ordered sequences due to significant differences in the relative energy of different rotational isomeric states along the polymer chain [39]. A detailed description of this molecular/configurational/conformational structural disorder is difficult to establish by other techniques. However, Raman is an excellent technique for such analysis. Simulation techniques combined with detailed Raman studies can resolve the various hypotheses presented in literature (see Figure 11.7). Given the relatively high compliance of the amorphous region in semicrystalline PLA, understanding the changes of these disordered chains is crucial to better assess the relationship between structure and properties during processing.

For disordered PLA structures, the vibrational spectra observed may be complex due to changes in both band frequency and shape. The vibrational mode can be quite complicated because the contributions of various internal coordinates to the character of the vibration may change significantly compared to an ordered unique crystalline chain, as shown above. Changing the frequency and character of skeletal vibrations provides tremendous insight into the structure of polymers in solution or melt [40]. Rather than analyzing one specific chain conformation, as for an ordered chain, the analysis of vibrational spectra from disordered chains requires the analysis of the vibrational spectrum associated with a conformational distribution. An extremely large number of possible conformers exist for disordered chains, each with a unique spectrum often only slightly different from one another. However, the isotropic Raman spectrum [29] for the disordered state of various polymers can be simulated as a composite of contributions from the ensemble of chains generated by a Monte Carlo procedure that assigned both a conformation and a probability of occurrence for each chain [40, 41]. The generation of these chains containing specific

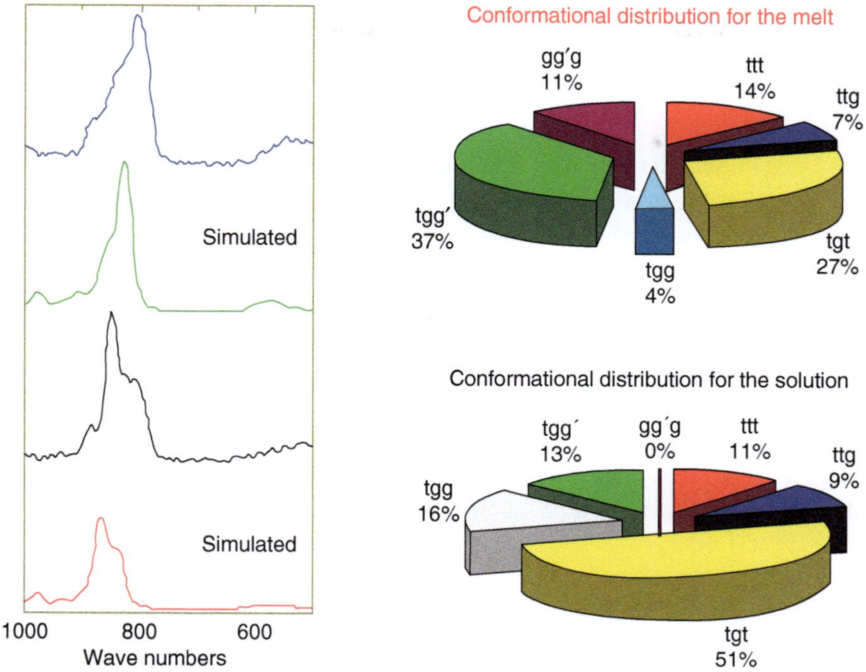

Figure 11.7 Simulated and experimental Raman data of poly(ethylene oxide) solution and melt [26].

configurational defect content is not as difficult as first envisioned. Flory developed methods for determining the probability of finding specific conformers along the chain, which have been incorporated into simulation programs. Polymeric chains containing both conformational and configurational defects can now be analyzed with confidence [40]. In the case of poly(ethylene oxide) (PEO), different conformational distributions were clearly found for melt and solution, as shown in Figure 11.7. The differences in segmental dynamics may influence their transport properties for possible applications in battery technologies. The results of the analysis of disordered PLA chains are shown below.

11.2.3 Description of Anisotropic PLA – Polarized Spectra

The segmental orientation and dichroic ratio of various samples of PLA will be described here. Vibrational spectroscopy, especially infrared, has provided simple and rapid dichroic measurements to assess polymer anisotropy (see Figure 11.10). Polarization, a vector quantity, is the key characteristic of an electromagnetic wave used in both infrared absorption and Raman scattering. For polymers, infrared or Raman techniques measure the polarized intensity of vibrational bands after being either absorbed (infrared) or scattered (Raman) relative to the incident polarized radiation. Dichroic ratio is simply a measured quantity that is characteristic of segmental orientation. This ratio, D (Eq. (11.1)), is the relative band intensity of the signal with polarization parallel to the incident beam relative to the polarized signal orthogonal to the polarization of the incident infrared beam.

$$D = \frac{A_{\parallel}}{A_{\perp}} \tag{11.1}$$

Following Fraser's derivation, the dichroism is then related to the second moment of the orientation function by Eq. (11.2) [42, 43]. This second moment can range from −0.5 to 1.0. A value of −0.5 represents orientation orthogonal to the deformation axis. 1.0 represents perfect orientation with respect to the deformation axis. A zero value is found for a totally randomly orientation.

$$\langle P_2 \rangle = \frac{3\langle \cos^2\theta \rangle - 1}{2} = \frac{D-1}{D+2} \tag{11.2}$$

It should be emphasized that infrared measures the change in dipole moment or the polarizability change associated with molecular structures. Even localized vibrations such as C=O stretching in PLA have a transition dipole that may be at an angle relative to the bond axis. To correct for this difference, a modified expression should be used (Eq. (11.3))

$$\frac{3\langle \cos^2\theta \rangle - 1}{2} = \left(\frac{D-1}{D+2}\right)\left(\frac{2\cot^2\alpha + 2}{2\cot^2\alpha - 1}\right) \tag{11.3}$$

where α is the angle between the reference chain axis and the transition moment direction. Most analysis simply employs the dichroic ratio to express the degree of orientation achieved in the polymer. As expressed above, it is often of additional interest to assess the orientation achieved as a function of deformation. Eloquent expressions have been derived to correlate the segmental orientation achieved with the draw ratio [44–47]. However, these expressions often employ the affine deformation model or rubber elasticity model as approximations, making them difficult to implement in actual polymeric systems.

Raman effect was first reported in 1928 [48], and its use in materials science was not realized until advances were made in laser technology in the 1960s that could provide highly polarized radiation. The polarized or depolarized scattered Raman radiation collected is especially suitable for the analysis of segmental orientation, as proposed earlier [30]. Raman techniques have been used online to elucidate the effects of changing processing parameters on morphology development in uniaxial and biaxial polymers [5, 6]. Using depolarized scattering data, even disordered samples such as amorphous polymers, polymer solutions, or melts can be characterized [49]. Since we are dealing with a vector quantity in both infrared and Raman, usually only the second moment of the orientation distribution function is of interest in our orientation analysis [43]. The polarized infrared radiation can be easily generated in many ways, and the degree of polarization achieved in most experiments is not exceptionally high. In contrast, the polarization of the polarized laser beam generally used in Raman instruments can be as high as 10 000 to 1. However, given the optical constraints of both infrared and Raman instruments (large solid angle needed for high signal-to-noise ratio to obtain quality spectra), the polarized spectroscopic data simply cannot be extremely accurate. In most polymer deformation studies, these qualitative data are more than sufficient for analyzing the sample anisotropy.

11.3 Simulation Studies for Both Ordered and Disordered Structures

Both infrared and Raman spectra can be simulated complementing the experimental data obtained. This type of analysis has usually been accomplished by a number of techniques, from traditional normal coordinate analysis to more recent molecular dynamics or quantum mechanical approaches [50, 51]. In our laboratory, we use techniques that are based on classic normal vibrational calculation, using Wilson's GF matrices method and force constants transferred from small molecules [24]. Once vibrational spectra are calculated for well-defined structures, this analysis is also tremendously useful for analyzing disordered chains of the same type as the known structures. Rather than analyzing one specific chain conformation for an ordered chain, the analysis of disordered chains requires the vibrational spectrum of a conformational distribution, as shown above [40, 52].

To validate the simulation techniques, the calculated Raman spectra were compared to the experimental data obtained for crystalline PLA [19]. Only a few parameters are needed for the calculation of isotropic Raman scattering intensities below ~1000 cm^{-1}. Additivity in bond polarizability is always assumed. For PLA, structural parameters and force fields transferred from other polyesters were used to simulate vibrational spectra of crystalline PLA. The PLA monomer unit is shown schematically in Figure 11.8. Three skeletal bonds occur along the molecular chain in PLA: C—O(ester), O—C, and C—C. The C—O(ester) bond is always assumed to be trans due to conjugation with the C—O double bond. The O—C bond has two energy minima around −160° and −48°, and the C—C bond has two energy minima around 160° and −73°. Since 160° is close to 180°, the 160° angles are termed trans (t). In the same way, −48° and −73° are close to −60° and are described as gauche (g).

The observed Raman and infrared vibrations shown in Figures 11.9 and 11.10 can be assigned on the basis of their frequencies and polarization characteristics [23, 24]. The dichroic ratios and polarized and/or depolarized Raman data are obtained as described above. For a uniaxially oriented PLA sample, the *A* modes of a 3_1 helix are infrared-active and have transition moments parallel to the chain axis. In contrast, the *E* modes are vibrations with transition moments perpendicular to the chain axis. The analysis of optical activity associated with a helix also revealed that *A* modes are polarized in Raman spectra; *E* modes are depolarized vibrations.

The assignments of crystalline vibrations based on polarization and group theory analysis are reliable [18]. It is then important to establish that experimental

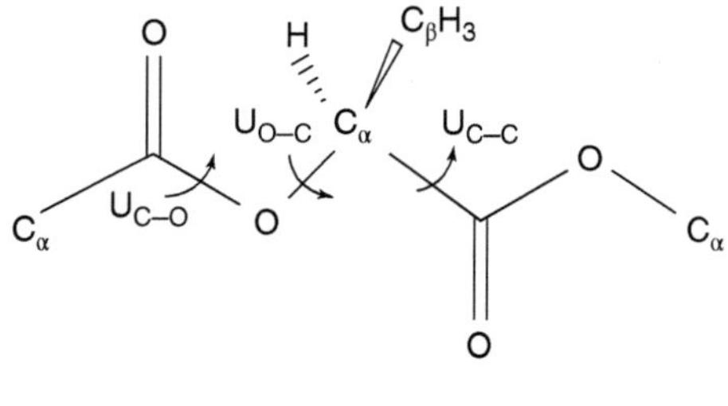

Figure 11.8 Schematic drawing of backbone structure of a PLA repeat. Source: Based on Kang et al. [19].

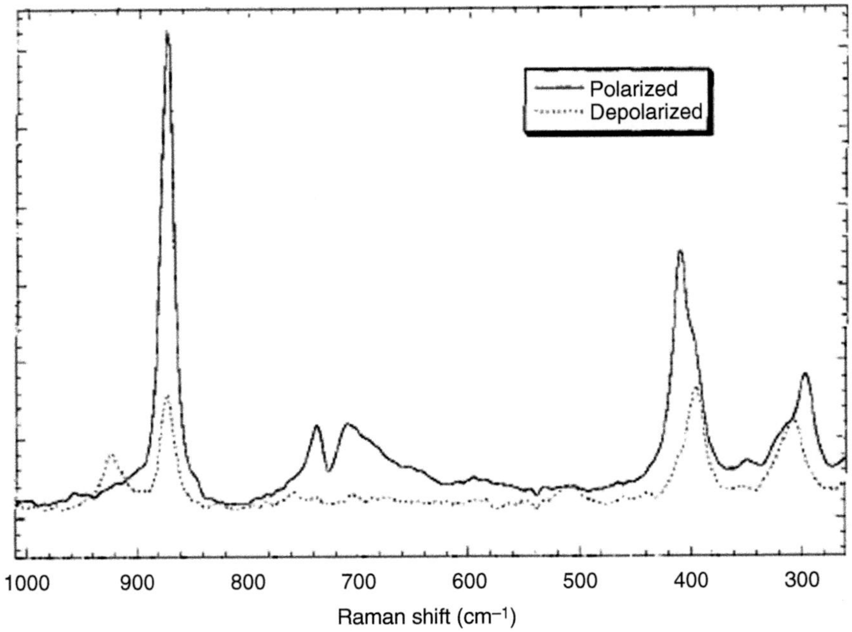

Figure 11.9 Polarized Raman spectra of PLA. The drawing direction of the sample is parallel to laser polarization direction. Polarized spectra are obtained by collecting scattered light along the direction of sample drawing; depolarized spectra are collected perpendicular to drawing direction. Source: Based on Kang et al. [19].

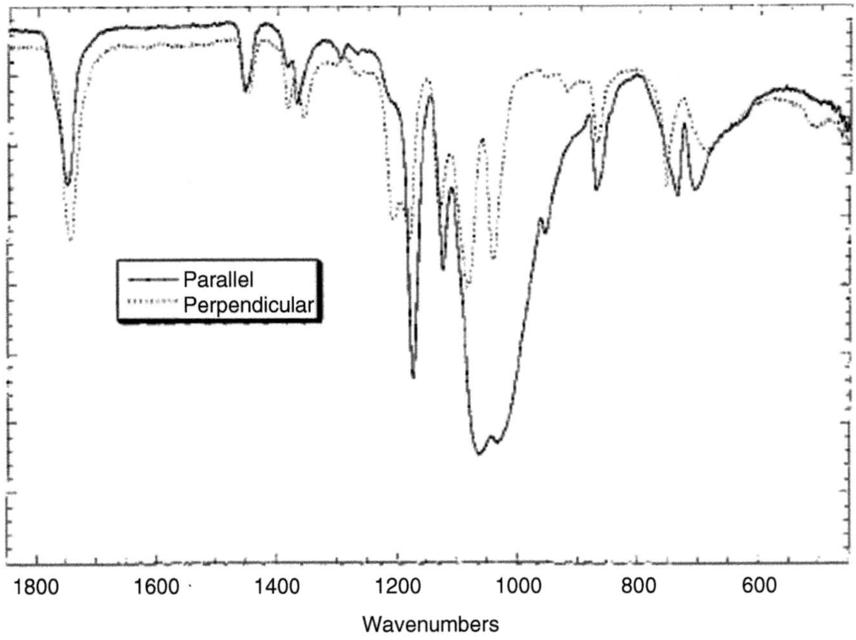

Figure 11.10 Polarized infrared spectra of drawn poly(lactic acid) film taken by the ATR technique. Source: Based on Kang et al. [19].

spectroscopic analysis, in conjunction with simulation studies, has the capacity to differentiate the various helices predicted. For the first structural analysis, without considering intermolecular interactions, the chain conformation corresponding to the four minima in the energy contour map has the C—C, C—O bonds tt, tg, gt, and gg. These correspond to a 2_1, 5_1, 3_1, and 4_1 helix, respectively [34]. The relative energies of the minima suggest that the 3_1 helix is the most favorable and the 4_1 helix is the second most favorable. According to the contour map presented in earlier studies, a completely random structure would contain 55% gt conformation and 37% gg conformation. It can then be assumed that the Raman band feature of such a chain would include 55% 3_1 helix (pure gt conformation) band feature and 37% 4_1 helix (pure gg conformation) band feature, respectively.

Furthermore, it is known that bands in the 920, 720, 520, 412, and 398 cm^{-1} regions are most sensitive to changes in structural order. Figure 11.11 shows the simulated spectra of the four helical structures. The four possible helices suggested by the conformational energy map can be readily differentiated. For a 4_1 helix, the 870 cm^{-1} band is shifted to 900 cm^{-1} and a strong band appears around 550 cm^{-1}. For the 2_1 and 5_1 helices, bands have also been calculated to exist between 500 and 600 cm^{-1}, but the 400 cm^{-1} band should be absent. The features expected for the 2_1 and 5_1 helices are quite different from the actual infrared and Raman observations, as shown in Figures 11.9 and 11.10. Since tt and tg conformations have very high energy according to this model, it is not surprising to find a very low probability of having a 2_1 or 5_1 helix. However, as the gg conformation has a very low energy, there is a

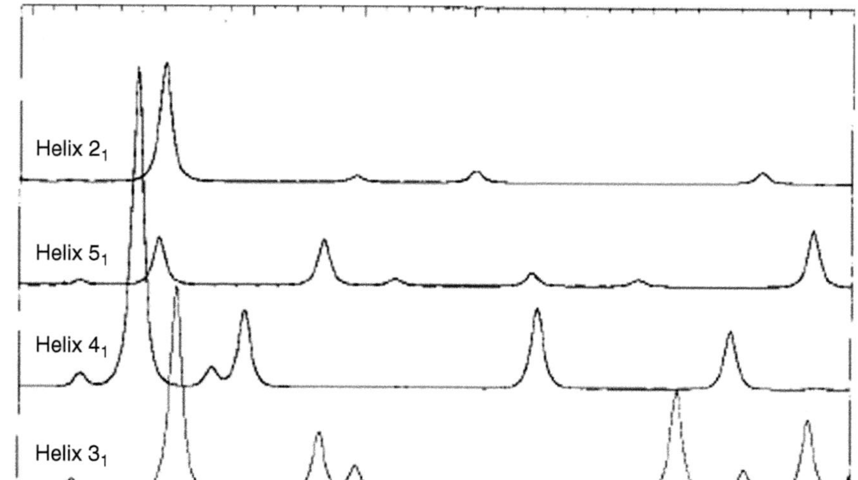

Figure 11.11 Calculated spectra of four helices (2_1, 3_1, 4_1, and 5_1) predicted in Brant's RIS model. Source: Based on Kang et al. [19].

probability of finding a 4_1 helix or gg conformation. The 550 cm^{-1} band can be used as an identification band for such a structure. This band has never been observed in any of the vibrational spectra obtained for PLA. It is then possible to conclude that the gg conformation may have much higher energy than the original conformational analysis suggests. These results, combining simulation with experiment, clearly establish the validity of using such simulation methods to determine the low probability of finding a PLA helical structure that is different from the 3_1 helix.

Since 3_1 and 10_3 helices are so similar in structure, our second simulation analysis employed a 3_1 helix for simplicity. The simulated isotropic Raman spectrum for a 3_1 helix together with the experimental Raman data is shown in Figure 11.12. The *A* modes are marked with an asterisk in the experimental spectrum. Virtually all the experimental data are reproduced accurately by the normal coordinate analysis. This agreement in both band position and intensity suggests that the transferred structural parameters, force field, and intensity parameters are sufficiently accurate for structural analysis. The differences in the conformational bands clearly show that only the simulated spectrum for the 3_1 helix matches the experimental data obtained.

As mentioned above, the application of vibrational spectroscopy, especially Raman scattering, to the analysis of disordered chains is an area of strong interest. The marked increase in computing performance over the past couple of decades has served as a prime impetus for progress in simulation research. With increased computing capabilities, simulation methods have become much more useful tools

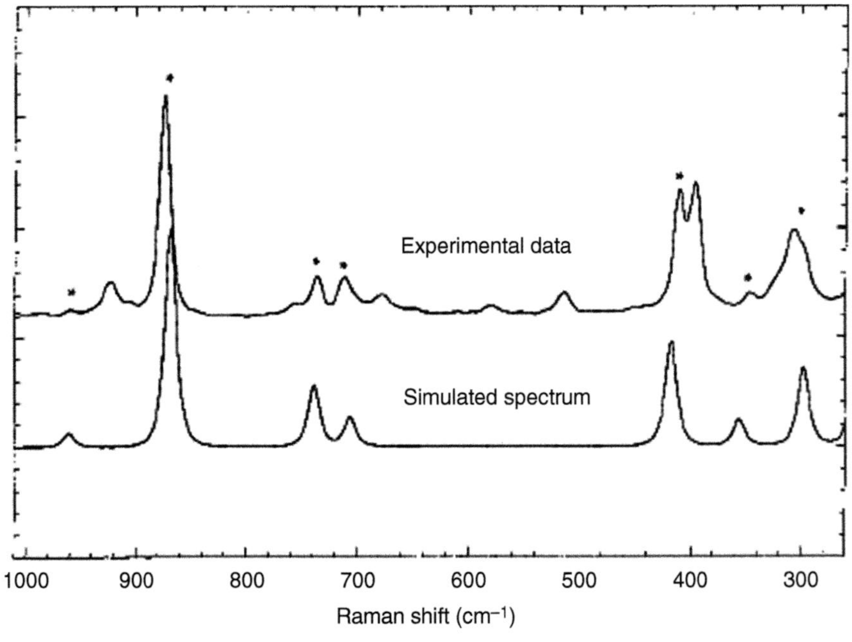

Figure 11.12 Measured and simulated (*A* modes only) Raman spectra for a 3_1 helix. Source: Based on Kang et al. [19].

in spectroscopy. For samples involving well-defined structures and dependable band assignments, the normal coordinate analysis described above works well. However, when a broad distribution of a large variety of conformations is involved, as in disordered structures, this method is difficult to apply with confidence. It takes some effort to simulate Raman spectra for all possible chain conformations associated with disordered chains, as seen above. In addition to the normal coordinate approach described above, other techniques to simulate disordered chains have also been developed and refined. From the vibrational dispersion curves obtained, it is possible to obtain the vibrational density of states which reflect the possible existence of vibrational bands [53]. Attempts to take intensity into consideration in this type of approach, however, have proven extremely difficult. Molecular dynamics has also been employed to simulate the vibrational spectrum of disordered chains. Using such techniques, correlation functions obtained may be transformed to obtain vibrational frequencies. Direct quantum mechanical calculations can also be carried out. However, this method is seldom utilized in polymer structural studies because of the substantial computation power required even for relatively small model compounds.

The success of the validation experiment rests with the ability to draw distinctions between different types of PLA helices and the ability to simulate experimental data for crystalline PLA accurately. Once that was established, it was possible to conduct a Raman analysis to examine the disordered structures of PLA [20]. Simulated spectra were obtained for an ensemble of chains generated by several rotational isomeric state models presented earlier [33, 34]. Again, when additivity in bond polarizability is assumed, only a few parameters are needed to calculate the isotropic Raman scattering intensities below 1000 cm^{-1}. The rotational isomeric states model can then be expressed as statistical weight matrices with two states [20]. The third rotational freedom is hindered due to the conjugation around the ester group. The experimental data for amorphous PLA were obtained by quenching a thin PLA film into liquid nitrogen, and the Raman data were obtained using a home-built sample holder fitted to the spectrometer.

To simulate a disordered PLA chain, an ensemble of 2000 chains was generated. It was then possible to examine the energy levels of the two degrees of rotational freedom (see Figure 11.8) that would yield the most satisfactory fit to the experimental Raman data for disordered chains. A detailed analysis of the simulated and experimental spectra was very instructive. On the basis of the simulation and experimental results of disordered chains, it was concluded that a distribution contained 70% of gt rotational isomers in the amorphous state. The conformation gg should be less than 20%. The other two chain conformations, tg and tt, are approximately 5% each.

A detailed analysis of the differences between simulated and experimental spectra showed that the rotational isomeric state reported in earlier studies leads to a chain that is much more flexible than the experimental data suggest. The rotational isomeric states governing the chain conformation distribution of PLA yielded a low characteristic ratio of 4.5, suggesting an extremely flexible chain. Even polyethylene has a characteristic value of 7 [39]. This new conformational distribution based on Raman analysis yields a much stiffer PLA chain, consistent with other experimental data. The brittle nature of both PLLA and PDLA is also inconsistent with a polymer possessing such a flexible backbone [54–56]. Light scattering measurements of PLA in acetonitrile solution also suggested a much higher characteristic value of 11.8 [55]. By convoluting the synthetic spectrum expected for an ensemble of disordered PLA chains calculated by various rotational isomeric states models, it was possible to create an accurate conformational model for PLA that matched all experimental data and processing requirements.

11.4 Analysis of Conformational Changes in PLA During Deformation

The physical properties of polymers can be significantly altered by processing methods such as blowing, extrusion, or drawing. For example, an isotropic polyethylene bead can have a modulus as low as a few GPa. Yet, when processed into a fiber without conformational defects, its modulus can be 300 GPa or higher [57, 58]. Piezoelectric polymers such as poly(vinylidene fluoride) cannot function well without the specific orientation of the β form polymer conformation produced by mechanical deformation and electric field poling. Since the most crucial aspects of PLA are the mechanical and thermal properties achieved through processing, it is very important to have a comprehensive understanding of its deformation behavior in uniaxial (e.g. fiber spinning) or uniplanar (e.g. biaxial film drawing) deformation processes [21, 59–61].

Virtually all starting PLA feedstocks are amorphous or have a low degree of crystallinity. PLA has a relatively slow quiescent crystallization rate compared to most polyolefins, such as polyethylene or polypropylene [62]. However, its strain-induced crystallization rate is surprisingly high [21, 63, 64]. This aspect of PLA generates considerable interest regarding its structural transformation during deformation. For instance, the undesirable high degree of crystallinity may form rapidly in the initial stages of deformation (uniaxial or biaxial), thereby reducing the efficiency of later processing stages and making the overall process difficult to control. In addition, if the final crystallinity is too high, the processed goods are brittle, resulting in poor impact strength. Another problem associated with deformed products is their dimensional instability [5, 6].

Deformation-induced structural changes in PLA necessarily involve the characterization of amorphous regions. While the degree of crystallinity changes with deformation, it should be emphasized that the conformational distribution in the amorphous regions may also change. The participation of crystalline–amorphous interphase has also been found to affect the mechanical properties ultimately achievable [65]. There are many techniques that are capable of elucidating the crystalline portion of semicrystalline polymers. Techniques such as X-ray diffraction are extremely capable of probing long-range order in polymer crystals. As mentioned above, vibrational spectroscopy, on the other hand, detects localized structures at the molecular scale. This is particularly useful for characterizing conformational distribution and the changes associated with amorphous chains during processing.

Among the four possible ways to describe the statistical conformational distribution of a PLA repeat unit (tt′t, tg′t, tt′g and tg′g; see the definitions of these rotational isomeric states above), the tg′t conformer was shown to dominate the conformation of amorphous PLA chains [19, 20]. As shown above, it is possible to accurately describe the vibrational transitions in both ordered (crystalline) and disordered (melt, solution, and amorphous phase of semicrystalline) PLAs [21]. Based on the study of a large number of PLA samples, it was revealed that the relative intensity of a pair of Raman-active vibrations (1044 and 1128 cm^{-1}) was an excellent indicator of sample deformation. The Raman spectra in the 1200–900 cm^{-1} region for different biaxial draw ratios are shown in Figure 11.13. As the figure shows, when the intensities were normalized to the 1044 cm^{-1} band, the relative intensity of the 1128 cm^{-1} band decreased with elongation. The assignments of the two bands are clear from normal coordinate analysis and are assigned to skeletal deformation and stretching. Because of their dependence on the chain backbone, it is expected that these two modes are sensitive to deformation-induced changes in conformation. The rationality of this method was supported by a series of isotropic Raman spectra simulations associated with different chain conformational distributions [21]. The amorphous chains with radii longer than the statistical average of a normal chain have a lower intensity of the 1128 cm^{-1} band. In contrast, the coils with more compact chains exhibit a higher intensity for this band.

Furthermore, the ratio of the two band intensities, R, is extremely useful as an indicator of changes in conformational distribution of PLA when deformed. To employ this ratio in chain conformation analysis, it is necessary to relate the intensity ratio to the number of tg′t conformations [21]. As shown in the simulation studies, the intensity ratio of the two peaks is mainly due to differences in the number of tg′t and tg′g conformers. Based on this observation, a two-phase model was used to calibrate the indicator ratio. Assuming a two-component model, the ratio of the two peaks (1044 and 1128 cm^{-1}) can then be related to the relative fraction of the tg′t conformer, $f_{tg't}$, using the Raman data from a series of annealed PLA crystals of different degrees of crystallinity, as shown in Figure 11.14.

Based on DSC measurements, the degree of crystallinity for various deformed PLA samples can be calculated easily and compared to the associated spectroscopic

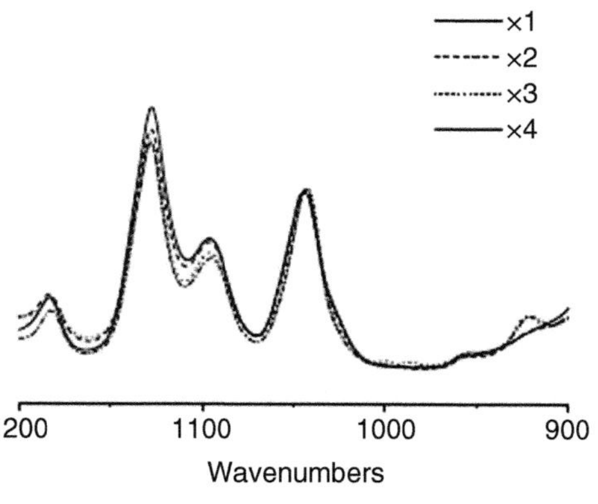

Figure 11.13 Experimental Raman spectra of biaxially stretched films. Source: Yang et al. [21].

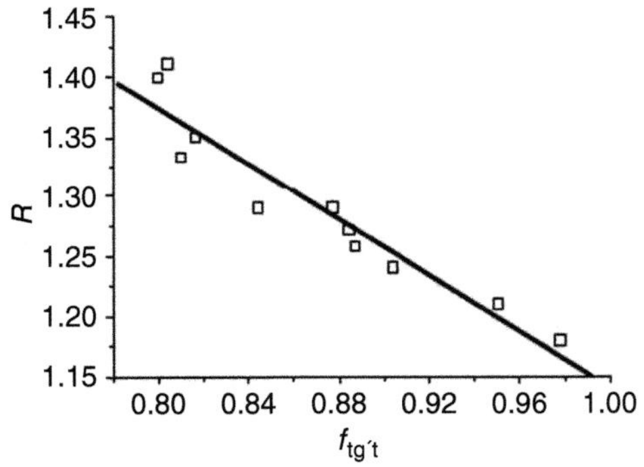

Figure 11.14 Plotting of the spectral intensity ratio R vs. $f_{tg't}$ for annealed samples. Source: Yang et al. [21].

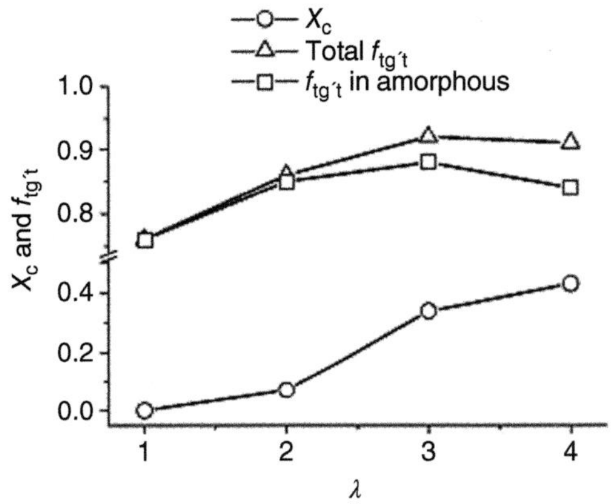

Figure 11.15 The percentage of tg′t conformation and the crystallinities of the biaxially stretched samples vs. draw ratios. Source: Yang et al. [21].

measurements. These amorphous PLA films were stretched at 80 °C (above the sample T_g) at a rate of 100% per second. Crystallinity increases quickly as a function of draw ratio. An R value of 1.425 was obtained for the undrawn sample. Much lower R values were measured for the deformed samples. Using the calibration curve shown in Figure 11.15, it was then possible to obtain the total contents of $f_{tg't}$ for the deformed samples drawn to different draw ratios. The experimental results are consistent with the fact that the elongated samples have a higher number of tg′t conformations than the undrawn sample. It is also clear that the degree of crystallinity in PLA samples can increase dramatically as a function of draw ratio. This behavior is quite different from that of other polyesters [66, 67].

One high-value application of PLA is in the role of biodegradable cardiovascular or peripheral stents (see Figure 11.1). The merit of this type of stent, in contrast to metallic ones, is that it will eventually degrade in the body and will not need to be extracted after use [13]. Cardiovascular stents or scaffolds (small tubular objects) simply cannot function without the biaxial orientation and controlled degree of crystallinity achieved during the extrusion process [13]. Vibrational spectroscopy, both infrared and Raman, can be used to analyze the degree of anisotropy and the crystalline phase in the processed samples.

The manufacturing process of PLA-based stents involves the extrusion of small diameter, "thick"-walled PLA tubes (wall thickness ~2 mm) into larger tubes with thinner walls. These tubes undergo an elongation along the tube axis and expansion perpendicular to this axis. This elongation and thinning of the walls is equivalent to the process of blowing a polymer plug into a full bottle for beverages. As shown above, the changes in sample morphology and segmental orientation can be extremely dramatic in PLA. Unlike other polymers, PLA exhibits rapid crystallization as a function of deformation [21]. This increase in the degree of crystallinity is easily verified using both infrared and Raman spectroscopy, as the crystalline features of PLA are well documented. What is more difficult is the assessment of segmental orientation as a function of deformation in these extruded tubes. Usually it is not possible to measure biaxial orientation using either infrared or Raman spectroscopy. However, if two vibrational bands with orthogonal transition moments can be found (such as the A and E modes in the 400 cm^{-1} region), it is then possible to follow the changes in the biaxially oriented samples, as in the walls of these expanding tubes. As shown in Figure 11.16 below, the spectra obtained for PLA show that crystallization increases rapidly as the tube elongates and expands [21, 68]. Raman technique also has the spatial resolution to measure the biaxial orientation within the tube wall. As we have demonstrated, it is possible to measure differences in the degree of orientation from the inner wall to the outer wall, even for the extremely thin walls of an individual stent [68].

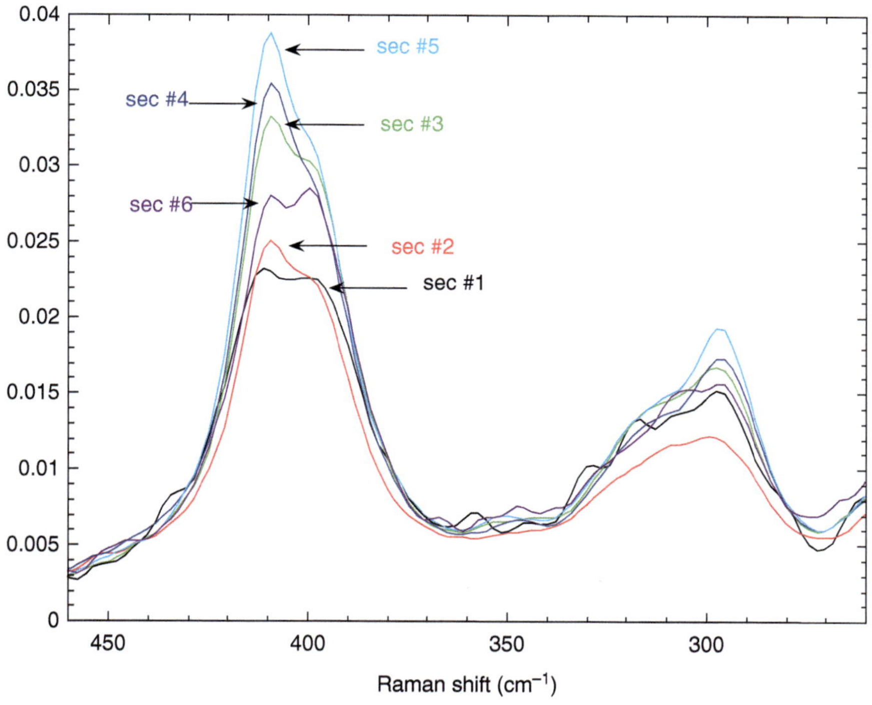

Figure 11.16 The increase in the degree of crystallinity of Raman bands in the 400 cm^{-1} region. The 412 cm^{-1} (A mode) band is the crystalline band. The 397 cm^{-1} band is the amorphous band. Deformation increases from sections #1 to #5 (initial plug to the highly deformed regions at the bottom). Sources: Based on Kang [68], Yang et al. [21].

The 400 cm^{-1} region shown in Figure 11.16 contains an *A* polarized Raman-active mode. There is also an *E* depolarized mode in this region. The 397 cm^{-1} band is shown to be an amorphous band. There are other vibrations in the 300 and 800 cm^{-1} region, which are insensitive to changes in the degree of crystallinity [5]. The degree of orientation in various drawn PLA films was determined by measuring the relative intensity of the polarized and depolarized components and the shift in the relative band positions of the bands in the 400 cm^{-1} region [5]. The bandwidth and specific band frequency correlated well with the degree of sample crystallinity. These aspects make the Raman method useful for the online monitoring of orientation and crystallinity in commercial processes. A molecular understanding of heat setting to reduce film shrinkage was also proposed. Raman data from films that were heat-set while physically constrained suggest that amorphous-phase relaxation does not occur during heat setting. Rather, the level of crystallinity increases substantially, indicating that the cause of low shrinkage in the films was due to crystalline network formation rather than amorphous-phase relaxation.

11.5 Aging Behavior in PLA

Vibrational spectroscopy has also contributed significantly to an increased understanding of the changing structural aspects of PLA. Research interest in a bio-based PLA has always been grounded on the potential that this polymer can be commercialized for a number of applications, especially in the low-temperature food-packaging industry [4]. High-volume applications, such as field covers requiring large areas to control temperature and humidity experienced by various crops, are also being developed around the world. It is well known that the condensed structure of PLA can change as a function of temperature and time, especially with moisture present [14]. Depending on processing conditions and thermal profile, two crystalline structures for PLA can be found [63]. One form, the pseudo-orthorhombic a structure (a = 10.6, b = 6.1, and c = 28.8 A) contains two chains in the unit cell and is found at relatively low drawing temperatures and/or low hot-draw ratios. At elevated temperatures, a second structure is formed. For this structure, an orthorhombic unit cell is proposed (a = 10.31, b = 18.21, and c = 9.00 Å) containing six chains. The chain conformations of these two structures are left-handed 10_3 and 3_1 helices, respectively. Calculations show that both conformations have approximately the same energy (see Figure 11.17). Therefore, the preference for one of the two structures is determined by packing considerations [69]. Vibrational spectroscopy, particularly Raman, was instrumental in establishing the dipolar intermolecular interactions necessary to stabilize these structures [12, 18, 70, 71]. This proves it to be an ideal technique to characterize the transformation from one form to another in the condensed phase.

Empirical simulation methods have improved substantially over the last few years. These calculations, in conjunction with Raman experiments, were able to determine the conformational distortion of the various crystalline forms of PLA and their changes as a function of time and temperature. It is known that a structure formed at low temperatures, possessing a substantial number of defects,

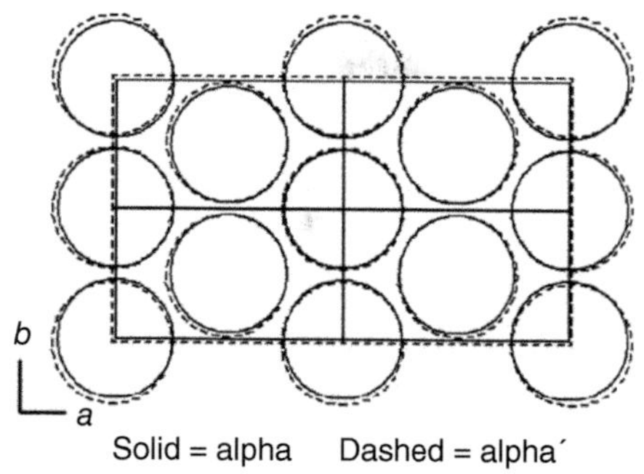

Figure 11.17 Unit cells of the α and α' structures from literature. Source: Kalish et al. [69].

can transform into a more perfect structure at elevated temperatures [19–21, 37]. Because of the similarity in helical diameters and the associated packing constraints of an overall helical structure, the chain conformation can only be approximated by slight deviations from the standard tgt monomer conformation of a PLA [11, 12]. Since it is known that imperfections exist in PLA, various models were generated with different magnitudes of random departures from the expected dihedral angles, keeping the helical parameters similar. Raman-active vibrations in the 700 cm^{-1} region that are sensitive to changes in chain conformation were then simulated (see Figure 11.18). The model showing the best fit to experimental data contains a majority (80%) tgt-10_3 sequences with randomly distributed 3_1 segments, shown schematically in Figure 11.19. The departure of the skeletal dihedral angle and the mechanism of order formation were also established, thus establishing that the α' phase is a metastable structure containing mainly the 10_3 helix, with some sequences reminiscent of the disordered structure (see Figure 11.19). The percentage of tgt-3_1 defects is shown to be ~30%. The α' phase can be transformed to the α phase at elevated temperatures (>120 °C). It was concluded that the transformation from metastable α' to α phase is associated with removal of conformational disorder that disrupts specific interchain interactions typically observed in the well-ordered α phase. In this case, vibrational spectroscopy provided the molecular mechanism (removal of isolated helical defects) for the transformation to take place as a function of time and temperature.

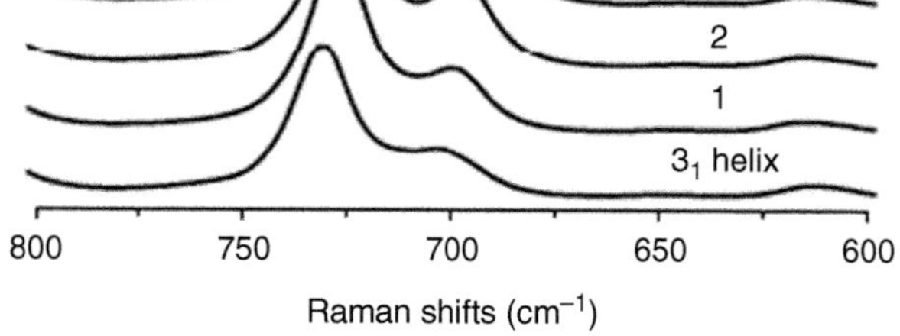

Figure 11.18 Simulated Raman spectra with the variable helix model [11, 12]. Source: Kalish et al. [12].

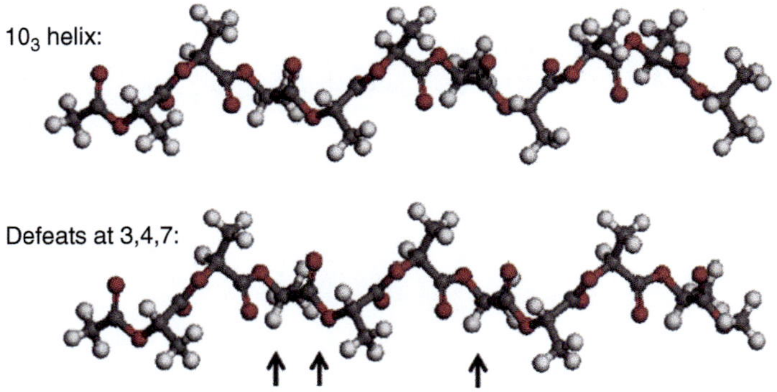

Figure 11.19 Molecular models of a 10_3 PLA helix chain containing 30% 3_1 defects. The arrows show where the conformational defects occur [11, 12]. Source: Kalish et al. [12].

11.6 Conclusion

The goal of this section is to demonstrate the merits of using vibrational spectroscopy, both infrared absorption and Raman scattering, to characterize the molecular structure of PLA. This polymer is fashionable because our society is rapidly realizing the need to control plastic wastes. Because it can decompose easily, PLA can potentially be used to manufacture both high-value and high-volume commercial products. Vibrational spectroscopy exhibits the potential to characterize PLA conformation in both crystalline and amorphous states. Vibrational spectroscopy has also demonstrated its merit in following changes in PLA structure during processing. Even the aging behavior of PLA at different temperatures and at different relative humidity can be quantified as a function of time. Both the successes of experimental and simulation techniques have been covered in this review, though it should be noted that many remain to be carried out.

Acknowledgment

Professors Hsu and Yang acknowledge years of continuous and generous support from several industrial partners, especially Abbott Vascular, Santa Clara, CA, and Saint Gobain, Northborough, MA. We also acknowledge the financial support from US Department of Agriculture. Lastly, the editing support of Ms. Margaret Gyorgy is greatly appreciated.

References

1 Manome, A., Okada, S., Uchimura, T., and Komagata, K. (1998). The ratio of L-form to D-form of lactic acid as a criteria for the identification of lactic acid bacteria. *J. Gen. Appl. Microbiol.* 44: 371–374.

2 Hamad, K., Kaseem, M., Yang, H.W. et al. (2015). Properties and medical applications of polylactic acid: a review. *Express Polym. Lett.* 9: 435–455.
3 Langer, R. and Tirrell, D.A. (2004). Designing materials for biology and medicine. *Nature* 428: 487–492.
4 Auras, R., Harte, B., and Selke, S. (2004). An overview of polylactides as packaging materials. *Macromol. Biosci.* 4: 835–864.
5 Smith, P.B., Leugers, A., Kang, S. et al. (2001). An analysis of the correlation between structural anisotropy and dimensional stability for drawn poly(lactic acid) films. *J. Appl. Polym. Sci.* 82: 2497–2505.
6 Smith, P.B., Leugers, A., Kang, S.H. et al. (2001). Raman characterization of orientation in poly(lactic acid) films. *Macromol. Symp.* 175: 81–94.
7 Garlotta, D. (2001). A literature review of poly(lactic acid). *J. Polym. Environ.* 9: 63–84.
8 Nampoothiri, K.M., Nair, N.R., and John, R.P. (2010). An overview of the recent developments in polylactide (PLA) research. *Bioresour. Technol.* 101: 8493–8501.
9 Faruk, O., Bledzki, A.K., Fink, H.P., and Sain, M. (2012). Biocomposites reinforced with natural fibers: 2000–2010. *Prog. Polym. Sci.* 37: 1552–1596.
10 Siro, I. and Plackett, D. (2010). Microfibrillated cellulose and new nanocomposite materials: a review. *Cellulose* 17: 459–494.
11 Kalish, J.P., Aou, K., Yang, X.Z., and Hsu, S.L. (2011). Spectroscopic and thermal analyses of α′ and α crystalline forms of poly(L-lactic acid). *Polymer* 52: 814–821.
12 Kalish, J.P., Zeng, X.G., Yang, X.Z., and Hsu, S.L. (2011). A spectroscopic analysis of conformational distortion in the α′ phase of poly(lactic acid). *Polymer* 52: 3431–3436.
13 Kossuth, M.B., Perkins, L.E.L., and Rapoza, R.J. (2016). Design principles of bioresorbable polymeric scaffolds. *Interventional Cardiol. Clin.* 5: 349–355.
14 Vyavahare, O., Ng, D., and Hsu, S.L. (2014). Analysis of structural rearrangements of poly(lactic acid) in the presence of water. *J. Phys. Chem. B* 118: 4185–4193.
15 Rathi, S.R., Chen, X., Coughlin, E.B. et al. (2011). Toughening semicrystalline poly(lactic acid) by morphology alteration. *Polymer* 52: 4184–4188.
16 Rathi, S.R., Coughlin, E.B., Hsu, S.L. et al. (2014). Maintaining structural stability of poly(lactic acid): effects of multifunctional epoxy based reactive oligomers. *Polymers* 6: 1232–1250.
17 Rathi, S.R., Ng, D., Coughlin, E.B. et al. (2014). Effects of molecular architecture on the stereocomplex crystallization in poly(lactic acid) blends. *Macromol. Chem. Phys.* 215: 320–326.
18 Aou, K. and Hsu, S.L. (2006). Trichroic vibrational analysis on the α-form of poly(lactic acid) crystals using highly oriented fibers and spherulites. *Macromolecules* 39: 3337–3344.
19 Kang, S.H., Hsu, S.L., Stidham, H.D. et al. (2001). A spectroscopic analysis of poly(lactic acid) structure. *Macromolecules* 34: 4542–4548.
20 Yang, X.Z., Kang, S.H., Hsu, S.L. et al. (2001). A spectroscopic analysis of chain flexibility of poly(lactic acid). *Macromolecules* 34: 5037–5041.

21 Yang, X.Z., Kang, S.H., Yang, Y.N. et al. (2004). Raman spectroscopic study of conformational changes in the amorphous phase of poly(lactic acid) during deformation. *Polymer* 45: 4241–4248.

22 Herzberg, G. (1945). *Infrared and Raman Spectra*. Princeton, NJ: D. Van Nostrand Company.

23 Turrell, G. (1972). *Infrared and Raman Spectra of Crystals*. New York: Academic Press.

24 Wilson, E.B. Jr., Decius, J.C., and Cross, P.C. (1955). *Molecular Vibrations: The Theory of Infrared and Raman Vibrational Spectra*. New York: Dover Publications, Inc.

25 Colthup, N.B., Daly, L.H., and Wiberly, S.E. (1990). *Introduction to Infrared and Raman Spectroscopy*. Boston: Academic Press, Inc.

26 Hsu, S.L. (2018). Vibrational Spectroscopy. In: *Encyclopedia of Polymer Science and Technology*. Wiley https://doi.org/10.1002/0471440264.pst381.pub2.

27 Hsu, S.L. (2000). Raman spectroscopic studies of polymer structure. In: *Raman Scattering in Materials Science* (eds. W.H. Weber and R. Merlin), 369–445. Heidelberg: Springer.

28 Snyder, R.G. (1971). Raman scattering activities for partially oriented molecules. *J. Mol. Spectrosc.* 37: 353–365.

29 Long, D.A. (2002). *The Raman Effect: A Unified Treatment of the Theory of Raman Scattering by Molecules*. New York: Wiley.

30 Porto, S.P.S. (1966). Angular dependence and depolarization ratio of the Raman effect. *J. Opt. Soc. Am.* 56: 1585.

31 Aleman, C., Lotz, B., and Puiggali, J. (2001). Crystal structure of the α-form of poly(L-lactide). *Macromolecules* 34: 4795–4801.

32 Sasaki, S. and Asakura, T. (2003). Helix distortion and crystal structure of the alpha-form of poly(L-lactide). *Macromolecules* 36: 8385–8390.

33 Brant, D.A., Tonelli, A.E., and Flory, P.J. (1969). The configurational statistics of random poly(lactic acid) chain. II. Theory. *Macromolecules* 2: 228–234.

34 Okihara, T., Tsuji, M., Kawaguchi, A. et al. (1991). Crystal structure of sterocomplex of poly(L-lactide) and poly(D-lactide). *J. Macromol. Sci. Part B Phys.* 30: 119–140.

35 Tonelli, A.E. and Flory, P.J. (1969). The configurational statistics of random poly(lactic acid) chain. I. Experimental results. *Macromolecules* 2: 225–227.

36 Kang, S.H., Zhang, G.Z., Aou, K. et al. (2002). Raman and light scattering analysis of poly(lactic acid) flexibility. *Abstr. Pap. Am. Chem. Soc.* 224: U473–U473.

37 Kang, S.H., Zhang, G.Z., Aou, K. et al. (2003). An analysis of poly(lactic acid) with varying regio regularity. *J. Chem. Phys.* 118: 3430–3436.

38 Chen, X.L., Kalish, J., and Hsu, S.L. (2011). Structure evolution of α′-phase poly(lactic acid). *J. Polym. Sci., Part B: Polym. Phys.* 49: 1446–1454.

39 Flory, P.J. (1969). *Statistical Mechanics of Chain Molecules*. New York: Interscience Publishers.

40 Yang, X., Su, Z., Wu, D. et al. (1997). Raman analysis of a conformational distribution of poly(ethylene oxide) and its model compound in the liquid state. *Macromolecules* 30: 3796–3802.

41 Hallmark, V.M., Bohan, S.P., Strauss, H.L., and Snyder, R.G. (1991). Analysis of the low-frequency isotropic Raman spectrum of molten isotactic polypropylene. *Macromolecules* 24: 4025–4032.

42 Fraser, R.D.B. (1953). The interpretation of infrared dichroism in fibrous protein structures. *J. Chem. Phys.* 21: 1511.

43 Roe, R.J. and Krigbaum, W.R. (1964). Orientation distribution function of statistical segments in deformed polymer networks. *J. Appl. Phys.* 35: 2215–2219.

44 Addadi, L., Weinstein, S., Gati, E. et al. (1982). Resolution of conglomerates with the assistance of tailor-made impurities. Generality and mechanistic aspects of the "rule of reversal". A new method for assignment of absolute configuration. *J. Am. Chem. Soc.* 104: 4610.

45 Hanyu, A. and Stein, R.S. (1991). Segmental orientation and infrared dichroism of model bimodal elastomeric networks. *Makromol. Chem. Macromol. Symp.* 45: 189–203.

46 Marrinan, H.J. (1959). The infrared dichroism of a stretched polymer. Part I. The theoretical treatment of rubberlike polymers. *J. Polym. Sci.* 39: 461–468.

47 Shindo, Y., Read, B.E., and Stein, R.S. (1968). Study of orientation of polyvinyl chloride films by means of birefringence and infrared visible and ultraviolet dichroism. *Makromol. Chem. - Macromol. Chem. Phys.* 118: 272.

48 Raman, C.V. and Krishnan, U.S. (1928). *Nature* 121: 501.

49 Savage, J.D., Wang, Y.K., Corbett, M. et al. (1992). An analysis of Raman active low frequency bands observed for syndiotactic polystyrene. *Macromolecules* 25: 3164–3169.

50 Barone, V., Improta, R., and Rega, N. (2008). Quantum mechanical computations and spectroscopy: from small rigid molecules in the gas phase to large flexible molecules in solution. *Acc. Chem. Res.* 41: 605–616.

51 Tobias, D.J., Stern, A.C., Baer, M.D. et al. (2013). Simulation and theory of ions at atmospherically relevant aqueous liquid-air interfaces. *Ann. Rev. Phys. Chem.* 64: 339–359.

52 Snyder, R.G. (1992). Chain conformation from direct calculation of Raman spectra of *n*-alkanes. *J. Chem. Soc., Faraday Trans.* 88: 1823.

53 Zbinden, R. (1964). *Infrared Spectroscopy of High Polymers*. New York: Academic Press.

54 Grijpma, D.W., Penning, J.P., and Pennings, A.J. (1994). Chain entanglement, mechanical properties and drawability of poly(lactide). *Colloid. Polym. Sci.* 272: 1068–1081.

55 Joziasse, C.A.P., Veenstra, H., Grijpma, D.W., and Pennings, A.J. (1996). On the chain stiffness of poly(lactide)s. *Macromol. Chem. Phys.* 197: 2219–2229.

56 Wu, S. (1992). Control of intrinsic brittleness and toughness of polymers and blends by chemical structure: a review. *Polym. Int.* 29: 229–247.

57 Asahina, M. and Enomoto, S. (1962). Elastic Moduli of oriented polymers. II. $(-A_1-A_2,)_n$-type polymers. *J. Polym. Sci.* 59: 101.

58 Shimanouchi, T., Enomoto, S., and Asahina, M. (1962). Elastic moduli of oriented polymers. I. Simple helix, polyethylene, polytetrafluoroethylene, and a general formula. *J. Polym. Sci.* 59: 93.

59 Miyata, T. and Masuko, T. (1998). Crystallization behavior of poly(L-lactide). *Polymer* 39: 5515–5521.

60 Miyata, T., Yachida, T., and Masuko, T. (1998). Uniaxial orientation behavior of polylactide films. *Sen-i Gakkaishi* 54: 262–270.

61 Postema, A.R. and Pennings, A.J. (1989). Study on the drawing behavior of poly(L-lactide) to obtain high strength fibers. *J. Appl. Polym. Sci.* 37: 2351–2369.

62 Schmidt, S.C. and Hillmyer, M.A. (2001). Polylactide stereocomplex crystallities as nucleating agents for isotactic polylactide. *J. Polym. Sci., Part B: Polym. Phys.* 39: 300–313.

63 Hoogsteen, W., Postema, A.R., Pennings, A.J. et al. (1990). Crystal-structure, conformation, and morphology of solution-spun poly(L-lactide) Fibers. *Macromolecules* 23: 634–642.

64 Lee, J.K., Lee, K.H., and Jin, B.S. (2001). Structure development and biodegradability of uniaxially stretched poly(L-lactide). *Eur. Polym. J.* 37: 907–914.

65 Cicero, J.A., Dorgan, J.R., Garrett, J. et al. (2002). Effects of molecular architecture on two-step, melt-spun poly(lactic acid) fibers. *J. Appl. Polym. Sci.* 86: 2839–2846.

66 Ajji, A., Guevremont, J., Cole, K.C., and Dumoulin, M.M. (1996). Orientation and structure of drawn poly(ethylene terephthalate). *Polymer* 37: 3707–3714.

67 Ma, J.P., Yu, L., Chen, S.C. et al. (2019). Structure-property evolution of poly(ethylene terephthalate) fibers in industrialized process under complex coupling of stress and temperature field. *Macromolecules* 52: 565–574.

68 Kang, S. (2003). Fundamentals of poly(lactic acid) microstructure, crystallization behavior, and properties. Doctoral, University of Massachusetts (Amherst).

69 Kawai, T., Rahman, N., Matsuba, G. et al. (2007). Crystallization and melting behavior of poly(L-lactic acid). *Macromolecules* 40: 9463–9469.

70 Aou, K. (2007). Effect of molecular structure on the thermal stability of amorphous and semicrystalline poly(lactic acid). University of Massachusetts Amherst.

71 Aou, K., Hsu, S.L., Kleiner, L.W., and Tang, F.W. (2007). Roles of conformational and configurational defects on the physical aging of amorphous poly(lactic acid). *J. Phys. Chem. B* 111: 12322–12327.

Part II

Topical Polymers Studied by Spectroscopy

12

Probing Molecular Events in Self-Healable Polymers

Qianhui Liu, Lei Li, and Marek W. Urban

Clemson University, Department of Materials Science and Engineering, Clemson 29634, USA

12.1 Introduction

Often inspired by behavior of biological systems and driven by the desire to create sustainable materials, autonomous self-healing became one of scientifically exciting and technologically important areas. Typically measured by mechanical property recovery after damage–repair cycle, self-healing is a complex process, which involves macromolecular rearrangements from nano- to micro- and macro-length scales. Although mechanical analysis serves as a practical assessment tool, which for ideal self-healing material should result in the same properties before damage and after autonomous repair, it does not reveal molecular events responsible for self-healing. This is particularly critical in the context of several self-healing mechanisms, which are often chemically and/or physically diverse, yet macroscopic outcomes may be similar. As shown in Figure 12.1, over the last couple of decades, several physical, and/or chemical self-healing approaches have been developed for polymeric materials [1]. They include, but are not limited to, covalent rebonding [2–5] or dynamic supramolecular restructuring [6, 7] and coordination chemistries [8, 9], van der Waals (vdW) [10] and ionic interactions [11, 12], encapsulation of reactive compounds [13] or incorporation of superparamagnetic nanoparticles [14], and embedding electrically conductive additives [15]. These self-healing processes can be easily assessed visually or physically, but it is not trivial to identify macromolecular rearrangements that govern physical recovery. The main challenge is how to detect local molecular events, which result in macroscopic recoveries. Although electron paramagnetic resonance (EPR) and nuclear magnetic resonance (NMR), in particular nuclear Overhauser effect spectroscopy (NOESY) and correlated spectroscopy (COSY) 2D NMR are capable of detecting through-space and through-bond interactions, a lack of spatial resolution requires severe delocalized damage to detect chemical changes during damage–repair cycle [16]. A promising unexplored approach for monitoring self-healing events in polymers is dynamic nuclear polarization (DNP) NMR spectroscopy, if applied to the detection of conformational changes of dangling chain ends generated

Spectroscopic Techniques for Polymer Characterization: Methods, Instrumentation, Applications, First Edition. Edited by Yukihiro Ozaki and Harumi Sato.

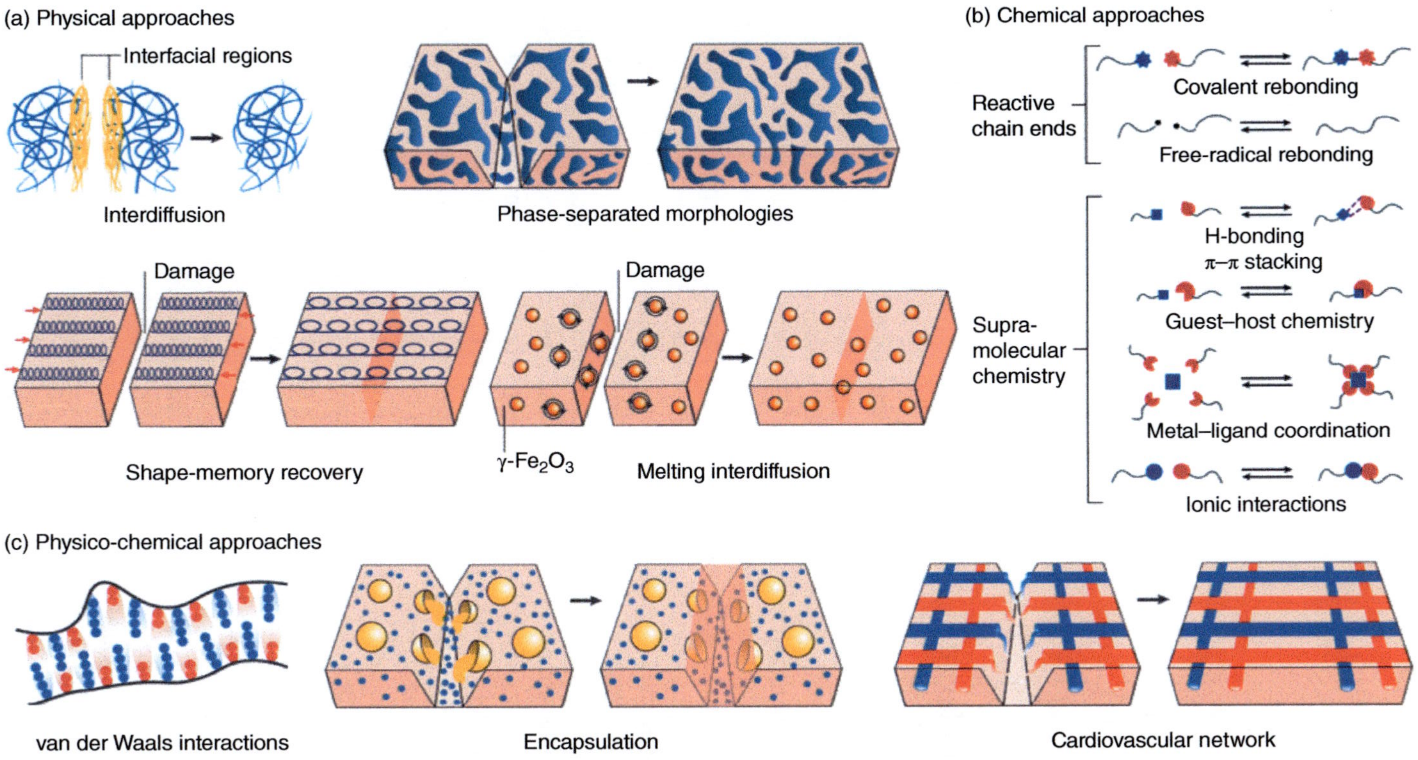

Figure 12.1 Self-healing mechanisms. (a) Physical self-healing processes include interdiffusion of polymer chains, introduction of phase-separated morphologies, shape-memory effects, and incorporation of active nanoparticles into a polymer matrix. (b) Chemical processes to facilitate self-healing involve either introducing reactive chain ends or supramolecular chemistries. (c) A combination of physical and chemical processes is achieved by enhancement of vdW interactions, encapsulating nanocapsules or microcapsules containing reactive liquids to heal a wound, or by mimicking cardiovascular architectures composed of hollow fibers filled with reactive chemicals to heal a polymer matrix. *Source:* Wang and Urban [1].

during damage. Increasingly important are computational approaches to predict or explain observed self-healing processes. To model fracture and self-healing, most models adopted continuum damage approaches in which cracking and healing are considered as degradation and recovery of materials' properties. On the other hand, cohesive-zone approaches treat damage as a discrete event allowing explicit modeling of crack evolution. For detecting local molecular events, sensitive and localized spectroscopic tools are infrared (IR) and Raman spectroscopy, but to identify mass transport during damage–repair cycle, atomic force microscopy (AFM) may be useful if acquisition times are fast enough to capture self-healing events. Although each approach exhibits spatial and spectroscopic limitations, this chapter highlights recent advances in IR, including internal reflection infrared images (IRIRIs) [17] and Raman imaging as well as AFM and molecular dynamics (MDs) simulations that have led to new advances in self-healing mechanisms and ultimately the development of self-healing in commodity polymers. Specifically, the role of microphase-separated heterogeneities, entropy-driven shape-memory effects (SMEs), free radical and cationic recombination, and vdW interactions in the development of new self-healing polymers is discussed along with the visual assessment of damage–repair cycle by embedding color-changing chemical sensors.

12.2 Microphase Separation

Wounds of *Delosperma cooperi* leaves can be closed by tissue bending or contraction due to the stored elastic stress built-in within the heterogeneous structure during growth. Inspired by this event, thermoplastic polyurethanes (TPUs) synthesized from polycaprolactone diol (PCL-diol), 1,4-butanediol (BDO), hydroxyl-modified spiropyran (SP), and hexamethylene diisocyanate (HDI) (Figure 12.2A) were drawn into microphase (PURP) (Figure 12.2B1) and nanophase-separated fibers (PURM) (Figure 12.2B2), where the phase separation was observed using IRIRI and AFM [18]. As shown in Figure 12.2C, the IR images of 1685 and 1724 cm^{-1} bands due to the PUR and PCL C=O vibrations, respectively, were collected from the fiber cross-sections (Figure 12.2C2, C3). Each cross-section exhibits microphase separation, where the micron-size PUR-rich domains are distributed within the PCL-rich matrix. The band at 1535 cm^{-1} (Figure 12.2C4) corresponds to urethane–urethane H-bonding amide II, and its intensity decreases in the PUR-poor domains. Figure 12.2D1–D3 shows a gradual decrease of the urethane C=O (1685 cm^{-1}) intensities when going from the center of PUR-rich domain outward toward the PCL-rich domain, and illustrate the presence of micron-scale interphase between the two domains. In contrast, the phase separation of PURM is indistinguishable by IRIRI (Figure 12.2E1, E2) due to the limited spatial resolution (~1 μm), but the presence of two glass transition temperatures (T_g) (differential scanning calorimetry [DSC]) as well as AFM phase images (Figure 12.2E3) indicates nanoscale phase separation.

Interestingly enough, these chemically identical PURP and PURM polymers exhibit different SMEs and self-healing characteristics. A cut on PURP closes within

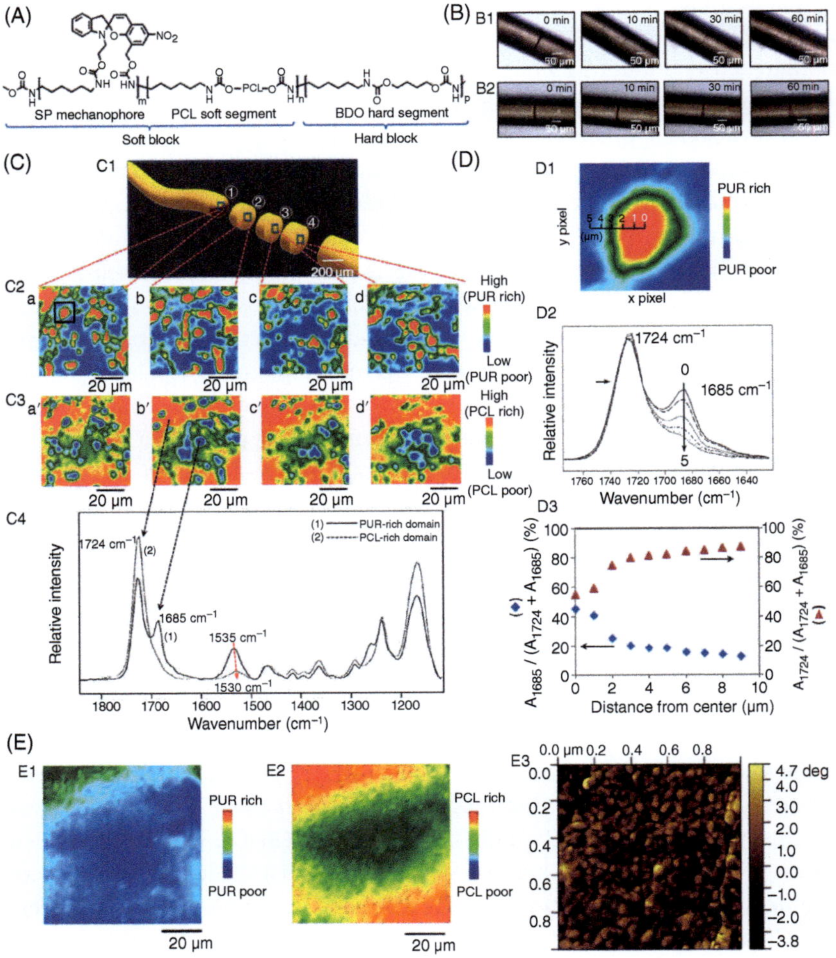

Figure 12.2 (a) chemical structure of synthesized thermoplastic polyurethane inspired by *Delosperma cooperi* leaves; (B1) optical images of self-healable PURP fibers; (B2) optical images of self-healing for PURM fibers; (C1) Schematic representation of PURP cross-section of a 200-mm thick fiber. (C2, C3) IR images of the 1685 and 1724 cm^{-1} bands due to PUR and PCL C=O vibrations, respectively; (C4) Corresponding IR spectra recorded from PUR-rich and PCL-rich domains indicated by black arrows; (D1) IR image of the 1685 cm^{-1} band due to PUR C=O vibration; (D2) IR spectra of C=O vibration, where traces 0–5 were collected from the corresponding areas shown in (D1) marked by 0–5, respectively; (D3) Percentage of the 1724 cm^{-1} (ester C=O) and 1685 cm^{-1} (urethane C=O) band areas (A_{1724} and A_{1685}) within 1780–1640 cm^{-1} region plotted as a function of the distance from the center of PUR-rich domain shown in (D1). The percentage is calculated based on integrated areas of the deconvoluted bands. (E1) IR images of PURM cross-section recorded by tuning into C=O vibrations of polyurethane groups at 1686 cm^{-1}; (E2) IR images of PURM cross-section recorded by tuning into C=O vibrations of PCL ester groups at 1724 cm^{-1}; (E3) AFM image of nanophase-separated PURM. *Source:* Yang et al. [18].

10 minutes whereas the PURM virtually do not self-heal (Figure 12.2b), because PURP exhibits higher shape recovery ratio and the degree of crystallinity. These observations formulated hypothesis that chain slippage of urethane segments from PUR-rich to PCL-rich domains occurs during deformation and therefore inhibits PCL chain packing in PURM. On the contrary, the distribution of PUR- and PCL-rich domains in PURP is less disrupted because microscale phase-separated interfacial regions act as stable junction points preventing chain slippage during mechanical deformation.

To test this hypothesis, IR spectra of PURP and PURM cross-sections were extracted from IR images and the urethane amide II bands were analyzed. The bands at 1523 and 1539 cm^{-1} due to the formation of urethane–ester (PCL) and urethane–urethane H-bonding were analyzed. 2D-FT-IR correlation analysis of PURP and PURM bands, shown in Figure 12.3A, was obtained using an asynchronous 2D-FT-IR analysis governed by the following relationships: $\Psi(v1, v2) = \frac{1}{\pi}[y1(v1)y2(v2) - y2(v1)y2(v2)]$ (where $v1$ and $v2$ represent two wavenumbers, and $y1$ and $y2$ represent IR intensities in the first and second spectra, respectively). In this study, spectra of undamaged specimens were assigned as $y1$, and the spectra collected after perturbation (shape memory cycle [SMC] or self-healing) were designated as $y2$. Going from $y1$ to $y2$, if the intensity ratio of the two bands at $v1$ vs. $v2$ changes, two asymmetrical cross-peaks with respect to the diagonal will be observed. The presence of positive cross-peak indicates the intensity ratio $y2(v2)/y2(v1) > y1(v2)/y1(v1)$ and a negative cross-peak implies $y2(v2)/y2(v1) < y1(v2)/y1(v1)$ [19]. As shown in Figure 12.3A1, two antisymmetric cross-peaks along the diagonal are detected, indicating the two types of H-bonding: urethane–urethane (1539 cm^{-1}) and urethane–ester (1523 cm^{-1}). Positive cross-peaks after recovery denote the increased urethane–urethane H-bonding with respect to urethane–ester counterpart for PURP after the first SMC. In contrast, the opposite trend is observed for PURM (Figure 12.3A2), where the negative cross-peaks exhibit the decrease of urethane–urethane H-bonding with respect to the urethane–ester. Along the same lines, the amide II bands of PURP and PURM, at the initial state and upon self-healing, were analyzed. Similar trends were observed that after self-healing, urethane–urethane H-bonding interactions increase and decrease for PURP and PURM, from the positive cross-peak in Figure 12.3A3 and the negative cross-peak in A4, respectively.

In summary, micron-scale damage–repair cycle and macroscopic shape memory recovery are governed by the same molecular events: the increase of urethane–urethane H-bonding in microphase-separated PURP copolymers and further phase separation at the healing temperatures as the two phases are immiscible. In contrast, the increase of urethane–ester interactions in nanophase-separated PURM shows that PUR blocks are partially dispersed in PCL-rich domains after SMC and self-healing, indicating that nanosized hard-segment crystals resulting from urethane–urethane H-bonding dissociate during deformation. For the microphase-separated PURP (Figure 12.3B1), PUR-rich domains and the interphase inhibit PUR blocks from disengagement under mechanical forces. As crystalline PCL blocks unfold, amorphous PCL segments elongate and therefore reduce

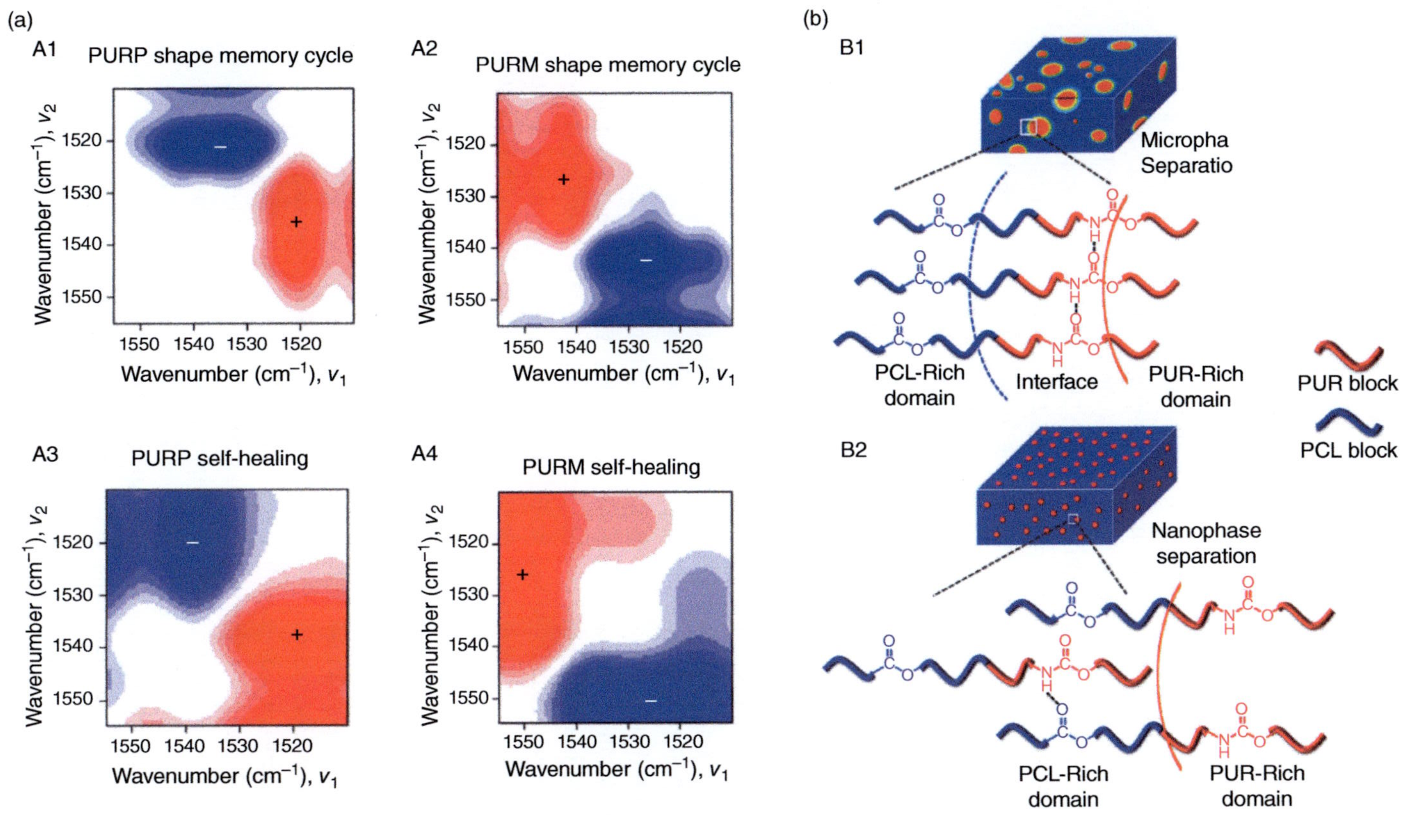

Figure 12.3 Asynchronous 2D-FT-IR correlation spectra of PURP (A1) and PURM (A2) before and after the first SMC; asynchronous 2D-FT-IR correlation spectra of PURP (A3) and PURM (A4) before damage and after self-healing; schematic illustrations of molecular events in micro- and nanophase-separated PURP (B1) and PURM (B2) copolymers upon self-healing. *Source:* Yang et al. [18].

entropy and store energy. When temperature increases, macromolecular chains in PCL-rich domains gain mobility and return to the randomized state with maximum entropy. The stored energy is released, and shape recovery occurs, which promotes wound closure. On the contrary, for nanophase-separated copolymers (Figure 12.3B2), PUR blocks slip from PUR-rich domains to PCL-rich phase upon mechanical deformation, resulting in energy dissipation and consequently the decrease in shape recovery and wound closure. In summary, the combination of IR imaging and 2D IR correlation spectroscopy revealed that modulating the length scale of polymer/copolymer heterogeneity provides a unique combination of tools for elucidating mechanisms of self-healing.

12.3 Entropically Driven Self-Healing

Mechanically robust TPUs capable of autonomous self-healing under ambient conditions were also developed, and two self-healing mechanisms were identified: viscoelastic shape memory driven by conformational entropic energy for polymers with high molecular weight, and surface energy/tension-induced reduction of newly generated surface areas for polymers with low molecular weight [20]. One of the unique features is that mechanical damage of higher M_w (~72 kDa) TPU fibers vanishes within 40 minutes, while lower M_w (~45 kDa) does not self-heal within 24 hours (Figure 12.4A). To understand how shape memory may impact self-healing, viscoelastic length transitions (VLTs) were determined by measuring strain as a function of temperature for TPUs with various molecular weights (Figure 12.4B). These studies showed that the strain increases to ε_{max} attributed to viscoelastic changes near T_g and decreases to ε_{min} due to conformational entropy of T_g-based SMEs. Molecular entanglements act as junction points for shape-memory-induced recovery and prevent chain slippage and flow, which diminish energy storage upon deformation. As molecular weight decreases, the % of strain recovery and released conformational entropic energy density (ΔS_R) decrease due to the reduced storing and releasing of conformational entropic energy. The stored entropic energy (ΔS_s), ε_{max}, and the stress at ε_{max} are plotted on the polymer shape memory prediction plane and show that TPUs with M_w of 72 and 180 kDa possess balanced σ and ε storage capabilities and high ΔS_s, whereas lower M_w (45, 32, and 22 kDa) do not fall on the plane due to poor entropic recovery (Figure 12.4C). 3D-laser microscope surface profiling (Figure 12.4D) was utilized to follow the progression of wound closure. As shown, the wound edges are pushed up and out from the center upon damage but return to their initial states during repair due to shape recovery and reestablished contact at the interface.

Molecular events responsible for wound closure for 72 and 45 kDa TPU were examined by IRIRI. Additionally, the analysis of urethane amide I, II, and III bands at 1695, 1533, and 1241 cm^{-1}, respectively, shows no covalent bond cleavage or reformation upon damage or repair, but reversible intensity changes due to conformational rearrangements. Furthermore, H-bonding dissociation inside the wound is manifested by the increase of the 1724 cm^{-1} band corresponding to

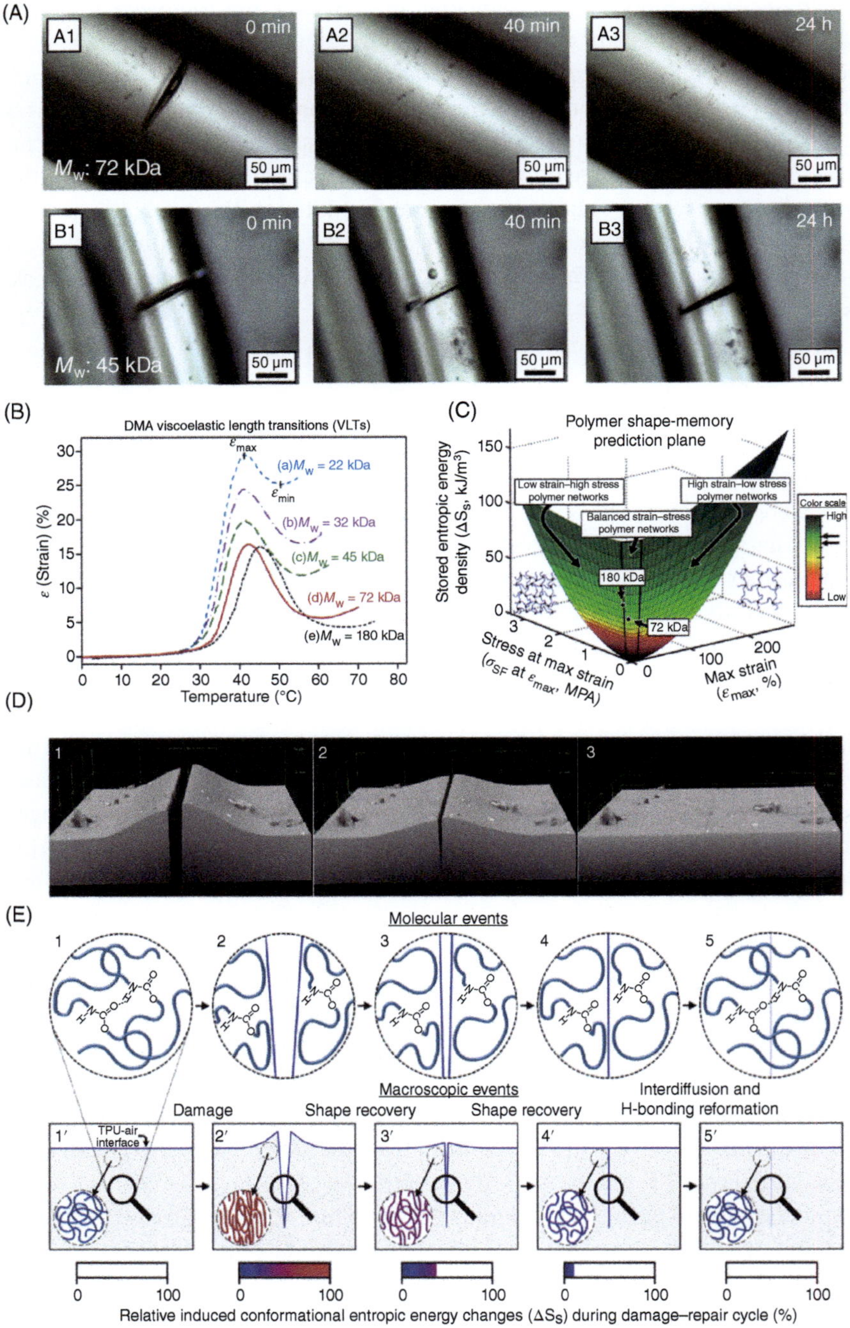

Figure 12.4 (A) Optical images of damaged TPU fibers: $M_W \approx 72$ kDa (A1–A3), and $M_W \approx 45$ kDa (B1–B3); (B) Dynamic mechanical analysis (DMA) strain (ε) curves showing the VLT measured as a function of TPU molecular weight: $M_W \approx 22$, 32, 45, 72, and 180 kDa; (C) 180 and 72 kDa TPU VLTs maximum strain (ε_{max}), stress at maximum strain (σ_{SF} at ε_{max}), and stored entropic energy densities (ΔS_S) on the DMA shape-memory prediction plane; (d) 3D-laser microscope surface profiles of 72 kDa TPU during wound closure; (e) Schematics of molecular level and macroscopic events during entropy driven self-repair. *Source:* Hornat and Urban [20]. Licensed under CC BY-4.0.

non-H-bonding C=O groups and the decrease of the 1683 cm^{-1} band denoting H-bonding C=O. 2D-FT-IR correlation spectroscopy was utilized to examine band intensity changes for H-bonding and free urethane groups under damage–repair cycle. Similar to the previous studies, synchronous 2D-FT-IR correlation spectra were calculated (Figure 12.5, A1–A3 and B1–B3) and show that upon both damage (A1–A3) and repair (B1–B3), H-bonding urethane amide I, II, and III bands (1682, 1540, and 1244 cm^{-1}, respectively) display positive cross-peaks, whereas free urethane amide I, II, and III bands (1719, 1505, and 1224 cm^{-1}) exhibit negative cross-peaks, indicating that H-bonding and non-H-bonding bands have opposite responses throughout the damage–repair cycle. On the other hand, asynchronous spectra (Figure 12.5, A4–A6 and B4–B6) show that upon damage (A4–A6) bands associated with free urethane increase relative to H-bonding ones, exhibiting positive cross-peaks (1683, 1716 cm^{-1}), (1537, 1508 cm^{-1}), and (1244, 1225 cm^{-1}). However, during repair (B4–B6), free urethane bands decrease relative to H-bonding ones, signified by negative cross-peaks (1683, 1716 cm^{-1}), (1537, 1508 cm^{-1}), and (1244, 1225 cm^{-1}).

In summary, the visual observations and spectroscopic analyses demonstrate that for polymers with sufficient molecular weight for entropic-recovery, self-repair consists of two events (Figure 12.4E): (i) H-bonding decreases and chain conformation changes in the wound, so the conformational entropy decreases near damaged areas, and restoring forces responsible for shape recovery generated; (ii) the entropic force brings the two cut surfaces back into contact, and H-bonding reformation and chain diffusion facilitate rebonding of the interface to restore mechanical integrity. This mechanism is ineffective for low M_w polymers lacking sufficient entanglements to resist chain slippage/flow upon damage. Nevertheless, low M_w polymers may exhibit surface tension-driven self-healing, where polymer flows to fill and repair the damage. The healing rate is a function of molecular weight and increases as molecular weight decreases (22 kDa ~18 days, 32 kDa ~100 days, 45 kDa TPU is not fully repaired after 100 days) because viscosity resists flow. Self-healing may not occur in a reasonable period if viscosity is too high. However, in real applications, the surface-tension-driven self-repair and viscoelastic creep should be balanced.

12.3.1 Free Radical and Cationic Recombination

Oxetane-substituted chitosan-polyurethane (OXE-CHI-PUR) crosslinked networks capable of self-healing under UV irradiation were developed by copolymerizing polyethylene glycol (PEG) diol with trifunctional HDI and the OXE-CHI macromonomer, under the catalysis of dibutyltin dilaurate (DBTDL) (Figure 12.6a) [21]. The isocyanate–polyol and OXE-CHI crosslinking reactions result in PUR and polyurea (PUA) formation (Figure 12.6b). Using IRIRI, the origin of molecular and macroscopic processes inside a mechanically generated scratch during UV-initiated self-repair was elucidated. Four linkages: (1) –NHCONH–, (2) –C–O–C– between CHI and CHI, (3) –C–O–C– between CHI and OXE, and (4) –C–O–C– of the OXE ring, highlighted in blue in Figure 12.6b, were discovered to be primarily responsible for UV-induced self-repair. These linkages undergo bond cleavage under

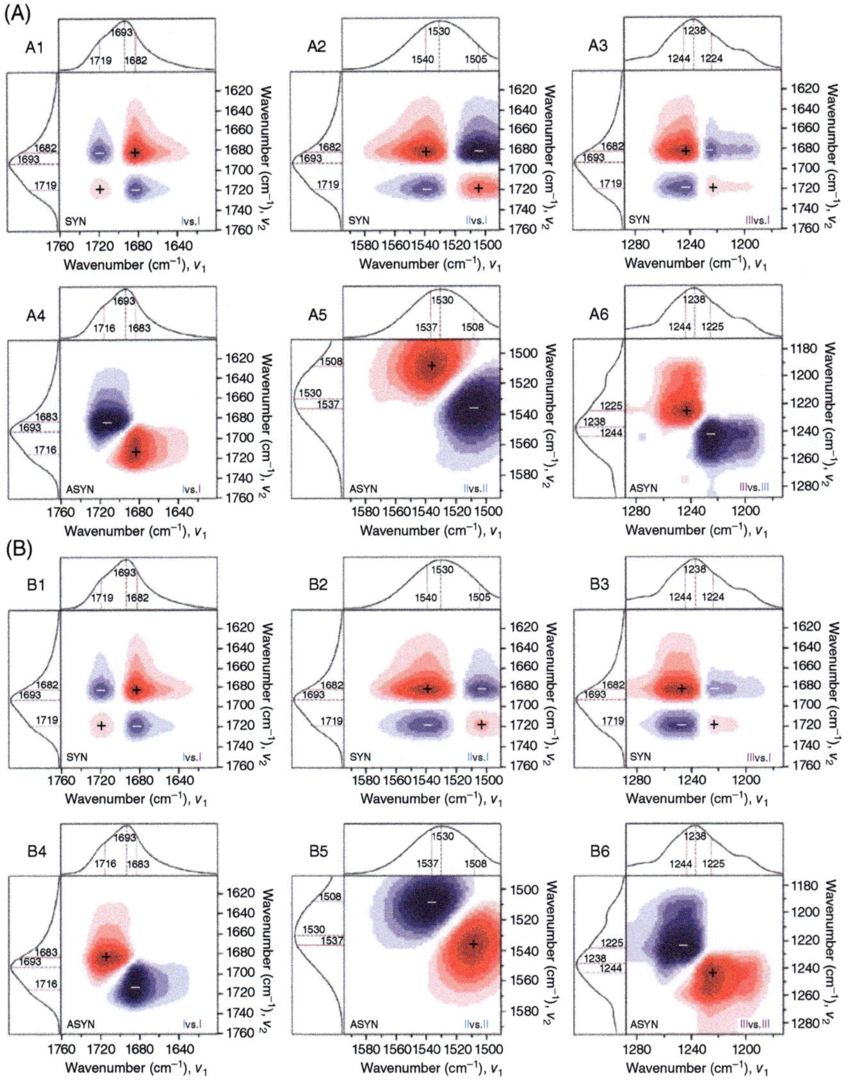

Figure 12.5 2D-FT-IR spectra of TPUs upon damage (A) and repair (B); Synchronous (A1–A3 and B1–B3, using the average spectrum as the reference spectrum) and asynchronous (A4–A6 and B4–B6, using no reference spectrum). (A1 and B1) 1760–1600 cm^{-1} (urethane amide I) vs. 1760–1600 cm^{-1} (urethane amide I); (A2 and B2) 1595–1490 cm^{-1} (II) vs. 1760–1600 cm^{-1} (i); (A3 and B3) 1289–1173 cm^{-1} (III) vs. 1760–1600 cm^{-1} (I). (A4 and B4) 1760–1600 cm^{-1} (I); (A5 and B5) 1595–1490 cm^{-1} (II); (A6 and B6) 1289–1173 cm^{-1} (III). (Red = positive, blue = negative). *Source:* Hornat and Urban [20]. Licensed under CC BY-4.0.

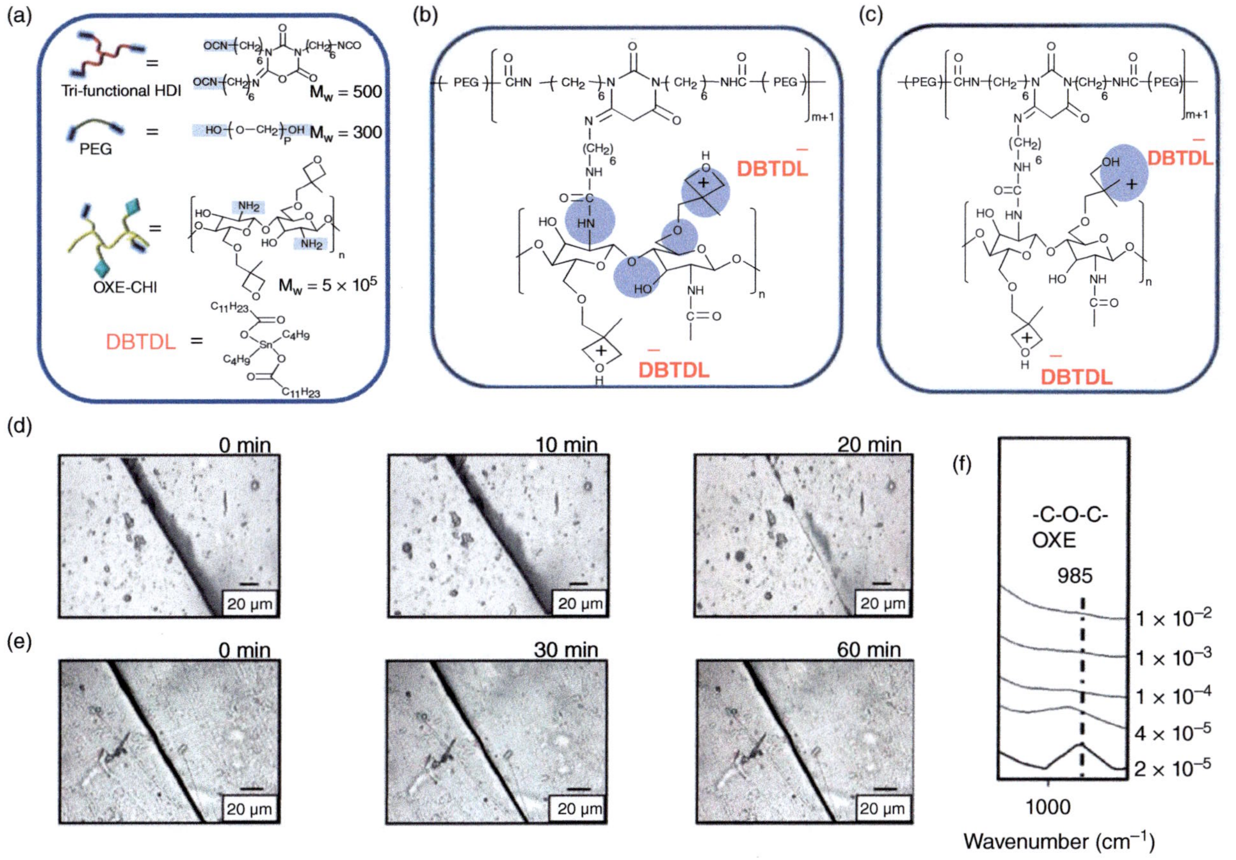

Figure 12.6 (a) monomers and catalyst utilized in the synthesis of OXE-CHI-PUR networks; (b) synthesized OXE-CHI-PUR networks with four linkages highlighted in blue responsible for UV-induced self-healing; (c) OXE ring opening as one of the proposed self-healing mechanisms; Scratch on OXE-CHI-PUR networks self-heals with (d) 2×10^{-5} and (e) 4×10^{-5} mol of DBTDL added; (f) intensity of −C−O−C− vibration of OXE as a function of the content of DBTDL. *Source:* Ghosh et al. [21].

mechanical damage and exposure to UV radiation. Free radicals are generated in linkages (1)–(3), and a carbocation forms in (4) where the OXE ring opening reaction propagates cationically (Figure 12.6c).

Furthermore, the role of DBTDL concentration levels and acidity on OXE ring opening reaction was assessed by optical microscopy and IR spectroscopy. In OXE-CHI-PUR networks (HDI: PEG: OXE-CHI = 1.0 : 1.33 : 1.17 × 10^{-4}, molar ratio), when DBTDL concentration is 2 ×10^{-5} mol, OXE-CHI-PUR networks exhibit self-healing behavior, but its excess inhibits this process (Figure 12.6d,e). This is because the synthesis of PUR in the presence of DBTDL involves the formation of cationic intermediates, and the excess of DBTDL results in the premature OXE ring opening, inhibiting self-healing. Figure 12.6f illustrates IR spectra of OXE-CHI-PUR networks containing various contents of DBTDL, and the band at 985 cm^{-1} corresponding to –C–O–C– stretching vibrations of OXE is detected only for 2 × 10^{-5} DBTDL mole content, and diminishes to the minimum at higher molar concentrations. Thus, only when the DBTDL stoichiometry is adjusted to catalyze crosslinking reactions of PEG-HDI and CHI-HDI, OXE rings remain in their dormant state as oxonium ions, which polymerize cationically after damage and UV irradiation. Protic environments also impact self-healing characteristics and in particular acidic pH values inhibit self-healing.

One of the questions that is always of interest is the mass flow during damage–repair cycle. AFM may serve this purpose. Figure 12.7A1–A3 illustrates topographical AFM images of OXE-CHI-PUR networks before (A1) and after (A2) mechanical damage and after UV exposure ($A2^1$, $A2^2$, A3). The open (a) and closed (b) circles represent mechanical damage and UV-induced repair areas, and undamaged areas as controls, respectively. In Figure 12.7A2, a groove forms along the side bank resulting from the mechanical damage. Upon UV exposure, polymer flow occurs from the bottom of the scratch ($A2^1$, $A2^2$), and the material self-heals with formation of a scar (A3). The heights of the surface topological features are shown in Figure 12.7, A1′–A3′. The surface has no height difference before damage, whereas the scratch results in a mass flow and formation of side banks. After UV irradiation, the heights of the side banks decrease, and the frontal growth within the damage appears, followed by scratch closure. During this process, thermal–mechanical properties change within scratches because of physical network rearrangement. Thermal expansion d(Δl) of the network normalized to $d(\Delta I)/dT$ values for undamaged, damaged, and self-repair areas (A1″–A3″, respectively) as a function of temperature shows that the peak value corresponds to T_g of the network. The T_g inside scratches drops from 61 to 56 °C, and the width of the transition increases. These observations indicate lower molecular weight and broader molecular weight distribution resulting from the bond cleavage. Upon UV exposure, the T_g shifts to 62 °C, due to enhanced crosslink density and self-repair. The shoulder at 102 °C corresponds to the formation of PUA, which was also detected spectroscopically. For reference purpose, the same analysis was performed on PUR without OXE-CHI, and no self-healing was observed. The decrease of T_g (61–59 °C) was observed in damaged area, but upon UV exposure no increase was observed, indicating marginal or no crosslinking reactions.

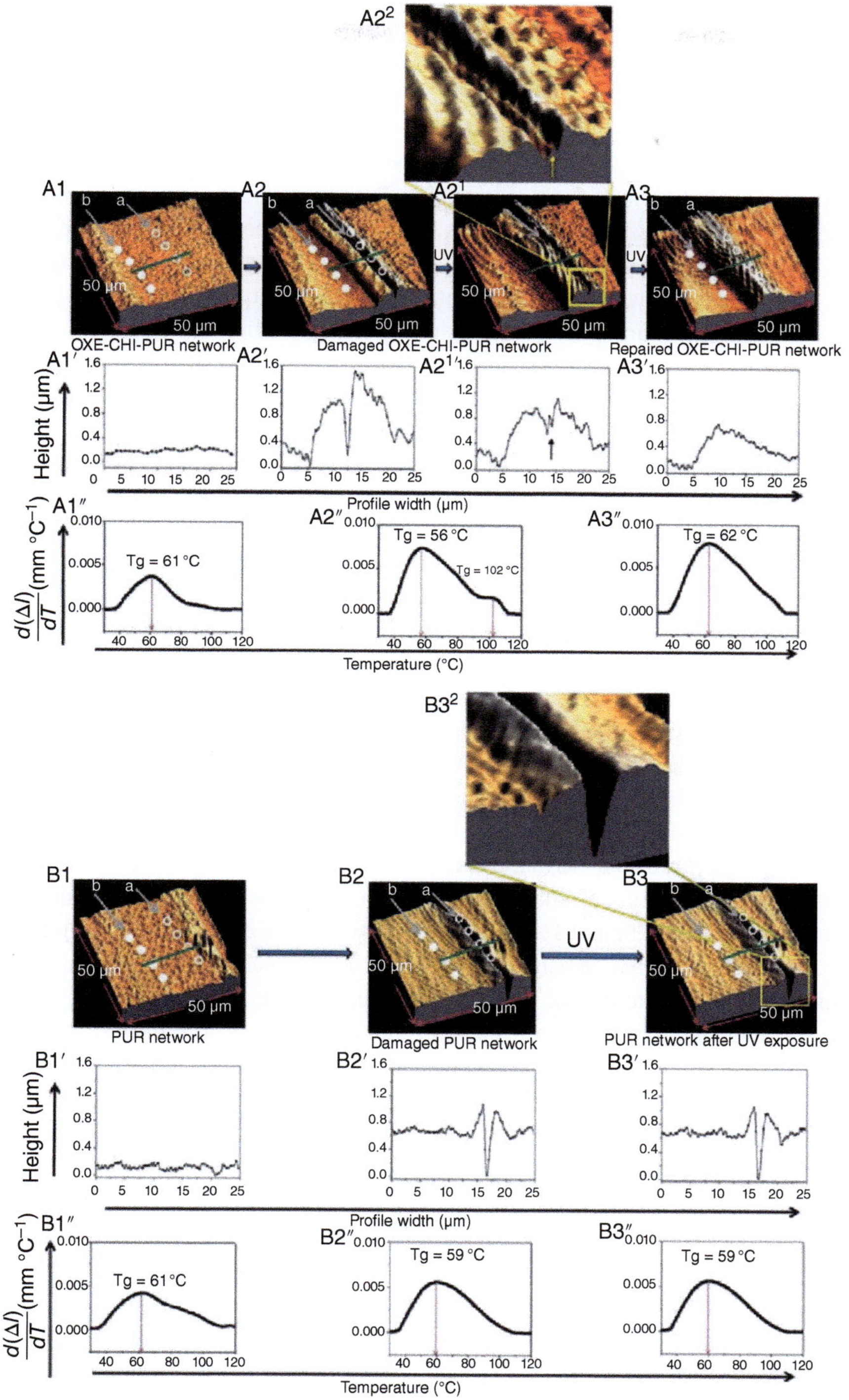

Figure 12.7 AFM images of undamaged (A1), damaged (A2), UV-exposed ($A2^1$, $A2^2$) and repaired (A3) OXE-CHI-PUR networks, and a plot of the height vs. width profile of undamaged (A1′), damaged (A2′), UV exposed ($A2^{1\prime}$), and repaired (A3′) networks, and $d(\Delta l)/dT$ vs. temperature (A1″–A3″) for the same. AFM images of undamaged (B1), damaged (B2), and UV-exposed (B3, $B3^2$) PUR networks and a plot of the height vs. width profile of undamaged (B1′), damaged (B2′), and UV exposed (B3′) networks, and $d(\Delta l)/dT$ vs. temperature (B1″–B3″) for the same. *Source:* Ghosh et al. [21].

12.3.2 Van der Waals Interactions

One of the attractive features of polymers is that vdW forces are always present. For example, vdW forces between the molecules and the surface of a gecko's foot allow the gecko to grip the wall. Although advanced materials functionalized with self-healing capability have been developed in the past several decades, only recently the use of vdW forces that enable multiple recovery upon mechanical damage without external intervention has been utilized in commodity copolymers [10]. For example, [p(MMA/nBA)] and its derivatives are capable of self-healing upon mechanical damage when alternating random copolymer topologies are employed. To decode molecular features during damage–repair cycle, MD simulations combined with precise copolymerization were utilized as a tool to elucidate and illustrate the effects of enhanced vdW interactions on self-healing.

To determine the key features responsible for self-healing, cohesive energy densities (CEDs), copolymer conformations of p(MMA/nBA), and end-to-end distances (r) with different compositions were determined. As shown in Figure 12.8a, the CED values exhibit maxima for self-healable copolymer compositions (45/55 to 50/50 MMA/nBA monomer molar ratios) due to higher interchain packing density compared to non-self-healable ranges I and III.

To explore the effects of monomer sequences and the contribution of vdW forces on self-healing, model MD simulations on pentads were conducted and showed that alternating BMBMB/BMBMB pair (1 : 1) exhibits the highest cohesive energy (CE_p) values compared to more "blocky" combinations, supporting experimental evidence that block copolymer topologies have adverse effects on self-healing. As shown in Figure 12.8b, interdigitated morphologies of preferentially alternating copolymers exhibit energetically and directionally favorable vdW interactions, which upon damage return to their initial conformations.

Another unique feature is that the copolymer chains within range II exhibit preference in extended helix-like conformations, which are energetically favorable with an equilibrium vdW density of 1.96×10^5 kJ m^{-3}. Helix-like conformations resulting from the alternating monomer sequences are the contributing factor to the higher CEDs and stronger vdW interchain forces. In contrast, block pMMA-*b*-pnBA copolymers are not self-healable, and their CED values are lower. Thus, key-and-lock configurations formed between adjacent copolymer chains (Figure 12.8b) are favorable due to alternating/random copolymer topologies, which are responsible for self-healing.

It is also useful to consider copolymer flexibility parameter (f_{eq}) defined as $f_{eq} = r_{max}/[r_{eq}^2(l\ (2 - f))]$, which represents fraction of covalent bonds capable of bending out from the collinear direction, where r_{max} is fully extended chain length, r_{eq} is the end-to-end distance obtained from MD simulations, and l is the length of the repeating unit. When polymer chains reach their equilibrium status (f_{eq}), the copolymers in self-healable region II exhibit smallest chain flexibility. In another word, if chains are deformed by external forces, energy will be stored, and the distorted conformations will act as mechanical springs to facilitate return to their original state. Again, the formation of key-and-lock junctions between

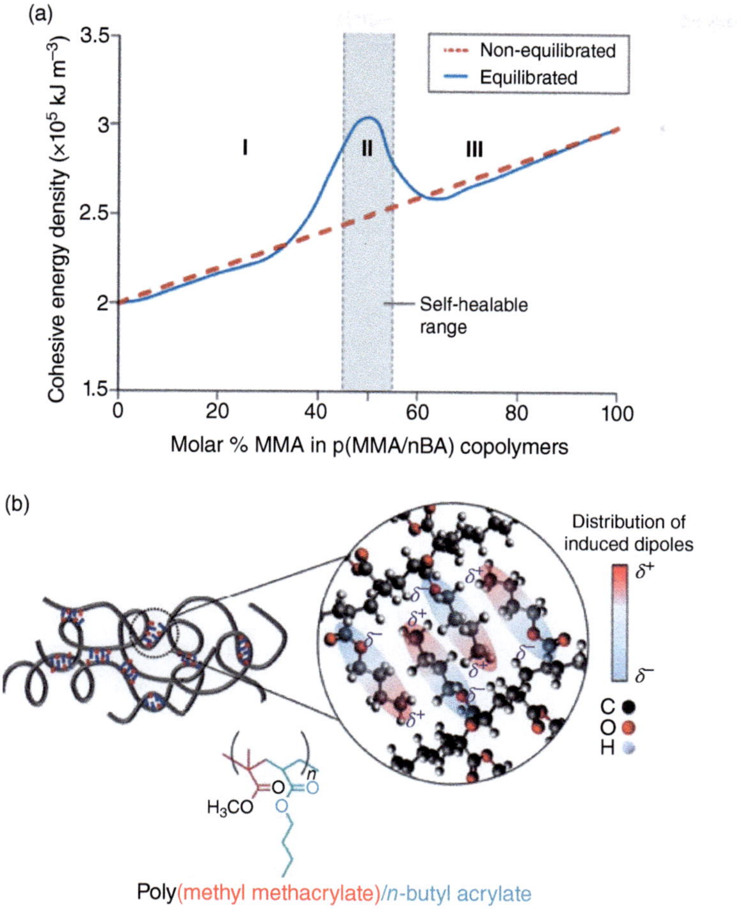

Figure 12.8 (a) Cohesive energy density (CED) as a function of molar ratio for non-equilibrated and equilibrated p(MMA/nBA) copolymers. *Source:* Wang and Urban [1]; (b) self-healing through vdW forces with key-and-lock configuration in p(MMA/nBA) copolymers.

neighboring chains driven by alternating/random monomer sequences ultimately leads to p(MMA/nBA) recovery.

12.3.3 Chemical Sensing of Damage–Repair Cycle

Nature has the unique ability of rebuilding damages in an effort to maintain living functions. For instance, a skin cut triggers blood clotting and color changes, which parallel wound healing. Inspired by this example, poly(methyl methacrylate/*n*-butylacrylate/1,3-dihydro-1,3,3-trimethylspiro[2H-indole-2,3′-[3H]-naphth[2,1-*b*][1,4]-oxazine]-2-amino-2 methylacrylate) [p(MMA/nBA/SNO)] was prepared, which is capable of sensing mechanical damage by color change and self-healing (Figure 12.9A) [22]. Mechanical damage and visible light irradiation result in the conversion between the colorless spironapthoxazine (SNO) moiety and the

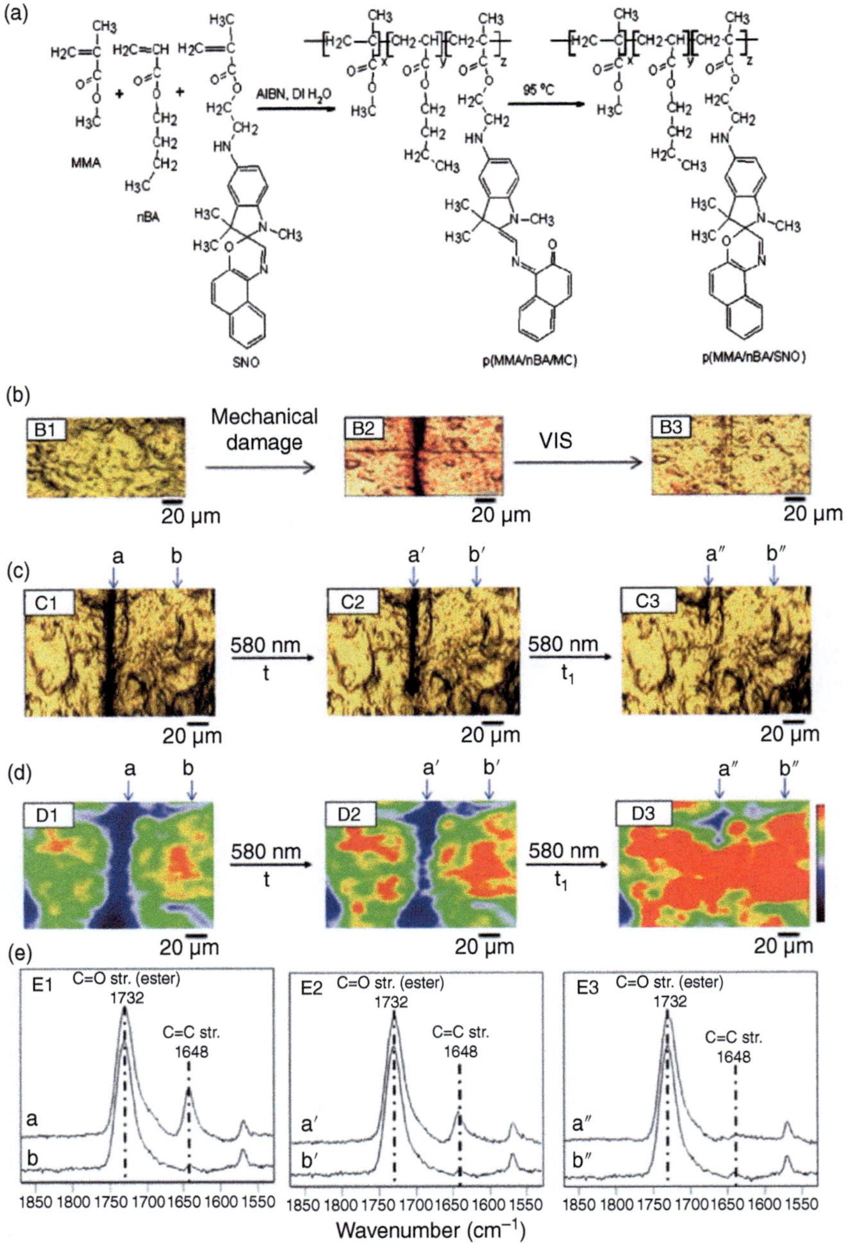

Figure 12.9 (A) Copolymerization of methyl methacrylate (MMA), *n*-butyl acrylate (nBA), and 2-[(1,3,3-trimethyl-1,3-dihydrospiro[indole-2,3′-naphtho[2,1-*b*][1,4]oxazin]-5-yl)amino] ethyl-2-methylacrylate (SNO) monomers; (B) optical images of p(MMA/nBA/SNO) films: undamaged (B1), mechanically damaged (B2), and self-healed after exposure to VIS radiation (B3); (C) optical images of p(MMA/nBA/SNO) films: mechanically damaged (C1), exposure to VIS radiation for 30 minutes (C2) and 60 minutes (C3); (D) Raman images of p(MMA/nBA/SNO) film: mechanically damaged (D1), exposure to VIS radiation for 30 minutes (D2) and 60 minutes (D3); (E) Raman spectra of p(MMA/nBA/SNO) film: mechanically damaged (E1), exposure to VIS radiation for 30 minutes (E3) and 60 minutes (E3). *Source:* Ramachandran et al. [22].

red merocyanine (MC) group due to ring opening/closing reactions. As shown in Figure 12.9B, the colorless polymer (SNO) turns reddish (MC) upon mechanical damage and turns back to the undamaged and colorless state (SNO) upon visible (VIS) irradiation. Optical images of the self-healing process of p(MMA/nBA/SNO) scratched film after exposure to VIS for 30 and 60 minutes are shown in Figure 12.9C. The damaged and undamaged areas are labeled a, a′, a″ and b, b′, b″, respectively. The repair process was monitored using Raman imaging, where intensity changes of the conjugated C=C band of the ring-opened MC groups at 1648 cm^{-1} were measured before and after VIS exposure for 30 and 60 minutes (Figure 12.9D). Traces a, a′, and a″ in Figure 12.9E illustrate Raman spectra collected from areas a, a′, and a″, whereas traces b, b′, and b″ are obtained from the undamaged areas b, b′, and b″. Using the 1732 cm^{-1} band due to C=O vibrations of methacrylate components of p(MMA/nBA/SNO) as a normalization band, the appearance of the 1648 cm^{-1} band in trace a denotes SNO to MC conversion, whereas the absence of this band on traces b, b′, and b″ indicates that ring-opening does not occur when there is no mechanical damage. Figure 12.9E shows that the 1648 cm^{-1} band diminishes after 30 minutes' exposure (trace a′) and disappears after 60 minutes' exposure to VIS (trace a″), manifesting the MC to SNO conversion. Transitions between MC and SNO groups play a major role in color changes, and their functions as a repairing entity of the entire p(MMA/nBA/SNO) copolymer can be determined by computer simulation.

In damaged p(MMA/nBA/MC) scratches, intermolecular H-bonding of MC groups of the neighboring chains parallels the formation of extended backbone conformations. Nevertheless, upon exposure to VIS radiation, H-bonding is disturbed by the ring closure (MC to SNO transition), which ultimately leads to localized backbone collapse facilitated by the free volume due to MC to SNO localized conversion, as shown in Figure 12.10A. This simulated process was analyzed using Raman spectroscopy. In Figure 12.10B, traces a and b were collected before damage and after repair upon exposure to VIS radiation, respectively. The bands at 1648 and 1408 cm^{-1} in trace a are due to C=C stretching vibrations of the conjugated structure of the open ring MC, and N–H–O bending vibrations attributed to the secondary amine groups, respectively. Upon repair, the conjugated MC form is converted to SNO, and the H-bonded secondary amine groups of the neighboring MC units are dissociated, as manifested by the intensity decrease of the 1408 cm^{-1} band (trace b). Schematics of the non-H-bonded molecular structures before mechanical damage and after self-healing, as well as the H-bonded structure in damaged status are shown in Figure 12.10C. The presence of the closed ring in SNO results in a collapsed state of the copolymer backbone, which facilitates self-healing by pulling entanglements to fill the volume from prior damage. In summary, Raman spectroscopy combined with MD simulation are able to probe the color changes and self-healing characteristics of p(MMA/nBA/SNO) copolymers, which originate from the transition between SNO and MC groups and the corresponding polymer backbone rearrangements due to H-bonding.

In summary, this chapter outlines how an interplay of vibrational spectroscopic tools to probe self-healing at molecular levels and simulations can lead to new

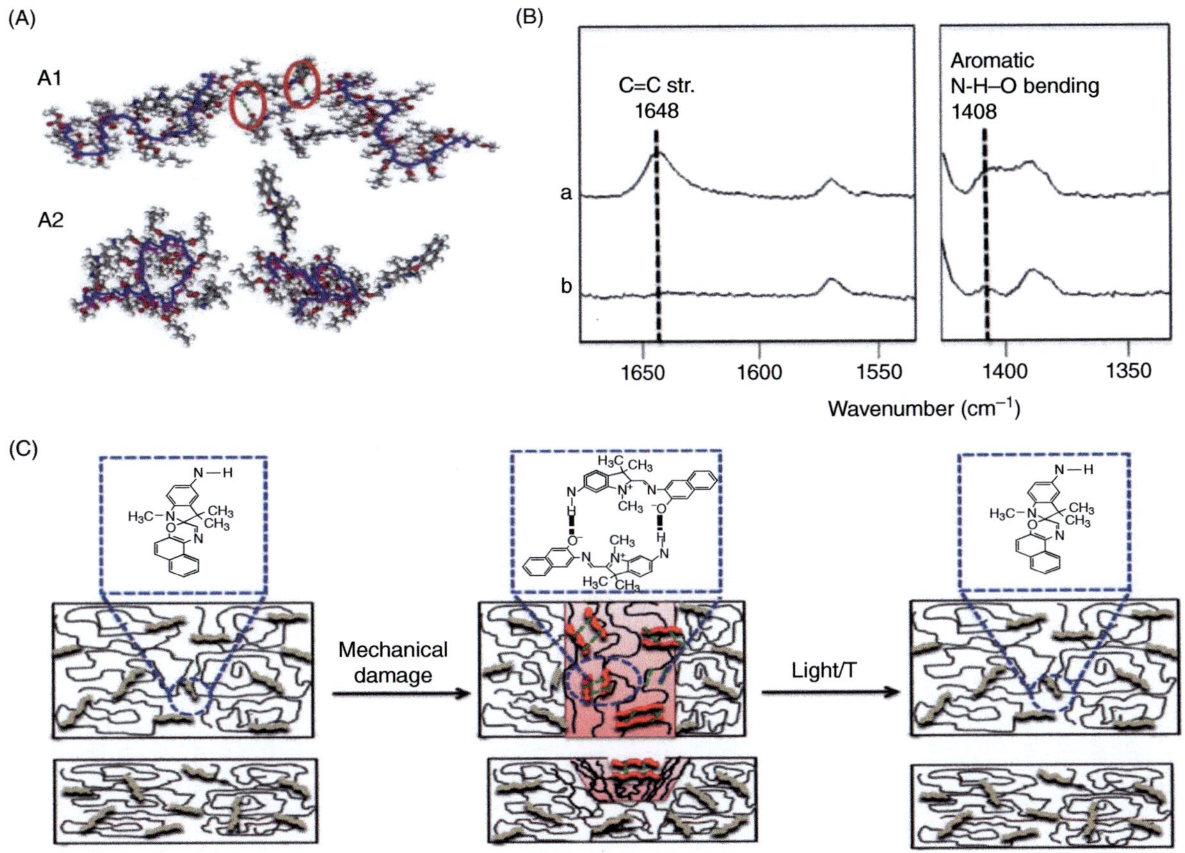

Figure 12.10 (A) Molecular modeling simulations of p(MMA/nBA/SNO) films: mechanically damaged (A1) and exposure to VIS radiation (A2); (B) Raman spectra of p(MMA/nBA/SNO) films: mechanically damaged (a) and exposure to VIS radiation (b); (C) schematic representation of p(MMA/nBA/SNO) films: undamaged, damaged, and exposure to VIS radiation. *Source:* Ramachandran et al. [22].

advances in self-healable polymers. While chemical imaging (IR, Raman) enables detection of localized molecular events associated with self-healing, 2D correlation spectroscopy further amplifies spectroscopic analysis outcomes. On the other hand, AFM, if sufficiently fast, is capable of detecting mass flow and thermal changes during damage–repair cycle if a thermocouple is used as an AFM tip. MD simulations offer significant advantages in modeling self-healing at nanoscale, especially when MD input uses the outcome of experimental data. Combination of these techniques offers unique capabilities in advancing limited knowledge about molecular events that lead to microscopic self-healing. While visualization via optical and electron microscopy as well as mechanical testing are useful and essential tools in monitoring damage–repair cycle, in an effort to advance the field of stimuli-responsive materials in general, and self-healing polymers in particular, molecular probes with sufficiently high spatial and spectroscopic resolutions are needed.

Acknowledgments

This work was partially supported by the Department of Energy under the Award EE0008827, National Science Foundation under Award DMR 2003005, and J. E. Sirrine Foundation Endowment at Clemson University.

References

1 Wang, S. and Urban, M.W. (2020). Self-healing polymers. *Nat. Rev. Mater.*: 1–22.
2 Chen, X. et al. (2002). A thermally re-mendable cross-linked polymeric material. *Science* 295: 1698–1702.
3 Yang, Y. and Urban, M.W. (2014). Self-repairable polyurethane networks by atmospheric carbon dioxide and water. *Angew. Chem. Int. Ed.* 53: 12142–12147.
4 Ghosh, B. and Urban, M.W. (2009). Self-repairing oxetane-substituted chitosan polyurethane networks. *Science* 323: 1458–1460.
5 Imato, K. et al. (2012). Self-healing of chemical gels cross-linked by diarylbibenzofuranone-based trigger-free dynamic covalent bonds at room temperature. *Angew. Chem.* 124: 1164–1168.
6 Cordier, P., Tournilhac, F., Soulié-Ziakovic, C., and Leibler, L. (2008). Self-healing and thermoreversible rubber from supramolecular assembly. *Nature* 451: 977–980.
7 Yan, X. et al. (2018). Quadruple H-bonding cross-linked supramolecular polymeric materials as substrates for stretchable, antitearing, and self-healable thin film electrodes. *J. Am. Chem. Soc.* 140: 5280–5289.
8 Holten-Andersen, N. et al. (2011). pH-induced metal-ligand cross-links inspired by mussel yield self-healing polymer networks with near-covalent elastic moduli. *Proc. Natl. Acad. Sci. U.S.A.* 108: 2651–2655.

9 Rao, Y.-L. et al. (2016). Stretchable self-healing polymeric dielectrics cross-linked through metal–ligand coordination. *J. Am. Chem. Soc.* 138: 6020–6027.

10 Urban, M.W. et al. (2018). Key-and-lock commodity self-healing copolymers. *Science* 362: 220–225.

11 Das, A. et al. (2015). Ionic modification turns commercial rubber into a self-healing material. *ACS Appl. Mater. Interfaces* 7: 20623–20630.

12 Cao, Y. et al. (2017). A transparent, self-healing, highly stretchable ionic conductor. *Advanced Materials* 29: 1605099.

13 White, S.R. et al. (2001). Autonomic healing of polymer composites. *Nature* 409: 794.

14 Corten, C.C. and Urban, M.W. (2009). Repairing polymers using oscillating magnetic field. *Adv. Mater.* 21: 5011–5015.

15 Williams, K.A., Boydston, A.J., and Bielawski, C.W. (2007). Towards electrically conductive, self-healing materials. *J. Roy .Soc. Interf.* 4: 359–362.

16 Davydovich, D. and Urban, M.W. (2020). Water accelerated self-healing of hydrophobic copolymers. *Nat. Commun.* 11: 5743.

17 Otts, D.B., Zhang, P., and Urban, M.W. (2002). High fidelity surface chemical imaging at 1000 nm levels: internal reflection IR imaging (IRIRI) approach. *Langmuir* 18: 6473–6477.

18 Yang, Y., Davydovich, D., Hornat, C.C. et al. (2018). Leaf-inspired self-healing polymers. *Chem* 4: 1928–1936.

19 Noda, I. (1993). Generalized two-dimensional correlation method applicable to infrared, Raman, and other types of spectroscopy. *Appl. Spectrosc.* 47: 1329–1336.

20 Hornat, C.C. and Urban, M.W. (2020). Entropy and interfacial energy driven self-healable polymers. *Nat. Commun.* 11: 1028.

21 Ghosh, B., Chellappan, K.V., and Urban, M.W. (2011). Self-healing inside a scratch of oxetane-substituted chitosan-polyurethane (OXE-CHI-PUR) networks. *J. Mater. Chem.* 21: 14473–14486.

22 Ramachandran, D., Liu, F., and Urban, M.W. (2012). Self-repairable copolymers that change color. *RSC Adv.* 2: 135–143.

13

Recent Application of Vibrational Spectroscopy to Conjugated Conducting Polymers

Yukio Furukawa

Waseda University, Graduate School of Advanced Science and Engineering, Department of Chemistry and Biochemistry, 3-4-1 Okubo, Shinjuku-ku, Tokyo, 169-8555, Japan

13.1 Introduction

The 2000 Nobel Prize in Chemistry was awarded to Drs. A.J. Heeger, A.G. MacDiarmid, and H. Shirakawa for the discovery and development of electrically conductive polymers, such as polyacetylene, polythiophene (PT), and polypyrrole (http://nobelprize.org/prizes/chemistry/2000/summary). The chemical structures of some conducting polymers are shown in Figure 13.1. They have conjugated π electrons, exhibiting semiconductor behavior. When a polymer film is doped with an electron acceptor, such as iodine or $FeCl_3$, or an electron donor, such as Na, it exhibits high electrical conductivity. Upon acceptor doping, the polymer chains become positively charged and form complexes with dopant anions such as I_3^- or $FeCl_4^-$. Upon donor doping, the polymer chains become negatively charged and form complexes with donor cations such as Na^+. The electrical conductivity σ can be expressed as the product of the carrier density n, the charge of a carrier q, and the mobility of the carrier μ. The carrier density and mobility are important quantities for characterizing polymer semiconductors. These organic compounds are applicable as the active materials of thin-film transistors [1], light-emitting diodes [2], and solar cells [3].

Primary polymer films are insoluble in organic solvents. These insoluble polymers have poor applications. Polymers with large substituents have been synthesized that are soluble in organic solvents. Soluble polymers have been used as active materials for organic polymer electronic devices. In particular, polymer solar cells [4] have attracted much attention because renewable solar energy is expected to be used instead of fossil fuels to solve the global warming problem. Polymer semiconductors have advantages in producing flexible devices via low-cost fabrication processes such as printing and solution casting. Regioregular poly(3-hexylthiophene) (P3HT) (Figure 13.1c) is one such conjugated polymer that demonstrates high performance in polymer transistors and solar cells; indeed, a variety of polymers with complex chemical structures demonstrate high performance in polymer electronic devices. The electrical properties of conjugated polymers

Spectroscopic Techniques for Polymer Characterization: Methods, Instrumentation, Applications, First Edition. Edited by Yukihiro Ozaki and Harumi Sato.

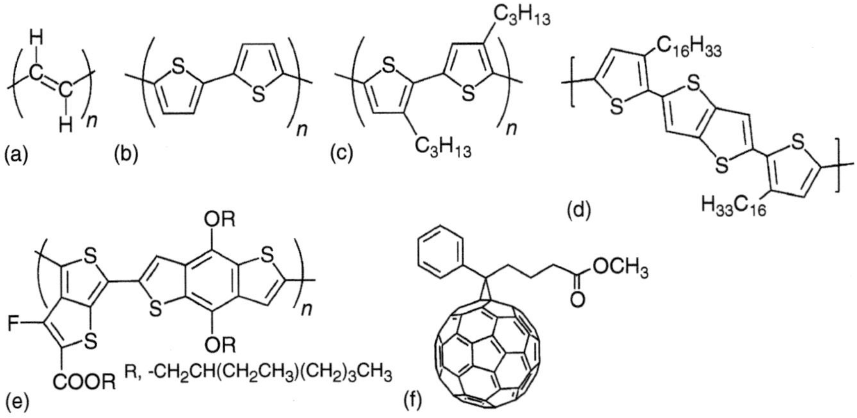

Figure 13.1 Chemical structures of conjugated polymers: (a) *trans*-polyacetylene; (b) polythiophene; (c) regioregular poly(3-hexylthiophene) (P3HT); (d) poly(2,5-bis (3-hexadecylthiophen-2-yl)thieno[3,2-*b*]thiophene) (PBTTT-C16); (e) poly[[4,8-bis [(2-ethylhexyl)oxy]benzo[1,2-*b*:4,5-*b*′]dithiophene-2,6-diyl][3-fluoro-2-[(2-ethylhexyl) carbonyl]thieno[3,4-*b*]thiophenediyl]] (PTB7); (f) [6,6]-phenyl-C_{61}-butyric acid methyl ether (PCBM).

are considered to be associated with the crystallinity of the polymers. Poly(2,5-bis (3-hexadecylthiophen-2-yl)thieno[3,2-*b*]thiophene) (PBTTT-C16) is a conjugated polymer with two thiophene rings with long side chains and one thienothiophene ring as a repeating unit, as shown in Figure 13.1d. This polymer has a large positive carrier mobility owing to its high crystallinity and is used in transistors because of its good solubility.

A new type of transistor fabricated with a semiconductor as the active layer and an electrolyte as the gate dielectric is called an electrolyte-gated transistor (EGT) [5]. The electrolyte compounds contain ionic liquids. In this review, a transistor fabricated with a polymer semiconductor and an ionic liquid as the gate dielectric [6] is called an ionic-liquid-gated transistor (ILGT). An ILGT has the advantages of a high drain current and a low operating gate voltage because of the high charge density ($\sim 10^{21}$ cm^{-3}) induced by the high capacitance of the ionic-liquid gate dielectric. Application of a gate voltage can lead to the formation of carriers in the semiconductor channel region upon electrochemical and/or electrostatic doping. Positive and negative carriers are generated by negative and positive gate voltages, respectively. The charge density and doping level can be obtained as a function of the gate voltage over a wide range. The electrical conductivity can be obtained as a function of the doping level using the relation between the drain current and the drain voltage. The carrier mobility can also be obtained. Thus, an ILGT is very useful for studying the electrical properties of polymer semiconductors.

It is distinctive of conjugated polymers that two kinds of carriers exist: singly charged polarons (charge $\pm e$, spin 1/2) and doubly charged bipolarons (charge $\pm 2e$, no spin) [7]. The formation of a polaron or a bipolaron on a polymer chain is associated with structural changes. These structural changes can be detected using infrared (IR) and Raman (vibrational) spectroscopy. Doping-induced infrared active vibrational (IRAV) bands were observed and explained using the effective conjugation coordinate theory proposed by Zerbi [8]. Although carriers give rise to strong IRAV bands, polarons and bipolarons were not identified using infrared spectroscopy. We presented a review concerning the Raman characterization of polarons and bipolarons that are formed upon chemical doping [9]. The Raman spectra of polyaniline [10], poly(*p*-phenylenevinylene) [11–13], poly(*p*-phenylene) [14, 15], polythiophene (PT) [16], P3HT [17], and PBTTTC16 [18] have been analyzed within the context of polarons and bipolarons.

Raman spectroscopy imaging has emerged as a powerful tool for the characterization of materials and the spatial distribution of chemical species in various research fields, such as biology, medicine, pharmacology, and materials science, with a spatial resolution of approximately 1 μm. In polymer electronic devices, the carrier distribution in an active area is essential to the function of devices. Raman imaging studies of electrochemical memories [19] and ILGTs [20] were reported. It is important to study polarons and bipolarons in detail to improve the performance of polymer electronic devices.

Solar cells fabricated with conjugated polymers are flexible, light, and environment friendly. The power conversion efficiencies of polymer solar cells were low. However, recently, a high 18% power conversion efficiency was reported [21]. In polymer semiconductors, the diffusion lengths of excited states are on the order of 10 nm, which is very short. Thus, a blend between a polymer and an acceptor, such as [6,6]-phenyl-C_{61}-butyric acid methyl ether (PCBM), is used as an active material. Such a blend is called a bulk heterojunction. Recently developed femtosecond stimulated and inverse Raman spectroscopy is useful for studying the ultrafast dynamics of bulk heterojunction films [22, 23]. Time-resolved Fourier-transform infrared (FT-IR) spectroscopy is useful for studying the microsecond dynamics of photogenerated long-lived carriers [24].

In this review, first, we present the electronic and vibrational properties of conjugated polymers, polarons, and bipolarons. Second, we present studies on ILGTs fabricated with P3HT and PBTTT-C16. We can determine the relation between the doping levels, carrier types (polarons or bipolarons), and carrier mobilities using the ILGT configurations. Third, we present femtosecond dynamics of excited states and polarons using near-infrared (NIR) inverse Raman spectroscopy and microsecond dynamics of long-lived polarons using time-resolved FT-IR spectroscopy.

13.2 Carriers

Polarons and bipolarons serve as carriers in polymers with a nondegenerate ground state, such as polythiophene (Figure 13.1b) [25–28]. Upon acceptor doping, the polymer chains become positively charged and form complexes with dopant anions such as I_3^- or $FeCl_4^-$. When an electron is removed from a polymer chain during this process, the positive charge $+e$ becomes localized over several thiophene rings, and structural changes occur in this region. This charge carrier is called a positive polaron; it has charge $+e$ and spin 1/2. Positive polarons correspond to radical

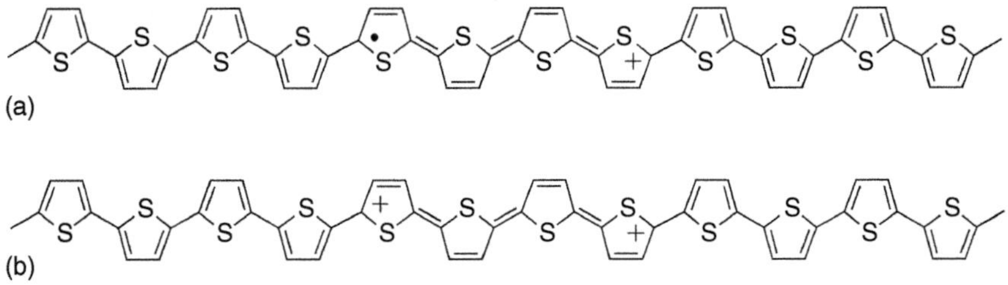

Figure 13.2 Chemical structures of (a) a positive polaron and (b) a positive bipolaron in polythiophene.

cations in chemical terminology. When an electron is subsequently removed from a positive polaron, a positive bipolaron, with charge $+2e$ and no spin, is formed. The bipolaron is a spinless carrier. Positive bipolarons correspond to dications. The chemical structures of positive polarons and positive bipolarons in P3HT are schematically illustrated in Figure 13.2. In the first stage of doping, positive polarons are formed. As the doping level increases, positive bipolarons are formed [29]. The doping level at the polaron-to-bipolaron transition is called the critical doping level x_c, which depends on the kind of conjugated polymer. In the case of donor doping, negative polarons (charge $-e$, spin 1/2) and negative bipolarons (charge $-2e$, no spin) are formed. Negative polarons and negative bipolarons correspond to radical anions and dianions, respectively.

The electronic structures based on one-electron theory [30] for a neutral polymer, a positive polaron, and a positive bipolaron are schematically shown in Figure 13.3. In an infinite polymer chain, the interaction between repeating units leads to the formation of the valence band (VB) and the conduction band (CB) as shown in Figure 13.3a. A bandgap exists. For a polaron or a bipolaron, two localized electronic levels ($\pm\varepsilon$ for the polaron and $\pm\varepsilon'$ for the bipolaron) form within the bandgap, as shown in Figure 13.3b,c. A positive polaron is expected to have the following intragap transitions since one electron is removed from the $-\varepsilon$ level:

(1) P_1 transition: $-\varepsilon$ level $\leftarrow$ valence band
(2) P_2 transition: $+\varepsilon$ level $\leftarrow$ $-\varepsilon$ level
(3) P_3 transition: $+\varepsilon$ level $\leftarrow$ valence band and conduction band $\leftarrow$ $-\varepsilon$ level

A positive bipolaron is expected to have the following two intraband transitions because two electrons are removed from the $-\varepsilon'$ level:

(1) BP_1 transition: $-\varepsilon'$ level $\leftarrow$ valence band
(2) BP_3 transition: $+\varepsilon'$ level $\leftarrow$ valence band

The P_3 transition of positive polarons and the BP_3 transition of positive bipolarons are very weak or forbidden. Thus, polarons should exhibit two strong absorption bands from the VIS region to the IR region, whereas bipolarons should exhibit one strong band. Shimoi et al. [31] theoretically showed the two-band feature of the polaron and the one-band feature of the bipolaron in the presence of electron–electron interactions.

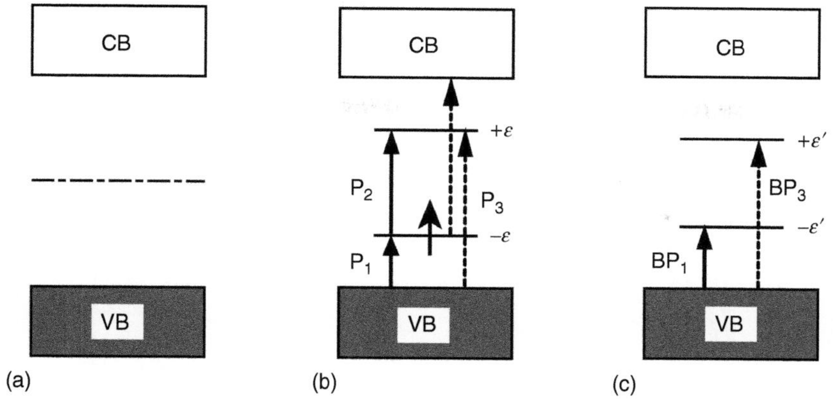

Figure 13.3 Schematic electronic structures. (a) Neutral polymer; (b) positive polaron; (c) positive bipolaron. CB, conduction band; VB, valence band.

A radical cation (anion) and a dication (dianion) of short oligomers correspond to a positive (negative) polaron and a positive (negative) bipolaron, respectively. Optical absorption studies of short oligomers have provided support for this paradigm, revealing two strong bands, P_1 and P_2, for polarons and a single strong band, BP_1, for bipolarons. The assignments of the observed optical absorptions were previously controversial regarding the intensities of the forbidden bands, which were described in a previous review [32].

A polaron can be detected via electron spin resonance (ESR) spectroscopy because a polaron has spin 1/2. By contrast, a bipolaron cannot be detected via ESR because a bipolaron has no spin. However, a singlet polaron pair also has no spin. It is difficult to distinguish a bipolaron from a singlet polaron pair using ESR alone.

13.3 Optical Absorption Spectra Upon Chemical Doping

13.3.1 P3HT

Figure 13.4 shows the changes in the VIS/NIR absorption spectrum of a P3HT film that occur upon $FeCl_3$ solution doping [17]. The spectrum of the neutral P3HT film is presented as spectrum 4a. The absorption at 520 nm is assigned to the π–π* interband transition of a conjugated polymer chain. Absorption spectra 4b–4e represent the spectra of the P3HT film at various doping levels. The doping level increases from 4b to 4e. As the doping level is increased, initially, the 520-nm absorption band of P3HT decreases, and new absorption bands appear at 788 nm (1.57 eV) and in the NIR range; subsequently, the 788-nm absorption band decreases slightly, and the NIR absorption band increases slightly. Upon additional doping, no further spectral changes are observed. The 788-nm band is assigned to the P_2 transition of positive polarons, and the NIR absorption is assigned to the P_1 transition. A comparison of spectra 4d and 4e reveals a decrease in the 788-nm absorption band and a simultaneous increase in the NIR absorption band. This spectral change originates from a

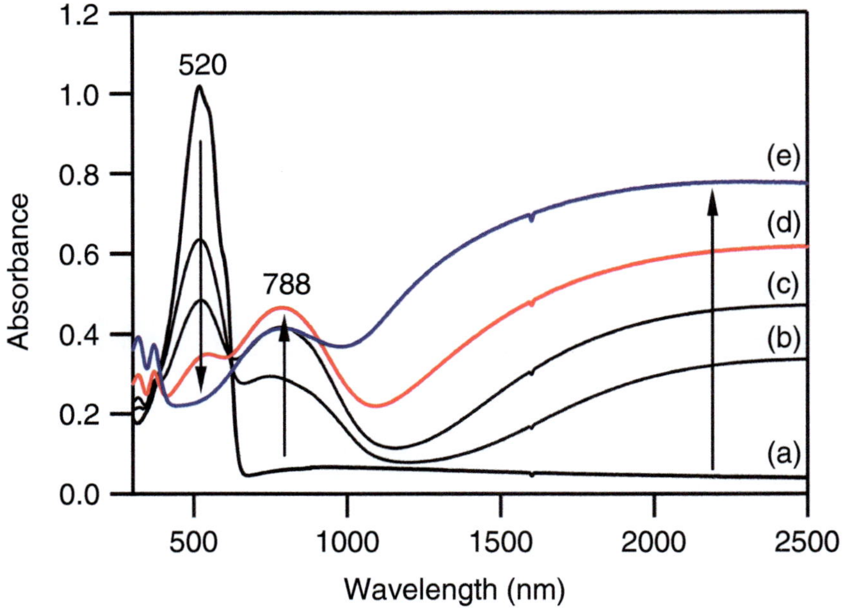

Figure 13.4 Changes in the VIS/NIR absorption spectrum of a P3HT film doped with an acetonitrile solution of $FeCl_3$. Source: Yamamoto and Furukawa [17].

decrease in the number of positive polarons and an increase in the number of positive bipolarons.

Figure 13.5 shows the changes in the VIS/NIR absorption spectrum of a P3HT film that occur upon $FeCl_3$ vapor doping [17]. As the doping proceeds, the absorption band at 520 nm associated with the π–π* transition decreases. Meanwhile, a new band appears at 790 nm (1.57 eV) and subsequently disappears. The strength of the absorption in the NIR range, however, increases monotonically. It should be noted that the 790-nm band disappears in the case of vapor doping, whereas this band persists during solution doping. This result suggests that the maximum doping level that can be achieved through vapor doping is higher than that achievable through solution doping. In the case of vapor doping, positive polarons form during the initial stage of doping, as evidenced by the increasing strength of the 790-nm P_2 transition. However, during the latter stage of doping, positive bipolarons form; at the maximum doping level, only positive bipolarons exist, as indicated by the absence of the 790-nm P_2 absorption band and the increase in the NIR absorption band. As the $FeCl_3$ doping proceeds, the intensity of the 520-nm absorption decreases. This originates from the decreases in the energy level densities of the VB and the CB.

13.3.2 Poly(2,5-bis(3-hexadecylthiophene-2-yl)thieno[3,2-*b*]thiophene) (PBTTT-C16)

The optical absorption spectral changes of an as-prepared PBTTT-C16 film upon $FeCl_3$ vapor doping are shown in Figure 13.6 [17]. The dopant content increases

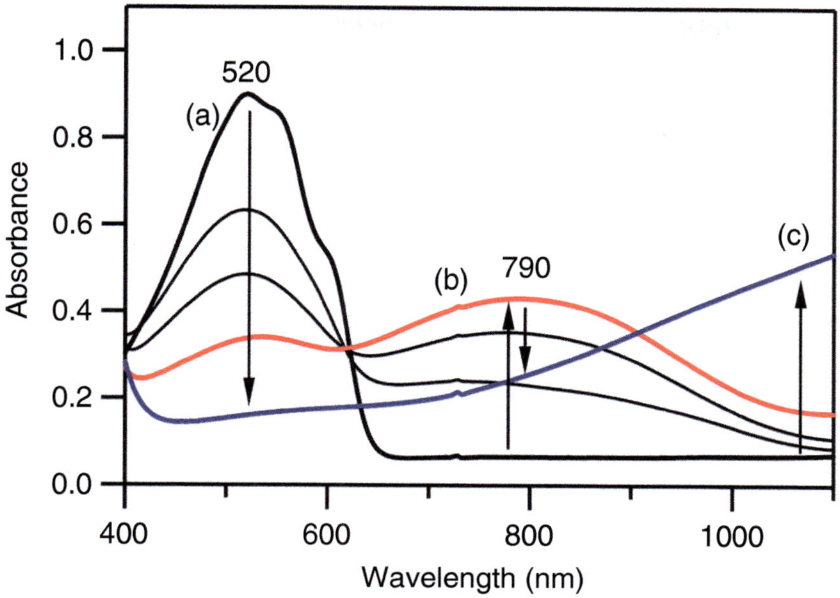

Figure 13.5 Changes in the VIS–NIR absorption spectrum of a P3HT film doped with $FeCl_3$ vapor: (a) neutral P3HT film; (b) after 200 minutes; (c) after 20 hours. Source: Yamamoto and Furukawa [17].

Figure 13.6 Optical absorption spectral changes of a PBTTT-C16 film upon $FeCl_3$ vapor doping. The sharp bands below 0.5 eV originate from water vapor. Source: Enokida and Furukawa [18].

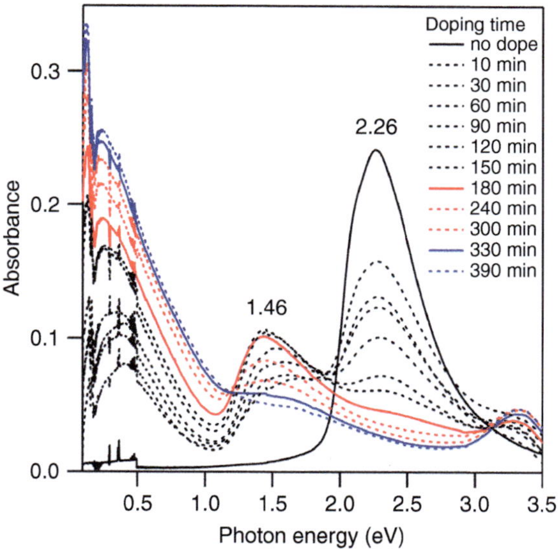

with increasing doping time. The pristine PBTTT-C16 film has an absorption maximum at 2.26 eV, which can be assigned to the transition from the VB to the CB. Upon $FeCl_3$ doping, two new bands appear at approximately 0.41 and 1.71 eV. The intensities of these bands increase and the peak positions redshift as the doping time increases. These two bands are assigned to the P_1 and P_2 transitions, respectively, originating from a polaron. After 150 minutes, the intensity of the P_2 band at 1.46 eV reaches a maximum; the P_1 band intensity shows a maximum at 0.24 eV. As the doping time increases further, the intensity of the band at 1.46 eV decreases with a concomitant increase in the intensity of the band at 0.24 eV. These spectral changes are attributed to the formation of a bipolaron from two polarons. At 330 minutes, the band at 1.46 eV almost disappears, whereas a band appears at 0.23 eV. This band is assigned to the BP_1 transition of bipolarons. These results indicate that the dominant carriers are polarons until 150 minutes, polarons and bipolarons between 180 and 300 minutes, and bipolarons between 330 and 390 minutes.

13.4 Raman Spectra of Positive Polarons and Bipolarons Generated Upon Chemical Doping

13.4.1 P3HT

The Raman spectra of the neutral P3HT film, positive polarons formed via oxidative doping with a CH_3CN solution of $FeCl_3$, and positive bipolarons formed via oxidative doping with gaseous $FeCl_3$ are presented in Figure 13.7a–c, respectively [33].

In the Raman spectrum of P3HT, the strong band observed at 1448 cm^{-1} is assigned to the C=C stretching vibration of thiophene rings. The 1383-cm^{-1} band is assigned to the C–C stretching vibration of thiophene rings. In the Raman spectrum of the positive polarons, the band at 1412 cm^{-1} is assigned to the CC stretching vibration of polarons, corresponding to the 1448-cm^{-1} C=C stretching mode of neutral P3HT. The lower wavenumber for the positive polaron bands is consistent with

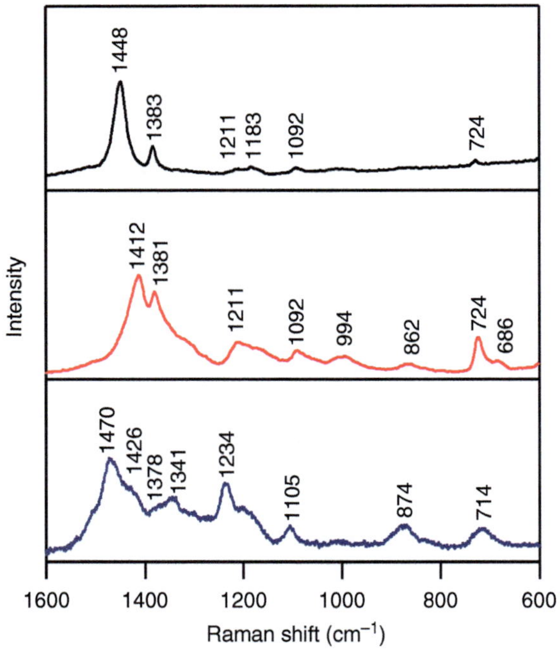

Figure 13.7 Raman spectra of (a) neutral P3HT, (b) chemically formed positive polarons, and (c) chemically formed positive bipolarons under excitation at 830 nm. Source: Yamamoto and Furukawa [33].

the extraction of an electron from the bonding orbital of the conjugated thiophene rings. The 1381-cm^{-1} band of positive polarons is assigned to the CC stretching mode, corresponding to the 1383-cm^{-1} CC stretching mode of neutral P3HT.

The Raman spectrum of the positive bipolarons is different from that of the positive polarons. The strongest band for positive bipolarons is observed at 1470 cm^{-1} in Figure 13.7a, whereas the strongest band for positive polarons is observed at 1412 cm^{-1} in Figure 13.7b, and the strongest band for neutral P3HT is observed at 1448 cm^{-1} in Figure 13.7a. The wavenumber of the bipolaron band is higher than that of the band for neutral P3HT, whereas the wavenumber of the polaron band is lower than that of the band for neutral P3HT. The high wavenumber of the bipolaron band (1470 cm^{-1}) is quite unusual and may be explained as follows [17]: The Raman spectrum of the positive bipolarons (excited at 830 nm) is resonant with the BP_1 bipolaron transition, whereas the corresponding spectrum of the positive polarons is resonant with the P_2 transition. These spectra reflect different resonance effects. Thus, the vibrational mode of the 1470-cm^{-1} band is likely different from the modes of the 1448-cm^{-1} band of neutral P3HT and the 1412-cm^{-1} band of the positive polarons. Different vibrational modes are most likely observed under different resonant conditions.

The 1412-cm^{-1} band (ring C=C stretching vibration), 1381-cm^{-1} band (CC stretching vibration), and 724-cm^{-1} band (CSC ring deformation) can be used as markers of polarons. The 1470-cm^{-1} band (CC stretching vibration) and 1234-cm^{-1} band (inter-ring CC stretching vibration) can be used as markers of bipolarons. The 1448-cm^{-1} band (C=C stretching vibration) can be used as a marker of neutral P3HT.

13.4.2 PBTTT-C16

The 785-nm excited Raman spectra of the PBTTT-C16 film were obtained simultaneously with the optical absorption spectral measurements. The Raman spectral changes of the film upon $FeCl_3$ doping are shown in Figure 13.8 [18]. The excitation wavelength of 785 nm (1.58 eV) is located within the P_2 absorption band attributable to the presence of polarons. It should be noted that the intensity of a Raman band attributable to polarons will be enhanced by the resonance Raman effect. In other words, the cross section of a band attributable to polarons is larger than that of bands attributable to a pristine polymer and bipolarons with excitation at 785 nm. The pristine film shows bands at 1490, 1461, 1417, and 1393 cm^{-1}. The band at 1393 cm^{-1} is assigned to a stretching mode of the thienothiophene ring [34].

The optical absorption spectra described above indicate that polarons are the dominant carriers up to 150 minutes, the formation of bipolarons begins at 180 minutes, and bipolarons are the dominant carriers from 330 to 390 minutes. On the basis of these results, one can assign the Raman bands that appear upon $FeCl_3$ doping. Figure 13.8 shows the large Raman spectral changes that are observed upon $FeCl_3$ doping. At 30 minutes, i.e. when polarons are dominant, new bands clearly appear at 1457, 1432, 1385, 1208, 744, and 692 cm^{-1}. These bands are attributed to polarons. All the bands observed at 150 minutes are also attributed to polarons. The bands at 1457 and 1385 cm^{-1} show small downward shifts. The band at 1385 cm^{-1} can be assigned

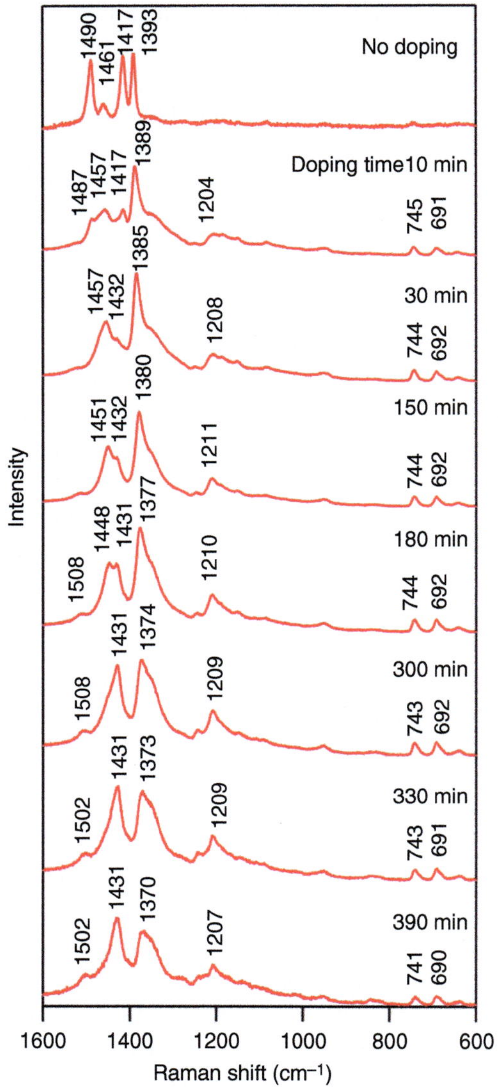

Figure 13.8 Raman spectra of neutral PBTTT-C16 upon $FeCl_3$ doping. Excitation wavelength, 785 nm. Source: Enokida and Furukawa [18].

to a stretching mode of the thienothiophene ring [34]. After 180 minutes, the intensity of the band at 1431 cm^{-1} largely increases. At 390 minutes, i.e. when bipolarons are dominant, the bands observed at 1502, 1431, 1370, 1207, 741, and 690 cm^{-1} can be attributed to bipolarons. The intensity of the band at 1431 cm^{-1} is stronger than that of the band at 1370 cm^{-1}. The band at 1370 cm^{-1} is assignable to a stretching mode of the thienothiophene ring.

Raman spectral features in the range between 1500 and 1300 cm^{-1} can be used to identify the types of carriers in PBTTT-C16. Polarons have the strongest band at 1385–1380 cm^{-1} and a strong band at 1457–1451 cm^{-1} with a 1432-cm^{-1} shoulder. On the other hand, bipolarons have the strongest band at 1431 cm^{-1} and a strong band at 1373–1370 cm^{-1}.

13.5 Carriers and Electrical Properties Based on ILGTs

13.5.1 ILGTs

The device structure of an ILGT is depicted in Figure 13.9, whereas various device structures of ILGTs have been reported [5]. The chemical structures of some ionic liquids are shown in Figure 13.10. The source and drain electrodes were formed by the vacuum deposition method using a shadow mask. A 5 nm adhesion layer of Ni was deposited followed by 50 nm of Au on a glass substrate. A P3HT film was prepared from a chlorobenzene solution of P3HT (24 mg ml^{-1}) using the spin-coating method. The thickness of the P3HT film was 1.0×10^2 nm. The width and length of

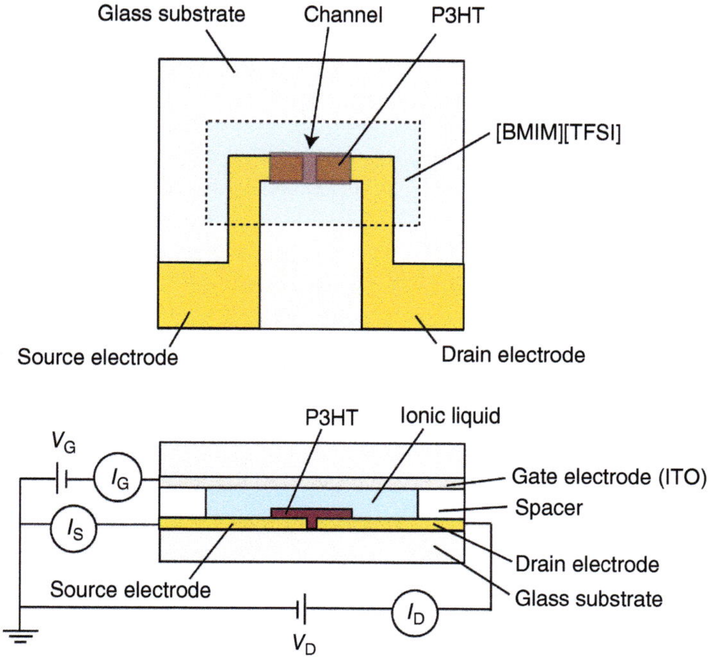

Figure 13.9 Device structure of the ILGT fabricated with P3HT and [BMIM][TFSI]. V_G, gate voltage; I_G, gate current; V_D, drain voltage; I_D, drain current; I_S, source current. Source: Yamamoto and Furukawa [33].

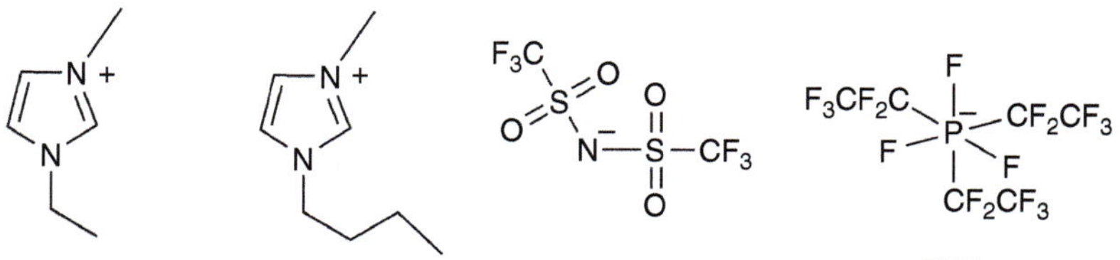

EMIM BMIM TFSI FAP

Figure 13.10 Chemical structures of some ionic liquids.

the channel of the transistor were 6.0×10^2 and 7.14×10 μm, respectively. The ionic liquid 1-butyl-3-methylimidazolium bis(trifluoromethylsulfonyl)imide ([BMIM][TFSI]) was used as the electrolyte. An indium-tin-oxide (ITO)-coated glass substrate was used as the gate electrode. The thickness of the [BMIM][TFSI] layer was set to 1.25×10^2 μm using a polyethylene terephthalate spacer.

Using the ILGT configuration, one can obtain the carrier mobilities as follows [33, 35]: The type of carriers, polarons or bipolarons, can be determined using Raman spectroscopy. A positive polaron has charge $q = e$, and a bipolaron has charge $q = 2e$. The injected charge density n (cm^{-3}) at each V_G can be obtained by integration of the gate current I_G; the doping level x mol% per π electron at V_G can be calculated from the n value and density of the film. The electrical conductivity σ ($S\,cm^{-1}$) at each V_G can be calculated from I_D, V_D, and the size of the channel. The carrier mobility μ ($cm^2\,V^{-1}\,s^{-1}$) can be calculated from σ, n, and q because $\sigma = nq\mu$.

13.5.2 Raman Spectra of ILGTs Fabricated with P3HT

Figure 13.11 illustrates the relation between $-I_D$ and $-V_G$ at $V_D = -100$ mV for an ILGT fabricated with P3HT and [BMIM][TFSI] [33]. When V_D is constant, $-I_D$ is proportional to the electrical conductivity of P3HT. $-I_D$ increases at 1.3 V and exhibits a quasi-plateau between 1.8 and 2.7 V. It then begins to drop at $-V_G = 2.8$ V. Paulsen and Frisbie [6] reported that $-I_D$ exhibits a peak at $-V_G = 2.0$ V. It should be noted that the value of V_G depends on the size and structure of the electrochemical transistor. For both results, $-I_D$ decreases at high $-V_G$ values. This drop is detrimental to the transistor. We studied the origin of this drop using Raman spectroscopy and mobilities obtained from electrochemical measurements.

The 785-nm excited Raman spectra of an ILGT with the ITO/[BMIM][TFSI]/P3HT structure as a function of $-V_G$ from 0.0 to 3.5 V are presented in Figure 13.12. The drain electrode is connected to the source electrode. The 785-nm laser line is located within the 780-nm absorption band of the polarons. The spectrum at 0.0 V is attributed to neutral P3HT. Based on the Raman spectra of polarons and bipolarons generated by $FeCl_3$ chemical doping (oxidation), the spectra at other voltages can be attributed to electrochemically oxidized species, polarons or bipolarons. The Raman

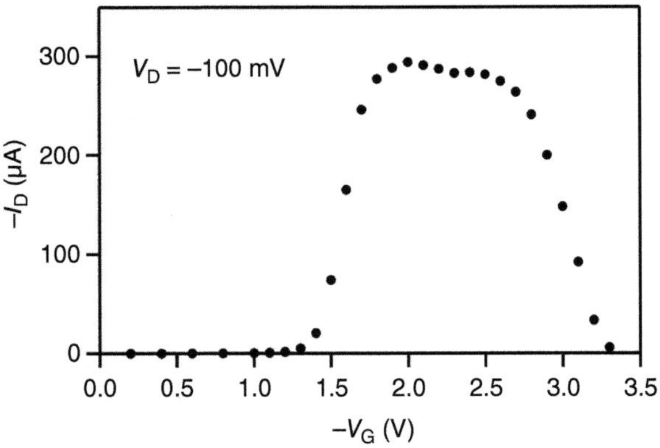

Figure 13.11 Relation between $-I_D$ and $-V_G$ at $V_D = -100$ mV for a P3HT ILGT. Source: Yamamoto and Furukawa [33].

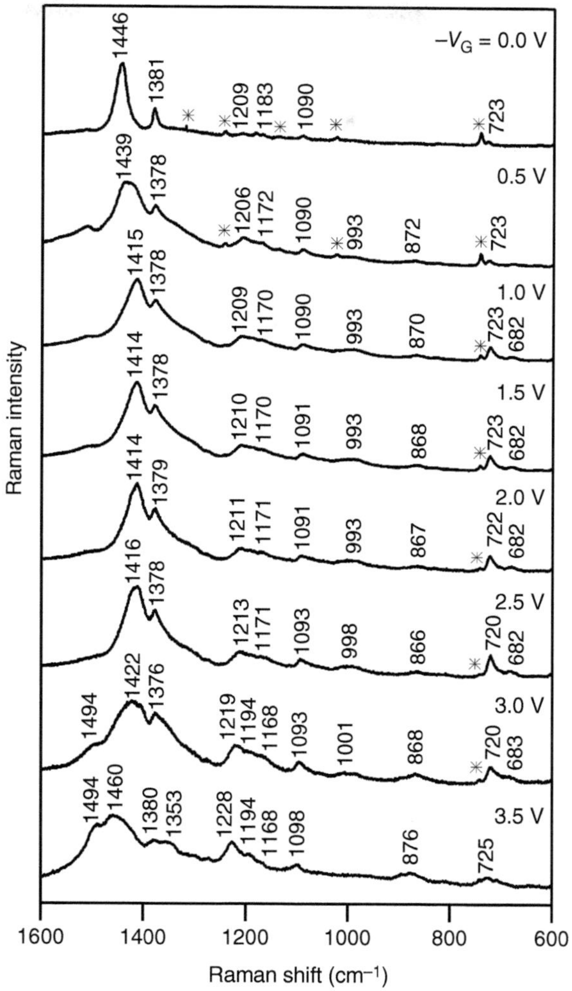

Figure 13.12 Raman spectra of the ITO/[BMIM][TFSI]/P3HT structure as a function of $-V_G$. The excitation wavelength was 785 nm. The bands with * originate from [BMIM][TFSI]. Source: Yamamoto and Furukawa [33].

spectra of the neutral P3HT film, positive polarons formed via oxidative doping with a CH_3CN solution of $FeCl_3$, and positive bipolarons formed via oxidative doping with gaseous $FeCl_3$ are presented in Figure 13.7a–c, respectively. The 1412-cm^{-1} band (CC stretching vibration), 1381-cm^{-1} band (CC stretching vibration), and 724-cm^{-1} band (CSC ring deformation) can be used as markers of polarons [17]. The 1470-cm^{-1} band (CC stretching vibration) and 1234-cm^{-1} band (interring CC stretching vibration) can be used as markers of bipolarons [17]. The 1448-cm^{-1} band (C=C stretching vibration) can be used as a marker of neutral P3HT.

The observed spectra at $-V_G$ = 1.0–2.5 V can be attributed to positive polarons because these spectra are essentially similar to the spectrum shown in Figure 13.7b. A small wavenumber upshift from 1414 to 1416 cm^{-1} and a broadening of the 1416-cm^{-1} band are observed in the spectrum at $-V_G$ = 2.5 V. These features likely reflect the formation of a small number of bipolarons. The observed spectrum at $-V_G$ = 3.5 V can be attributed to bipolarons because the spectrum is essentially similar to that shown in Figure 13.7c. The spectrum at $-V_G = 0.5$ V can be attributed to a superimposition of the bands of neutral P3HT and polarons. The spectrum at $-V_G = 3.0$ V can be attributed to a superimposition of the polaron and bipolaron bands.

The doping level per π electron x (mol% per π electron) was calculated from the incorporated charges. Positive polarons are formed below $x = 6.75$ mol% ($V_G = -2.5$ V, $n = 1.1 \times 10^{21}$ cm^{-3}). At $x = 9.25$ mol% ($V_G = -3.0$ V, $n = 1.5 \times 10^{21}$ cm^{-3}), positive polarons and bipolarons coexist. Thus, $6.75 < x_c < 9.25$ holds, where x_c is the critical concentration of the polaron-to-bipolaron transition. The coexistence can be explained by considering that polarons are stable at structural defects in the P3HT film, i.e. for low crystallinity of the P3HT film. The conductivity and mobility of polarons are shown in Figure 13.13. The highest mobility of polarons is 0.31 cm^2 V^{-1} s^{-1} at $x = 3.75$ mol%. The mobility of positive bipolarons is 2.5×10^{-3} cm^2 V^{-1} s^{-1}. This is much smaller than that of polarons.

13.5.3 Raman Spectra of ILGTs Fabricated with PBTTT-C16

ILGTs fabricated with unannealed and annealed PBTTT-C16 films and [BMIM][TFSI] were studied [18]. The relation between $-I_D$ and $-V_G$ at $V_D = -0.01$ V for the ILGT fabricated with an unannealed PBTTT-C16 film is shown in Figure 13.14 [18]. $-I_D$ increases gradually in the $-V_G$ range between 0.0 and 1.0 V. On the other hand, $-I_D$ increases rapidly from 1.0 to 1.4 V and shows a plateau from 1.4 to 2.0 V. In this case, the current drop that was reported for the P3HT ILGT is not observed. The increase in $-I_D$ indicates the formation of carriers in the channel region. Electrochemical measurements provide us with information about carrier formation in the channel. However, the types of generated carriers cannot be determined by electrochemical measurements.

Figure 13.15 shows the V_G dependence of the 785-nm excited Raman spectrum of the unannealed ILGT [18]. The observed spectra originate from the PBTTT-C16

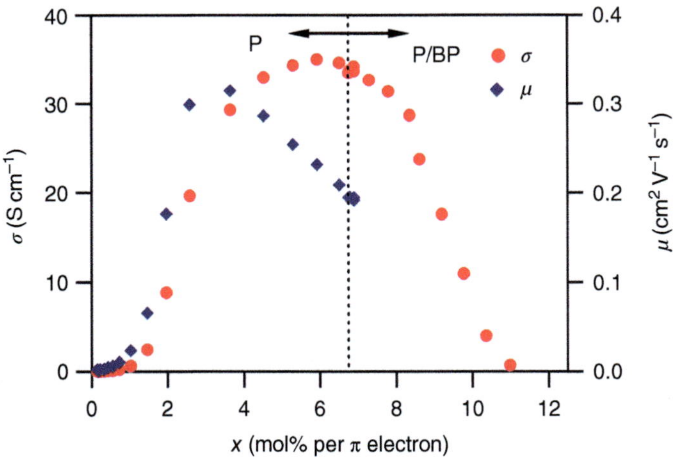

Figure 13.13 Conductivity and mobility as a function of doping level.

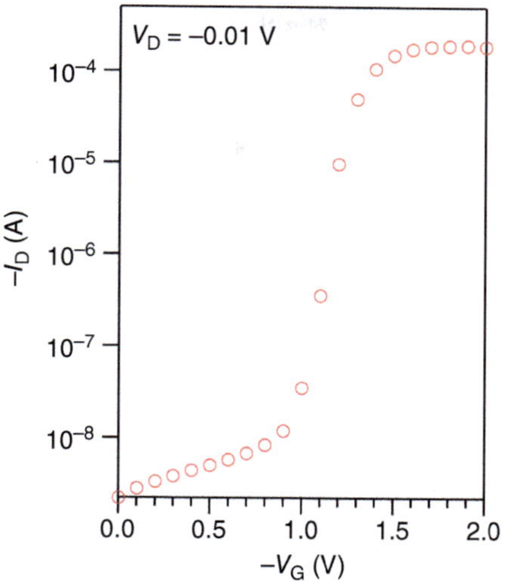

Figure 13.14 Relation between $-I_D$ and $-V_G$ at $V_D = -0.01$ V for an unannealed PBTTT-C16 ILGT. Source: Enokida and Furukawa [18].

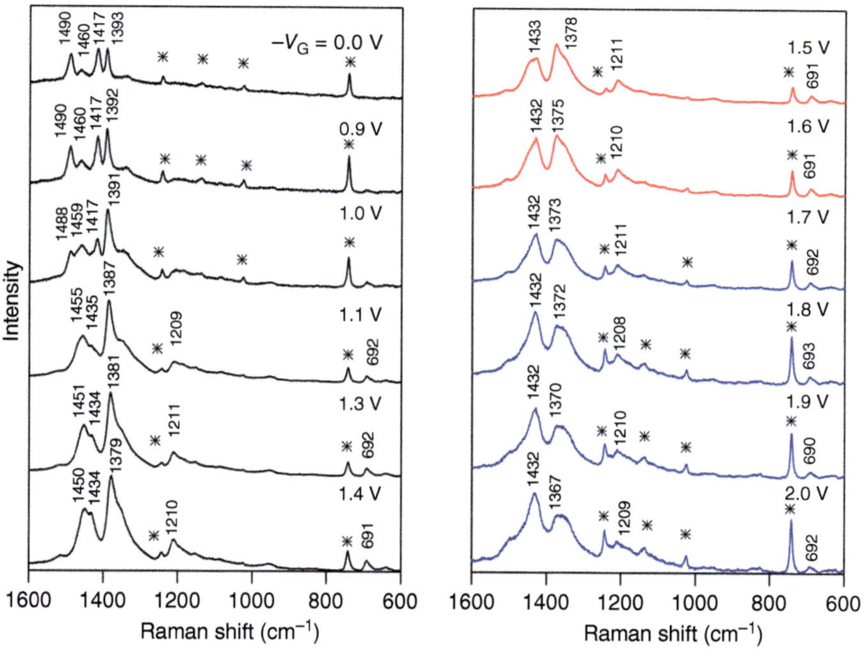

Figure 13.15 Raman spectral changes of an unannealed PBTTT-C16 ILGT upon application of $-V_G$. Bands with * are attributed to [BMIM][TFSI]. Excitation wavelength, 785 nm. Source: Enokida and Furukawa [18].

channel region of the device. The type of carrier can be identified on the basis of the spectra of polarons and bipolarons generated upon $FeCl_3$ doping. The spectrum at $V_G = 0.0$ V is attributed to the neutral PBTTT-C16 polymer. At $-V_G = 0.9$ and 1.0 V, the intensity of the 1392-cm^{-1} band becomes strong. These observed spectra are attributed to polarons and the neutral polymer. The spectra between $-V_G = 1.1$ and

1.4 V are attributable to polarons. For $-V_G = 1.4$ and 1.7 V, the intensity at approximately 1432 cm^{-1} increases. The spectra at $-V_G = 1.5$ and 1.6 V can be attributed to a mixture of polarons and bipolarons. The spectra above $-V_G = 1.7$ V originate from bipolarons. Accordingly, polarons are the dominant carriers below $-V_G = 1.4$ V, polarons and bipolarons coexist between $-V_G = 1.5$ and 1.6 V, and bipolarons are the dominant carriers above $-V_G = 1.7$ V.

The charge density n (cm^{-3}) and doping level per π electron x (mol% per π electron) were calculated from the measured injected charge. The σ values were calculated from the observed I_D values and are plotted against x in Figure 13.16. The μ values of polarons were obtained from the data below $x = 5.5$ mol% ($n = 6.6 \times 10^{20}$ cm^{-3}, $-V_G = 1.4$ V). The μ values of bipolarons were obtained from the data above $x = 11$ mol% ($n = 1.3 \times 10^{21}$ cm^{-3}, $-V_G = 1.7$ V), although a small number of polarons may exist. These μ values are also plotted in Figure 13.16 with the σ values.

σ increases gradually from $x = 0$ to $x = 1.2$ mol% and then rapidly above $x = 1.5$ mol%. The Raman results show that polarons are generated below 5.5 mol%; bipolarons begin to form from polarons at $x = 7.3$ mol% and are the dominant carriers above $x = 11$ mol%. Thus, the critical doping level (x_c) for the polaron-to-bipolaron transition is approximately equal to 7.3 mol%. Above $x = 11$ mol%, σ shows a plateau at approximately 193 S cm^{-1}. The μ values of polarons strongly depend on x, and they increase with increasing x. The highest μ value of polarons is 1.0 cm^2 V^{-1} s^{-1} at $x = 5.5$ mol% ($n = 6.6 \times 10^{20}$ cm^{-3}). The μ values of bipolarons moderately depend on x. The highest μ value of bipolarons is 0.92 cm^2 V^{-1} s^{-1} at $x = 11$ mol% ($n = 1.3 \times 10^{21}$ cm^{-3}).

The study of the ILGT fabricated with annealed PBTTT-C16 and [BMIM][TFSI] showed that the critical doping levels (x_c) for the polaron-to-bipolaron transition are 7.3 and 3.4 mol% for the unannealed and annealed films, respectively. This result indicates that x_c is apparently lowered by annealing. The defects in the polymer film favor polarons. The anion size effect on the critical doping level x_c in

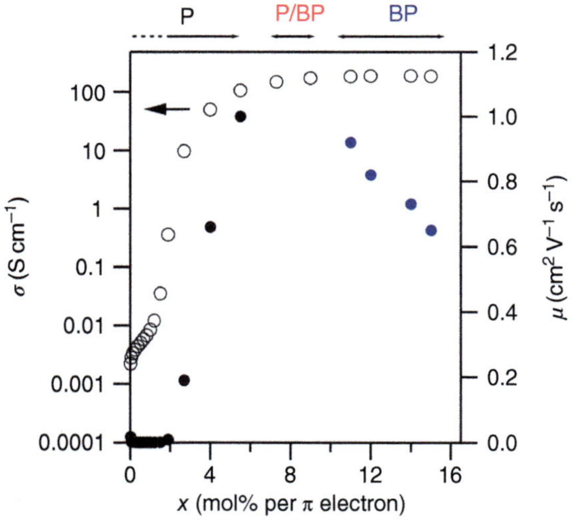

Figure 13.16 σ and μ versus x plots for an unannealed PBTTT-C16 ILGT. Source: Enokida and Furukawa [18].

PBTTT-C16 was studied using ILGTs fabricated with 1-ethyl-3-methylimidazolium bis(trifluoromethylsulfonyl)-imide ([EMIM][TFSI]) or 1-ethyl-3-methylimidazolium tris(pentafluoroethyl)trifluorophosphate ([EMIM] [FAP]) [36]. The volumes of [TFSI] and [FAP] anions were calculated using the density functional theory (DFT) method based on the B3LYP/6-31++G(d,p) level. The calculated volumes of [TFSI] and [FAP] anions are 223 and 310 Å^3, respectively. Raman studies of the ILGTs indicated that the x_c values are 4.5 and 12 mol% for [EMIM][TFSI] and [EMIM][FAP], respectively. The large anion FAP impedes bipolaron formation even at a high doping level.

13.6 Carrier Mobilities

The μ values of positive polarons and bipolarons are listed in Table 13.1. It should be noted that the mobilities of bipolarons may be contributed to by a small number of polarons. The highest μ value of polarons in the unannealed PBTTT-C16 ILGT is on the same order as that in the annealed ILGT. The highest μ value of bipolarons in the annealed PBTTT-C16 ILGT is larger than that in the unannealed ILGT. The crystallinity of the annealed film is greater than that of the unannealed film. A higher crystallinity leads to an improvement in the highest μ value. The highest μ value of bipolarons in PBTTT-C16 is 1.2 cm^2 V^{-1} s^{-1}. This mobility of bipolarons in PBTTT-C16 is quite large compared to that in P3HT.

13.7 Raman Images in the Channel Region

For a conventional transistor, I_D is plotted against V_D for a V_G value, which is called the output characteristics. In the low V_D region, I_D is approximately proportional to

Table 13.1 Highest mobilities of positive polarons and bipolarons.

Polymer	Carrier	$-V_G$ (V)	x (mol% per π electron)	μ (cm^2 V^{-1} s^{-1})	References
PBTTT-C16 (unannealed)	Polaron	1.4	5.5	1.0	[18]
	Bipolaron	1.7	11	0.92	[18]
PBTTT-C16 (annealed)	Polaron	1.1	2.7	1.0	[18]
	Bipolaron	1.4	4.8	1.2	[18]
		1.5	5.2	1.2	[18]
PBTTT-C16	—	1.0	2.9	3.4	[37]
P3HT	Polaron	1.7	3.75	0.31	[33]
	Bipolaron	3.3	11	2.5×10^{-3}	[33]
P3HT	—	1.25	9	0.5	[35]

V_D, which is called the linear region. On the other hand, beyond the "pinch-off" voltage, I_D shows saturation in the high V_D range, which is called the saturation region. This feature can be explained by pinch-off theory [38]. Pinch-off means depletion in the carrier density in the channel near the drain electrode due to the application of a large V_D. It is important to clarify the carrier densities in the channel for polymer transistors. In an ILGT, a negative bias on the gate electrode can lead to positive carrier generation in the semiconductor film between the source and drain electrodes. The characteristics of the ILGT depend on the density, distribution, and mobility of carriers.

A Raman imaging study of the carrier density distribution in the P3HT channel in an ILGT was reported [20]. The ionic liquid [EMIM][TFSI] was used as the electrolyte (gate dielectric) in the ILGT. The I_D values are plotted as a function of V_D at $V_G = -1.0$ V for a P3HT ILGT in Figure 13.17. As V_D increases, I_D increases linearly in the range between 0 and −0.4 V, which is the linear region. Quasi-pinch-off is observed at −0.4 V. In the range between −0.4 and −1.1 V, I_D shows a slow increase, which is similar to the saturation region. According to the theory of inorganic semiconductor transistors, the linear and saturation behaviors originate from the carrier distribution in the channel of the transistor [38]. Thus, it is important to clarify the carrier density distribution in the channel to understand the working mechanism of an ILGT.

The intensity of the 1415-cm^{-1} band attributed to positive polarons (carriers) is proportional to the carrier density. Figure 13.18 shows Raman images of the intensities of the polaron band at $V_G = -1.0$ V and $V_D = -0.1$, −0.5, −1.0, and −1.5 V. The carrier density distribution is clearly observed in the Raman images. In these images, the carrier density has an almost uniform distribution along the horizontal electrode direction, whereas it shows changes along the vertical direction from the source electrode to the drain electrode. Thus, it is important to analyze the cross sections of the images from the source electrode to the drain electrode.

Figure 13.19 shows the cross sections of the Raman images along the source electrode to the drain electrode at the center of the electrodes. Near the source and drain electrodes, slow changes in the Raman intensity (carrier density) are observed. These

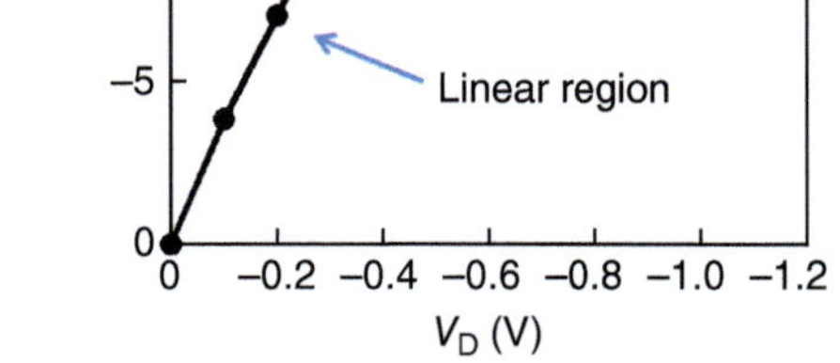

Figure 13.17 I_D–V_D characteristics at $V_G = -1.0$ V. Source: Wada et al. [20].

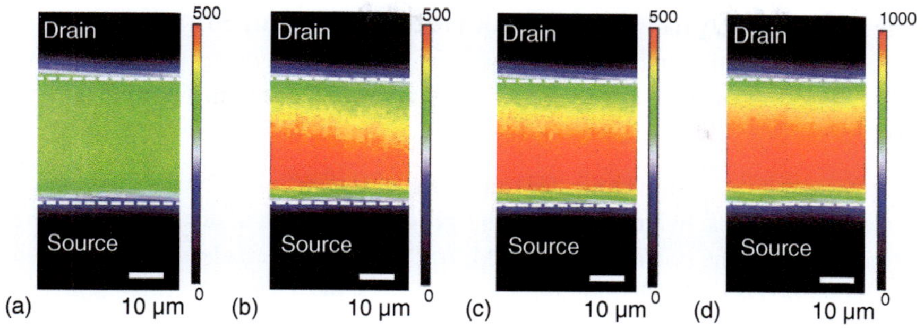

Figure 13.18 Raman images of the channel region of a transistor at $V_G = -1.0\,V$: (a) $V_D = -0.1\,V$; (b) $V_D = -0.5\,V$; (c) $V_D = -1.0\,V$; (d) $V_D = -1.5\,V$. The excitation wavelength was 785 nm. Source: Wada et al. [20].

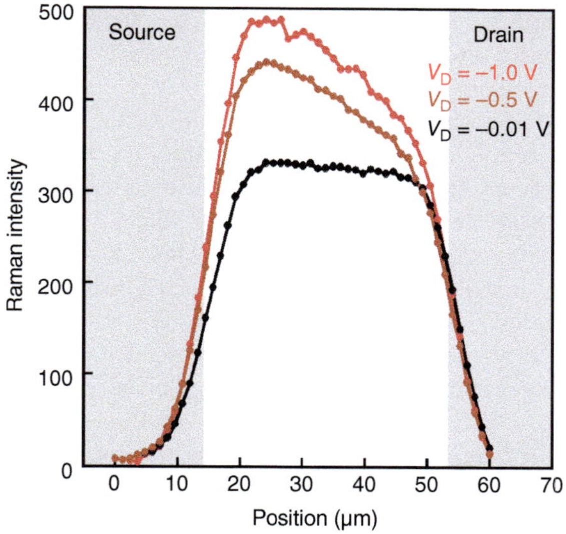

Figure 13.19 Cross sections of the Raman images in the channel region at $V_G = -1.0\,V$. Source: Wada et al. [20].

are probably due to the deterioration in the spatial resolution and dull edges of the metal electrodes. If the edges of the electrodes are perpendicular to the substrate, then one can calculate the spatial resolution of the measurements. According to Sommer's criteria [39], the spatial resolution is defined as the ordinate distance that corresponds to 5% and 95% of the peak height; thus, the spatial resolution is 9.4 μm. Since a normal objective lens was used in this study, refraction at the glass substrate was not compensated for. At $V_D = -0.1\,V$, the carrier density is almost uniform. As V_D increases, the carrier density near the source electrode increases, and the carrier density near the drain electrode is almost the same as that at $V_D = -0.1\,V$. At $V_D = -0.5$ and $-1.0\,V$, gradual decreases in the Raman intensity toward the drain electrode are observed. According to pinch-off theory, bleaching, almost zero carrier density, will be observed near the drain electrode. However, no bleaching is observed. Thus, the carrier density distribution in the channel of an ILGT is different from that of inorganic semiconductor transistors.

13.8 Carrier Dynamics in Bulk Heterojunction Films

13.8.1 Photoexcitation Dynamics on Femto- and Picosecond Time Scales

After photoexcitation, singlet excited states are generated on polymer chains in a polymer–PCBM bulk heterojunction. The generated excited states migrate to the interface between the polymer and PCBM. At the interface, an electron is removed from the excited state by PCBM, and a pair of a positive carrier (positive polaron) and a negative carrier (the anion of PCBM) is formed. Geminate recombination may occur. Then, some of the positive and negative carrier pairs separate into a free positive carrier and a free negative carrier. Some of these long-lived positive and negative carriers recombine to return to neutral species on micro- to millisecond time scales. The other part of the long-lived positive carriers and negative carriers reach the electrodes in the device. Thus, the solar cell can work. Understanding the charge separation and transport processes is necessary for improving the efficiency and durability of solar cells.

Initial photochemical processes, the formation of singlet excited states and charge separation, were studied using femtosecond time-resolved NIR inverse Raman spectroscopy [23]. Figure 13.20 shows the time-resolved NIR inverse Raman spectra from a P3HT:PCBM bulk heterojunction film. The samples were photoexcited with an energy density of 60 μJ cm^{-2} for the 480 nm actinic pump pulse. The wavelength of the Raman pump pulsed laser was 1190 nm, which is located within the absorptions attributable to singlet excited states and positive polarons. The spectrum at −0.36 ps is attributed to neutral P3HT. The spectra at 0.20 and 50 ps are mainly attributed to singlet excited states and positive polarons, respectively. The observed spectra were fitted with a linear combination of the three components representing the neutral P3HT, singlet excited states, and positive polarons. The following conclusions were made: (i) Singlet excited states are formed at the moment of photoexcitation, and

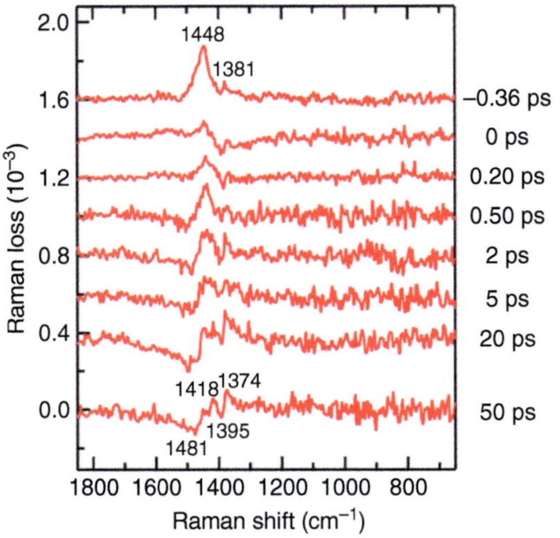

Figure 13.20 Femtosecond time-resolved NIR inverse Raman spectra from a P3HT:PCBM bulk heterojunction film. Source: Takaya et al. [23].

the decay time constant of the singlet excited states is estimated to be 0.08 ps. (ii) The rise time constants of positive polaron formation are estimated to be 0.3 and 10 ps. Thus, there are two pathways for polaron formation. Recently, solar cells fabricated with donor–acceptor type polymers, such as poly[*N*-9′-heptadecanyl-2,7-carbazole-*alt*-5,5-(4′,7′-di-2-thienyl-2′,1′,3′-benzothiadiazole)] (PCDTBT) [22] or poly[[4,8-bis[(2-ethylhexyl)oxy]benzo[1,2-*b*:4,5-*b*′]dithiophene-2,6-diyl][3-fluoro-2-[(2-ethylhexyl)carbonyl]thieno[3,4-b]thiophenediyl]] (PTB7) [40], exhibited high power conversion efficiencies. Positive polarons are formed within 0.3 and 1 ps after photoexcitation for PCDTBT-PCBM and PTB7-PCBM bulk heterojunction films, respectively.

13.8.2 Microsecond Recombination Dynamics of Long-Lived Carriers

Infrared spectroscopy is also a powerful tool for studying time-dependent phenomena because it provides information about their temporal evolution as well as the molecular structures of transient species. The dynamics of photogenerated carriers in the bulk heterojunction of a polymer solar cell are key to clarifying the transport of photogenerated carriers and improving the performance of the solar cell. The microsecond recombination dynamics of photogenerated carriers in a P3HT:PCBM bulk heterojunction were studied using time-resolved infrared spectroscopy [41].

Microsecond time-resolved infrared spectra of a P3HT:PCBM blend film were obtained in the temperature range between 78 and 120 K using a system based on a step-scan FT-IR spectrometer (JASCO FT/IR-6800) equipped with a photovoltaic HgCdTe detector and a pulsed Nd:YAG laser (Continuum Superlight II-10). The time resolution of the system was 10 μs. The wavelength of the laser light was 532 nm, the repetition rate was 10 Hz, and the pulse width was 5 ns. Photoinduced infrared absorption was measured by the FT-IR difference spectrum method.

The temporal evolution of the photoinduced infrared absorption spectrum from a P3HT:PCBM (1 : 1) bulk heterojunction film is shown in Figure 13.21. The observed

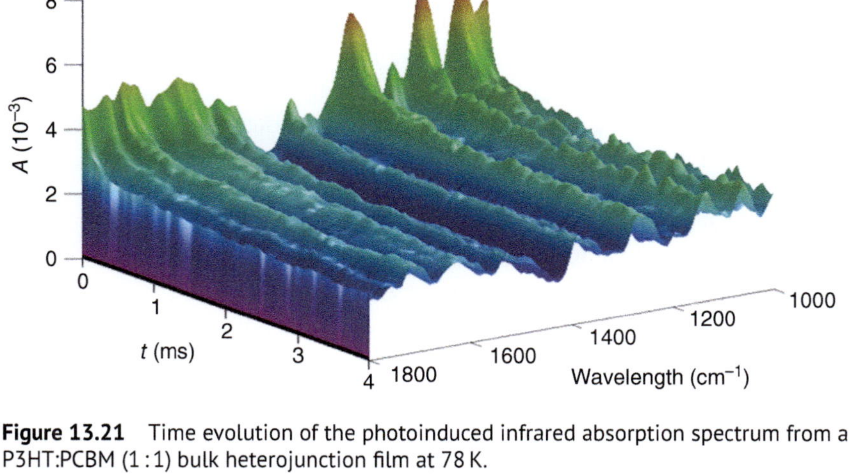

Figure 13.21 Time evolution of the photoinduced infrared absorption spectrum from a P3HT:PCBM (1 : 1) bulk heterojunction film at 78 K.

spectrum is attributed to photogenerated positive polarons. The spectrum shows complicated features, which are attributed to the interaction between broad electronic absorption and vibrational transitions [42]. No bands attributable to the radical anion of PCBM are found because the absorption coefficient of the polaron band is much larger than that of the radical anion of PCBM. Assuming that a positive polaron and a radical anion of PCBM (a negative carrier) are recombined via the bimolecular process, the decay dynamics of polarons can be expressed by the bimolecular reaction. The data observed at 78 K were successfully fitted with two bimolecular processes by the least-squares method. The obtained rate constants are 1.6×10^{-12} and 1.2×10^{-13} $cm^3\ s^{-1}$. The two rate constants probably originate from amorphous and crystalline regions of the bulk heterojunction.

13.9 Conclusions

We present a review of recent applications of Raman and infrared spectroscopy to conjugated conducting polymers and their devices. Raman spectroscopy is a powerful tool for studying charge carriers, polarons and bipolarons in polymers. Using the ILGT configuration, one can elucidate the doping-level dependences of the electrical conductivity, carrier type, and mobility. It is remarkable that positive bipolarons show high mobilities, c. $1.2\ cm^2\ V^{-1}\ s^{-1}$, in PBTTT-C16 films. Femtosecond time-resolved stimulated and inverse Raman spectroscopy provides us with the formation process of polarons. Time-resolved FT-IR spectroscopy gives us the microsecond recombination dynamics of long-lived polarons. These methods are useful for the research on the polymer materials used for solar cells.

References

1 Burroughes, J.H., Jones, C.A., and Friend, R.H. (1988). New semiconductor device physics in polymer diodes and transistors. *Nature* 335: 137–141.
2 Burroughes, J.H., Bradley, D.D.C., Brown, A.R. et al. (1990). Light-emitting diodes based on conjugated polymers. *Nature* 347: 539–541.
3 Sariciftci, N.S., Smilowitz, L., Heeger, A.J., and Wudl, F. (1992). Photoinduced electron transfer from a conducting polymer to buckminsterfullerene. *Science* 258: 1474–1476.
4 Hedley, G.J., Ruseckas, A., and Samuel, I.D.W. (2017). Light harvesting for organic photovoltaics. *Chem. Rev.* 117: 796–837.
5 Kim, S.H., Hong, K., Xie, W. et al. (2013). Electrolyte-gated transistors for organic and printed electronics. *Adv. Mater.* 25: 1822–1846.
6 Paulsen, B.D. and Frisbie, C.D. (2012). Dependence of conductivity on charge density and electrochemical potential in polymer semiconductors gated with ionic liquids. *J. Phys. Chem. C* 116: 3132–3141.
7 Baeriswyl, D., Campbell, D.K., and Mazumdar, S. (1992). An overview of the theory of π-conjugated polymers. In: *Springer Series in Solid-State Sciences*, vol. 102 (eds. N.J. Everall, J.M. Chalmers and P.R. Griffiths), 7–134. Wiley.

8 Zerbi, G. (2007). Vibrational spectroscopy of conducting polymers: theory and perspective. In: *Vibrational Spectroscopy of Polymers: Principles and Practice* (eds. N.J. Everall, J.M. Chalmers and P.R. Griffiths), 487–536. Wiley.

9 Furukawa, Y. (2007). Vibrational spectroscopy of conducting polymers: fundamental and applications. In: *Vibrational Spectroscopy of Polymers: Principles and Practice* (eds. N.J. Everall, J.M. Chalmers and P.R. Griffiths), 537–556. Wiley.

10 Furukawa, Y., Ueda, F., Hyodo, Y. et al. (1988). Vibrational spectra and structure of polyaniline. *Macromolecules* 21: 1297–1305.

11 Sakamoto, A., Furukawa, Y., and Tasumi, M. (1992). Resonance Raman characterization of polarons and bipolarons in sodium-doped poly(*p*-phenylenevinylene). *J. Phys. Chem.* 96: 3870–3874.

12 Sakamoto, A., Furukawa, Y., and Tasumi, M. (1994). Resonance Raman and ultraviolet to infrared absorption studies of positive polarons and bipolarons in sulfuric-acid-treated poly(*p*-phenylenevinylene). *J. Phys. Chem.* 98: 4635–4640.

13 Sakamoto, A., Furukawa, Y., and Tasumi, M. (1997). Spectroscopic studies on the radical-cation dimer of a model compound of poly(*p*-phenylenevinylene). Similarities between the dimer and the state of positive polarons in the sulfuric-acid-treated polymer. *J. Phys. Chem. B* 101: 1726–1732.

14 Furukawa, Y., Ohtsuka, H., and Tasumi, M. (1993). Raman studies of polarons and bipolarons in sodium-doped poly-*p*-phenylene. *Synth. Met.* 55: 516–523.

15 Honda, K., Furukawa, Y., Furuya, K. et al. (2002). Density functional theory study on the Raman spectra of negative polarons and negative bipolarons in Na-doped poly(*p*-phenylene). *J. Phys. Chem. A* 106: 3587–3592.

16 Yokonuma, N., Furukawa, Y., Tasumi, M. et al. (1996). Electronic absorption and Raman studies of BF_4^--doped polythiophene based on the spectra of the radical cation and dication of α-sexithiophene. *Chem. Phys. Lett.* 255: 431–436.

17 Yamamoto, J. and Furukawa, Y. (2015). Electronic and vibrational spectra of positive polarons and bipolarons in regioregular poly(3-hexylthiophene) doped with ferric chloride. *J. Phys. Chem. B* 119: 4788–4794.

18 Enokida, I. and Furukawa, Y. (2019). Doping-level dependent mobilities of positive polarons and bipolarons in poly(2,5-bis(3-hexadecylthiophen-2-yl)thieno[3,2-*b*]thiophene) (PBTTT-C16) based on an ionic-liquid-gated transistor configuration. *Org. Electron.* 68: 28–34.

19 Kumar, R., Pillai, R., Pekas, N. et al. (2012). Spatially resolved Raman spectroelectrochemistry of solid-state polythiophene/viologen memory devices. *J. Am. Chem. Soc.* 134: 14869–14876.

20 Wada, Y., Enokida, I., Yamamoto, J., and Furukawa, Y. (2018). Raman imaging of carrier distribution in the channel of an ionic liquid-gated transistor fabricated with regioregular poly(3-hexylthiophene). *Spectrochim. Acta, Part A* 197: 166–169.

21 Liu, Q., Jiang, Y., Jin, K. et al. (2020). 18% efficiency organic solar cells. *Sci. Bull.* 65: 272–275.

22 Provencher, F., Bérubé, N., Parker, A.W. et al. (2014). Direct observation of ultrafast long-range charge separation at polymer-fullerene heterojunctions. *Nat. Commun.* 5: 4288–4298.

23 Takaya, T., Enokida, I., Furukawa, Y., and Iwata, K. (2019). Direct observation of structure and dynamics of photogenerated charge carriers in poly(3-hexylthiophene) films by femtosecond time-resolved near-IR inverse Raman spectroscopy. *Molecules* 24: 431.

24 Furukawa, Y., Seto, K., Nakajima, K. et al. (2012). Infrared and Raman spectroscopy of organic thin films used for electronic devices. *Vib. Spectrosc.* 60: 5–9.

25 Su, W.P. and Schrieffer, J.R. (1980). Soliton dynamics in polyacetylene. *Proc. Natl. Acad. Sci. U.S.A.* 77: 5626–5629.

26 Brazovskiĭ, S.A. and Kirova, N.N. (1981). Excitons, polarons, and bipolarons in conducting polymers. *Sov. Phys. JETP Lett.* 33: 4–8.

27 Bishop, A.R., Campbell, D.K., and Fesser, K. (1981). Polyacetylene and relativistic field theory models. *Mol. Cryst. Liq. Cryst.* 77: 253–264.

28 Brédas, J.L., Chance, R.R., and Silbey, R. (1981). Theoretical studies of charged defect states in doped polyacetylene and polyparaphenylene. *Mol. Cryst. Liq. Cryst.* 77: 319–332.

29 Shimoi, Y. and Abe, S. (1994). Competition between polarons and bipolarons in nondegenerate conjugated polymers. *Phys. Rev. B* 50: 14781–14784.

30 Fesser, K., Bishop, A.R., and Campbell, D.K. (1983). Optical absorption from polarons in a model of polyacetylene. *Phys. Rev. B* 27: 4804–4825.

31 Shimoi, Y., Abe, S., and Harigaya, K. (1995). Theory of optical absorption in doped conjugated polymers. *Mol. Cryst. Liq. Cryst.* 267: 329–334.

32 Furukawa, Y. (1996). Electronic absorption and vibrational spectroscopies of conjugated conducting polymers. *J. Phys. Chem.* 100: 15644–15653.

33 Yamamoto, J. and Furukawa, Y. (2016). Raman characterization and electrical properties of poly(3-hexylthiophene) doped electrochemically in an ionic liquid-gated transistor geometry. *Org. Electron.* 28: 82–87.

34 Francis, C., Fazzi, D., Grimm, S.B. et al. (2017). Raman spectroscopy and microscopy of electrochemically and chemically doped high-mobility semiconducting polymers. *J. Mater. Chem. C* 5: 6176–6184.

35 Ando, Y., Ito, H., Watanabe, S., and Kuroda, S. (2012). Variable-range hopping conduction in ion-gel-gated electrochemical transistors of regioregular poly(3-hexylthiophene). *J. Phys. Soc. Jpn.* 81: 114721.

36 Enokida, I. and Furukawa, Y. (2019). Effect of anions on bipolaron formation in ionic-liquid-gated transistors fabricated with poly(2,5-bis(3-hexadecylthiophen-2-yl)thieno[3,2-*b*]thiophene) (PBTTT-C16). *Chem. Lett.* 23: 498–501.

37 Harada, T., Ito, H., Ando, Y. et al. (2015). Signature of the insulator–metal transition of a semicrystalline conjugated polymer in ionic-liquid-gated transistors. *Appl. Phys Express* 8: 021601.

38 Sye, S.M. (2002). *Semiconductor Devices, Physics and Technology*, 2e, 187–193. Wiley.

39 Sommer, A.J., Tisinger, L.G., Marcott, C., and Story, G.M. (2001). Attenuated total internal reflection infrared mapping microspectroscopy using an imaging microscope. *Appl. Spectrosc.* 55: 252–256.

40 Takaya, T., Enokida, I., Furukawa, Y., and Iwata, K. (2020). Near-infrared resonance stimulated Raman study of short-lived transients in PTB7 films. *Vib. Spectrosc.* 106: 103011.

41 Furukawa, Y. and Oki, N. (2019). Microsecond dynamics of photogenerated carriers in a regioregular poly(3-hexylthiophene): PCBM blend film by time-resolved infrared spectroscopy. *Tenth International Conference on Advanced Vibrational Spectroscopy* (7–12 July 2019), Auckland, New Zealand.

42 Jiang, X.-M., Österbacka, R., Korovyanko, O. et al. (2002). Spectroscopic studies of photoexcitations in regioregular and regiorandom polythiophene films. *Adv. Funct. Mater.* 12: 587.

14

Vibrational Spectroscopy for Fluoropolymers and Oligomers

Takeshi Hasegawa

Kyoto University, Institute for Chemical Research, Gokasho, Uji, Kyoto, 611-0011, Japan

14.1 Perfluoroalkyl-Containing Compounds

14.1.1 Molecular Conformation on Phase Diagram

The molecular structure of a perfluoroalkyl (R_f) group is mostly characterized by the helical conformation about the molecular axis as found in Figure 14.1, which is largely different from normal hydrocarbons having a planar-zigzag conformation [1]. The van der Waals radius of fluorine is ca. 20% larger than that of hydrogen, which makes a linear arrangement of CF_2 groups difficult along the R_f chain. As a result of the steric repulsion, the R_f group has to be twisted to have the helical conformation [1]. This R_f-specific unique conformation, collaborated with a fairly strong dipole moment along the C—F bond [2], yields various material and spectroscopic characters [3, 4].

The twisting rate of the helical conformation depends on a phase diagram as shown in Figure 14.2 [5–11] measured on polytetrafluoroethylene (PTFE) that is known as Teflon®. PTFE is widely used in daily life due to its unreplaceable unique material characters represented by the water-and-oil repellency, which is discussed in Section 14.1.3. At an early stage, only the three phases of I through III were known, and the phase IV was additionally found by Hirakawa and Takemura by using the ultrasonic waves, thermal expansion, differential thermal analysis, and X-ray diffraction techniques [6]. At ambient pressure, the phases I, II, and IV appear on the abscissa axis. Phase III is, on the other hand, available only for a very high pressure, in which the PTFE molecule is known to take the planar zigzag conformation denoted by the point group of D_{2h} [12] with the orthorhombic unit cell inclined at 35° to the *b* axis [7]. The rest phases commonly have the helical conformation with a different twisting rate. Phase II has a helix with a twisting angle of 180° over 13 carbons (or 12 C—C bonds). On the other hand, Phases I and IV have a helix with a twisting angle of 180° over 15 carbons [1, 9, 11], which are all determined by the X-ray diffraction technique. The boundary temperature between the phases II and IV at ambient pressure is found at 19 °C (292 K) [5, 12]. On the other hand, Phase I above 30 °C is known to have a less helicity [13].

Spectroscopic Techniques for Polymer Characterization: Methods, Instrumentation, Applications, First Edition. Edited by Yukihiro Ozaki and Harumi Sato.

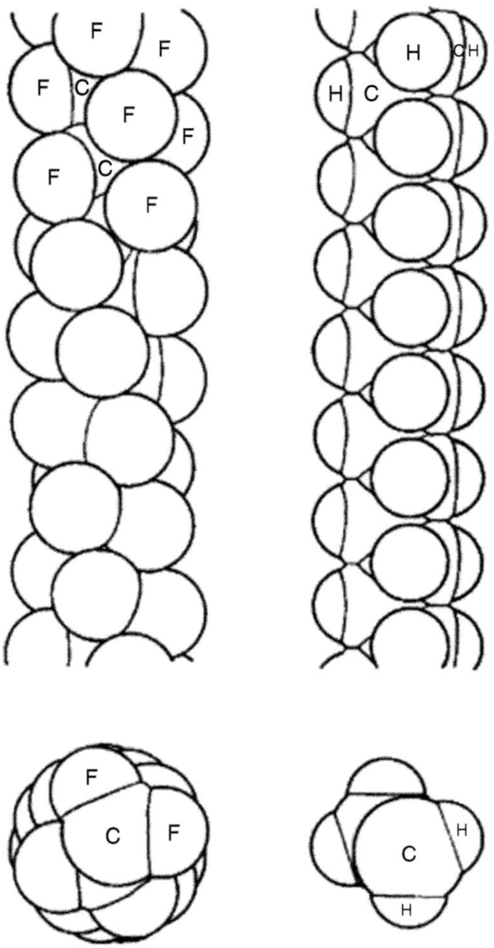

Figure 14.1 Schematics of R_f chain (left) and normal alkyl chain (right). Source: Bunn and Howells [1].

Molecular packing and conformation can be analyzed by using Raman spectroscopy. Koenig and Boerio measured Raman spectra of PTFE and its homologues of an oligomer compound, $C_{14}F_{30}$ and $C_{16}F_{34}$, as a function of temperature and pressure [12, 14]. They discussed a strong band at 575 cm^{-1}. As found in Figure 14.3, the band is dominant at 16 °C in Phase II, whereas the counterpart band at 595 cm^{-1} develops at 29 °C in Phase IV. In fact, the bands at 575 and 595 cm^{-1} are major bands in cold and heated PTFE, respectively. In appearance, therefore, we have an impression that they are normal modes of the 13_6 and 15_7 helices, respectively.

This hypothesis, however, is denied through a precise discussion of Raman and infrared (IR) spectra of the compounds. They focus on a similarity of the thermal behavior of the two bands to those of IR bands at 636 and 625 cm^{-1} [12], which is readily explained by an increase of thermal defects. Therefore, the bands of 575 and 595 cm^{-1} are not assigned to a fundamental normal mode, but they should be associated with defects. In fact, ten Raman bands *except* three bands involving the 575

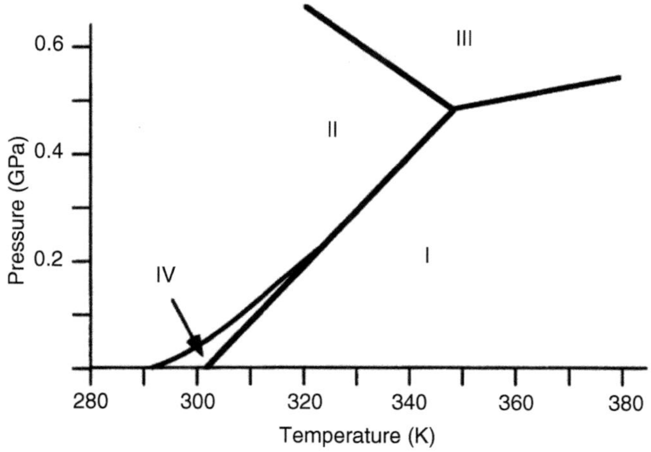

Figure 14.2 Phase diagram of PTFE. Source: Clark [5].

Figure 14.3 Raman spectra of PTFE at different temperatures. Source: Koenig and Boerio [14].

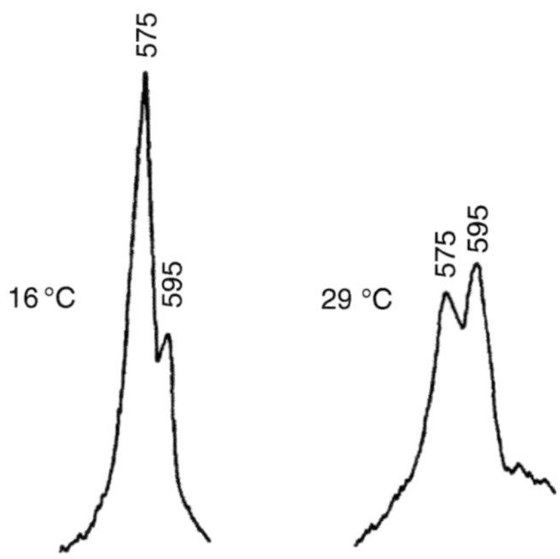

and 595 cm^{-1} bands are readily assigned to the strong A_1 mode, and the weak E_1 and E_2 modes, by analysis of intensity and depolarization [11].

Regardless, extension of this band shift can still be useful for identification of Phase III. The phase boundaries are readily revealed by using Raman spectroscopy. As described in the paper by Wu and Nicol [10], Figure 14.4 apparently shows a first-order phase transition at the boundary of Phases II and III. As a result, the Raman band appears at a highly shifted position of c. 620 cm^{-1} at a very high pressure, which is a useful good marker of Phase III [9].

Here, we have to characterize the Raman bands of 575 and 595 cm^{-1} in terms of normal modes. According to Hannon et al. [15], the homologue of $C_{14}F_{30}$ is crystallized on cooling down to −70 °C where the band at 575 cm^{-1} is readily resolved, which is similar to that found at 19 °C of PTFE. The Raman band at 575 cm^{-1} is not dissolved from the neighboring band at ambient temperature, whereas it appears apparently at −70 °C. The band is thus assigned to the crystal phase. Here, we have to note that the *gauche* conformer occurs for both crystalline and amorphous

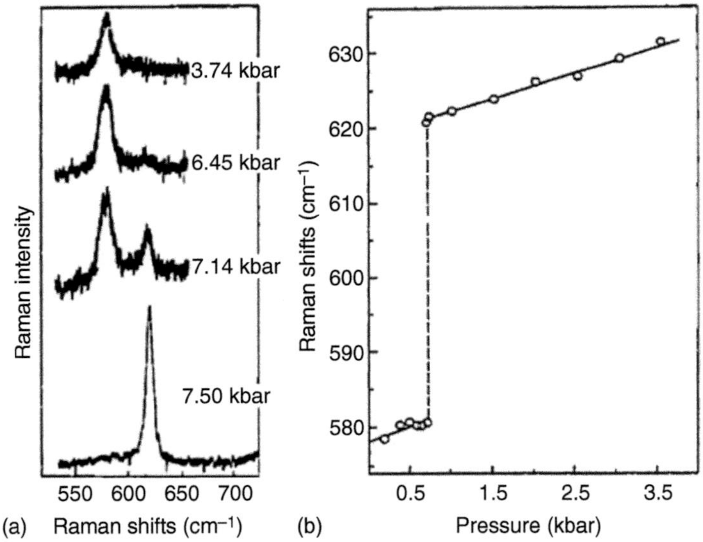

Figure 14.4 Raman shift at the boundary of Phases II and III. Source: Wu and Nicol [10].

fluorocarbons at a high temperature, in which the *gauche* conformation is on the carbon linkage of the backbone. After the energy diagram as a function of conformation calculated by De Santis et al. [16], the band at 575 cm^{-1} is finally assigned to the CF_2 deformation vibration band with a trans-planar linkage in the backbone, whereas the band at 595 cm^{-1} is for the same mode adjoining a *gauche* linkage in the backbone.

In this manner, we find that the Raman bands at 575 and 595 cm^{-1} are due to the structural stress on the backbone of the molecule. In this sense, the characteristic band of c. 620 cm^{-1}, which is associated with the all-trans zigzag conformation specific to Phase III, has an unusually high stress on the backbone of an R_f group. This is a decisively different point from a normal hydrocarbon that has the most relaxed state for the all-trans zigzag conformation [16].

14.1.2 Molecular Vibration of an R_f Group

To discuss the IR and Raman spectra of R_f-containing compounds further, elemental knowledge of molecular vibration specific to an R_f group must be considered: the mass of fluorine is larger than that of carbon. This is a totally different point from a normal hydrocarbon, in which the mass of hydrogen is much smaller than that of carbon. As a result, a normal mode of an R_f group has a unique picture where the fluorine atoms relatively stay unmoved, but instead the carbon atoms are vibrated as a coupled oscillator, since they are directly connected with each other.

An R_f-group specific character of IR spectra is found in Figure 14.5. The IR spectra are on a series of myristic acids having an R_f group at the tail end with a different length (denoted as MA-$R_f n$), and the spectral range between 1000 and 1400 cm^{-1} is attributed to the CF-related modes [11–15, 18], which means that the CH part of the

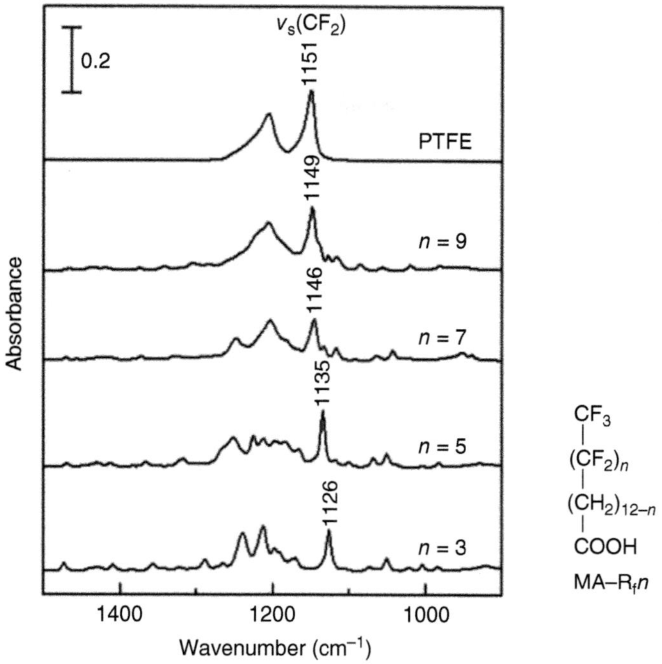

Figure 14.5 (a) The C–F stretching vibration region of IR spectra of a compound series of MA-R$_f$*n*. The sample involves a unit of $CF_3(CF_2)_{n-}$. Source: Shimoaka et al. [17].

molecule can be excluded from the discussion. The most intensive band is of a vibrational mode with the name of "CF_2 symmetric stretching vibration ($\nu_s(CF_2)$)" mode [11] exhibiting a unique higher wavenumber shift with an increase of the chain length. If we perform the same measurements on hydrocarbons, the band shows a lower wavenumber shift [19], which is quite understandable since a longer chain has a higher reduced mass giving a lower wavenumber. Therefore, the higher wavenumber shift on the R$_f$-containing compounds seems unusual, which requires another mechanism than the change of reduced mass.

The $\nu_s(CF_2)$ mode has a name of "CF_2 symmetric stretching vibration." As a matter of fact, however, the vibration is highly coupled with the chain-stretching vibration [15]. In such a case of the skeletal-stretching vibration, Raman spectra should be referred.

Raman spectra of the same compounds as used for IR spectra are presented in Figure 14.6. In the Raman spectra, the $\nu_s(CF_2)$ mode appears in a very much lower wavenumber region than the IR spectra. This is because the Raman active mode (A_g; $\delta = 0$) has a largely different from the IR-active mode (B_{2u}; $\delta = \pi$) in terms of the phase shift, δ, between the adjacent CF_2 units, if the chain skeleton has the all-trans zigzag conformation with the point group of D_{2h} [20].

Note that the molecular conformation of an R$_f$ group also depends on the chain length, if the length is very short. Here, the length, n, is counted by the number of CF_2 groups. If $n = 3$, the skeleton does not have the helical conformation, but it has the planar structure, which can be denoted by D_{2h} as found for a normal alkyl group.

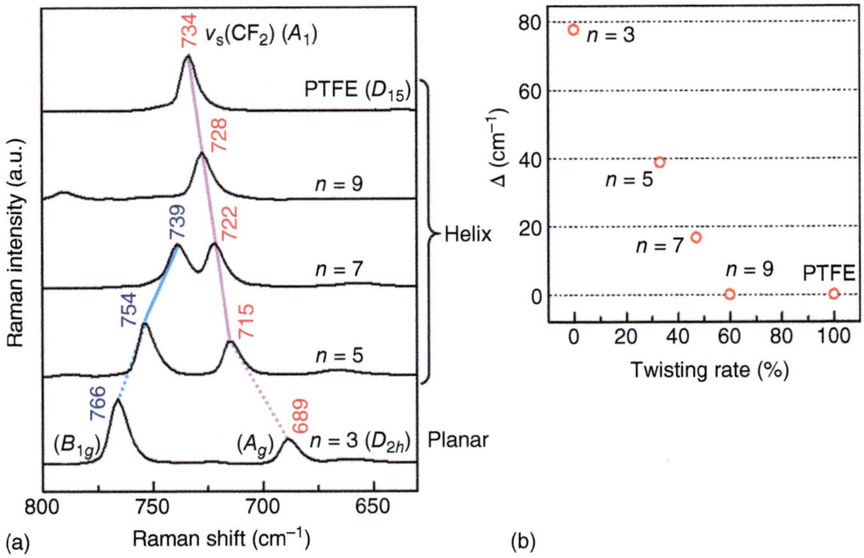

Figure 14.6 (a) The C–F stretching vibration region of Raman spectra of the same compounds as used for the IR measurements presented in Figure 14.5, (b) The shift depends on the twisting rate. Source: Shimoaka et al. [17].

For a molecule having a repeat unit such as $(CF_2)_2$, the factor group analysis is employed on D_{2h}, and we have the irreducible representation of vibrational modes [20]:

$$\Gamma^{\text{vib}} = 3A_g + A_u + 2B_{1g} + 2B_{1u} + B_{2g} + 2B_{2u} + 2B_{3g} + B_{3u}$$

Since the factor group analysis obeys the $3N - 4$ law [20], the number of atoms in the repeat unit, $N = 6$, is put in the law to have 14 modes. According to the character table of D_{2h}, five modes go to IR-active modes, while eight modes go to Raman-active modes. The rest one mode is inactive for both IR and Raman. When we focus on the wavenumber range of 1000–1400 cm^{-1}, only two modes having the A_g and B_{1g} representations are Raman active. In short, *two peaks* appear in this region of the Raman spectrum, if the chain skeleton has the point group of D_{2h}.

On the other hand, if the length is $n = 5$ or longer, the conformation is known to have D_{15} (i.e. the 15_7 helix; see Figure 14.7) for the temperature of Phase IV involving ambient temperature (Figure 14.2).

The same procedure as used for D_{2h} is performed for D_{15} [20, 21], and the irreducible representation is obtained as:

$$\Gamma^{\text{vib}} = 4A_1 + 3A_2 + 8E_1 + 9E_2 + 9(E_3 + E_4 + E_5 + E_6 + E_7)$$

This is simplified in terms of IR and Raman spectroscopy to be:

$$\Gamma^{\text{IR}} = 3A_2 + 8E_1$$

$$\Gamma^{\text{Raman}} = 4A_1 + 8E_1 + 9E_2$$

$$\Gamma^{\text{Inactive}} = 9(E_3 + E_4 + E_5 + E_6 + E_7)$$

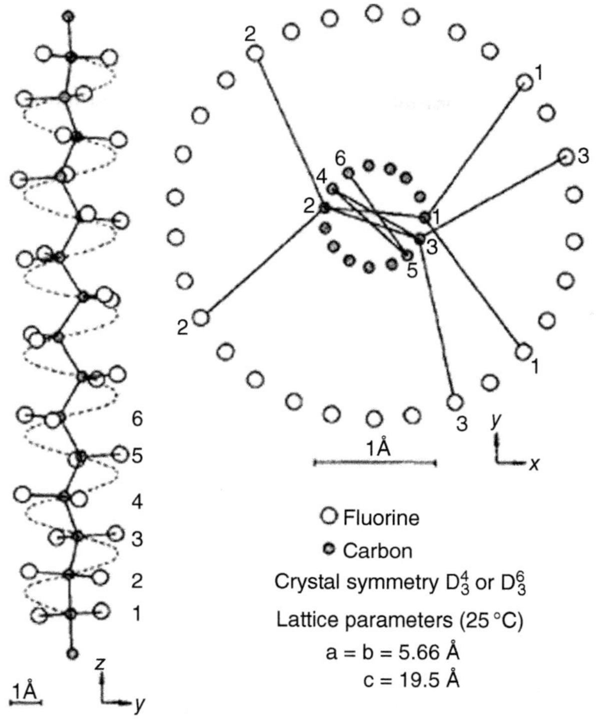

Figure 14.7 The helical conformation of an R_f group with D_{15}. Source: Piseri et al. [13].

In the range of 1000–1400 cm^{-1} of a Raman spectrum, only A_1 and E_1 modes are active for Raman and IR spectroscopy, respectively. In short, for Raman spectra, two peaks appear for D_{2h}, while a single peak appears for D_{15}.

This is apparently found in the Raman spectra in Figure 14.6a. The $\nu_s(CF_2)$ band appears as doublet peaks for $n = 3$, while it has a single peak for $n = 9$. In other words, the splitting width decreases with an increase of the chain length, and the width is getting zero to have a singlet peak. In this manner, with respect to vibrational spectroscopy, the conformational change is a continuous change. The extreme limit is, of course, found for PTFE: a relatively sharp single peak is found at a highest wavenumber position. This is a big hint for considering the R_f-specific higher wavenumber shift found in the IR spectra (Figure 14.5).

In quite a similar manner, the factor group analysis is performed for IR spectra in the 1000–1400 cm^{-1} region, a simple result is obtained: for both D_{2h} and D_{15}, a single peak (B_{2u} and E_1, respectively) appears. In other words, IR spectra detect only one band of the two bands, and the other band is missing because of the selection rule. The intrinsic phenomenon is the conformational change from D_{2h} to D_{15} as apparently found in the Raman spectra, but IR spectra detect only one side of the splitting, which gives us an impression that the band position is higher shifted, although the reduced mass is increased on the increase of the chain length.

These phenomena are specific to R_f groups, and they are not found for hydrocarbons. In this manner, when we analyze IR and Raman spectra of R_f-containing compounds, we have to pay special attention to these unique characteristics.

14.1.3 The SDA theory

The R_f-specific material characters are theoretically understandable based on fundamental elemental characters of fluorine. The theoretical framework is known as the stratified dipole-arrays (SDA) theory [3, 4]. Since the SDA theory is the basis not only for material character, but also for spectroscopic one, the theory is briefly summarized here.

The most simple and famous polymeric material of an R_f compound is PTFE. PTFE has a lot of unique material characters represented by the water-and-oil repellency, low solubility to most of the organic solvents, high melting point, and low electric permittivity. These characters cannot be reproduced by any other compounds having no fluorine, and therefore R_f-containing compounds are recognized to be quite useful for applications in daily life. The unique characters, however, were very difficult to understand in terms of molecular structure. Above all, the water-and-oil repellency (the "fluorous" property) sounds strange, since we are mostly accustomed to a binary system of water and oil. In the binary system, a compound soluble to one of the solvents, it is insoluble to the other, on which the fluorous property sounds strange. The SDA theory, however, readily answers the questions in a uniformed manner.

The SDA theory focuses on a fact that the material character of a single molecule is totally different from that of a bulk matter. To bridge the single molecule and the bulk matter, molecular interaction must carefully be considered. In the past (and even now), the molecular interaction between molecules exhibiting a hydrophobic character was believed to be driven by London's dispersion force. This misunderstanding was led by an inappropriate replacement of "dispersion force" by "hydrophobic interaction." As shown by a simple model calculation on London's equation [2], however, the interaction between R_f groups is mostly due to the dipole–dipole interaction (i.e. the orientation effect) that is a member of van der Waals forces. The three essential factors of van der Waals forces are theorized by the following equations [22].

$$\overline{U}_{\text{orientation}} = -\frac{1}{3R^6}\frac{\mu^4}{k_B T}(1 + 3\cos^2\theta) \tag{14.1}$$

$$\overline{U}_{\text{induction}} = -\frac{2}{R^6}\mu^2\alpha \tag{14.2}$$

$$\overline{U}_{\text{dispersion}} = -\frac{3}{4R^6}h\nu_0\alpha^2 \tag{14.3}$$

The C—F bond has the molecular polarizability, α, of 0.555 that is apparently less than 0.652 for the C—H bond (for the unit, see the reference [2]). In addition, the C—F bond has the permanent dipole moment, μ, of 1.39 that is much larger than 0.40 for the C—H bond. Since Eqs. (14.1) and (14.3) are functions of μ^4 and α^2, respectively, the differences are very much enhanced, and as a result, the dipole–dipole interaction is the dominant factor for spontaneous aggregation of

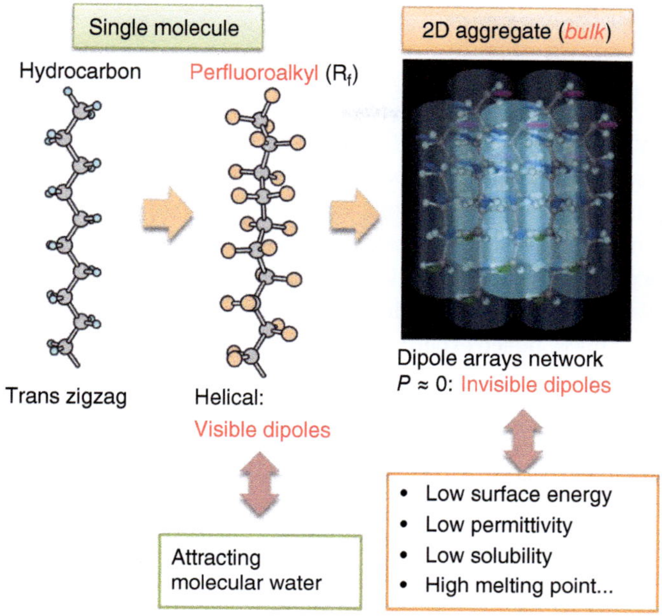

Figure 14.8 Schematic view of the SDA theory. Source: Hasegawa [4].

R_f-containing molecules. (Equation (14.2) doesn't contribute significantly as mentioned by London himself [22].)

Note that, at the moment, "a single R_f group" does not exhibit a hydrophobic property [23]. The hydrophobic character is given only by the bulk matter that is a molecular aggregate via the dipole–dipole interaction.

To understand the aggregation of R_f-containing molecules to be a bulk matter, the following two points are fundamental reasons.

(1) The C—F bond has a large permanent dipole moment along the bond.
(2) An R_f group has a helical conformation about the molecular axis as stated in Section 14.1.1.

The R_f-containing molecules are spontaneously aggregated to mostly have a hexagonal packing because of the two reasons cooperatively (Figure 14.8). In the "bulk" matter, molecules are strongly attracted with each other by the dipole–dipole interaction, but it is nearly invisible because of averaging the dipole arrays with various directions. This simple schematic elucidates all the unique characteristic comprehensively [4].

On the other hand, a single molecule of an R_f-containing compound exhibits totally different characters: molecular water (not a droplet) is attracted by the single molecule, whereas it is not dissolved by bulk water and it stays at the air/water interface [23]. In this sense, the single R_f group does not exhibit hydrophilic or hydrophobic, either, which is called a "dipole interactive" character.

14.2 Spectroscopy for R_f Compounds

14.2.1 ROA analysis of R_f Compounds

An R_f group is relatively stiff, and we thus don't have to take care about a conformation change as found for hydrocarbons. As shown by the SDA theory in the previous section, R_f-containing molecules are spontaneously aggregated two-dimensionally, mostly in a hexagonal manner, due to the dipole–dipole interaction and the helical conformation cooperatively. If this molecular scheme were true, the R_f groups should have a perfectly perpendicular stance in the two-dimensional aggregate as illustrated in Figure 14.8. This is indeed experimentally confirmed by using an X-ray diffraction technique [24] and the pMAIRS technique [25, 26]. The pMAIRS technique equipped with FT-IR is being widely accepted as a useful and powerful technique typically for analyzing molecular orientation in a thin film. When a thin film deposited on a solid substrate is subjected to pMAIRS, two spectra of in-plane (IP) and out-of-plane (OP) are simultaneously obtained from an identical sample. Since pMAIRS is uniquely designed to have the IP and OP spectra in a common ordinate scale, the orientation of each transition moment can easily be calculated quantitatively. Here, we have to note that the R_f groups are spontaneously aggregated in a two-dimensional manner generating a thin film. The molecular aggregate is thus subjected to pMAIRS, and the molecular orientation is discussed.

The pMAIRS spectrum of a single-monolayer Langmuir–Blodgett (LB) film of MA-R_f9 deposited on a silicon substrate is presented in Figure 14.9.

The pMAIRS-IP and -OP spectrum selectively shows the surface-parallel and surface-perpendicular components of a transition moment, respectively. The key band of the spectrum is the band at 1344 cm^{-1}, which is attributed to the symmetric CF_3 stretching vibration ($\nu_s(CF_3)$) band. Note that this very high-wavenumber position is for "symmetric" stretching of the CF_3 group, and this is another impressive R_f-specific assignment [4, 20], which is largely different from the common sense to a hydrocarbon. The pMAIRS spectrum shows that the $\nu_s(CF_3)$ band appears in the OP spectrum only, and it is invisible in the IP one. Since the transition moment of the $\nu_s(CF_3)$ mode is parallel to the molecular axis [4, 20, 28], the OP-oriented appearance of the $\nu_s(CF_3)$ band directly indicates that the R_f group has a perpendicular stance to the surface of the two-dimensional molecular aggregate.

The perpendicular orientation in the spontaneously aggregated R_f molecules is induced by the SDA packing as found in Figure 14.8. Each R_f group has a helical conformation about the axis, and it takes right- and left-handed helices. As found in the solid line in Figure 14.10, the right- and left-handed helices are separated by a very small barrier (1.59 kJ mol^{-1}), which means that they are in equilibrium in a solution (atrop isomers).

The SDA-packing model, however, implies that a molecular domain should comprise one-side oriented helices; otherwise, the molecules cannot be aggregated spontaneously. This means that a "seed" molecule having a right-handed (or left-handed) helix induces the same-handed helices to be aggregated. If this chemical model is true, the aggregate should have an apparent optical activity, since it has a chirality

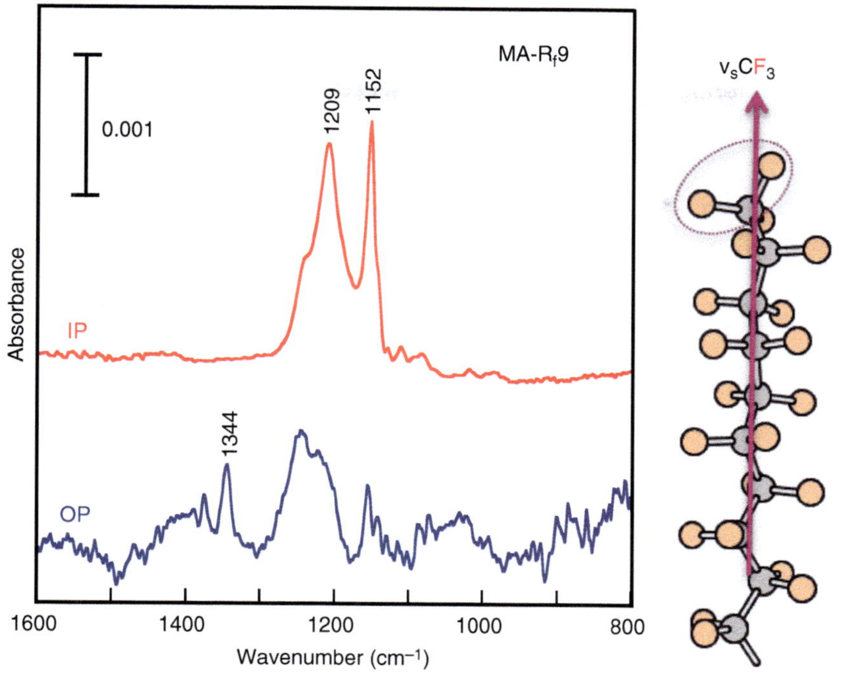

Figure 14.9 Infrared pMAIRS spectrum of a single-monolayer LB film of MA-R_f9 deposited on a silicon substrate [27]. The $\nu_s(CF_3)$ mode has a parallel transition moment to the molecular axis. Source: Hasegawa [4].

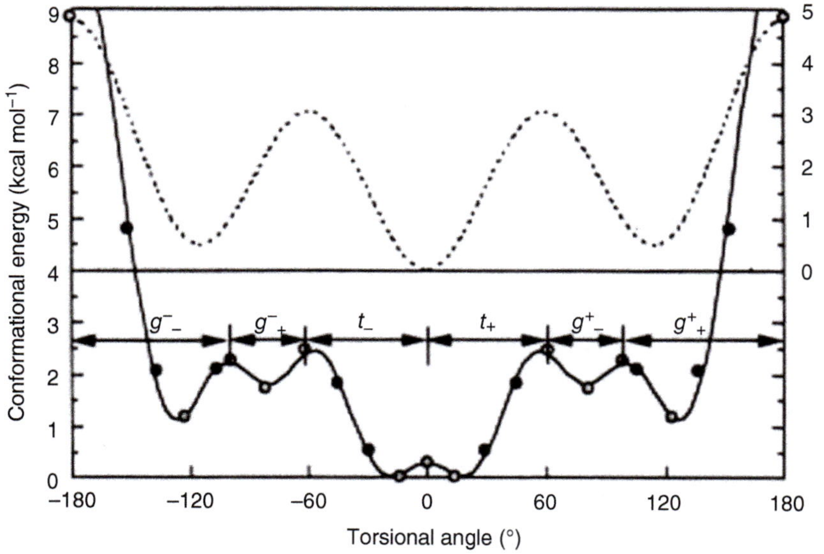

Figure 14.10 Conformational energy against the torsion angle for a hydrocarbon (dotted line) and an R_f group with a helix of 15_7 (solid line). Source: Smith et al. [29].

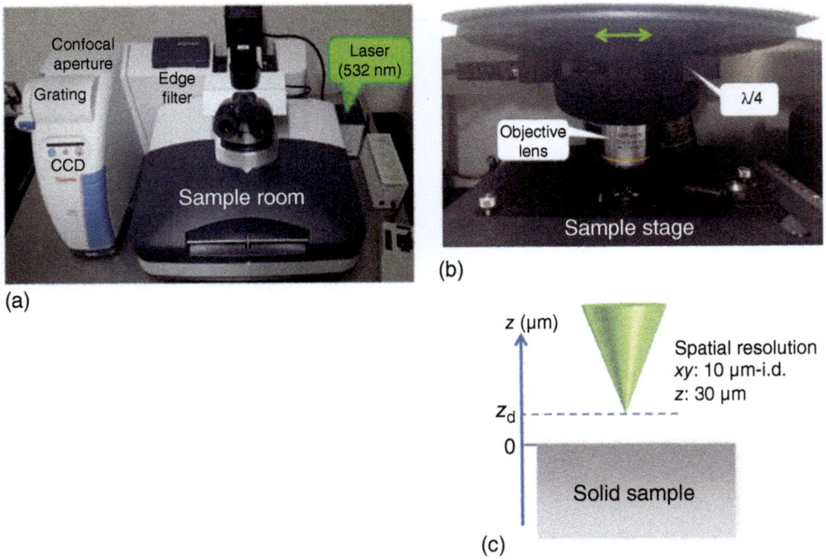

Figure 14.11 (a) The ROA measurement setup built on a Raman imaging microscope. (b) A 1/4 wavelength plate is put in the light path. (c) The light focus is made above the sample surface. Source: Shimoaka et al. [30].

due to the homogeneous aggregation of the identical-handed helices. If the right- and left-handed helices are equally mixed in the aggregate, the aggregate would make achiral.

To experimentally confirm these theoretical predictions, Raman optical activity (ROA) was employed. ROA is one of the chirality-sensitive vibrational spectroscopic techniques. Another famous one is vibrational circular dichroism (VCD) that is built on FT-IR. Both techniques are mostly employed for analysis of a liquid solution. In our study, a molecular aggregate that is solid at ambient temperature is analyzed. For that purpose, a Raman imaging microscope was chosen for building a ROA setup, since our Raman microscope has a very accurate stage-positioning system with a reproducible positioning accuracy of 100 nm in both x and y directions (Figure 14.11b), which is suitable for obtaining right and left circular polarized Raman spectra, I_R and I_L, respectively, at the same position. An ROA spectrum, $I_{ROA} \equiv I_R - I_L$, is obtained by the difference between the two circular polarized Raman spectra. The ROA spectrum, in general, has 1000-times weaker intensity than Raman spectra. Therefore, very accurate and reproducible measurements are definitely required.

According to Yamamoto et al. [31], the intensities of both Raman and ROA signals are enhanced by a decrease of the intermolecular distance between two chiral molecules, and ROA in fact shows an apparently larger enhancement than Raman. When the distance between the two chiral molecules is less than 0.5 nm, in particular, a drastic increase of the enhancement is found. In addition, when a molecule is surrounded by four molecules, for example, ca. 40-times' enhancement is readily explained. Fortunately, in our case, the sample is solid, in which the molecules are

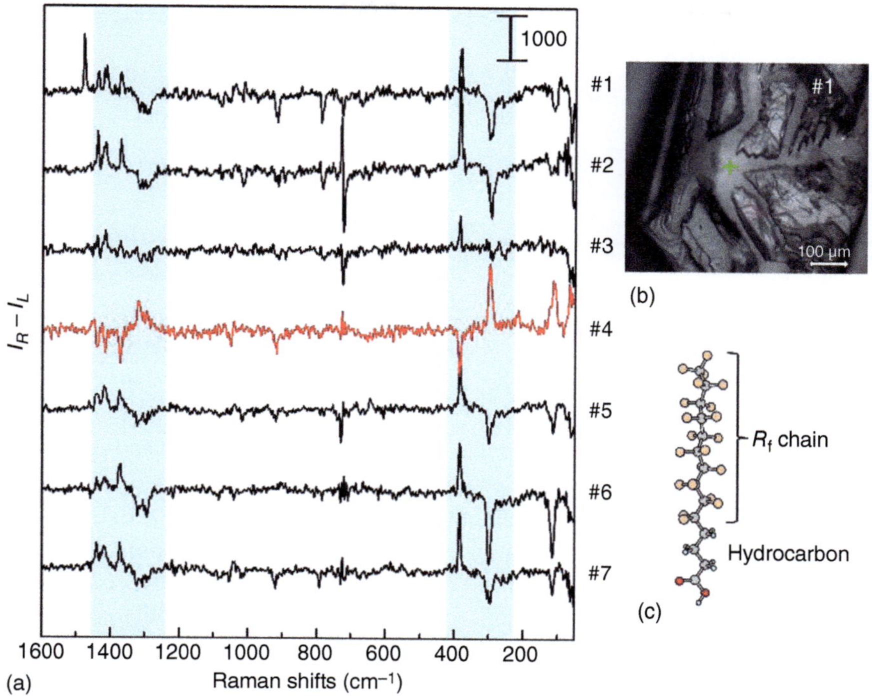

Figure 14.12 (a) ROA spectra of seven crystals of MA-R_f9. (b) Visual image of a crystal surface. (c) Molecular structure of MA-R_f9. Source: Shimoaka et al. [30].

closely packed, which readily enhances the ROA intensity. In this manner, fortunately, the ROA spectra of MA-R_f9 (see Figure 14.12c) crystals were readily measured as presented in Figure 14.12a.

The ROA spectra show an apparent positive and negative ROA-specific peaks, which implies that the crystals have chirality as predicted. Of another note is that only one spectrum (the fourth spectrum from the top) shows an inverted pattern: this crystal comprises reverse-wound molecules. Since a seed crystal has right-wound molecules, for example, the seed induces homogeneous windings because of the atrop isomers with a very shallow energetic barrier.

14.2.2 Surface Modes of Phonon and Polariton

Another important topic of R_f compounds for vibrational spectroscopy is the phonon and polariton modes [32, 33] found in molecular aggregates involving both crystal [34–36] and amorphous polymeric matters [37, 38]. IR and Raman spectra of organic matters mostly consist of normal-mode bands and their related combination bands. On closer inspection, however, some significant bands are found unusual, which cannot be explained by a concept of the normal mode. If such an abnormal band is inappropriately recognized as a normal-mode band, the chemical discussion would go in a wrong direction yielding an incorrect conclusion. To prevent the analytical

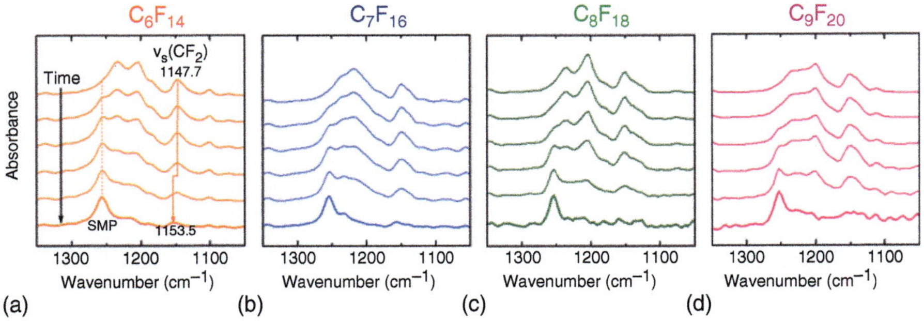

Figure 14.13 IR ATR spectra of C_nF_{2n+2} ($n = 6$–9) with time during an evaporation process. Source: Fukumi et al. [39].

risk, understanding of the "surface modes" appeared with the phonon and polariton is necessary.

Figure 14.13a presents time-dependent IR ATR spectra of liquid of C_6F_{14} at an ambient condition, which is indeed an evaporation process. The ordinate scale is normalized for a better visualization of the spectral shape change. Of interest is that, at the last moment of the evaporation, the spectrum has a largely different shape from that of the former spectra. The strong bands at both 1234 and 1205 cm^{-1} disappear, and a new band at 1255 cm^{-1} develops instead.

When the same experiment was performed for other compounds having different length, very similar changes were found irrespective of even and odd of the chain length (Figure 14.13b–d). Since the phenomena are not found for normal hydrocarbons at all [39], the newly appeared spectrum at the last moment should be attributed to an R_f group.

As apparently shown in the SDA theory (Section 14.1.3), the R_f-containing molecules are aggregated mostly via the dipole–dipole interaction, which can be ignored for hydrocarbons. The dipole–dipole interaction combines intermolecular vibrations to have a lattice vibration, if the vibration propagates for a long distance. This physical situation can be theorized by the "top-down" schematic using the dielectric relative permittivity, $\varepsilon_r(\omega)$, that is a function of the angular frequency, ω, as shown by Eq. (14.4) [20].

$$\varepsilon_r(\omega) = \varepsilon_{r,\infty} + \frac{(\varepsilon_{r,0} - \varepsilon_{r,\infty})\omega_{TO}{}^2}{\omega_{TO}{}^2 - \omega^2 - i\gamma\omega} \tag{14.4}$$

A simple case for a NaCl crystal is shown in Figure 14.14a [39]. Here, the relative permittivity at a high frequency and at a static state are set to $\varepsilon_{r,\infty} = 2.31$ and $\varepsilon_{r,0} = 5.90$, respectively [40].

In an ionic crystal, the vibration propagates for a long distance, which is interpreted as "zero damping," i.e. $\gamma = 0$. Therefore, only the real part remains in Eq. (14.4). As found in the figure, the permittivity has a negative region in the frequency region between ω_{TO} and ω_{LO}. When this function is put in $c^2k^2 = \omega^2\varepsilon_r(\omega)$, then the dispersion relation is obtained as Figure 14.14b. Besides, the reflectivity is also calculated by using the permittivity to have Figure 14.14c. As for $\gamma = 0$, the

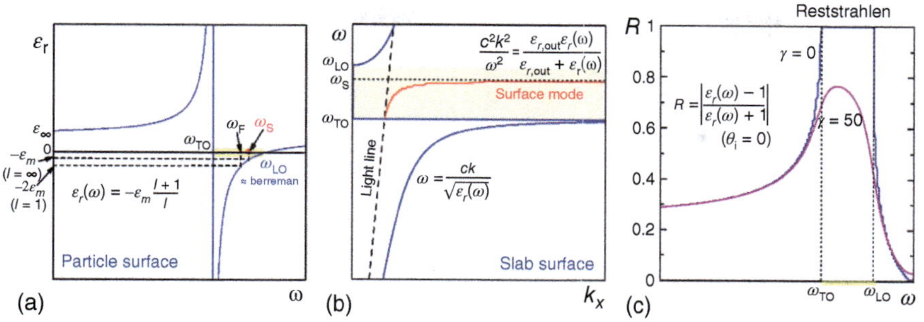

Figure 14.14 (a) Electric relative permittivity of a crystal of NaCl involving the surface modes represented by the Fröhlich mode (ω_F), (b) the dispersion relation on the permittivity, and (c) the reflectivity on the permittivity.

incident light is 100% reflected between ω_{TO} and ω_{LO}, which is known to be the Reststrahlen band.

In an organic material, however, the situation largely changes to have a finite damping factor. For example, for $\gamma = 50$, the Reststrahlen band is changed to have a round shape (Figure 14.14c), which is similar to an absorption peak of a normal mode.

To figure out a hidden Reststrahlen band, the surface modes are focused on. The surface modes are roughly categorized into three types: (i) Berreman's mode of a continuous thin film (slab), (ii) surface polariton mode on a slab, and (iii) surface modes of particles (SMP). The Berreman mode can be distinguished by using the pMAIRS technique, since the mode appears as a pure OP mode of a thin film. Next, the surface polariton mode of a slab is a nonradiative mode, and it requires a waveguide-coupling technique for actual measurements, which can be excluded because we don't employ the coupling technique [32, 33].

The third type depends on the radius of the particle. When the particle size is large, the collective vibrational mode accompanies time retardation, which is a polariton mode. In addition, the modes also depend on the degree of the multipole modes. In any case, however, as found in Figure 14.14a, all the surface modes have a limited frequency region between ω_F and ω_S, both of which are apparently less than ω_{LO}. This rule can be quite useful to find the hidden Reststrahlen band.

The top and the bottom spectra of Figure 14.13 are replotted in Figure 14.15a. By using the attenuated total-reflection (ATR) spectrum of the thick sample at the initial stage, the transverse-optic (TO) and longitudinal optic (LO) functions are calculated by applying the Kramers–Kronig (KK) relations [20] as presented in Figure 14.15b. When the two functions are compared to the two measured spectra (Figure 14.15a), the TO function is close to the spectrum of the thick sample. On the other hand, the SMP band exhibits an apparently lower wavenumber position than the calculated LO peak by about 10 cm^{-1}. By referring to the rule of the third type above, the very thin sample should be a discontinuous film of particles. To confirm this discussion, a similar measurement was performed by using the s-polarization. As shown in Figure 14.15c, the spectra reproduce the unpolarized results. In this manner, the

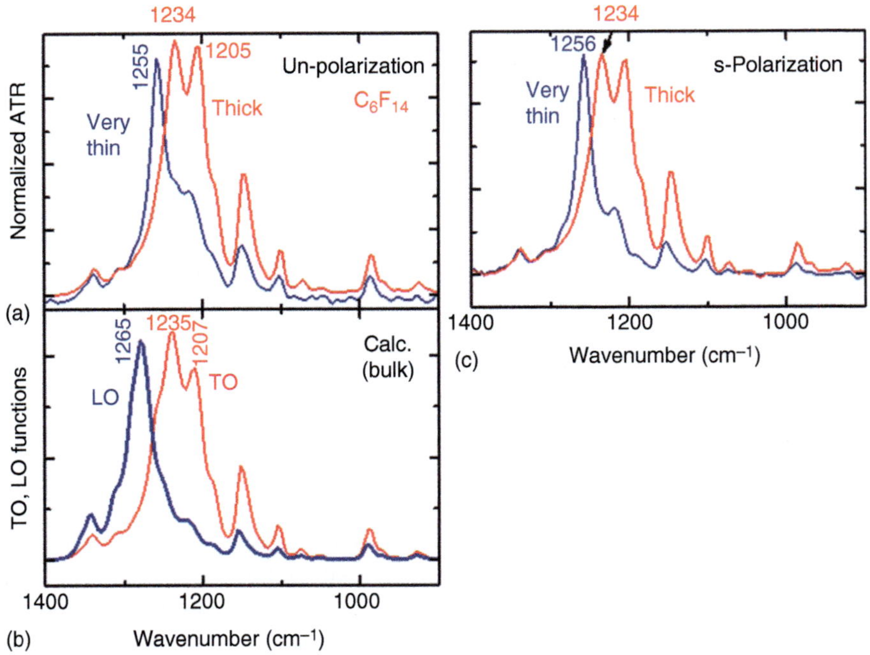

Figure 14.15 (a) The unpolarized IR ATR spectra of thick and very thin C_6F_{14}. (b) Calculated TO and LO functions from the ATR spectrum of the thick sample using the KK relations. (c) A similar measurement to (a) but using the s-polarization. Source: Fukumi et al. [39].

unusual spectral change is found to be induced by the SMP peaks of molecular aggregates of R_f compound.

The CF_2 symmetric stretching vibration ($\nu_s(CF_2)$) band exhibits another interesting change. As found in Figure 14.16a, this peak shows an apparent band shift to a higher wavenumber when the SMP mode appears. Since the $\nu_s(CF_2)$ band is known to be sensitive for molecular packing, this higher-wavenumber shift means the molecular packing is made loose when the continuous film is broken to yield the particle aggregates. This trend is commonly found for C_6F_{14}–C_8F_{18} (Figure 14.16a–c).

Of interest is that C_9F_{20} does not show the shift (Figure 14.16d) even when the SMP peak appears. According to the SDA theory (Section 14.1.3), the chain length of C_9F_{20} corresponds to seven CF_2 groups ($n = 7$), which is a critical length to exhibit a much stronger intermolecular aggregation. This very clear difference implies that the C_9F_{20} molecular aggregates do not come apart even when the surface morphology is largely changed from the continuous to discontinuous film exhibiting the SMP peak.

14.2.3 Summary and Perspective

R_f-containing compounds require a change of fundamental concept of both material chemistry and vibrational spectroscopy. The intrinsically big change is induced

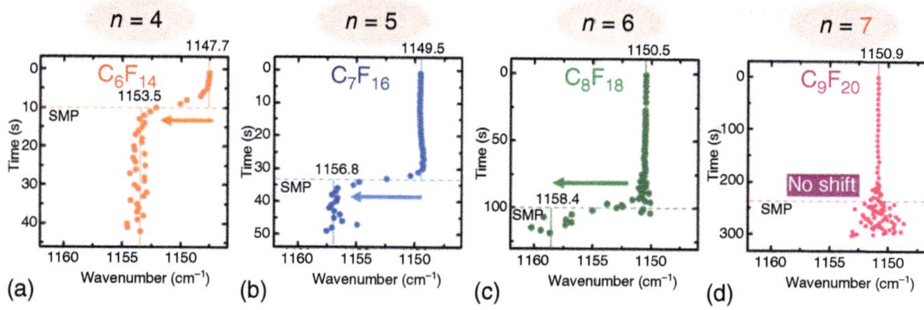

Figure 14.16 Wavenumber plots with time for the same R_f alkanes as used for Figure 14.13. Source: Fukumi et al. [39].

mostly by only two factors of fluorine: (i) the radius of fluorine is larger than that of hydrogen, and (ii) the electronegativity of fluorine is largest of all the elements as represented by the SDA theory. This implies an important fact that nonfluorine compounds cannot be a replacement of an R_f-containing compound. Since the fluorous property is a result of a weak dispersion effect and a strong dipole–dipole interaction, both a small molecular polarizability and a large electronegativity are necessary. Since the molecular polarizability becomes large with a radius of element, we should go up in the periodic table. In a similar manner, a larger electronegativity is found in the right side of the table. As a result, fluorine is the unique solution to yield the fluorous property, and no other element can be a replacement.

At the moment, the material properties are readily understood by the two major factors of fluorine. If we discuss further the molecular interaction of R_f-containing molecules with another compound, electron orbitals should also be taken into account, which is the next step of our study.

References

1 Bunn, C.W. and Howells, E.R. (1954). Structures of molecules and crystals of fluorocarbons. *Nature* 174: 549–551.

2 Hasegawa, T. (2015). Understanding of the intrinsic difference between normal- and perfluoro-alkyl compounds toward total understanding of material properties. *Chem. Phys. Lett.* 627: 64–66.

3 Hasegawa, T., Shimoaka, T., Shioya, N. et al. (2014). Stratified dipole-arrays model accounting for bulk properties specific to perfluoroalkyl compounds. *ChemPlusChem* 79 (10): 1421–1425.

4 Hasegawa, T. (2017). Physicochemical nature of perfluoroalkyl compounds induced by fluorine. *Chem. Rec.* 17 (10): 903–917.

5 Clark, E.S. (1999). The molecular conformations of polytetrafluoroethylene: forms II and IV. *Polymer* 40 (16): 4659–4665.

6 Hirakawa, S. and Takemura, T. (1969). Transitions and phases of polytetrafluoroethylene. *Jpn. J. Appl. Phys.* 8 (6): 635–641.

7 Nakafuku, C. and Takemura, T. (1975). Crystal structure of high pressure phase of polytetrafluoroethylene. *Jpn. J. Appl. Phys.* 14 (5): 599–602.
8 Flack, H.D. (1972). High-pressure phase of polytetrafluoroethylene. *J. Polym. Sci. Part A-2: Polym. Phys.* 10 (9): 1799–1809.
9 Rabolt, J.F., Piermarini, G., and Block, S. (1978). Raman spectroscopic evidence for conformational deformation in the high pressure phase of polytetrafluoroethylene. *J. Chem. Phys.* 69 (6): 2872–2876.
10 Wu, C.-K. and Nicol, M. (1973). Raman spectra of high pressure phase and phase transition of polytetrafluoroethylene (teflon). *Chem. Phys. Lett.* 21 (1): 153–157.
11 Koenig, J. and Boerio, F. (1969). Raman scattering and band assignments in polytetrafluoroethylene. *J. Chem. Phys.* 50 (7): 2823–2829.
12 Brown, R. (1964). Vibrational spectra of polytetrafluoroethylene: effects of temperature and pressure. *J. Chem. Phys.* 40 (10): 2900–2908.
13 Piseri, L., Powell, B.M., and Dolling, G. (1973). Lattice dynamics of polytetrafluoroethylene. *J. Chem. Phys.* 58 (1): 158–171.
14 Koenig, J. and Boerio, F. (1970). Raman scattering and thermal defects in polytetrafluoroethylene. *J. Chem. Phys.* 52 (8): 4170–4171.
15 Hannon, M., Boerio, F., and Koenig, J. (1969). Vibrational analysis of polytetrafluoroethylene. *J. Chem. Phys.* 50 (7): 2829–2836.
16 De Santis, P., Giglio, E., Liquori, A.M., and Ripamonti, A. (1963). Stability of helical conformations of simple linear polymers. *J. Polym. Sci. Part A* 1 (4): 1383–1404.
17 Shimoaka, T., Sonoyama, M., Amii, H. et al. (2017). Study of perfluoroalkyl chain-specific band shift in infrared spectra on the chain length. *J. Phys. Chem. A* 121 (44): 8425–8431.
18 Rabolt, J.F. and Fanconi, B. (1978). Raman scattering from finite polytetrafluoroethylene chains and a highly oriented TFE-HFP copolymer monofilament. *Macromolecules* 11 (4): 740–745.
19 Uno, T., Machida, K., and Miyajima, K. (1968). Infrared spectra of finite chain molecules – II n-alkyltrimethylammonium bromides. *Spectrochim. Acta Part Mol. Spectrosc.* 24 (11): 1749–1763.
20 Hasegawa, T. (2017). *Quantitative Infrared Spectroscopy for Understanding of a Condensed Matter*. Tokyo: Springer.
21 Higgs, P.W. (1953). The vibration spectra of helical molecules: infra-red and Raman selection rules, intensities and approximate frequencies. *Proc. Royal Soc. Lond. Ser. Math. Phys. Sci.* 220 (1143): 472–485.
22 London, F. (1937). The general theory of molecular forces. *Trans. Faraday Soc.* 33: 8–26.
23 Shimoaka, T., Tanaka, Y., Shioya, N. et al. (2016). Surface properties of a single perfluoroalkyl group on water surfaces studied by surface potential measurements. *J. Colloid Interface Sci.* 483: 353–359.
24 Honda, K., Morita, M., Otsuka, H., and Takahara, A. (2005). Molecular aggregation structure and surface properties of poly(fluoroalkyl acrylate) thin films. *Macromolecules* 38 (13): 5699–5705.

25 Hasegawa, T. (2002). A novel measurement technique of pure out-of-plane vibrational modes in thin films on a nonmetallic material with no polarizer. *J. Phys. Chem. B* 106 (16): 4112–4115.

26 Hasegawa, T. (2007). Advanced multiple-angle incidence resolution spectrometry for thin-layer analysis on a low-refractive-index substrate. *Anal. Chem.* 79 (12): 4385–4389.

27 Hasegawa, T., Shimoaka, T., Tanaka, Y. et al. (2015). An origin of complicated infrared spectra of perfluoroalkyl compounds involving a Normal alkyl group. *Chem. Lett.* 44 (6): 834–836.

28 Hasegawa, T. and Shioya, N. (2020). MAIRS: innovation of molecular orientation analysis in a thin film. *Bull. Chem. Soc. Jpn.* 93: 1127–1138.

29 Smith, G.D., Jaffe, R.L., and Yoon, D.Y. (1994). Conformational characteristics of poly(tetrafluoroethylene) chains based upon ab initio electronic structure calculations on model molecules. *Macromolecules* 27 (12): 3166–3173.

30 Shimoaka, T., Sonoyama, M., Amii, H. et al. (2019). Raman optical activity on a solid sample: identification of atropisomers of perfluoroalkyl chains having a helical conformation and no chiral center. *J. Phys. Chem. A* 123 (18): 3985–3991.

31 Yamamoto, S., Li, X., Ruud, K., and Bouř, P. (2012). Transferability of various molecular property tensors in vibrational spectroscopy. *J. Chem. Theory Comput.* 8 (3): 977–985.

32 Ruppin, R. and Englman, R. (1970). Optical phonons of small crystals. *Rep. Prog. Phys.* 33 (1): 149–196.

33 Agranovich, V.M. and Mills, D.L. (1982). *Surface Polaritons*. North-Holland Publishing Company.

34 Fuchs, R. and Review, K.K. (1965). Optical modes of vibration in an ionic crystal slab. *Physiol. Rev.* 140 (6A): 2076–2088.

35 Kliewer, K.L. and Fuchs, R. (1966). Optical modes of vibration in an ionic crystal slab including retardation I. Nonradiative region. *Physiol. Rev.* 144 (2): 495–503.

36 Kliewer, K.L. and Fuchs, R. (1966). Optical modes of vibration in an ionic crystal slab including retardation II. Radiative region. *Physiol. Rev.* 150 (2): 573–588.

37 Nagai, N., Okawara, M., and Kijima, Y. (2015). Infrared response of sub-Micron-scale structures of polyoxymethylene: surface polaritons in polymers. *Appl. Spectrosc.* 70 (8): 1278–1291.

38 Nagai, N., Okada, H., and Hasegawa, T. (2019). Morphology-sensitive infrared absorption bands of polymers derived from surface polaritons. *AIP Adv.* 9 (10): 105203.

39 Fukumi, A., Shimoaka, T., Shioya, N. et al. (2020). Infrared active surface modes found in thin films of perfluoroalkanes reveal the dipole–dipole interaction and surface morphology. *J. Chem. Phys.* 153: 044703.

40 Shanker, J., Agrawal, G.G., and Singh, R.P. (2006). The Clausius-Mossotti and Mott-Littleton theories of the static dielectric constant of alkali halide crystals. *Philos. Mag. B* 39 (5): 405–411.

15

Probing Structures of Conductive Polymers with Vibrational Spectroscopy

Jianming Zhang and Yuan Yuan

Qingdao University of Science & Technology, Key Laboratory of Rubber-Plastics, Ministry of Education/ Shandong Provincial Key Laboratory of Rubber-plastics, School of Polymer Science and Engineering, 53 Zhengzhou Road, Qingdao, 266042, China

15.1 Introduction

The utility of the vibrational spectroscopy, including infrared spectroscopy and Raman spectroscopy, stems from the specificity to different molecular groups and segmental constituents of polymers, which gives the technology a unique capacity to establish relation between the molecular structures and the properties. Along with the rapid technological development, vibrational spectroscopy has been widely used in probing hierarchical polymer structures as a powerful tool. In this chapter, we reviewed the applications and research progress in the fields of the structure characterization focused on conductive polymers by using vibrational spectroscopy in recent 20 years. The applications include the following six aspects: (i) the sensitive bands identification for polymer chain conformation, chain packing/aggregates, and doping state; (ii) the structural changes under the external fields, including temperature and electric field; (iii) structural dynamics; (iv) chemical composition/morphology analysis in conductive-polymer-based blends; (v) surface/interface molecular orientation; and (vi) structure and dynamics of charge carriers. Last, the development of vibration spectroscopy in polymer research is prospected.

15.2 Application of Vibrational Spectroscopy

15.2.1 Chain Packing/Aggregate Mode Identification

Infrared spectroscopy can be utilized to investigate the multiple chain packing in conjugated polymers such as poly(3-butylthiophene) (P3BT). Up to now, three types of polymorphs termed forms I, I′, and II have been identified in P3BT. The characteristic vibration in the C_b–H out-of-plane deformation mode $\delta(C_b\text{–H})$ at 860–790 cm^{-1} receives continuous attention because of its sensitivity to the

Spectroscopic Techniques for Polymer Characterization: Methods, Instrumentation, Applications, First Edition. Edited by Yukihiro Ozaki and Harumi Sato.

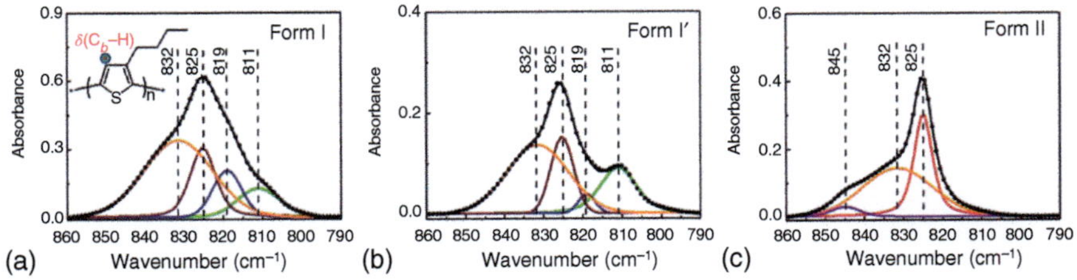

Figure 15.1 Room-temperature FTIR spectra of the δ(C$_b$–H) vibration mode for P3BT Form I (a), Form I′ (b), and Form II (c). Measured spectra are indicated by black filled circles and fitted spectra are shown by solid lines. Dashed lines, serving as guides to the eye, indicate peak positions. The inset shows the chemical structure of P3BT, where the out-of-plane vibration direction of the δ(C$_b$–H) mode is indicated by the blue–red circle. Source: Yuan et al. [1].

phase state in poly(3-alkylthiophenes) (P3ATs) [1–5]. By using the deconvolution analysis, several peaks are observed in the spectrum of the δ(C$_b$–H) mode for each of the P3BT modifications in Figure 15.1 [1]. It is proposed that the 811 and 819 cm^{-1} bands exhibited in forms I and I′ represent that two types of main-chain stacking structures coexist in the crystalline phase, while one ordered main-chain structure is probed in form II, exhibiting the sharp 825 cm^{-1} band. Additionally, it is considered a mesophase exists in forms I and I′, which could exhibit the broad band at 825 cm^{-1} because it shows distinct intensity-temperature dependence from the sharp 825 cm^{-1} band in form II at low temperature (below room temperature).

By using Raman spectroscopy, Gao et al. identified the C=C symmetric stretching Raman modes (ν_s(C=C)) of poly(3-hexylthiophene) (P3HT) in P3HT/[6,6]-phenyl-C$_{61}$-butyric acid methyl ester (PCBM) blend films [6]. As shown in Figure 15.2a, the ν_s(C=C) mode of the as-cast film shows contributions from the aggregated and unaggregated P3HT chains with characteristic Raman frequencies of ~1450 ($I_{C=C^{agg}}$) and ~1470 cm^{-1} ($I_{C=C^{un}}$) by spectrum deconvolution, respectively. Upon annealing, the relative concentration of the two species (defined as $R = I_{C=C^{agg}}/I_{C=C^{un}}$) increases as shown in Figure 15.2b. It is proposed that R value is associated with the relative density of the aggregated and unaggregated species. Resonance Raman imaging was further performed to spatially map morphology-dependent variations of R values (Figure 15.2c). From both R and frequency dispersion resonance Raman images of $I_{C=C^{agg}}$ and $I_{C=C^{un}}$ species, four types of P3HT chains with aggregation degree are identified and mapped in annealed blend films.

15.2.2 Conformation-Sensitive Bands Identification

Infrared spectra collected below room temperature can be used for uncovering the conformation characteristic of the conductive polymers such as P3BT. The order of the side-chains of P3BT polymorphs is evaluated from the profiles of the characteristic vibrations [1]. Cooling the samples from 20 to −100 °C helps to visualize spectral features more clearly. Because disorder usually causes larger line-widths in the infrared spectrum, the observation of significantly narrower line widths in form I′ reveals more ordered butyl side chains as compared to form I

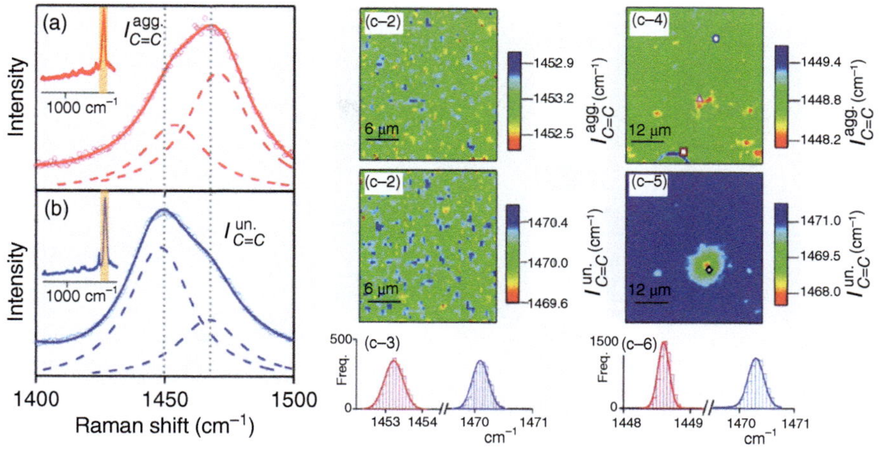

Figure 15.2 (a, b) Raman spectra of as-cast (a, red) and annealed (b, blue) blend films excited with 488 nm light show the v_s(C=C) band of P3HT represented by the shaded regions of the complete spectra shown as insets. The band is fitted with two Lorentzian functions (dashed traces) showing the relative contributions of both aggregated ($I_{C=C^{agg}}$) and unaggregated ($I_{C=C^{un}}$) components. (c-1 and c-2) $I_{C=C^{agg}}$ and $I_{C=C^{un}}$ center frequency dispersion images for P3HT/PCBM as-cast films, and (c-3) histograms of frequency components. (c-4 and c-5) $I_{C=C^{agg}}$ and $I_{C=C^{un}}$ center frequency images for P3HT/PCBM annealed films, and (c-6) histograms of frequency components. Source: Reproduced by permission of the American Chemical Society from Gao and Grey [6].

as shown in Figure 15.3a,b. In contrast, the CH_2-vibrations of form II as shown in Figure 15.3c reveal a band splitting into two components at 2862 and 2852 cm^{-1} ($\upsilon_s(CH_2)$) and at 2933 and 2920 cm^{-1} ($\upsilon_{as}(CH_2)$). Band splitting is even more visible in the second derivative spectra. The mode of the interdigitated butyl side chains is suggested therefore.

Raman spectroscopy is a powerful technique for investigating the conformation of conjugated polymers [7, 8], in which resonance Raman spectroscopy was employed by Gao et al. to identify the ordered and disordered conformers of poly(2,5-bis(3-tetradecylthiophen-2-yl)thieno[3,2-*b*]thiophene) (PBTTT) blended with PCBM in bulk heterojunction solar cells [7]. It is identified that the PBTTT thiophene ring symmetric C=C stretching mode (v_s(C=C)) consists of contributions both from ordered and disordered components as shown in Figure 15.4a. The vibration mode of the ordered conformer is located at 1489 cm^{-1} with the peak width of ~15 cm^{-1}, whereas that of the disordered one is located at 1500 cm^{-1} with the peak width of ~25 cm^{-1}. The relative amounts of two components are sensitive to PCBM loading, annealing, and excitation energy. Resonance Raman images of the annealed PBTTT/PCBM model solar cells as shown in Figure 15.4b confirm that the ordered PBTTT chains are most concentrated in PCBM-rich phases.

15.2.3 Doping-Sensitive Bands Identification

Raman spectroscopy and its imaging technique were used successfully to directly study polaron localization/relaxation with the added advantage of excellent spatially resolution in doped conducting polymers. For example, the polaronic

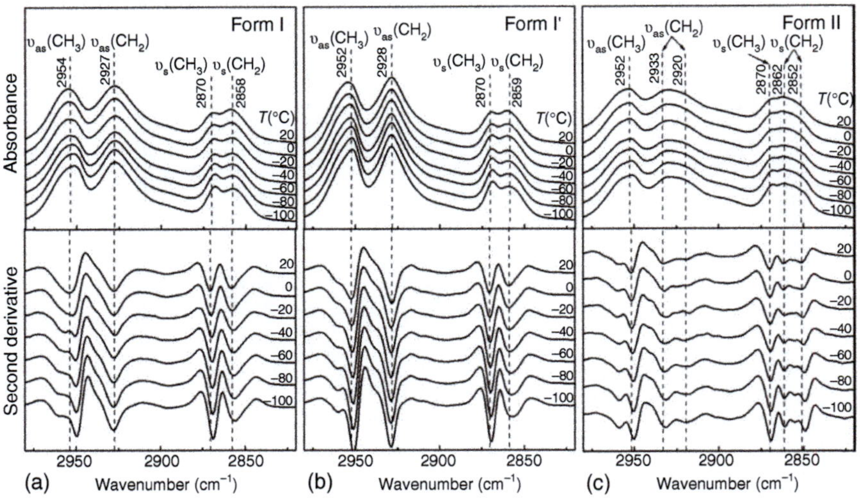

Figure 15.3 Temperature-dependent FTIR spectra of P3BT Form I (a), Form I′ (b), Form II (c), and their corresponding second derivatives in the region of 3000–2800 cm^{-1}. Dashed lines, serving as guides to the eye, indicate peak positions and their assignment along with wavenumbers determined from the spectra measured at room temperature. Source: Yuan et al. [1].

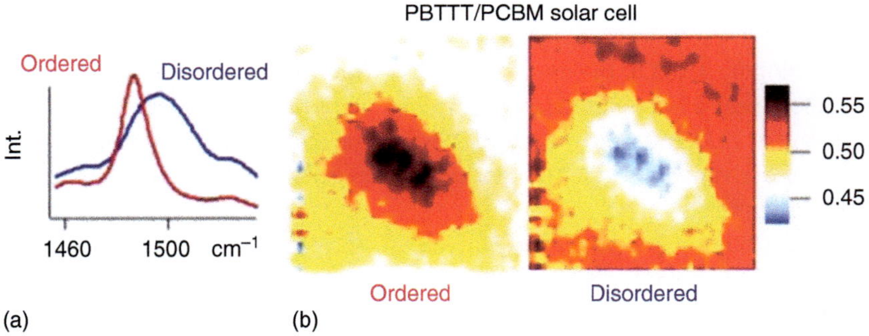

Figure 15.4 (a) Resonance Raman spectra of the ν_s(C=C) mode for as-cast PBTTT/PCBM blend thin films showing lineshapes of the ordered and disordered PBTTT forms. (b) Resonance Raman images of the integrated Raman intensity ratio of the ordered and disordered PBTTT species. Source: Reproduced by permission of the American Chemical Society from Gao et al. [7].

nature of PBTTT is investigated by Raman spectroscopy combined with the density functional theory (DFT) calculation [9]. As shown in Figure 15.5a, PBTTT exhibits a significant increase of the relative intensity of the ν(C=C) mode (as denoted by mode A) in thienothiophene upon electrochemical and chemical hole doping. This behavior indicates that the positive polaron of the doped polymer chain is mostly localized on the thienothiophene unit. Computed Raman spectra as shown in Figure 15.5b suggest that the observed intensity increase of the ν(C=C) mode is supported by the additional polarization of the polymer induced by the electric field. Furthermore, the observed changes in the intensity ratios of characteristic

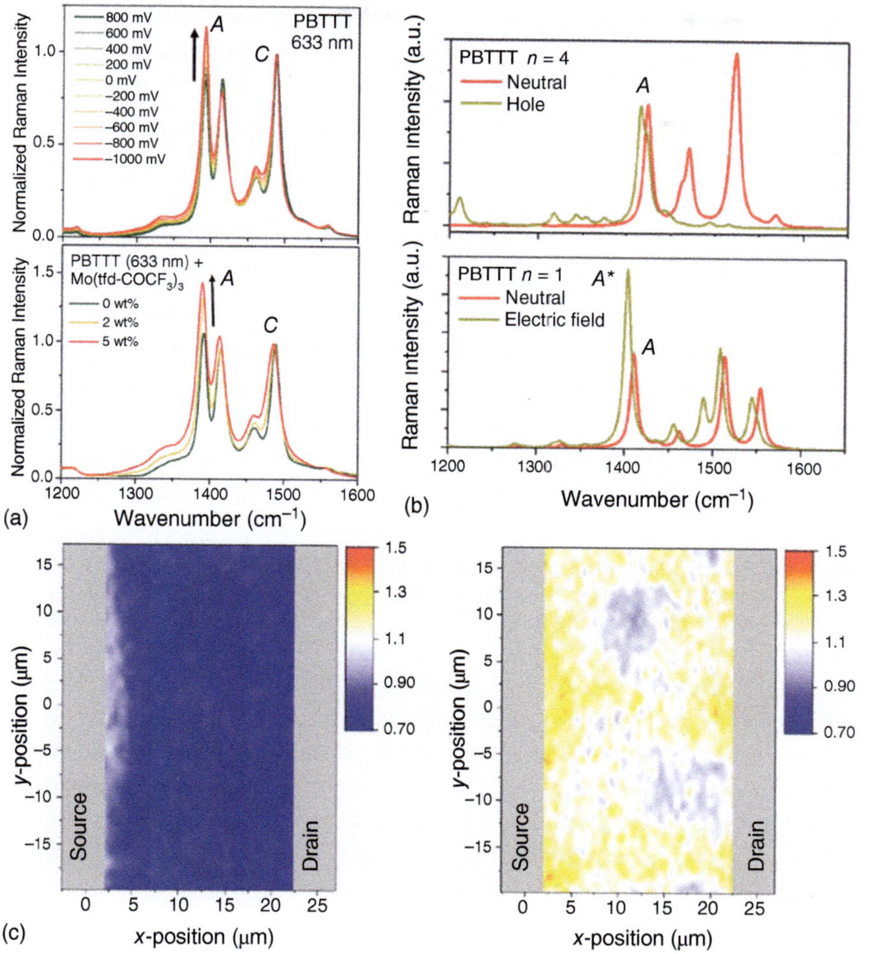

Figure 15.5 (a) Raman spectra of electrochemically doped PBTTT depending on applied gate voltage and doping levels with Mo(tdf-COCF$_3$)$_3$. Raman intensity normalized with respect to the 1488 cm^{-1} band (mode *C*). (b) DFT computed Raman spectra. The spectra of the neutral and cation species of PBTTT ($n = 4$) oligomers, and those of the neutral species and polarized species (i.e. applied electric field) of the PBTTT monomer are compared. (c) Raman maps of electrolyte-gated transistors with PBTTT (nanoribbon morphology) showing variation of charge carrier density with gate voltage visualized by the peak height ratio of mode *A*/mode *C* that is roughly proportional to the hole concentration. Source: Reproduced by permission of the Royal society of Chemistry from Francis et al. [9].

Raman bands, depending on the hole concentration, enable fast and simple in situ mapping of carrier densities in electrolyte-gated polymer transistors with high spatial resolution as shown in Figure 15.5c.

15.2.4 Thermally Induced Phase Transitions

The utility of infrared spectroscopy stems from its specificity to different molecular groups and segmental constituents of polymer systems, which gives this method a

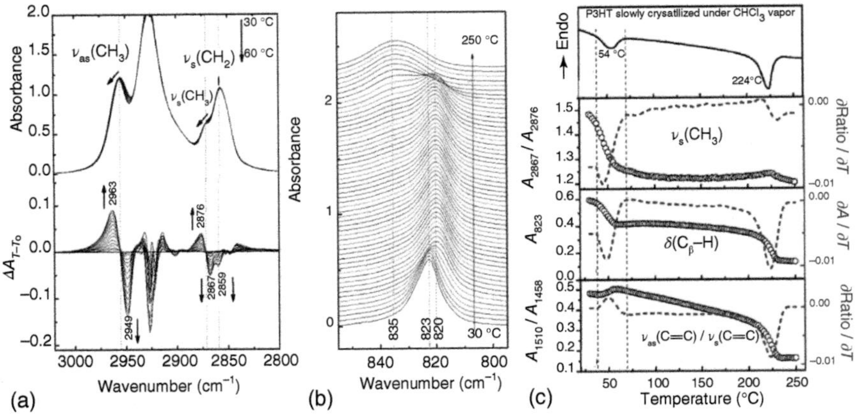

Figure 15.6 (a) FTIR spectral changes in the C–H stretching vibration region from 30 to 60 °C and the corresponding difference spectra obtained by the subtraction of the initial spectrum measured at 30 °C for slowly crystallized P3HT sample. (b) Temporal infrared spectral changes in the $\delta(C_\beta-H)$ band region during the heating process. (c) Differential scanning calorimetry (DSC) heating trace and the spectral changes associated with side-chain conformation $\nu_s(CH_3)$, main-chain conformation $\nu_{as}(C{=}C)/\nu_s(C{=}C)$ as well as chain packing $\delta(C_\beta-H)$. Source: Yuan et al. [2].

unique capacity to elucidate the molecular origin of phase transition phenomena [2, 10]. For example, temperature-dependent Fourier transform infrared spectroscopy (FTIR) was employed to explain the origin of the form I′-to-form I phase transition that occurred around 54 °C in P3HT crystal [2]. As seen from the original spectra in Figure 15.6a, a subtle peak shift toward higher frequencies can be discernible for the $\nu_s(CH_3)$ and $\nu_{as}(CH_3)$ bands of the P3HT side chains. Such spectral changes are presented more clearly in the corresponding difference spectra. The appearance of a positive peak located at 2963 cm^{-1} and a negative one located at 2949 cm^{-1} contributes to the peak shift of $\nu_{as}(CH_3)$ band. A similar situation can be found for the $\nu_s(CH_3)$ mode. This observation leads to the conclusion that the packing of end CH_3 groups in form I′ experiences an obvious disordering during the phase transition as revealed by the intensity ratio of A_{2867}/A_{2876} as shown in Figure 15.6c. With increasing temperature, the characteristic band of the main-chain packing located at 823 cm^{-1} decreases in intensity and shifts toward 820 cm^{-1} gradually as shown in Figure 15.6b,c, indicating a change in the π–π stacking modes of the conjugated backbone chain during the phase transitional process.

15.2.5 Structural Dynamics

Time-resolved infrared spectroscopy can be used to study the isothermal melt-crystallization kinetics of the conductive polymers [11]. Figure 15.7a shows the time-dependent spectral changes in the region of 850 to 800 cm^{-1} for poly(3-octylthiophene) (P3OT). The crystalline 821 cm^{-1} band gradually increases while the amorphous 835 cm^{-1} band in intensity during the whole isothermal crystallization process. The simultaneous exchange of these two bands in intensity

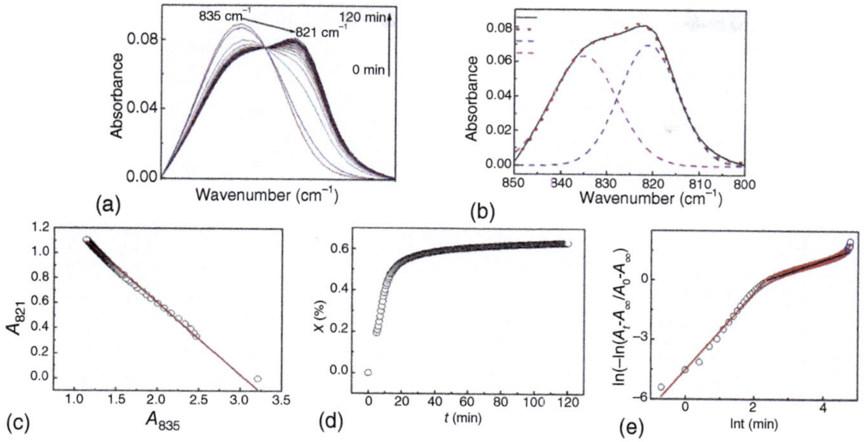

Figure 15.7 (a) Spectral changes in the region of 850 to 800 cm^{-1}during the melt-crystallization process of P3OT film at 165 °C. (b) An example for the curve fitting in spectral region of 850 to 800 cm^{-1}. (c) The plot of the intensity of the 821 cm^{-1} band as a function of the intensity of the 835 cm^{-1} band. (d) Calculated crystallinity of P3OT as a function of crystallization time. (e) Avrami plot of the crystallization dynamics of P3OT. Source: Wang et al. [11].

and the isosbestic point strongly suggests that the crystallization of P3OT is a first-order phase transition process without the intermediate phase.

For the two-phase system, it is possible to evaluate the changes in relative crystallinity of polymer. Based on the Beer–Lambert law and the definition of crystallinity, the crystalline degree of P3OT can be expressed as follows:

$$X_c = \frac{A_{821}}{\frac{k_1}{k_2} \times A_{835} + A_{821}} \tag{15.1}$$

where X_c is the crystallinity, A is the integrated peak area, which can be determined by spectrum deconvolution as shown in Figure 15.7b, and k is specific absorption coefficient. By linear fitting of the A_{821}/A_{835} plot as shown in Figure 15.7c, $k_1/k_2 = 0.57$ is derived. Then, the plot of P3OT crystallinity as a function of crystalline time can be depicted in Figure 15.7d. The well-known Avrami equation is used to analyze the isothermal crystallization process further. When infrared data are used, Avrami equation can be stated as follows:

$$\ln\left[-\ln\left(\frac{A_t - A_\infty}{A_0 - A_t}\right)\right] = \ln k + n \ln t \tag{15.2}$$

where A_t is the peak intensity at the crystallization time, A_0 and A_∞ are the initial and final peak intensities during isothermal crystallization, respectively, k is the crystallization rate constant, t is the crystallization time, and n is the Avrami exponent. The intensity change of the 821 cm^{-1} band with crystallization time is plotted according to Eq. (15.2). The result is shown in Figure 15.7e. It is interesting to find that two linear fittings are required, indicating two distinct growth stages for P3OT main-chain crystallization.

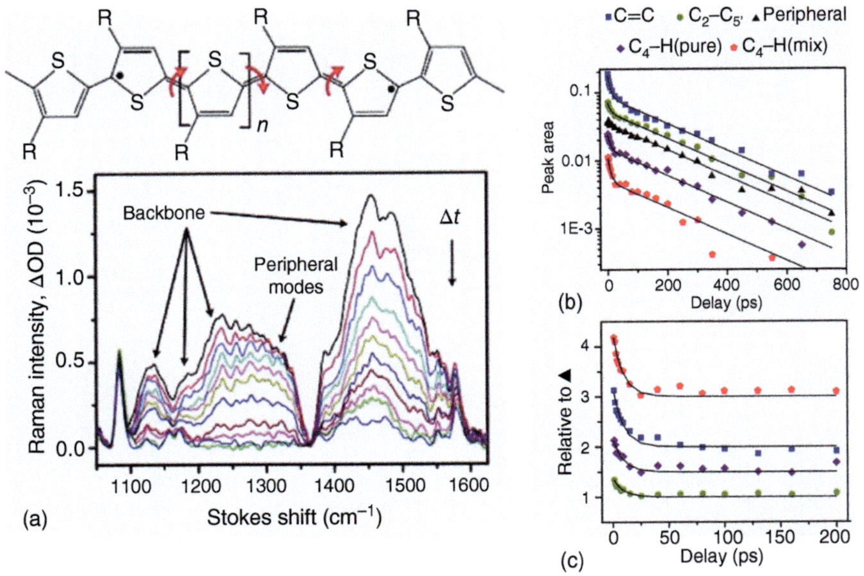

Figure 15.8 (a) Valence-bond Structures of the Quinoidal Excited State of P3HT and the time-resolved resonant-Raman spectra of P3HT in chlorobenzene photoexcited at 510 nm. (b, c) Time dependence of Raman band intensities in figure (a). Integrated intensities (b), black lines correspond with biexponential fits with constrained lifetimes of 9 ± 1 and 220 ± 20 ps. Relative change in feature intensities attributed to torsion-induced exciton conformational relaxation (c). Source: Yu et al. [12].

Time-resolved resonant Raman spectroscopy was used to investigate the conformational relaxation of P3HT in chlorobenzene on a time scale of picosecond [12]. As shown in Figure 15.8a, the characteristic vibrations along the thiophene backbone and peripheral modes, which are insensitive to the relaxation dynamics, are assigned according to the time dependence of excited-state Raman vibrations. The intensity changes of all the assigned Raman features with time are plotted in Figure 15.8b subsequently. The relative intensity dependence of Raman bands uncovers the evolution in bond lengths along the backbone resulting from increased conjugation as revealed by Figure 15.8c.

15.2.6 Chemical Composition/Morphology Analysis in Conductive-Polymer-Based Blends

Surface-enhanced Raman spectroscopy (SERS) is a powerful technique to offer the same potential selectivity and vertical resolution to the study of organic thin films. Razzell-Hollis et al used SERS to probe the interfacial composition and molecular order in P3HT:poly((9,9-dioctylfluorene)-2,7-diyl-alt-[4,7-bis(3-hexylthien-5-yl)-2,1,3-benzothiadiazole]-2′,2″-diyl) (F8TBT) blend thin films [13]. The spectra of neat P3HT and F8TBT films were firstly collected to identify the spectral variations that may result from changes in bulk (Raman) or interface (SERS) morphology as a result of annealing. Then, the spectra of the blend films under different sample

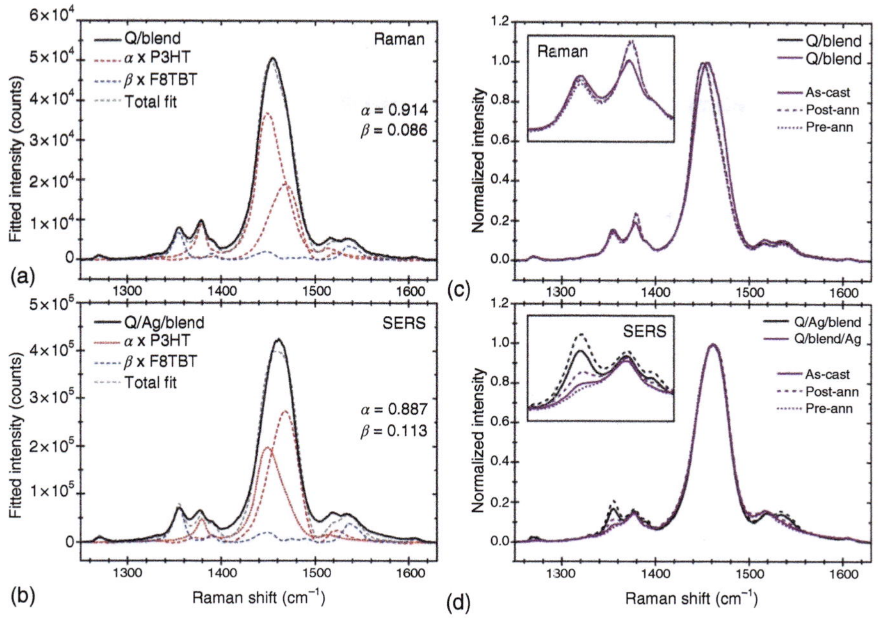

Figure 15.9 Raman (a) and SERS (b) spectra for an as-cast sample of quartz (quartz) Q/Ag/P3HT:F8TBT, fitted using RR-P3HT (as ordered fraction), RRa-P3HT (as disordered fraction) and F8TBT spectra to obtain relative contributions of P3HT (α) and F8TBT (β). Normalized Raman (c) and SERS (d) spectra for P3HT:F8TBT blends in five different sample configurations, with variation in the relative intensity of the F8TBT peak at 1356 cm^{-1} shown in each inset. Source: Razzell-Hollis et al. [13].

configurations were collected, and the spectrum deconvolution was performed to obtain the intensity values α for P3HT and β for F8TBT, which is essential to calculate the chemical composition. The results are shown in Figure 15.9. The authors suggest that it is possible to modify the interfacial compositions through thermal annealing, but it strongly depends on whether the films are annealed before or after evaporation of the metallic top contact. Preannealing increases the concentration of low-surface-energy P3HT at that top interface, whereas postannealing raises the concentration of high-surface-energy F8TBT.

Raman imaging technique was employed successfully to investigate the chemical composition in conductive polymers blended with others [14, 15] [16]. For example, Raman imaging was used to spatially map morphology-dependent vibrational signatures of poly[2,6-(4,4-bis-(2-ethylhexyl)-4*H*-cyclopenta[2,1-*b*;3,4-*b*′]dithiophene)-alt-4,7(2,1,3-benzothiadiazole)] (PCPDTBT) donor/acceptor moieties within the solar cell devices [14]. Spectrum analysis was performed on the BT symmetric in-plane C–H wagging mode (1269 cm^{-1}) and dithiophene C–C stretching mode (1423 cm^{-1}) as shown in Figure 15.10a. Figure 15.10b shows integrated Raman intensity images of the given modes for a PCPDTBT:PCBM solar cell device excited with high-energy (458 nm) and low-energy (647 nm) excitation. Raman images indicate a well-mixed morphology in devices with the blends of PCPDTBT and

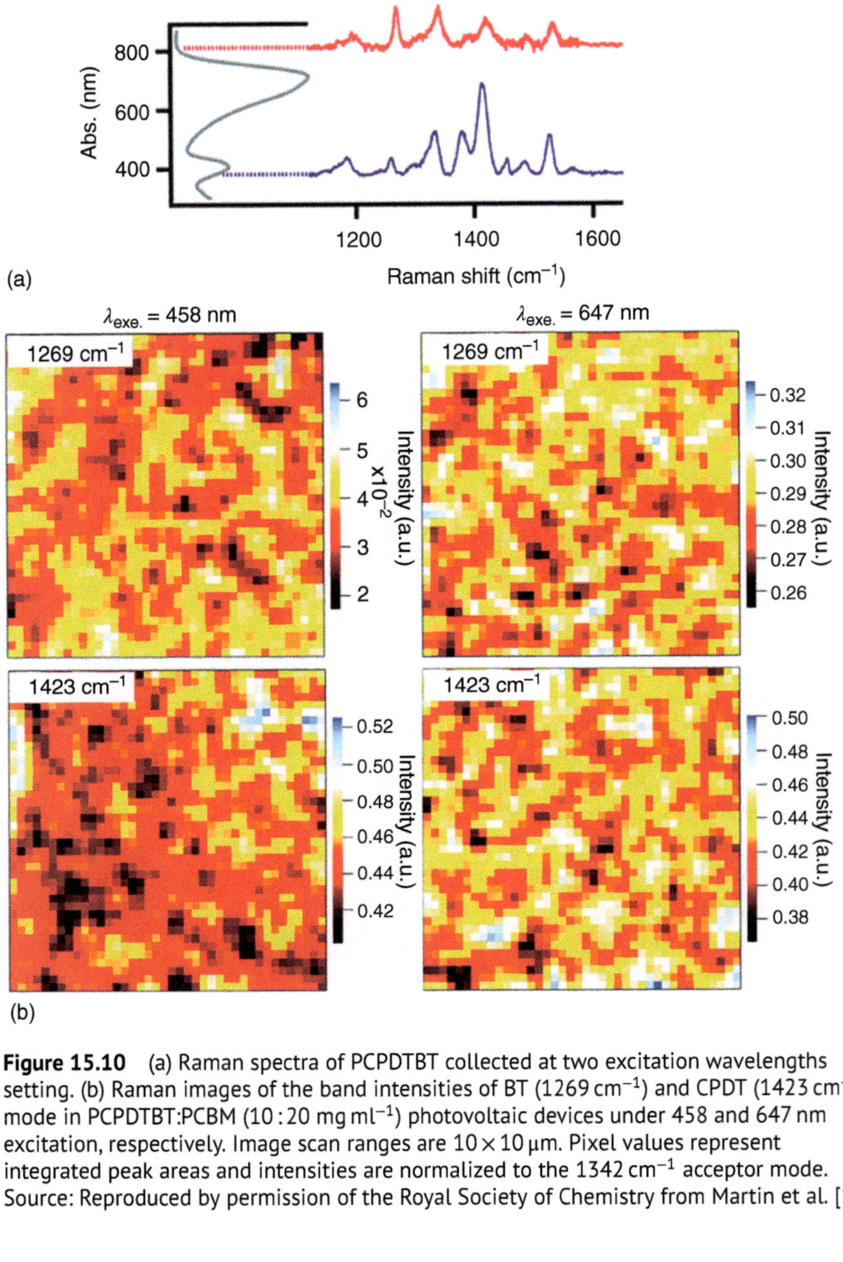

Figure 15.10 (a) Raman spectra of PCPDTBT collected at two excitation wavelengths setting. (b) Raman images of the band intensities of BT (1269 cm^{-1}) and CPDT (1423 cm^{-1}) mode in PCPDTBT:PCBM (10 : 20 mg ml^{-1}) photovoltaic devices under 458 and 647 nm excitation, respectively. Image scan ranges are 10 × 10 μm. Pixel values represent integrated peak areas and intensities are normalized to the 1342 cm^{-1} acceptor mode. Source: Reproduced by permission of the Royal Society of Chemistry from Martin et al. [14].

PCBM, which exhibit small-scale phase segregation at micrometer level with no apparent crystallite formation.

Scanning angle (SA) Raman spectroscopy (the schematic of which is shown in Figure 15.11A) is a versatile method applicable for studying molecular order in conductive polymers and their blend films [17, 18]. Enhanced Raman signals, film thickness, and molecular order can be achieved simultaneously. Whether the Raman scattering is generated within the polymer film can be determined

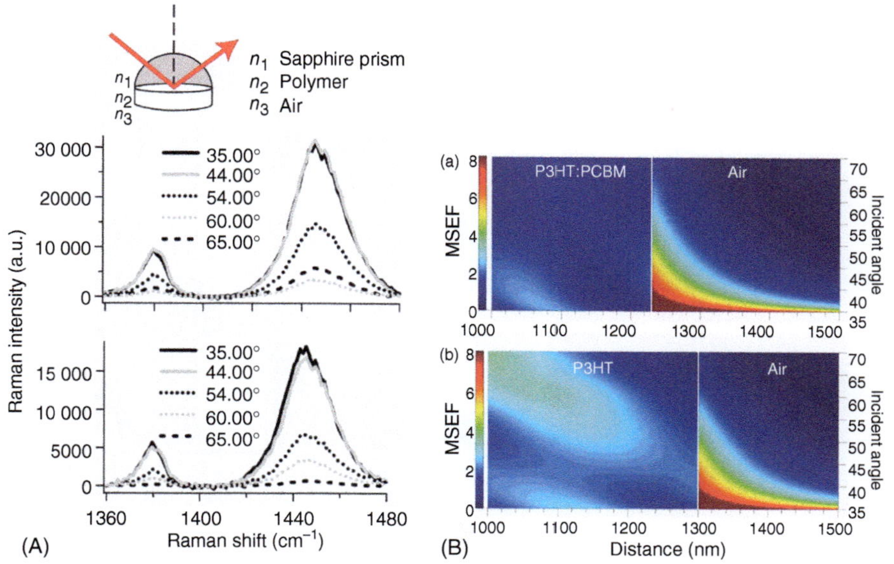

Figure 15.11 (A) Schematic of the SA Raman interface used to collect the data shown in (b) and (c) (a). SA Raman spectra at the indicated incident angles for (b) P3HT and (c) 1 : 1 P3HT:PCBM deposited on a sapphire substrate. (B) Calculated MSEF as a function of distance and incident angle for the interface: 0–1000 nm sapphire/1000–1230 nm P3HT:PCBM/1230–6000 nm air (a), 0–1000 nm sapphire/1000–1300 nm P3HT/1300–6000 nm air (b). The MSEF in the sapphire layer (0–1000 nm) and the majority of the air layer (greater than 1500 nm) are omitted for clarity. The calculated plots show the expected distance dependence of the experimental Raman signal in the *Z* direction. Source: Reproduced by permission of the American Chemical Society from Meyer et al. [17].

when combined with mean square electric field (MSEF) calculations furthermore. The film morphology of P3HT and P3HT:PCBM blends on several substrates was studied using SA Raman technique [17]. The SA Raman spectra collected at selected incident angles for films on sapphire are shown in Figure 15.11A. The Raman intensity of the P3HT film decreases as the SA increases from 35° to 60° and then increases at larger incident angles, whereas the Raman signal of the blend film decreases with increased incident angle. MSEF calculation as shown in Figure 15.11B reveals the distance dependence on the Raman signal in the *Z* direction, which aids in the Raman spectra interpretation. The expected Raman signal is proportional to the integrated MSEF across the entire polymer thickness, which agrees with the experimental Raman spectra. Furthermore, as the width of the thiophene C=C stretching mode (~1447 cm^{-1}) is sensitive to P3HT order, it is revealed that the molecular order varies depending on the underlying substrates.

15.2.7 Surface/Interface Molecular Orientation

Sum-frequency generation (SFG) vibrational spectroscopy is a second-order nonlinear spectroscopy in which tunable visible and infrared laser sources are spatially and temporally overlapped in a sample to produce light at the sum of their frequencies.

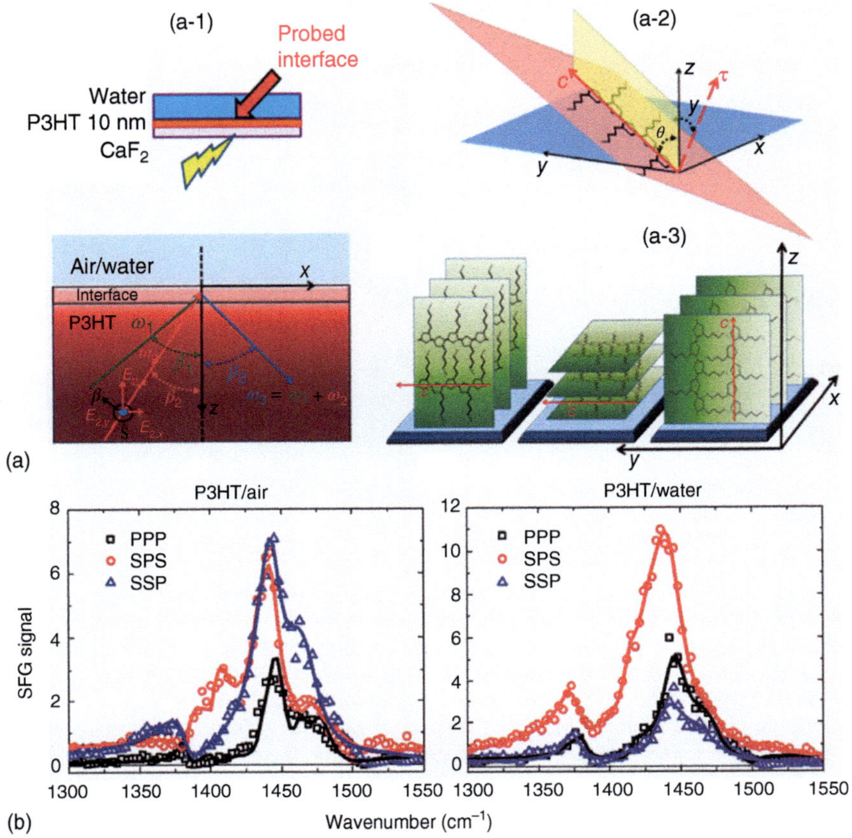

Figure 15.12 (a-1) Sketch of the considered interface and of the set-up geometry; (a-2) definition of lab (*xyz*) and molecular (*abc*) coordinates; the chain tilt angle *y*, defined as the angle between the axes *c* and *z*, and the twist angle *g*, defined as the angle formed between the torsion versor *t* and the vertical plane *cz*, are also shown; (a-3) sketch of prototypical molecular arrangement of P3HT; *c* indicates the orientation axis of the polymer chain. From left to right: edge-on, face-on, and vertical configurations. (b) SFG spectra measured for a thin film of P3HT in contact with air and with water for three different polarization sets. Results of the fitting to a superposition of vibrational resonances of the P3HT backbone are represented, for each polarization set, as solid lines. Source: Bellani et al. [19].

It is a surface-specific technique compatible with electrochemical/biological conditions and can be employed to study the molecular orientation at the water/air [19] or the metal [20] interfaces. As shown in Figure 15.12a, Bellani et al. used SFG spectroscopy to study the chain orientation of the P3HT thin film exposed to direct contact with water and air [19]. The ν_s(C=C) mode (~1440 cm^{-1}) as shown in Figure 15.12b exhibits the intensity dependence on the interface property and the polarization sets (see Figure 15.12a). This implies a more pronounced parallel orientation of the backbone with respect to the water interface that is a preferential "edge-on" configuration in the case of water and a "vertical" one in the case of air. Experimental results are corroborated by molecular dynamics simulations further.

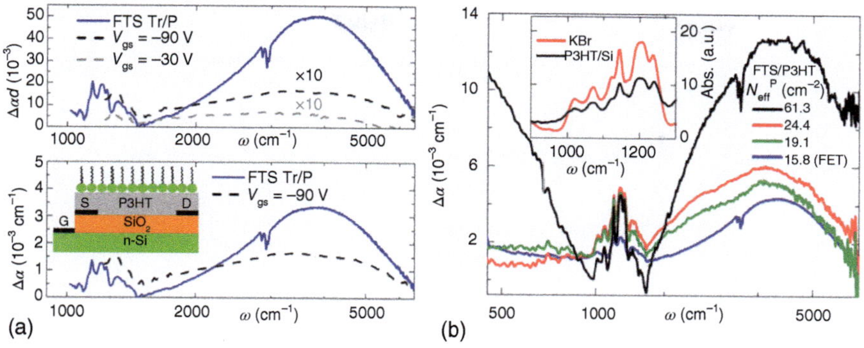

Figure 15.13 (a) IR response of P3HT film modified by electrostatic doping in an OFET structure and under exposure to FTS fumes. Top panel: mid-IR absorption $\Delta\alpha d = [1 - T(V_{gs})]/T(V_{gs} = 0\,V)$ for a device gated at $V_{gs} = -90\,V$ (black dashed curve) and $V_{gs} = -30\,V$ (gray dashed curve); $\Delta\alpha d = (1 - T_{tr}/T_p)$ for FTS modification with no applied bias (blue solid curve). Bottom panel: Change in the absorption coefficient $\Delta\alpha$ for the same device. Bottom inset: Schematic of FTS-treated OFET device. (b) Infrared spectroscopy (IR) absorption of four different P3HT-based structures extended down to 40 cm^{-1} to include the far-IR range of the spectrum. The blue curve is FTS-induced absorption of the field effect transistor (FET) device presented in figure (a). All other curves are two-probe (ungated) P3HT/Si structures doped with FTS. The curves are labeled according to their integrated polaron spectral weight and listed in order of polaron absorption strength. Inset: vibrational spectrum of an FTS-treated P3HT/Si sample and FTS-coated KBr substrate. Source: Khatib et al. [21].

15.2.8 Structure and Dynamics of Charge Carriers

In situ and ex situ vibrational spectroscopies exhibit the potential for the investigation of physical phenomena at the nanoscale occurring at the semiconductor–insulator interface in field-effect transistor devices [21–26]. Infrared measurements directly probe electronic excitations associated with injected charges. Using infrared spectroscopy, the low-energy excitations induced in fluoroalkyl trichlorosilane (FTS)-modified structures were characterized as shown in Figure 15.13 [21]. The three-terminal organic field-effect transistors (OFETs) serve as a reference system for characterizing the infrared absorption associated with FTS effects on the electronic transport in the P3HT active layer. Charges were first injected in a P3HT film by applying a gate voltage, forming a hole-accumulation layer in the polymer at the interface with SiO_2. The spectroscopic fingerprints of electrostatic charge injection with this gate-induced carrier density are observed as shown in Figure 15.13a, including the sharp peaks attributed to the active vibration modes located in the 1200–1400 cm^{-1} range as well as a broad band centered around 3500 cm^{-1} attributable to polaron absorption. The infrared data confirm that the large increase in conductivity in FTS-treated structures is based on doping via electron transfer. As shown in Figure 15.13b, such strong doping is indicated by the significantly enhanced polaron band and the strong far-infrared Drude-type absorption, which is characteristic of a free carrier response.

Time-resolved terahertz spectroscopy (TRTS) as the noncontact probes of conductivity is a powerful tool for interrogating the intrinsic properties of well-ordered

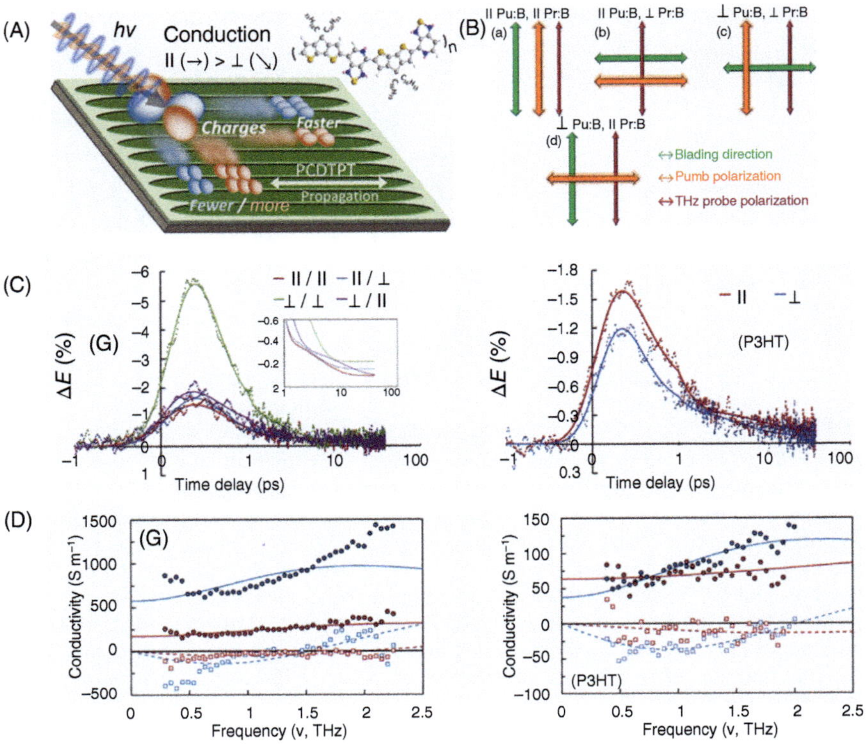

Figure 15.14 (A) Schematic of the polarization dependence of charge conduction in PCDTPT film. (B) Specifically, (a, b) parallel and (c, d) perpendicular pump:blade (Pu:B in the figure) and (a, d) parallel and (b, c) perpendicular probe:blade (Pr:B) directionalities are illustrated. Orientations of pump and THz probe beam polarizations relative to the blading direction (green) for four polarization configurations are shown in orange and red colour, respectively. (C) Integrated TRTS illustrated for (G)-PCDTPT and P3HT films for different excitation and probe polarizations. The key displays pump:blade/probe:blade and pump:probe polarization configurations. The inset shows the long time (tail) traces of smooth models fit through each dataset. PCDTPT TRTS plots were normalized to excitation via a 1.04 mJ cm^{-2}, 800 nm pulse. The time-delay axis for each plot is linear from −1 to 1 ps and logarithmic beyond 1 ps. (D) Peak frequency-dependent conductivity (points) and Drude–Smith model fits (lines) for a perpendicular pump:blade (G)-PCDTPT sample collected using 1.7 mJ cm^{-2}, 800 nm excitation and P3HT using 0.22 mJ cm^{-2}, 400 nm excitation. Source: Magnanelli et al. [27].

region in conductive polymer films [27, 28]. As shown in Figure 15.14A, for example, ultrafast, time-domain THz, and TRTS spectroscopies are employed to measure the carrier mobility and the charge generation efficiency in anisotropic film of poly[[1,2,5]thiadiazolo[3,4-*c*]pyridine-4,7-diyl(4,4-dihexadecyl-4Hcyclopenta[2,1-*b*:3,4-*b*′]dithiophene-2,6-diyl)[1,2,5]thiadiazolo[3,4-*c*]pyridine-7,4-diyl(4,4-dihexadecyl-4*H*-cyclopenta[2,1-*b*:3,4-*b*′]dithiophene-2,6-diyl)] (PCDTPT), and isotropic P3HT film [27]. Consecutive measurements were performed on the same spot of each sample under four polarization conditions as indicated in Figure 15.14B to separate excitation anisotropy from the anisotropic THz response. Polymer conduction anisotropies are compared from the TRTS spectra between PCDTPT and P3HT as shown in Figure 15.14C. Frequency-dependent TRTS data were further collected at fixed pump–probe time delays to differentiate between competing factors responsible for the observed anisotropies. Spectral analysis and modeling permit extraction of carrier conductivity, yields, and total charge (electron + hole) mobility. The results are shown in Figure 15.14D. Based on the spectral analysis, differences in the dynamics of the photo-induced charge carrier populations and their conductivities are revealed in PCDTPT, compared to nominally isotropic P3HT film. Polarization anisotropy can be observed as preferential charge conduction along the backbone propagation direction of PCDTPT (the schematic of which is shown in Figure 15.14A).

The structure and dynamics of photogenerated transients in conjugated polymers are allowed to be identified by vibrational spectroscopy [14, 16, 29–34]. Using femtosecond time-resolved resonance inverse Raman spectroscopy in the near-infrared region, Takaya et al. investigated the structure and dynamics of photogenerated transients in a P3HT:PCBM blend film, as well as in a pristine P3HT film [29]. To obtain the reference spectra of P3HT in the ground-state and positively charged excited states, the steady-state inverse Raman spectra of pristine and $FeCl_3$-doped films were recorded with the Raman pump and probe wavelength at 1190 and 900–1150 nm, respectively. The results are shown in Figure 15.15a, in which the characteristic bands of the thiophene ring are assigned. Femtosecond time-resolved near-infrared inverse Raman spectra as shown in Figure 15.15b,c were recorded for pristine P3HT and P3HT:PCBM blend films for observing the structure of photogenerated transients as well as their dynamics, respectively. The spectra, after the baseline subtraction, contain the inverse Raman bands of P3HT in the ground state as well as those of photogenerated transients.

As shown in Figure 15.16a–d, the transient near-infrared inverse Raman spectra at 0.2 and 50 ps were recorded with increased energy density of the actinic pump pulse

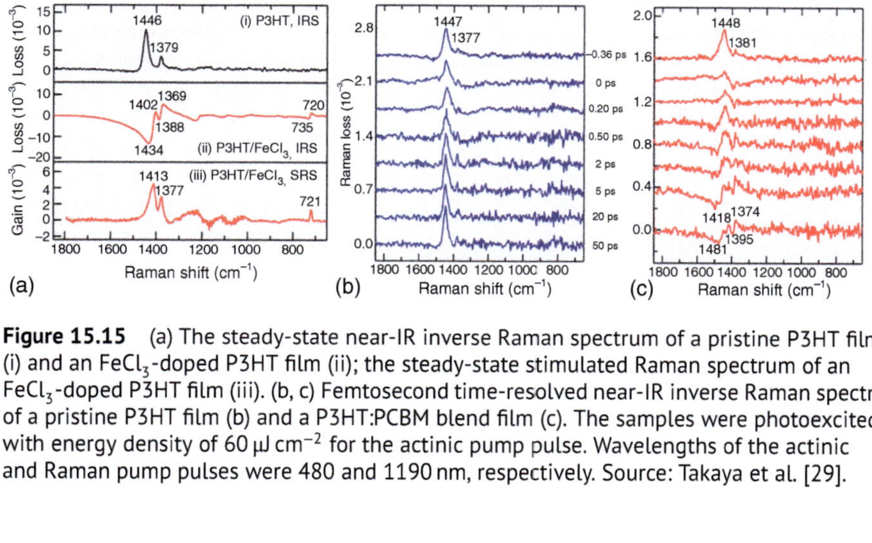

Figure 15.15 (a) The steady-state near-IR inverse Raman spectrum of a pristine P3HT film (i) and an $FeCl_3$-doped P3HT film (ii); the steady-state stimulated Raman spectrum of an $FeCl_3$-doped P3HT film (iii). (b, c) Femtosecond time-resolved near-IR inverse Raman spectra of a pristine P3HT film (b) and a P3HT:PCBM blend film (c). The samples were photoexcited with energy density of 60 μJ cm^{-2} for the actinic pump pulse. Wavelengths of the actinic and Raman pump pulses were 480 and 1190 nm, respectively. Source: Takaya et al. [29].

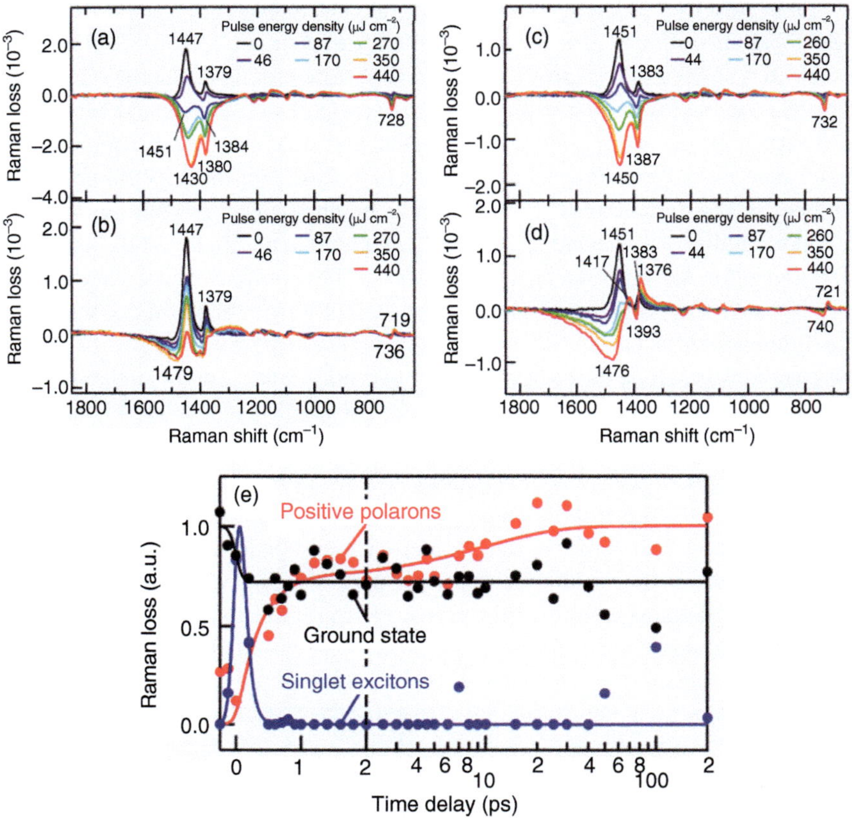

Figure 15.16 Pulse energy density dependence of transient inverse Raman spectra of (a) a pristine P3HT film at 0.20 ps; (b) a pristine P3HT film at 50 ps; (c) a P3HT:PCBM blend film at 0.20 ps; (d) a P3HT:PCBM blend film at 50 ps. (e) Time dependence of Raman loss for P3HT in the ground state, singlet exciton states, and positive polaron states in the blend film. Energy density of the actinic pump pulse was 60 μJ cm^{-2}. Filled circles represent the amplitude of the transients obtained by the least-squares fitting analysis of the time-resolved inverse Raman spectra with linear combinations of the spectra of the transients. Solid traces are the best-fitted curves obtained by the least-squares fitting analysis with an error function, an exponential function and an offset, two exponential functions for the ground state, the singlet exciton states, and the positive polaron states, respectively. Source: Takaya et al. [29].

from 0 to 4.4×10^2 μJ cm^{-2} further. Time dependence of Raman loss for P3HT in the ground state, singlet exciton states, and positive polaron states in the blend film is shown in Figure 15.16e. The transient inverse Raman spectrum of the pristine P3HT film at 50 ps uncovers the coexistence of neutral and charged excitations, whereas that of the blend film at 50 ps suggests the formation of positive polarons with a different structure from that in the $FeCl_3$-doped film. Time-resolved near-infrared inverse Raman spectra of the blend film indicate the absence of charge separation between P3HT and PCBM within the instrument response time of the spectrometer, while they indicate two independent pathways of the polaron formation with time constants of 0.3 and 10 ps.

15.2.9 Electric-Field-Induced Structural Changes

Experimental combined with theoretical study on the polarized Raman spectra provides a method for detecting the extension of the effective conjugation length and the structural homogeneity of conductive polymers [35]. A designed geometry as shown in Figure 15.17a was used to measure the spectroscopic properties of P3HT films within the channel directly. The vibrational modes coupled to the electronic transition in the 1350–1500 cm^{-1} region are analyzed, in which the most intense symmetric C_α–C_β stretching deformation (~1448 cm^{-1}, denoted as ν_1 band) and C_β–C_β stretching deformations (c. 1380 cm^{-1}, denoted as ν_2 band) existed where the aromatic thiophene ring is located.

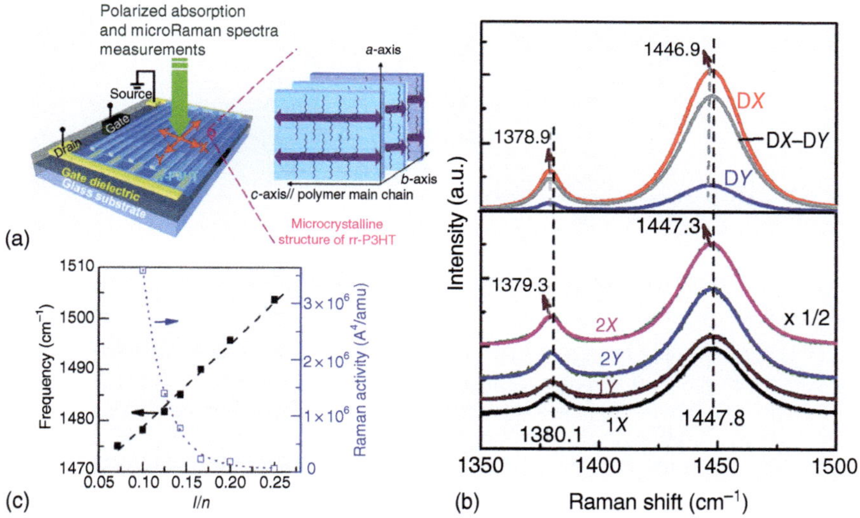

Figure 15.17 (a) Schematic diagram of the thin-film transistor configuration used in this study and polarized spectra measurements setup. *X*- and *Y*-directions are defined as parallel and perpendicular to the active channel of thin-film transistor, respectively. Inset: self-assembled lamellar-like crystal orientations of P3HT. (b) Comparison of polarized micro-Raman spectra of the P3HT films within the active channel of thin-film transistors (L/W = 10 μm/20 mm) before (denoted by 1) and after (denoted by 2) long-term operations (λ_{exc} = 532 nm). The *X*- and *Y*-directions are parallel and perpendicular to the channel, respectively. The solid curves represent the best fit with a Gaussian/Lorentzian function, while the thin lines to the experimental data. Traces D*X* and D*Y* are obtained by subtracting 1*X* from 2*X* and 1*Y* from 2*Y*, respectively. Trace D*X* – D*Y* is obtained by subtracting the D*Y* from D*X*. The dashed lines serve as guidelines. (c) Density functional theory calculated frequencies (left axis) and Raman activity (right axis) of the strongest symmetric C_α–C_β stretching (ν_1) band versus the reciprocal of the number of thiophene units (1/n) for oligo-(thiophene)s. The dashed and dotted lines are fitted to a linear and an exponential decay law, respectively. Source: Cheng et al. [35].

The effect of long-term operations on thin-film transistor along the X and Y directions of the Raman spectra from the P3HT films is shown in Figure 15.17b. Compared to the initial film, a slight red-shift of the v_1 and v_2 bands in both directions is observed. It suggests that the initial films have stable properties in structure with fine effective conjugation length. An enhanced Raman intensity after long-term operations is indicated also. According to Figure 15.17c, the increase of n, i.e. the effective

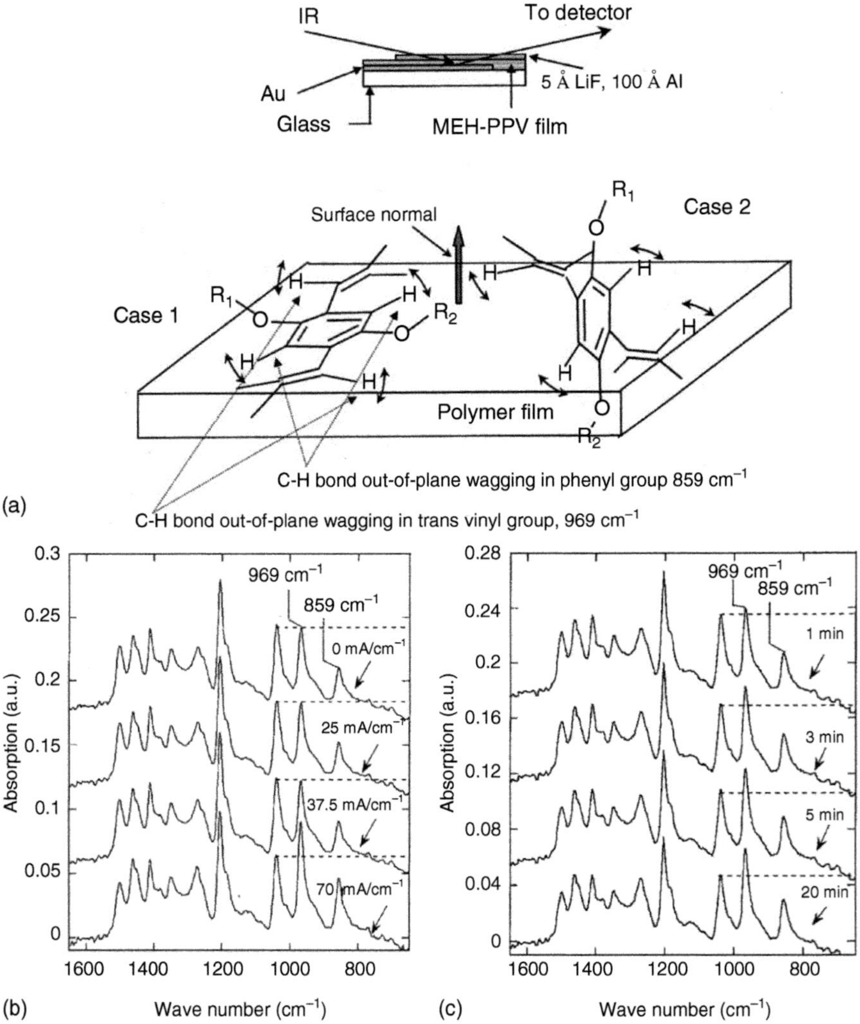

Figure 15.18 (a) Device structure and the schematic drawing of the in situ RA-FTIR setup. Au and semitransparent LiF/Al are used as the anode and cathode, respectively. The bottom part shows the two extreme examples for the orientation of conjugated chains inside the MEH-PPV films. (b) In situ RA-FTIR spectra taken 20 minutes after devices had been biased at 0, 25, 37.5, and 75 mA cm^{-2}, respectively. (c) In situ RA-FTIR spectra are taken at 1, 3, 5, and 20 minutes, while the device was being biased at 70 mA cm^{-2}. Source: Guo and Yang [36].

conjugation length, results in a rapid increase of Raman activity of the ν_1 band. Consequently, the increased intensity suggests an extended effective conjugation length in the X direction more than the Y direction. Compared to the initial films, the half-width of the bands measured in the operated film decreases more in the X direction than in the Y direction. The reduced half-width of the Raman bands indicates more structural homogeneity and better interchain order.

The reflection–absorption infrared spectroscopy (RAIR) enables in situ investigation of the correlation between device performance and the transformation of the molecular chain orientation in poly(2-methoxy-5-(2′-ethylhexyloxy)-1,4-phenylene vinylene) (MEH-PPV) [36]. The RAIR setup based on the device structure is drawn schematically in Figure 15.18a. In Figure 15.18b, the in situ RAIR spectra were measured after the device was biased for 20 minutes at different current densities. As the current density increases to 70 mA cm^{-2}, the intensities of the C–H in-phase out-of-plane wagging mode of the vinylene (969 cm^{-1} band) and phenylene groups (859 cm^{-1} band) increase. This enhanced absorption of p-polarized infrared may be contributed from the reoriented vibrating dipoles of the C—H bonds of these two groups, that is case 1 as shown in Figure 15.18a. A consistent observation on RAIR spectra is obtained as the bias time on devices is changed as shown in Figure 15.18c. The 969 and 859 cm^{-1} bands enhanced with increased time and finally reached a steady state.

15.3 Conclusion

In this chapter, probing structures of the conductive polymers with vibrational spectroscopy have been reviewed by illustrating many concrete studies. With the help of the vibrational spectroscopy, abundant information from the molecular conformation and the intermolecular interactions of macromolecular chains can be obtained. Although there are still some problems that limit the application of vibrational spectroscopy, for example, the poor spatial resolution in the depth of the sample, the application range is expected to be continuously extended with better understanding of advantages and limitations of this useful technique. More complicated structurally identification for polymer systems could be achieved by using multiequipment collaborative measurements, such as X-ray scattering, Raman, and infrared spectroscopy.

References

1 Yuan, Y., Shu, J., Kolman, K. et al. (2016). Multiple chain packing and phase composition in regioregular poly(3-butylthiophene) films. *Macromolecules* 49 (24): 9493–9506.

2 Yuan, Y., Zhang, J., Sun, J. et al. (2011). Polymorphism and structural transition around 54 °C in regioregular poly(3-hexylthiophene) with high crystallinity as revealed by infrared spectroscopy. *Macromolecules* 44 (23): 9341–9350.

3 Yuan, Y., Zhang, J., and Sun, J. (2011). Effect of solvent evaporation rate on order-to-disorder phase transition behavior of regioregular poly(3-butylthiophene). *Macromolecules* 44 (15): 6128–6135.

4 Zhang, T., Yuan, Y., Cui, X. et al. (2018). Impact of side-chain length on the phase structures of P3ATs and P3AT:PCBM films as revealed by SSNMR and FTIR. *J. Polym. Sci., Part B: Polym. Phys.* 56 (9): 751–761.

5 Yuan, Y., Shu, J., Liu, P. et al. (2015). Study on pi–pi interaction in H- and J-aggregates of poly(3-hexylthiophene) nanowires by multiple techniques. *J. Phys. Chem. B* 119 (26): 8446–8456.

6 Gao, Y. and Grey, J.K. (2009). Resonance chemical imaging of polythiophene/fullerene photovoltaic thin films: mapping morphology-dependent aggregated and unaggregated C=C species. *J. Am. Chem. Soc.* 131 (28): 9654–9662.

7 Gao, J., Thomas, A.K., Johnson, R. et al. (2014). Spatially resolving ordered and disordered conformers and photocurrent generation in intercalated conjugated polymer/fullerene blend solar cells. *Chem. Mater.* 26 (15): 4395–4404.

8 Winfield, J.M., Donley, C.L., Friend, R.H., and Kim, J.-S. (2010). Probing thin-film morphology of conjugated polymers by Raman spectroscopy. *J. Appl. Phys.* 107 (2): 024902.

9 Francis, C., Fazzi, D., Grimm, S.B. et al. (2017). Raman spectroscopy and microscopy of electrochemically and chemically doped high-mobility semiconducting polymers. *J. Mater. Chem. C* 5 (25): 6176–6184.

10 Wang, X.-J., Yuan, Y., Han, L.-l. et al. (2014). Effect of solvent annealing temperature on crystal modifications and phase transition behavior of regioregular poly(3-octylthiophene). *Chin. J. Polym. Sci.* 32 (9): 1158–1166.

11 Wang, X., Zhang, Y., Yuan, Y., and Zhang, J. (2014). One dimensional main-chain crystallization kinetics of poly(3-octylthiophenes) investigated by infrared spectroscopy. *Vib. Spectrosc.* 71: 1–5.

12 Yu, W., Zhou, J., and Bragg, A.E. (2012). Exciton conformational dynamics of poly(3-hexylthiophene) (P3HT) in solution from time-resolved resonant-Raman spectroscopy. *J. Phys. Chem. Lett.* 3 (10): 1321–1328.

13 Razzell-Hollis, J., Thiburce, Q., Tsoi, W.C., and Kim, J.S. (2016). Interfacial chemical composition and molecular order in organic photovoltaic blend thin films probed by surface-enhanced Raman spectroscopy. *ACS Appl. Mater. Interfaces* 8 (45): 31469–31481.

14 Martin, E.J.J., Bérubé, N., Provencher, F. et al. (2015). Resonance Raman spectroscopy and imaging of push–pull conjugated polymer–fullerene blends. *J. Mater. Chem. C* 3 (23): 6058–6066.

15 Badi, N., Khasim, S., and Roy, A.S. (2016). Micro-Raman spectroscopy and effective conductivity studies of graphene nanoplatelets/polyaniline composites. *J. Mater. Sci. - Mater. Electron.* 27 (6): 6249–6257.

16 Gao, J., Thomas, A.K., Yang, J. et al. (2015). Polythienylene–vinylene structure–function correlations revealed from resonance Raman spectroscopy and photocurrent imaging. *J. Phys. Chem. C* 119 (16): 8980–8990.

17 Meyer, M.W., Larson, K.L., Mahadevapuram, R.C. et al. (2013). Scanning angle Raman spectroscopy of poly(3-hexylthiophene)-based films on indium tin oxide, gold, and sapphire surfaces. *ACS Appl. Mater. Interfaces* 5 (17): 8686–8693.

18 James, D.T., Kjellander, B.K.C., Smaal, W.T.T. et al. (2011). Thin-film morphology of inkjet-printed single-droplet organic transistors using polarized Raman spectroscopy: effect of blending TIPS-pentacene with insulating polymer. *ACS Nano* 5 (12): 9824–9835.

19 Bellani, S., Porro, M., Caddeo, C. et al. (2015). The study of polythiophene/water interfaces by sum-frequency generation spectroscopy and molecular dynamics simulations. *J. Mater. Chem. B* 3 (31): 6429–6438.

20 Maia, F.C.B. and Miranda, P.B. (2015). Molecular ordering of conjugated polymers at metallic interfaces probed by SFG vibrational spectroscopy. *J. Phys. Chem. C* 119 (13): 7386–7399.

21 Khatib, O., Lee, B., Yuen, J. et al. (2010). Infrared signatures of high carrier densities induced in semiconducting poly(3-hexylthiophene) by fluorinated organosilane molecules. *J. Appl. Phys.* 107 (12): 123702.

22 Meyertholen, A.D., Li, Z.Q., Basov, D.N. et al. (2007). Concentration-dependent mobility in organic field-effect transistors probed by infrared spectromicroscopy of the charge density profile. *Appl. Phys. Lett.* 90 (22): 222108.

23 Li, Z.Q., Wang, G.M., Sai, N. et al. (2006). Infrared imaging of the nanometer-thick accumulation layer in organic field-effect transistors. *Nano Lett.* 6 (2): 224–228.

24 Figer, D.F. (2005). An upper limit to the masses of stars. *Nature* 434 (7030): 192–194.

25 Furukawa, Y., Yamamoto, J., Cho, D.-C., and Mori, T. (2004). Voltage-induced infrared spectra from polymer field-effect transistors. *Macromol. Symp.* 205 (1): 9–18.

26 Meana-Esteban, B., Lete, C., Kvarnström, C., and Ivaska, A. (2006). Raman and in situ FTIR-ATR characterization of polyazulene films and its derivate. *J. Phys. Chem. B* 110 (46): 23343–23350.

27 Magnanelli, T.J., Engmann, S., Wahlstrand, J.K. et al. (2020). Polarization dependence of charge conduction in conjugated polymer films investigated with time-resolved terahertz spectroscopy. *J. Phys. Chem. C* 124 (13): 6993–7006.

28 Vukmirović, N., Ponseca, C.S., Němec, H. et al. (2012). Insights into the charge carrier terahertz mobility in polyfluorenes from large-scale atomistic simulations and time-resolved terahertz spectroscopy. *J. Phys. Chem. C* 116 (37): 19665–19672.

29 Takaya, T., Enokida, I., Furukawa, Y., and Iwata, K. (2019). Direct observation of structure and dynamics of photogenerated charge carriers in poly(3-hexylthiophene) films by femtosecond time-resolved near-IR inverse Raman spectroscopy. *Molecules* 24 (3): 431.

30 Datko, B.D., Livshits, M.Y., Zhang, Z. et al. (2018). Unravelling the enigma of ultrafast excited state relaxation in non-emissive aggregating conjugated polymers. *Phys. Chem. Chem. Phys.* 20 (34): 22159–22167.

31 Magnanelli, T.J. and Bragg, A.E. (2015). Time-resolved Raman spectroscopy of polaron pair formation in poly(3-hexylthiophene) aggregates. *J. Phys. Chem. Lett.* 6 (3): 438–445.

32 Gao, J. and Grey, J.K. (2013). Resonance Raman overtones reveal vibrational displacements and dynamics of crystalline and amorphous poly(3-hexylthiophene) chains in fullerene blends. *J. Chem. Phys.* 139 (4): 044903.

33 Eigner, A.A., Jones, B.H., Koprucki, B.W., and Massari, A.M. (2011). Ground-state structural dynamics in doped and undoped polyaniline films probed by two-dimensional infrared vibrational echo spectroscopy. *J. Phys. Chem. B* 115 (16): 4583–4591.

34 Pensack, R.D., Banyas, K.M., and Asbury, J.B. (2010). Charge trapping in organic photovoltaic materials examined with time-resolved vibrational spectroscopy. *J. Phys. Chem. C* 114 (12): 5344–5350.

35 Cheng, H.-L., Lin, J.-W., Jang, M.-F. et al. (2009). Long-term operations of polymeric thin-film transistors: electric-field-induced intrachain order and charge transport enhancements of conjugated poly(3-hexylthiophene). *Macromolecules* 42 (21): 8251–8259.

36 Guo, T.-F. and Yang, Y. (2002). In situ study on the reorientation of polymer chains in operating polymer diodes. *Appl. Phys. Lett.* 80 (1): 148–150.

16

Weak Hydrogen Bonding in Biodegradable Polymers

Harumi Sato[1,2]

[1]*Kobe University, Graduate School of Human Development and Environment, Tsurukabuto 3-11 Nada, Kobe 657-8501, Japan*
[2]*Kobe University, Molecular Photoscience Research Center, Rokkodaicho 1-1, Nada, Kobe 657-8501, Japan*

16.1 Introduction

Hydrogen bonding, such as OH· · ·O and NH· · ·O, is well known since long ago. In the general definition of hydrogen bonds, the X—H· · ·A bond is formed between a donor X—H and an acceptor A [1–7]. For example, NH· · ·O=C hydrogen bonding is a well-known hydrogen bonding in various kinds of molecules such as polyamides, peptides, and proteins. On the other hand, CH· · ·O hydrogen bondings are not widely known, even though their existence has long been reported [8–19]. The C—H· · ·O hydrogen bonds have been found in proteins and nucleic acids, and their important functions are attracting considerable attention. Most important point of this CH· · ·O hydrogen bonding is that this hydrogen bond is a fairly weak interaction compared to other hydrogen bonds such as OH· · ·OH, OH· · ·O=C, and NH· · ·O=C. The modern definition of hydrogen bonds by Pimentel and McClellan is as follows: *hydrogen bond is said to exist when (i) there is evidence of a bond, and (ii) there is evidence that this bond sterically involves a hydrogen atom already bonded to another atom* [3]. According to them, hydrogen bonds should be defined phenomenologically rather than by their strength. Therefore, the CH· · ·O interaction is also a hydrogen bond, although not a unique one [3].

However, the CH· · ·O hydrogen bonds can be broken easily or weaken with slight changes in the molecular environment. This feature is advantageous because weak hydrogen bonding results in a reversible bond. In addition, biomaterials, such as proteins, have strong hydrogen bonding as in OH· · ·O and NH· · ·O along with weak hydrogen bondings as in CH· · ·O. It has been reported that weak CH· · ·O hydrogen bonds play important roles in stabilizing the structure of proteins and nucleic acids, in addition to the mutual recognition of protein–nucleic acids, enzyme–substrate interactions, and stabilization of biomolecular transition states [20–22]. The CH· · ·O hydrogens are hydrogen bonds commonly present in many materials. In this chapter, we present our research studies on CH· · ·O hydrogen bonds in biodegradable polymers. We have investigated the CH· · ·O hydrogen bonds

Spectroscopic Techniques for Polymer Characterization: Methods, Instrumentation, Applications, First Edition. Edited by Yukihiro Ozaki and Harumi Sato.

in biodegradable polymers using infrared (IR) spectroscopy, Raman spectroscopy, near-infrared (NIR) spectroscopy, far-infrared (FIR) spectroscopy, terahertz (THz) spectroscopy, wide-angle X-ray diffraction (WAXD), and small-angle X-ray scattering (SAXS) [23–42].

The CH· · ·O hydrogen bonding has been reported for various polymers such as poly(3-hydroxybutyrate) (PHB) [23–35], poly(3-hydroxybutyrate-*co*-3-hydroxyhexanoate (P(HB-*co*-3-HHx)) [24–26], poly(3-hydroxyvalerate) (PHV) [28], poly(3-hydroxybutyrate-*co*-3-hydroxyvalerate) (P(HB-*co*-HV)) [37], polyglycolic acid (PGA) [38–40], poly(ε-caprolactone) (PCL) [41], polyethylene terephthalate (PET) and polybutylene terephthalate (PBT) [42], and so on. The hydrogen bonds in these polymers stabilize the chain folding in the crystal lamella and may be responsible for the abnormally high melting temperature and crystallinity.

In this chapter, I introduce mainly IR and WAXD studies of weak hydrogen bondings in biodegradable polymers. FIR/THz, and low-frequency Raman studies and NIR studies of CH· · ·O hydrogen bondings are reported in Chapters 4 and 5, respectively.

16.2 Weak Hydrogen Bonding in Poly(3-hydroxybutyrate)

Biodegradable polymers are classified as follows: (i) polymers made by microorganisms, (ii) natural polymers made by plants, and (iii) synthetic polymers made by chemical synthesis as represented by polylactic acid. PHB (Figure 16.1), a representative biodegradable polymer derived from microorganisms, is known to have a peculiar crystal growth in which molecular chains are folded in the *a*-axis direction of the crystal lattice.

The crystal system of PHB is orthorhombic; the lattice constants are $a = 5.76$ Å, $b = 13.20$ Å, and $c = 5.96$ Å (fiber axis). The space group has double spiral symmetry with respect to all the crystal axes and has been reported to be $P2_12_12_1\text{-}D_2^4$ [43, 44]. It should be noted here that the O atom (C=O group) and the CH_3 group are oriented in the outward direction in the helix, facing each other. In addition, the distance

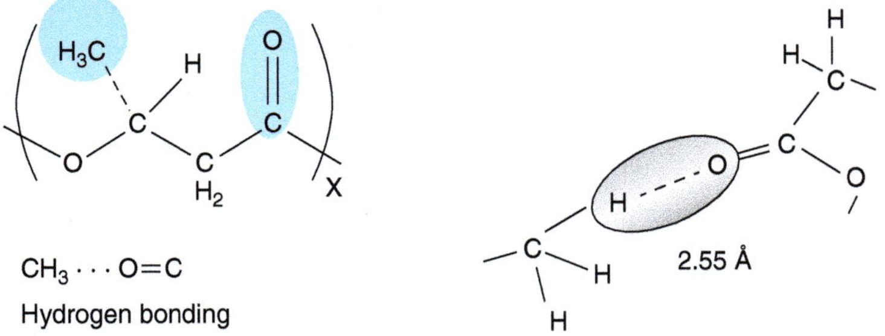

Figure 16.1 Chemical structure and its intermolecular hydrogen bonding of poly(3-hydroxybutyrate) (PHB). Source: Sato et al. [24].

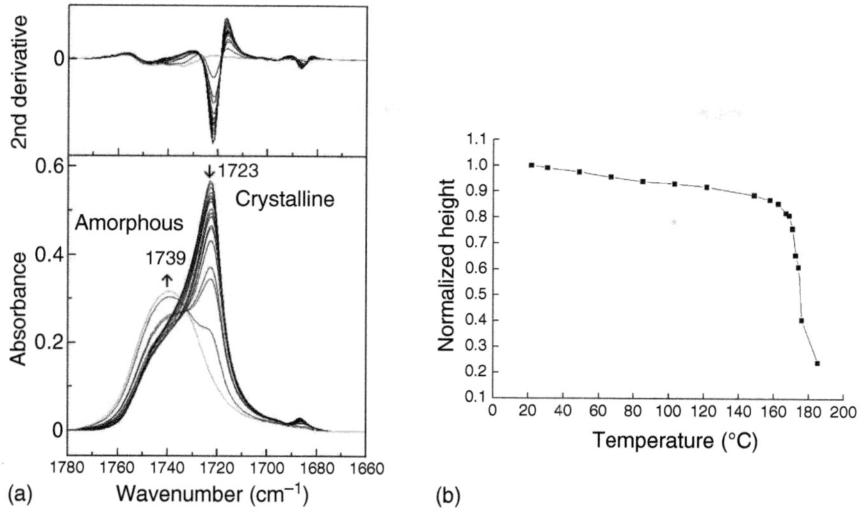

Figure 16.2 (a)Temperature-dependent IR spectral variations in the C=O stretching band region and its second-derivatives spectra. (b) Plot of the normalized peak height of the C=O stretching band at 1723 cm^{-1} of PHB. Source: (a) Sato et al. [24], (b) Based on Sato et al. [24].

between the O atom of the C=O group and the H atom of the CH_3 group is shortened. Estimating the distance of the CH···O=C bonds using the reported atom coordinates of the PHB crystal, the distance (2.55 Å) is shorter than the sum of the van der Waals separation between the O atom and H atom (2.72 Å). This suggests some interaction (CH···O=C hydrogen bond) between the O and H atoms. Furthermore, the temperature change measurement of X-ray diffraction showed that the temperature change in the lattice constant, *a*, of PHB was larger than that of *b*. This result also indicates that the CH···O=C hydrogen bond is a weak interaction, and the bonding weakens with an increase in temperature.

Figure 16.2a shows the temperature-dependent IR spectra variations in the C=O stretching vibration region (1660–1780 cm^{-1}) of PHB and their second derivative spectra [24]. Since the band at 1723 cm^{-1} decreases as the temperature and eventually disappears with high-temperature increase, this band is assigned to the crystalline C=O stretching band of PHB. The crystalline C=O stretching band of PHB at room temperature appears at 1723 cm^{-1}, which is lower by approximately 25 cm^{-1} compared to the frequency of the stretching band of a conventional free C=O group. That means the C=O group of PHB is involved in some intermolecular interactions. We ascribed this lower-frequency shift to the CH···O=C hydrogen bonding [24]. The normalized peak intensity of C=O stretching band at 1723 cm^{-1} changes little during heating process even after 160 °C, but it suddenly changes around its melting temperature [24] (Figure 16.2b). Therefore, the crystalline structure and intermolecular interaction of PHB are retained until close to the melting temperature.

Figure 16.3 shows the temperature dependence in the CH stretching vibration region (3050 to 2850 cm^{-1}) of the IR spectra of PHB and its second derivative spectra (3050 to 2950 cm^{-1}). In the original spectra, the band that appears as a

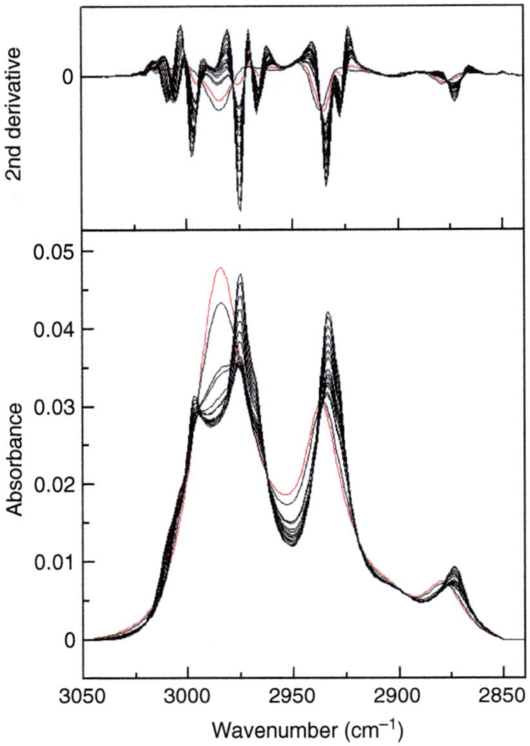

Figure 16.3 Temperature-dependent IR spectra variations and its second derivative spectra in the CH stretching region of PHB measured over a temperature range of 20–185 °C. Source: Sato et al. [24].

"shoulder" peak is observed in the second derivative spectra, and it gradually shifts to a lower wavenumber as the temperature increases. This band at 3009 cm^{-1} is considerably higher than the frequency of normal CH_3 degenerate expansion and contraction vibrations (3009 cm^{-1}), which is one of the evidence of the existence of the CH···O=C hydrogen bonds.

Moreover, in the C=O stretching region, the crystal and amorphous bands could be assigned, and the crystal structure collapse could be understood from the temperature-dependent IR spectra.

The evidence of the existence of the CH_3···O hydrogen bond of PHB is as follows: (i) C—H···O distance shorter than the van der Waals separation, 2.72 Å, and (ii) the appearance of a CH_3 asymmetric stretching band above c. 3000 cm^{-1}. In addition, the temperature dependence of the CH_3···O hydrogen-bonded band at 3009 cm^{-1} and lattice parameter also support the existence of CH_3···O hydrogen bonds. Spectra in the low-frequency region, such as the THz spectra, also show the presence of the C—H···O hydrogen bond of PHB and will be described later in this book.

16.3 Comparison Between Weak and Strong Hydrogen Bonds

It was found that PHB has a hydrogen bond, CH···O=C, which connects two helices and contributes to the stabilization of the crystal structure. Consequently, the crystal structure of PHB is maintained up to a melting point of 178 °C, which is close to the decomposition temperature of PHB (~190 °C), and the crystallinity is considered to be as high as approximately 60% to 70%. The wavenumber shift of the CH···O=C hydrogen-bonded band can be seen in the infrared spectra during the heating process, owing to the weak hydrogen bond. The band derived from the CH···O=C hydrogen bond of PHB appears in the C—H stretching vibration region. The band appears at an abnormally higher wavenumber region in the C—H stretching vibrational mode. This is because the H atom of the methyl group of PHB forms a hydrogen bond with the C=O group, and it is assumed that the hydrogen-bonded C—H group in the methyl group and vibration of the pseudo CH_2 group will appear. The band derived from the hydrogen bonds was observed as a shoulder. The strength of the CH···O=C hydrogen bond is very small (c. 1.22 kcal mol^{-1}), which is 1/4th of the conventional NH···O=C hydrogen bond strength (nylon). Considering that the internal rotational energy of ethane is c. 2.9 kcal mol^{-1}, the methyl group in the side chain of PHB will not be able to rotate freely at room temperature. At high temperatures, the methyl group in the side chain of PHB can rotate, and the CH···O=C hydrogen bond could be easily broken/connected (Figure 16.4).

In the polyethylene (PE) crystalline lamellae, which is the simplest structure, folding proceeds along the (110) plane direction. On the other hand, it has been reported that the chain folding of PHB shows alternate folding between the (110) and $(1\bar{1}0)$ planes in the opposite direction (*a*-axis) and not in the (110) plane direction. However, it is still unknown why PHB causes the molecular chain folding to proceed in the *a*-axis direction. In addition, as is well known, in the crystal structure of nylon 66, N—H···O=C hydrogen bonds present in the crystallographic *a*-axis direction stabilize the crystal structure. It is probable that the CH···O=C hydrogen bond plays a vital role in stabilizing the peculiar chain folding in PHB.

16.4 Difference in the Side Chain Length; PHB and PHV

Both PHB and PHV are semicrystalline bio-polyesters and have orthorhombic $P2_12_12_1$ unit cell with lattice parameters of $a = 5.76$ Å, $b = 13.20$ Å, and $c = 5.96$ Å for PHB [43, 44] and $a = 9.32$ Å, $b = 10.02$ Å, and $c = 5.56$ Å for PHV [45, 46]. PHB has a methyl side chain, whereas PHV has an ethyl side chain. The only difference between PHB and PHV is the number of C atoms in the side chain. PHB exhibits CH···O=C bonding between the methyl and the C=O groups (Figure 16.1). What happens to the hydrogen bonds in the case of PHV?

The CH···O=C hydrogen bond exists in the PHB crystalline structure, and it connects two molecular chains in the *a*-axis direction in the folded crystal (lamellar) structure. The number of CH···O=C hydrogen bonds depends on the thickness of the lamellar structure and the size of the two-dimensional spread. In other words, even though the CH···O=C hydrogen bond is weak, a large number of them will affect the crystal structure. Therefore, PHV with an ethyl group in the side chain was examined to determine whether CH···O=C hydrogen bonds exist in the crystalline structure, similar to PHB. It was found that in PHV, unlike PHB, intermolecular

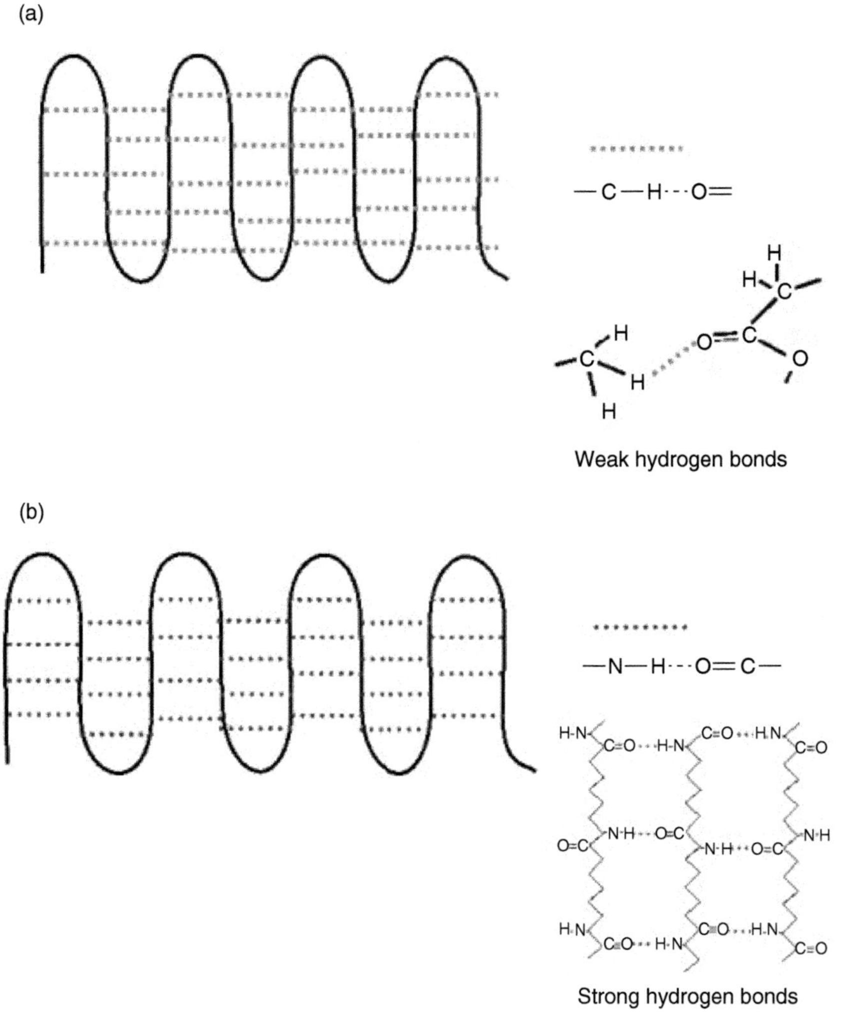

Figure 16.4 Weak hydrogen bonding of PHB and N—H· · ·O C hydrogen bonding network in nylon. (a) PHB and (b) nylon. Source: Sato et al. [26].

hydrogen bonds form between the methylene group of the main chain or side chain and the C=O group (Figure 16.5). In other words, the PHB has CH_3···O=C hydrogen bonds, while the PHV has CH_2···O=C hydrogen bonds. When the side chain length increases (C_3H_7), it can no longer form a crystal structure.

Furthermore, as described above, PHB has hydrogen bonds along the *a*-axis, and the folding direction of the lamella is also in the same direction of hydrogen bondings. In PHV, the hydrogen bonds exist in the (110) direction, and the chain folding direction is also the same. This suggests a relationship between the hydrogen bonds and the stability of the crystal (lamellar) structure.

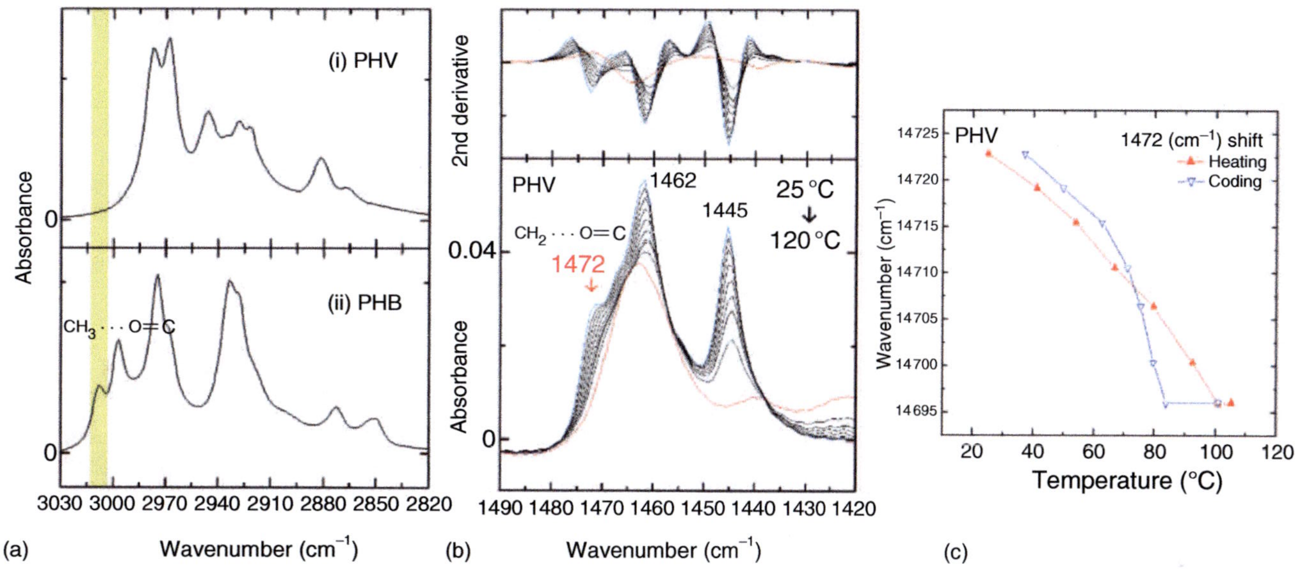

Figure 16.5 Comparison of (a) C–H stretching and (b) C–H deformation band regions of IR spectra of PHV and PHB. (c) Plots of the frequency of 1472 cm^{-1} band vs. temperature of PHV during the heating and cooling processes. Source: Sato et al. [28, 37].

16.5 Polyhydroxyalkanoate Copolymers

Polyhydroxyalkanoate copolymers, such as P(HB-*co*-HV) and P(HB-*co*-HHx), have been developed to improve the physical properties of these biodegradable homopolymers. By copolymerization, the crystallinity of PHB is reduced, and it shows more elastic properties.

In the case of P(HB-*co*-HV), co-crystallization occurs in the crystal lamellae, in which two different polymers exist as components in the same crystalline form [47, 48]. In the case of PHB-rich copolymers, it shows a PHB-type crystalline structure. PHV-rich copolymers show PHV-type crystalline structures. PHV has CH···O=C hydrogen bonds between the C=O group and one or two CH_2 groups in the main or side chains. Therefore, P(HB-*co*-HV) with low HV content has a PHB-type crystal structure with CH_3···O=C hydrogen bonds, whereas P(HB-*co*-HV) with high HV content has a PHV-type crystal structure with CH_2···O=C hydrogen bonds. The crystal structure of P(HB-*co*-HV) with HV content at approximately 50 mol% may have negligible intermolecular interactions between the C=O group and either CH_2 or CH_3 groups. The transition region from the PHB- to the PHV-type occurs at approximately 50 mol% HV content in the crystal structure of P(HB-*co*-HV). Therefore, the crystal structure of P(HB-*co*-HV) (HV contents ~50 mol%) collapses more easily than other P(HB-*co*-HV) copolymers during the heating process (Figure 16.6).

In P(HB-*co*-HHx), the HHx unit has a C_3H_7 side chain, which is longer than the methyl group. Therefore, the presence of the HHx unit inhibits the crystal structure of PHB. The number of CH···O=C hydrogen bonds along the *c*-axis (direction of lamellar thickness), based on the reported lamellar thickness [49, 50], is approximately eight or nine for PHB, and approximately three for P(HB-*co*-HHx) (HHx = 10.5 and 12 mol%). Since the lamella of P (HB-*co*-HHx) is thinner than

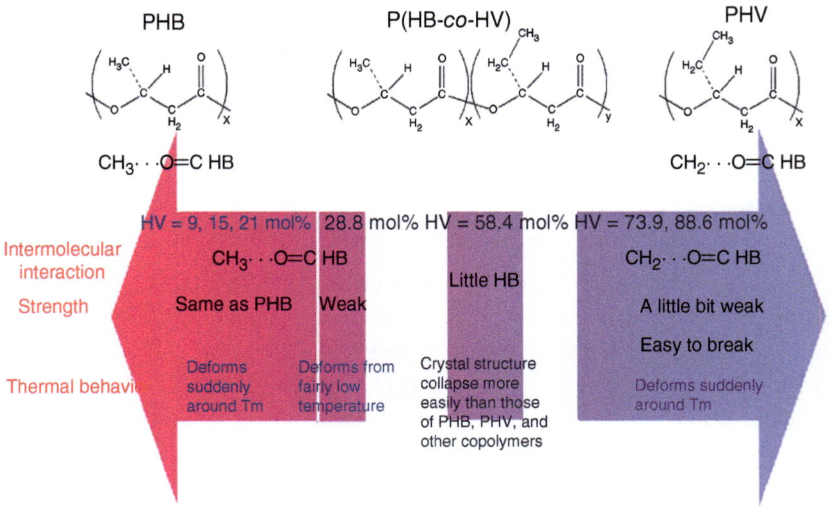

Figure 16.6 Intermolecular hydrogen bondings of P(HB-*co*-HV).

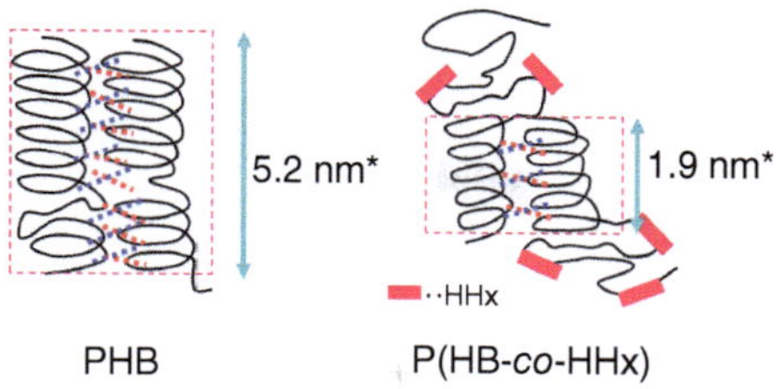

Figure 16.7 Schematic illustration of the lamella for PHB and P(HB-*co*-HV). *The lamella thicknesses were reported by Abe et al. [49]. Source: Based on Abe et al. [49].

that of PHB and more amorphous, the CH···O=C hydrogen bond in P(HB-*co*-HHx) appears to break more easily than that in PHB. However, it is assumed that polymer chains retain their lamellar structure even in copolymers with significant HHx content because of these hydrogen bonds. In other words, the CH···O=C hydrogen bonds stabilize the crystal structure of PHB. The P(HB-*co*-HHx) copolymer still has a crystalline structure, and it shows a lower crystallinity and flexibility compared to the PHB homopolymer, owing to weak hydrogen bonds (Figure 16.7).

16.6 Crystallization Process of PHB

Isothermal crystallization process of PHB with CH···O hydrogen bonding was studied by means of simultaneous measurements of time-resolved WAXD and SAXS combining with two-dimensional correlation spectroscopy (2DCOS) and multivariate curve resolution–alternating least squares (MCR–ALS) methods [32]. Hetero-2DCOS of WAXD and SAXS could analyze the time sequence of the development of WAXD and SAXS profiles, and the MCR–ALS analysis was applied to decompose the WAXD profiles into the intermediate and lamellar structures. The crystallization process of PHB was classified into regions I–III. In Region I, the density fluctuations were formed in the amorphous region firstly, then transformed into intermediate structures. Both the intermediate structure and lamellar crystal, which are created from intermediate structure, were developed in region II. In Region III, the intermediate structure decreased and the lamellar crystals increased with time [32] (Figures 16.8 and 16.9).

16.7 Other Kinds of CH···O Hydrogen Bonding

Here, we pay attention to hydrogen bonding in polyhydroxyalkanoates (PHAs) and other biodegradable polyesters. PGA is a petroleum-based plastic with a simple molecular structure and an abnormally high melting point, which is the highest among the aliphatic polyesters. Although PGA is a synthetic polymer, it shows biodegradability and is used in biomedical applications. The crystalline structure of PGA consists of an orthorhombic system with a zigzag conformation determined using X-ray diffraction by Chatani et al. [51] Chatani pointed out the short distances between the ester groups between the polymer chains of PGA using the results of X-ray crystallography, which result in an abnormally high melting point and

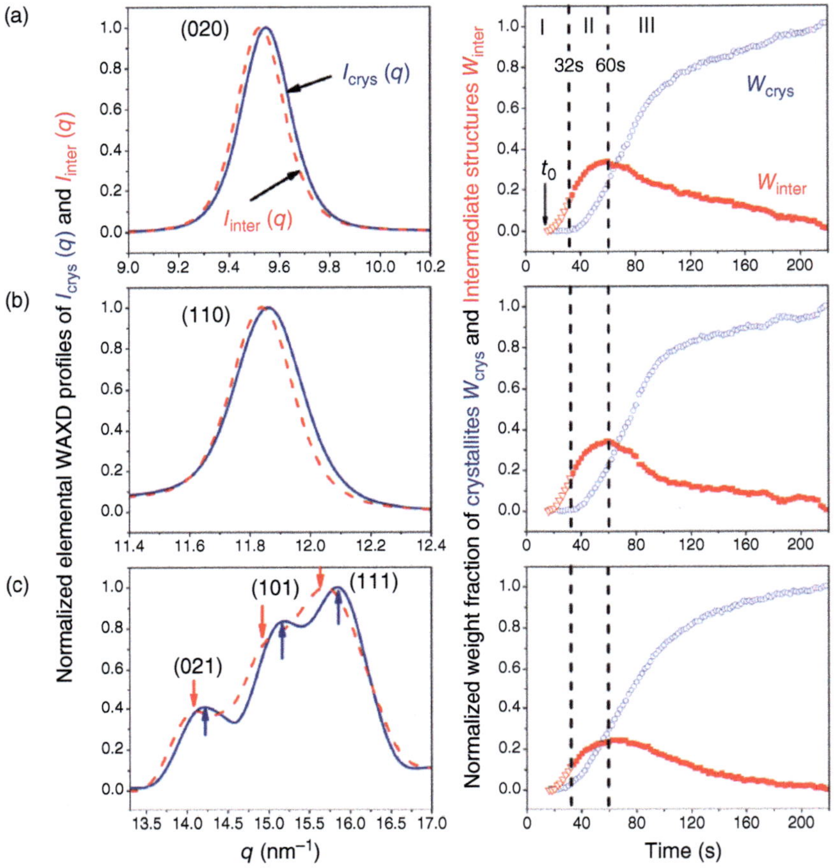

Figure 16.8 Decomposed elemental WAXD profiles of (020), (110), and multiple diffraction peaks composed of (021), (101), and (111) for the intermediate structures; I_{inter}(q) and the lamellar crystallites; I_{crys}(q) and time evolution of the weight fractions of the intermediate structures; W_{inter} and lamellar crystallites; W_{crys} during the isothermal crystallization process. Source: Guo et al. [32].

stabilize the crystal lattice. However, the details of the specific intermolecular interactions in PGA have not yet been explored. Using IR spectroscopy, WAXD, quantum mechanical, and natural bond orbital (NBO) calculations, the weak CH···O hydrogen bonding in PGA was explored. Since the crystal C=O stretching band appears at approximately 1740 cm^{-1} in the IR spectra of PGA, non-hydrogen bonding appears in the C=O group of PGA. Compared with PHB, which has hydrogen bonding in the C=O group, the C=O stretching band of PGA appeared at a high wavenumber. The distance between the O (ether) and H atoms estimated from the reported atom coordinates of PGA is shorter than the van der Waals separation, and the temperature-dependent IR spectral variations in PGA show the same trend between the CH_2 stretching band at 2992 cm^{-1} and the C–O–C stretching band at 1092 cm^{-1}. The result of NBO calculations also shows that there is a possibility to have an interaction between the O (ether) atom and H atom of the

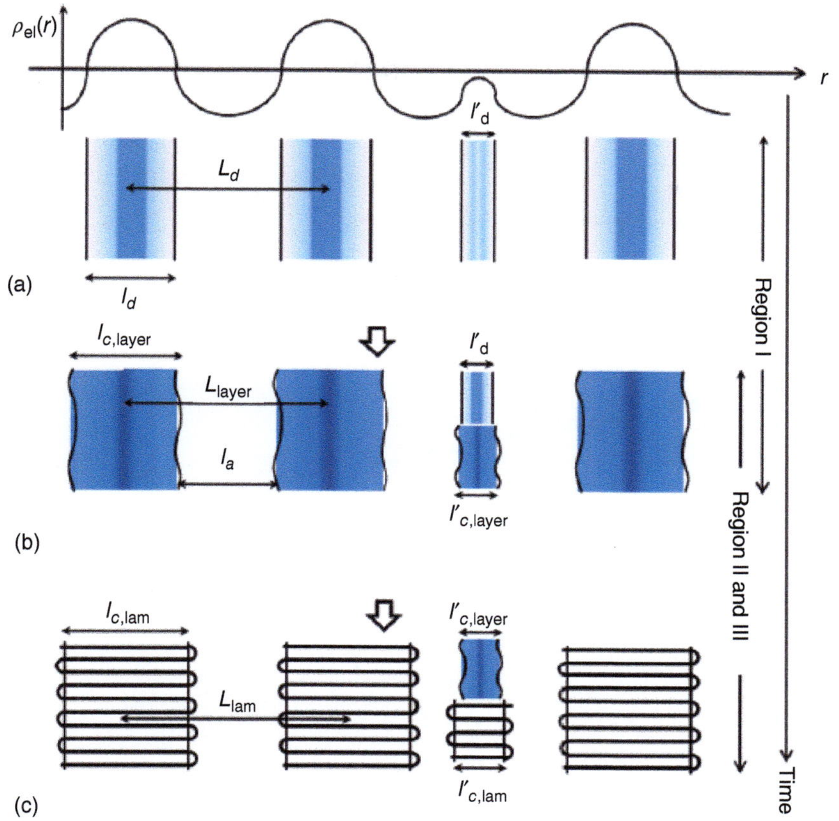

Figure 16.9 A schematic illustration of the multistep crystallization process of PHB studied by time-resolved simultaneous WAXD and SAXS measurements. Source: Guo et al. [32].

CH_2 group. PGA shows weak CH···O hydrogen bonding between the CH_2 group and ether oxygen atom along the (110) direction (Figure 16.10).

This intermolecular weak hydrogen bonding is possibly one of the origins of the very high melting point of PGA. PGA (CH···O (ether)), PHB (CH_3···O=C), and PHV (CH_2···O=C) have different types of weak CH···O hydrogen bonds, although all the CH···O hydrogen bonds play key roles in the chain folding and stabilizing the crystal lamellae. Temperature-dependent variation in the C=O stretching band region of PGA indicated that its intensity abruptly changes near its melting temperature, similar to PHB. However, the position of the crystalline C=O band is higher than that of PHB. This result implies that PGA might not have the CH···O=C hydrogen bond, similar to PHB (Figure 16.11).

As mentioned above, there are different types of weak CH···O hydrogen bonds in biodegradable polyesters, such as PHB, PHV, and PGA. The strength of the CH···O hydrogen bonds is determined mainly by the distance between the H and O atoms. In addition, the number of such weak CH···O hydrogen bonds in a repeat unit is also important for stabilizing the crystal lamellae. For example, PGA has one CH···O

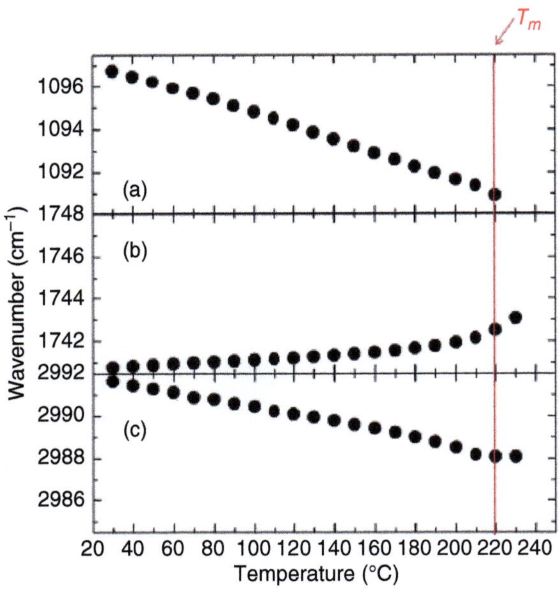

Figure 16.10 Plots of the wavenumbers of the (a) C–O–C (1092 cm^{-1}), (b) C=O (1740 cm^{-1}), and (c) CH_2 antisymmetric stretching (2992 cm^{-1}) bands vs. temperature. Source: Sato et al. [38].

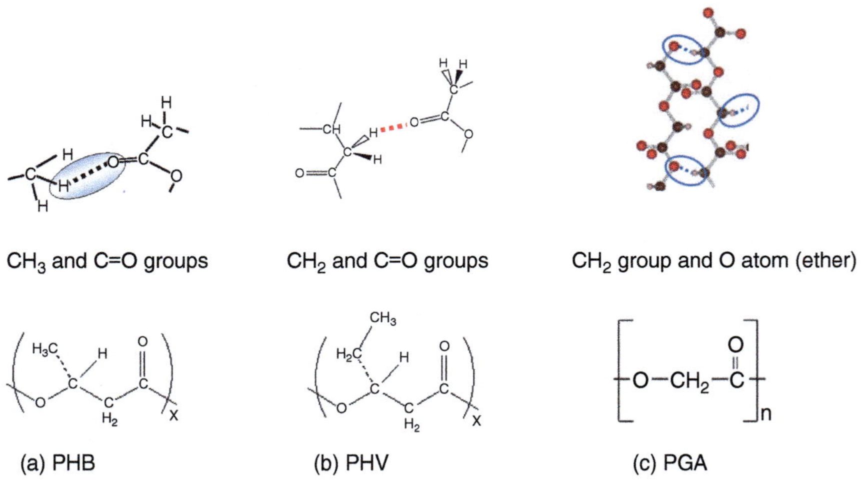

Figure 16.11 Chemical structure and the model of these intermolecular hydrogen bondings of (a) poly(3-hydroxybutyrate), (PHB), (b) poly(3-hydroxyvarelate) (PHV), (c) polyglycolic acid (PGA). Source: Based on Sato et al. [24].

hydrogen bond per repeat unit that links two adjacent parallel chains in the crystal lamellae. The repeat unit of PGA is relatively small compared to PHB and PHV. Therefore, the hydrogen bonding in total of the crystal lamellae of PGA will be relatively strong.

According to NBO calculations, a weak CH···O hydrogen bonding of approximately 4 kJ mol^{-1} exists between the ether oxygen and the C—H bond of the CH_2 group along the (110) plane in PGA. On the other hand, the quantum chemical

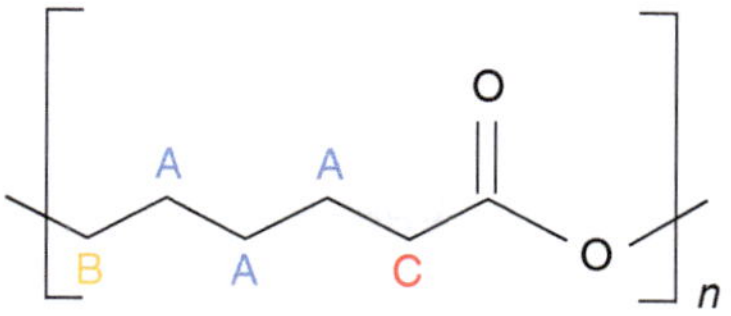

Figure 16.12 Chemical structure of PCL. "A" is the CH_2 adjacent to methylene groups, "B" is the CH_2 adjacent to ester group, and "C" is the CH_2 adjacent to carbonyl groups. Source: Funaki et al. [41].

calculations (QCC) of small molecules showed that the energy of $CH_3\cdots O{=}C$ hydrogen bonds is 4–6 kJ mol^{-1} [9]. Compared with the wavenumber of the CH···O hydrogen-bonded band in the CH stretching and vibrational region of small molecules, the energy of $CH_3\cdots O{=}C$ hydrogen bonds of PHB appears to be less than 4 kJ mol^{-1}. The CH···O hydrogen bond energy of PHB appears to be smaller than that of PGA.

Another biodegradable polyester with CH···O=C hydrogen bonding is PCL [41]. PCL has three types of weak CH···O=C intermolecular hydrogen bonding interactions, as determined by NBO calculations. The C=O stretching band of PCL appeared at 1724 cm^{-1}, close to the hydrogen-bonded band of PHB. In addition, many bands in the CH_2 stretching region may be caused by hydrogen bonding. The temperature-dependent THz spectra combined with the QCC of PCL support several types of hydrogen bonds in the crystal structure. (The THz spectra of PCL will be discussed in Chapter 4.) The lamellar thickness of PCL and PGA is approximately 6 nm, and the repeat unit length of PCL is more than double that of PGA. This indicates that the estimated number of hydrogen bonds in the lamellar thickness of PCL is smaller than that of PGA. This probably caused the significantly higher melting temperature of PGA than that of PCL (Figures 16.12 and 16.13).

16.8 Conclusions

The weak C—H···O hydrogen bonds of PHB, PHV, PHB copolymers, PGA, and PCL were studied by IR and Raman spectroscopy and WAXD in combination with QCC. It was found that there are intramolecular interactions (C—H···O=C hydrogen bonds) between the C=O groups in one helical structure and the CH_3 group in the other helical structure in PHB. The strength and thermal behavior of the C—H···O=C hydrogen bond were explored by frequency shifts of the CH_3 asymmetric stretching band at 3009 cm^{-1}. The C—H···O=C hydrogen bond is primarily responsible for the presence of the CH_3 stretching band above 3000 cm^{-1}, and the distance between the O and H atoms is less than that of the sum of their van der Waals separation. PHV has an ethyl group in the side chain and C—H···O=C hydrogen bonding between the CH_2 and the C=O groups. The wavenumber of the C=O stretching band and the thermal behavior of the CH_2 bending mode in the IR spectra are useful for investigating $CH_2\cdots O{=}C$ hydrogen bonding. In the case of a longer side chain in the PHA group, PHHx, which has a propyl side chain, a crystalline structure is not feasible. Therefore, P(HB-*co*-HV) and P(HB-*co*-HHx) show large differences in intermolecular hydrogen bonding for copolymers. P(HB-*co*-HHx) showed the same hydrogen bonding as PHB, while

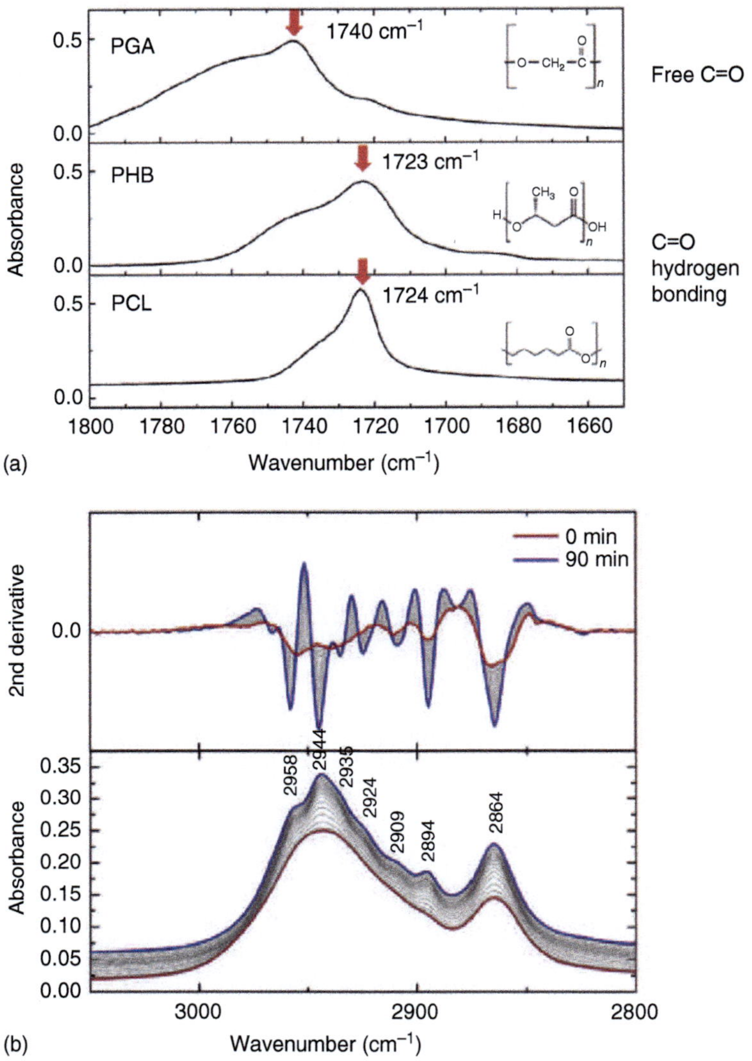

Figure 16.13 (a) IR spectra of PGA, PHB, and PCL cast film in the C=O stretching vibration region at room temperature. (b) IR spectra of PCL in the CH_2 stretching vibration region measured during isothermal crystallization at 48 °C (bottom) and their second derivatives (top). Source: Funaki et al. [41].

P(HB-*co*-HV) indicated two types of hydrogen bonding, PHB- and PHV-type. Another type of C—H· · ·O hydrogen bonding in the biodegradable polyester such as PGA and PCL is also discussed in this chapter. PGA has C—H···O hydrogen bonds between the O (ether) and CH_2 groups, and PCL shows at least three kinds of C—H···O hydrogen bonds with different CH_2 groups. The C—H···O hydrogen bonds have been identified in many polymers. The C—H···O hydrogen bonds play a crucial role in stabilizing the crystalline lamellae and folding of the molecular chain of these polymers. Although these hydrogen bonds are relatively weak, they are present in large numbers in the crystal lamellae, and it becomes a major force in the whole and has a significant effect on the physical properties of the polymer. This hydrogen bond helps to stabilize the crystal structure of the polymer, is related to thermal and mechanical properties, and affects the biodegradability of biodegradable polymers. It is possible that the physical properties of polymers can be controlled by controlling weak hydrogen bonds such as C—H···O.

References

1 Han, K.-L. and Zhao, G.-J. (eds.) (2010). *Hydrogen Bonding and Transfer in the Excited State*, vol. 1–2. Chichester, UK: Wiley.

2 Schuster, P., Zundel, G., and Sandorfy, C. (eds.) (1976). *The Hydrogen Bond. Recent Developments in Theory and Experiments*, Vols. 1–3,. North Holland, Amsterdam: North-Holland Publishing Company.

3 Pimentel, G.C. and McClellan, A.L. (1960). *The Hydrogen Bond*. San Francisco: W. H. Freenan and Co.

4 Gilli, G. and Gilli, P. (2009). *The Nature of the Hydrogen Bond: Outline of a Comprehensive Hydrogen Bond Theory*. Oxford: Oxford University Press.

5 Huettermann, A. (2019). *The Hydrogen Bond: A Bond for Life*. De Gruyter.

6 Hadži, D. (ed.) (1997). *Theoretical Treatments of Hydrogen Bonding*. Chichester: Wiley.

7 Grabowski, S.J. (ed.) (2006). *Hydrogen Bonding: New Insights*. Dordrecht: Springer.

8 Desiraju, G.R. and Steiner, T. (1999). *The Weak Hydrogen Bond*. Oxford: Oxford University Press.

9 Matsuura, H., Yoshida, H., Hieda, M. et al. (2003). Experimental evidence for intramolecular blue-shifting C—H· · ·O hydrogen bonding by matrix-isolation infrared spectroscopy. *J. Am. Chem. Soc.* 125 (46): 13910–13911.

10 Sutor, D.J. (1962). The C—H···O hydrpgen bond in crystals. *Nature* 195: 68–69.

11 Sutor, D.J. (1963). Evidence for the existence of C—H···O hydrpgen bond in crystals. *J. Chem. Soc.*: 1105–1110.

12 Hobza, P. and Havlas, Z. (2000). Blue-shifting hydrogen bonds. *Chem. Rev.* 100: 4253–4264.

13 Harada, T., Yoshida, H., Ohno, K., and Matsuura, H. (2002). Conformational stabilities of 1-methoxy-2-(methylthio)ethane and relevant intramolecular CH···O interaction studied by matrix-isolation infrared spectroscopy and density functional calculations. *Chem. Phys. Lett.* 362: 453.

14 Taylor, R. and Kennard, O. (1982). Crystallographic evidence for the existence of C—H···O, C—H···N, and C—H···Cl hydrpgen bonds. *J. Am. Chem. Soc.* 104: 5063–5070.

15 van der Veken, B.J., Herrebout, W.A., Szostak, R. et al. (2002). The nature of improper, blue-shifting hydrogen bonding verified experimentally. *J. Am. Chem. Soc.* 124: 11854–11855.

16 Musah, R.A., Jensen, G.M., Rosenfeld, R.J. et al. (1997). Variation in strength of an unconventional C—H to O hydrogen bond in an engineered protein cavity. *J. Am. Chem. Soc.* 119: 9083–9084.

17 Alabugin, I.V., Manoharan, M., Peabody, S., and Weinhold, F. (2003). Electronic basis of improper hydrogen bonding: a subtle balance of hyperconjugation and rehybridization. *J. Am. Chem. Soc.* 125: 5973–5987.

18 Scheiner, S. and Kar, T. (2002). Red- versus blue-shifting hydrogen bonds: are there fundamental distinctions? *J. Phys. Chem. A* 106: 1784–1789.

19 Desiraju, G.R. and Steiner, T. (1999). The weak hydrogen bond. In: *Structural Chemistry and Biology*. USA: Oxford University Press.

20 Jiang, L. and Lai, L.H. (2002). CH···O hydrogen bonds at protein-protein interfaces. *J. Biol. Chem.* 277: 37732–37740.

21 Castellano, R.K. (2004). Progress toward understanding the nature and function of CH···O interactions. *Curr. Org. Chem.* 8: 845–865.

22 Mandel-Gutfreund, Y., Margalit, H., Jernigan, R.L., and Zhurkin, V.B. (1998). A role for CH···O interactions in protein-DNA recognition. *J. Mol. Biol.* 277: 1129–1140.

23 Sato, H., Nakamura, M., Padermshoke, A. et al. (2004). Thermal behavior and molecular interaction of poly(3-hydroxybutyrate-*co*-3-hydroxyhexanoate) studied by wide-angle X-ray diffraction. *Macromolecules* 37: 3763–3769.

24 Sato, H., Murakami, R., Padermshoke, A. et al. (2004). Infrared spectroscopy studies of CH···O hydrogen bondings and thermal behavior of biodegradable poly(hydroxyalkanoate). *Macromolecules* 37: 7203–7213.

25 Sato, H., Mori, K., Murakami, R. et al. (2006). Crystal and lamella structure and C—H···O=C hydrogen bonding of poly(3-hydroxyalkanoate) studied by X-ray diffraction and infrared spectroscopy. *Macromolecules* 39: 1525–1531.

26 Sato, H., Murakami, R., Zhang, J. et al. (2005). Infrared spectroscopy and X-ray diffraction studies of C—H···O hydrogen bonding and thermal behavior of biodegradable poly(hydroxyalkanoate). *Macromol. Symp.* 230: 158–166.

27 Sato, H., Dybal, J., Murakami, R. et al. (2005). Infrared and Raman spectroscopy and quantum chemistry calculation studies of C—H/O hydrogen bondings and thermal behavior of biodegradable polyhydroxyalkanoate. *J. Mol. Struct.* 744–747: 35–46.

28 Sato, H., Ando, Y., Dybal, J. et al. (2008). Crystal structures, thermal behaviors, and C—H···O=C hydrogen bondings of poly(3-hydroxyvalerate) and poly(3-hydroxybutyrate) studied by infrared spectroscopy and X-ray diffraction. *Macromolecules* 41: 4305–4312.

29 Sato, H., Murakami, R., Mori, K. et al. (2009). Specific crystal structure of poly(3-hydroxybutyrate) thin films studied by infrared reflection–absorption spectroscopy. *Vib. Spectrosc.* 51: 132–135.

30 Guo, L., Sato, H., Hashimoto, T., and Ozaki, Y. (2010). FTIR study on hydrogen-bonding interactions in biodegradable polymer blends of poly(3-hydroxybutyrate) and poly(4-vinylphenol). *Macromolecules* 43: 3897–3902.

31 Guo, L., Sato, H., Hashimoto, T., and Ozaki, Y. (2011). Thermally induced exchanges of hydrogen bonding interactions and their effects on phase Structures of poly(3-hydroxybutyrate) and poly(4-vinylphenol) blends. *Macromolecules* 44: 2229–2239.
32 Guo, L., Spegazzini, N., Sato, H. et al. (2012). Multistep crystallization process involving sequential formations of density fluctuations, "intermediate Structures", and lamellar crystallites: poly(3-hydroxybutyrate) as investigated by time-resolved synchrotron SAXS and WAXD. *Macromolecules* 45: 313–328.
33 Suttiwijitpukdee, N., Sato, H., Zhang, J., and Hashimoto, T. (2011). Effects of intermolecular hydrogen bondings on isothermal crystallization behavior of polymer blends of cellulose acetate butyrate and poly(3-hydroxybutyrate). *Macromolecules* 44: 3467–3477.
34 Sato, H., Suttiwijitpukdee, N., Hashimoto, T., and Ozaki, Y. (2012). Simultaneous synchrotron SAXS/WAXD study of composition fluctuations, cold-crystallization, and melting in biodegradable polymer blends of cellulose acetate butyrate and poly(3-hydroxybutyrate). *Macromolecules* 45: 2783–2795.
35 Suttiwijitpukdee, N., Sato, H., Zhang, J. et al. (2011). Intermolecular interactions and crystallization behaviors of biodegradable polymer blends between poly(3-hydroxybutyrate) and cellulose acetate butyrate studied by DSC, FT-IR, and WAXD. *Polymer* 52: 461–471.
36 Yamamoto, S., Morisawa, Y., Sato, H. et al. (2013). Quantum mechanical interpretation of intermolecular vibrational modes of crystalline poly-(*R*)-3-hydroxydutyrate observed in low-frequency Raman and terahertz spectra. *J. Phys. Chem. B* 117: 2180–2187.
37 Sato, H., Ando, Y., Mitomo, H., and Ozaki, Y. (2011). Infrared spectroscopy and X-ray diffraction studies of thermal behavior and lamella Structures of poly(3-hydroxybutyrate-*co*-3-hydroxyvalerate) (P(HB-*co*-HV)) with PHB-type crystal structure and PHV-type crystal structure. *Macromolecules* 44: 2829–2837.
38 Sato, H., Miyada, M., Yamamoto, S. et al. (2016). C—H···O (ether) hydrogen bonding along the (110) direction in polyglycolic acid studied by infrared spectroscopy, wide-angle X-ray diffraction, quantum chemical calculations and natural bond orbital calculations. *RSC Adv.* 6: 16817–16823.
39 Yamamoto, S., Miyada, M., Sato, H. et al. (2017). Low-frequency vibrational modes of poly(glycolic acid) and thermal expansion of crystal lattice assigned on the basis of DFT-spectral simulation aided with a fragment method. *J. Phys. Chem. B* 121: 1128–1138.
40 Nishimura, F., Hoshina, H., Ozaki, Y., and Sato, H. (2019). Isothermal crystallization of poly(glycolic acid) studied by terahertz and infrared spectroscopy and SAXS/WAXD simultaneous measurements. *Polym. J.* 51: 237–245.
41 Funaki, C., Yamamoto, S., Hoshina, H. et al. (2018). Three different kinds of weak C—H···O=C inter- and intramolecular interactions in poly(ε-caprolactone) studied by using terahertz spectroscopy, infrared spectroscopy and quantum chemical calculations. *Polymer* 137: 245–254.
42 Yamamoto, Y., Hoshina, H., and Sato, H. (2021). Differences in intermolecular interactions and flexibility between poly(ethylene terephthalate) and

poly(butylene terephthalate) studied by far-infrared/terahertz and low-frequency Raman spectroscopy. *Macromolecules* 54: 1052–1062.

43 Cornibert, J. and Marchessault, R.H. (1972). Physical properties of poly-β-hydroxybutyrate. IV. Conformational analysis and crystalline structure. *J. Mol. Biol.* 71: 735–756.

44 Yokouchi, M., Chatani, Y., Tadokoro, H. et al. (1973). Structural studies of polyesters: 5. Molecular and crystal structures of optically active and racemic poly(β-hydroxybutyrate). *Polymer* 14: 267–272.

45 Marchessault, R.H., Debzi, E.M., Revol, J.F., and Steinbuhel, A. (1995). Single crystals of bacterial and synthetic poly(3-hydroxyvalerate). *Can. J. Microbiol.* 41: 297–302.

46 Yokouchi, M., Chatani, Y., Tadokoro, H., and Tani, H. (1974). Structural studies of polyesters. VII. Molecular and crystal structures of racemic poly(β-ethyl-β-propiolactone). *Polym. J.* 6: 248–255.

47 Yoshie, N., Asaka, A., and Inoue, Y. (2004). Cocrystallization and phase segregation in crystalline/crystalline polymer blends of bacterial copolyesters. *Macromolecules* 37: 3770–3779.

48 Yoshie, N., Saito, M., and Inoue, Y. (2001). Structural transition of lamella crystals in a isomorphous copolymer, poly(3-hydroxybutyrate-*co*-3-hydroxyvalerate). *Macromolecules* 34 (26): 8953–8960.

49 Abe, H., Doi, Y., Aoki, H., and Akehata, T. (1998). Solid-state structures and enzymatic degradabilities for melt-crystallized films of copolymers of (*R*)-3-hydroxybutyric acid with different hydroxyalkanoic acids. *Macromolecules* 31: 1791–1797.

50 Doi, Y., Kitamura, S., and Abe, H. (1995). Microbial synthesis and characterization of poly(3-hydroxybutyrate-*co*-3-hydroxyhexanoate). *Macromolecules* 28 (14): 4822–4828.

51 Chatani, Y., Suehiro, K., Okita, Y. et al. (1968). Structural studies of polyesters. I. Crystal structure of polyglycolide. *Makromol. Chem.* 113: 215–229.

Index

a

acceptor doping 367, 369
analytical technique 3, 117, 152, 217, 259, 261, 267
anti-Stokes Raman scattering, 189, 190
asymmetric poly(L-lactide)/ poly(D-lactide) (PLLA/PDLA) blend, isothermal crystallization 23–29
atomic force microscopy (AFM) 7, 40, 88, 90–92, 94, 96, 349, 350, 358, 359, 365
atomic force microscopy-infrared spectroscopy (AFM-IR) 7, 40, 88, 91, 92, 94
ATR–FUV spectroscopy 166
  graphene nanocomposites 172–176
  measurement 166–167
  nylons 167–172
  poly (ethylene glycol) 176–181
  poly(3-hydroxybutyrate) 172–176
attenuated total reflectance (ATR) technique 322
  absorption 166
  -FTIR spectroscopy 45–55
  spectrometer 165
Avrami equation 136, 139, 146–147, 419

b

band splitting 323, 415
Beer–Lambert law 45, 65, 419
biodegradable polyesters 117, 443, 445
biodegradable polymers 151, 261, 303, 307, 435–449
biological macromolecules 193
biopolymers 46, 197–203, 298
bipolarons 369
  doubly charged bipolarons 368
  negative bipolarons 370
  positive bipolarons 370, 372, 374, 375, 379, 380, 388
Born–Oppenheimer molecular dynamics (BOMD) 299–301, 303–305
1-butyl-3-methylimidazolium bis(trifluoromethylsulfonyl) imide ([BMIM][TFSI]) 377–379, 381, 382

c

Car–Parrinello molecular dynamics (CPMD) 301
carbon dot nanocomposite films 84, 86
carbon nanotubes (CNTs) 36, 40, 78, 81, 94

Spectroscopic Techniques for Polymer Characterization: Methods, Instrumentation, Applications, First Edition. Edited by Yukihiro Ozaki and Harumi Sato.

carrier density 367, 384, 385, 417, 425
carrier mobilities 369, 378, 383
carriers 368, 369–371, 374–384, 386–388, 413, 425–429
Cartesian coordinate tensor transfer (CCT) method 111–113
CF_2 symmetric stretching vibration ($\nu_s(CF_2)$) band 397, 399, 408
chain packing/aggregate mode identification 413–414
chain stretching vibration 397
charge carriers, structure and dynamics of 425–429
charge density 368, 378, 382
charge–transfer (CT) enhancement 9, 167, 176
chemical composition/morphology analysis 413, 420–423
chemometric techniques 149, 270–272
chromophore 82, 83, 165, 191–194, 199
cold-crystallization process 51, 53
conductive polymers 4, 7, 367, 413–431
conductive-polymer-based blends 413, 420–423
conformation-sensitive bands identification 414
crystallinity 7, 46, 51, 82, 88, 109, 117–120, 128, 132, 146, 151–153, 156, 158–160, 174–175, 278, 317–318, 334–338, 351, 368, 380, 383, 419, 436, 439, 442–443
crystallization 51
  in polymers under high-pressure or supercritical CO_2 61
  solvent-induced 51–52

d
deep-UV (DUV) 3, 9, 39, 40, 165, 166, 173–176, 192
degree of swelling 56, 66
density functional theory (DFT) 8, 16–20, 127, 166, 167, 169, 170, 174, 178, 179, 299, 383, 416, 417, 429
deoxyribonucleic acid (DNA) 193, 197–201, 203
2D gradient mapping method 269, 270
2D hetero-spectral correlation analysis 267
dication 370, 371
dichroism 129, 132, 259, 298, 328, 404
dielectric relative permittivity 406
dispersion force 181, 400
donor doping 367, 370
doping-induced infrared active vibrational (IRAV) bands 368–369
doubly charged bipolarons 368
3D SERS method 29
  symmetric nanoporous silver microparticles 28–34

e
effective thickness 66, 67
electric-field-induced structural changes 429–431
electrical conductivity 80, 367, 368, 378, 388
electrically conductive polymers 367
electrolyte-gated transistor (EGT) 368, 417
electromagnetic (EM) enhancement 9
electron spin resonance (ESR) spectroscopy 371

electronic transitions 9, 10, 125, 165, 166, 172–174, 176, 181, 193, 194, 209, 217
empirical simulation methods 338
entropically-driven self-healing 353–365
epoxy resins 76, 84
785-nm excited Raman spectra 375, 378

f

factor group analysis 398, 399
far-infrared (FIR) spectroscopy 8, 107
  intermolecular hydrogen bonds
    poly(3-hydroxybutyrate) 108
    poly(ethylene terephthalate) 115–117
    polybutylene succinate 113–115
    polycaprolactone 109–113
far ultraviolet (FUV) spectroscopy 3, 9–10, 39, 40, 165, 166, 176, 181
$FeCl_3$ doping 372, 373, 375, 376, 381
femtosecond time-resolved NIR inverse Raman spectroscopy 386
femtosecond time-resolved resonance inverse Raman spectroscopy 427
Fickian diffusion 52
finite damping factor 407
Flory–Huggins equation 46, 47
fluorescence spectroscopy 4, 76, 82–85
fluoroalkyl trichlorosilane (FTS)-modified structure 425
fluorocarbons 396
fluorous' property 400, 409
Fourier Transform Infrared (FTIR) spectroscopy 75, 183
  high-pressure or supercritical CO_2 55
    crystallization 61–67
  swelling and CO_2 sorption 65–67
  intermolecular interactions 47–51
  miscibility of polymer blends 46–47
Fraser's derivation 328

g

graphene 40, 78–80, 84, 88, 93–94
  nanocomposites 172–176
graphene oxide (GO) 78–80, 84

h

harmonic oscillator model 185
hydrogen bonding 435
  definition 435
  weak hydrogen bonding 435–449
hydrophobic interaction 400
hyperbranched epoxy/carbon dot nanomaterials 84
hyperspectral imaging (HSI) 278–283

i

imaging spectroscopy 40, 117
in-situ ATR-FTIR spectroscopic approach 66
in-situ FTIR spectroscopy 6, 45–68, 78, 89, 90, 96
infrared reflection–absorption spectroscopy (IRRAS) 262–264, 269
infrared (IR) spectroscopy 3–7, 39, 89, 91, 108, 126, 128, 227, 259, 261–262, 265, 267, 269, 302, 358, 369, 387–388, 399, 413, 436, 444
interfaces
  characterization 80–95

interfaces (*contd.*)
fluorescence spectroscopy 82–85
solid-state NMR spectroscopy 85–88
vibrational spectroscopy 88–95
interactions 76–80
interfacial region 75, 76, 80, 82, 83, 95, 228, 348, 351
intermolecular weak hydrogen bonding 445
internal reflection element (IRE) 165–167
inverse Raman spectroscopy 369, 386, 388, 427
ionic-liquid-gated transistor (ILGT) 368, 369, 377–385, 388
isothermal crystallization process 25, 26, 29, 51, 136, 418, 419, 443, 444
isothermal melt-crystallization kinetics 418

k
Kramers–Kronig (KK) relations 407, 408
Kramers–Kronig transformation (KKT) 166, 167

l
light-emitting diodes 367
linear low-density polyethylene (LLDPE) 21–23, 128
linear region 384
liquid NMA 167, 168
lithiumbis(trifluoromethanesulfonyl) imide (LiTFSI) 177, 179
London's equation 400
longitudinal acoustic modes (LAM) 5, 8
low-frequency Raman spectroscopy 6, 8, 107, 108, 113, 117, 120
lower critical solution temperature (LCST) 46, 55, 262, 265

m
macroscopic information 75, 82
macroscopic shape memory recovery 351
mean square electric field (MSEF) calculation 423
melt-crystallization process 51, 135, 419
microsecond time-resolved infrared spectra 387
mineral fillers 75
molecular conformation on phase diagram 393–396
molecular dynamics (MD) 297, 302–309, 333
molecular orbitals (MOs) 9, 172, 174, 175
molecular polarizability 7, 189, 190, 400, 409
molecular vibration 183
of an R_f group 396–399
multi-angle incidence resolution spectrum (MEIRS) 39
multiplicative scatter correction (MSC) 21–25
multiwall carbon nanotubes (MWCNTs) 34–39, 78, 80, 81, 93, 94
nanocomposites 34–39

n
nanospectroscopy 40
near-infrared (NIR) spectroscopy 8, 40, 125
analysis 126–128
applications 128–147
biodegradable polymer degradable process monitoring, 154–155

crystal evolution combined with chemometrics, 151–153
diffusion process monitoring 153–154
characteristics 126
imaging 148–151
linear low-density polyethylene 21–23
molecular orientation 128–134
principles 125
residual crystal amounet, at the extruder outlet nozzle 140–141
negative bipolarons 370
negative polarons 370
N-methylacetamide (NMA) 167, 168
nylons 16, 167–172

o

one-electron theory 370
optical absorption spectra upon chemical doping 371–374
organic field-effect transistors (OFETs) 425
organic solvents 55, 79, 367, 400
output characteristics 383
overtone bands 8, 21, 125, 134, 186
oxetane-substituted chitosan-polyurethane (OXE-CHI-PUR) 355, 357–359

p

partial least squares (PLS) regression analysis 21, 22, 25, 127, 153, 157–160
perfluoroalkyl-containing compounds
molecular conformation on phase diagram 393–396
molecular vibration of an R_f group 396–399
ROA analysis of R_f compounds 402–405
stratified dipole-arrays (SDA) theory 400–401
surface modes of phonon and polariton 405–408
[6,6]-phenyl-C_{61}-butyric acid methyl ether (PCBM) 368, 369, 386–388, 414–416, 422, 423, 427, 428
photoinduced infrared absorption spectrum 387
piezoelectric polymers 334
planer-zigzag conformation 393
PLLA crystallization 53, 64, 65
pMAIRS technique 402, 407
polarons 369
negative polarons 370
positive polaron 369–372, 374–376, 378–380, 383, 384, 386–388, 416, 428
polyacetylene 367, 368
poly(acrylic acid) (PAA) 10–16
poly(3-alkylthiophenes) (P3ATs) 414
polyaniline 369
poly[2,6-(4,4-bis-(2-ethylhexyl)-4H-cyclopenta[2,1-*b*:3,4-*b*′] dithiophene)-alt-4,7(2,1,3-benzothiadiazole)] (PCPDTBT) donor/acceptor moieties 421, 422
poly[[4,8-bis[(2-ethylhexyl)oxy]benzo [1,2-b:4,5-b′]dithiophene-2,6-diyl][3-fluoro-2-[(2-ethylhexyl) carbonyl]thieno[3,4-*b*]thio-phenediyl]] (PTB7) 368, 387
poly(2,5-bis(3-hexadecylthiophene-2-yl)thieno[3,2-b]thiophene) (PBTTT-C16) 368, 369, 372, 375, 381

poly(2,5-bis(3-tetradecylthiophen-2-yl)thieno[3,2-b]thiophene) (PBTTT) 368, 369, 372–376, 380–383, 388, 415–417
polybutadiene 61, 62, 86, 227, 244
polybutylene succinate (PBS) 107, 113–115
poly(3-butylthiophene) (P3BT) 413, 414, 416
polycaprolactone (PCL) 46, 47, 58, 59, 70, 107, 109–113, 117, 118, 119, 349–353, 436, 447, 448
poly(dimethylsiloxane) (PDMS) 66, 76–78, 80, 81, 84, 86, 90, 93, 129–134, 227
poly((9,9-dioctylfluorene)-2,7-diyl-alt-[4,7-bis(3-hexylthien-5-yl)-2,1,3-benzothiadiazole]-2′,2″-diyl) (F8TBT) blend thin films 420, 421
polyethylene (PE) 21–24, 132, 133, 227, 439
polyethylene glycol (PEG) 5, 49, 60, 176, 355
poly(ethylene terephthalate) (PET) 64, 65, 107, 115–117, 228, 436
poly(3-hexylthiophene) (P3HT) 367, 369, 371–374, 414
  P3HT/PCBM blend films 368, 369, 386–388, 414–416, 421–423, 427, 428
  P3HT:PCBM bulk heterojunction film 386
polyhydroxyalkanoates (PHAs) 443
  copolymers 442–443
poly(3-hydroxybutyrate) (PHB) 30, 32, 34, 46–48, 53–55, 107, 108, 128, 134–140, 172–176, 436–438, 443
  crsytallization process 443
  isothermal crystallization kinetics 134–140
poly(3-hydroxybutyrate-*co*-3-hydroxyhexanoate) (PHBH) 140–147
polyisoprene 87, 238, 240–244
poly(lactic acid) (PLA) 156, 157, 317, 318–328, 334–338
  aging behavior 338–340
  anisotropic 327–328
  morphological features 318
  ordered and disordered structures 329–334
  polarization 327–328
  vibrational features 321–323
polymer semiconductors 367–369
polymer solar cells 367, 369
polymer spectroscopy 3
polymer swelling 61, 65–67
polymer-filler system 75, 81, 83, 85, 90
polymeric hydrogels
  phase transition in
    thermo-sensitive 208–212
  water and polymer dynamics 212–215
  water confinement 204–208
poly(methyl methacrylate) (PMMA) 48–51, 56, 58, 76, 83, 227, 232, 234, 247, 361
poly(2-methoxy-5-(2′-ethylhexyloxy)-1,4-phenylene vinylene) (MEH-PPV) 430, 431
poly(*N*-9′-heptadecanyl-2,7-carbazole-alt-5,5-(4′,7′-di-2-thienyl-2′,1′,3′-benzothiadiazole) (PCDTBT) 387
poly(3-octylthiophene) (P3OT) 418–419
poly(*p*-phenylene) 369
poly(*p*-phenylenevinylene) 369
polypyrrole 367

polystyrene (PS) 33, 51, 56, 83, 227, 230–237, 238–240, 298
  non-solvent interface 230–237
  solid interface 238–240
  syndiotactic polystyrene (sPS) 51, 62, 64
poly(styrene-*co*-butadiene) rubber 244–250
polytetrafluoroethylene (PTFE) 307, 393–395, 397, 399, 400
poly[[1,2,5]thiadiazolo[3,4-c|i] pyridine-4,7-diyl(4,4-dihexadecyl-4Hcyclopenta [2,1-*b*:3,4-b′]dithiophene-2,6-diyl)[1,2,5]thiadiazolo[3,4-c] pyridine-7,4-diyl(4,4-dihexadecyl-4*H*-cyclo-penta[2,1-b:3,4-b′]dithiophene-2,6-diyl)] (PCDTPT) 426, 427
polythiophene (PT) 367–370
poly(vinyl acetate) 76
poly (vinyl methyl ether) (PVME) 56, 57
polyvinyl pyrrolidone (PVP) blend 33, 34, 60–61
positive bipolarons 370, 372, 374, 375, 379, 380, 388
positive polaron 369–372, 374–376, 378–380, 383, 384, 386–388, 416, 428
primary polymer films 367
principal component analysis (PCA) 11, 21, 22, 24, 127, 136, 139, 150, 152, 261, 262, 265, 270, 271–275, 284

q
quantum chemical calculations (QCC) 8, 40, 107, 127, 166, 167, 169, 170, 179, 181, 299, 447
quantum dots 83, 84

r
radical cation 371
Raman and infrared (IR) spectra 394
Raman effect 183–191, 328, 375
Raman imaging 23, 28, 369, 383–385
Raman optical activity (ROA) 298, 402–405
Raman scattering 5, 34–39, 78, 88, 91, 189–193, 298, 319–321, 327, 329, 332, 333, 340, 422
Raman spectra of ILGTs
  fabricated with P3HT 378–380
  fabricated with PBTTT-C16 380–383
Raman spectra of positive polarons and bipolarons generated upon chemical doping
  P3HT film 374–376
  PBTTT-C16 film 375–376
Raman spectroscopy 5, 6, 10, 23, 91, 374, 378, 380, 413
Rayleigh (or elastic) scattering 189
Rayleigh component 189
reflection-absorption infrared spectroscopy (RAIR) 431
regioregular poly(3-hexylthiophene) (P3HT) 368
regression coefficients (RC) 21–22, 25
resonance Raman (RR) scattering 4, 6, 7, 35, 191–193, 199, 201, 414, 416
Reststrahlen band 407
R_f-containing compounds 396–397, 399–401, 406, 408–409
Rydberg transitions 9, 166, 167, 173, 176

s
saturation region 384
scanning angle (SA) Raman spectroscopy 422

SDA-packing model 402
self-healing polymers 347
damage-repair cycle 361–365
mechanisms 348
microphase separation 349–353
Van der Waals interactions 360–361
self-modeling curve resolution (SMCR) 127
shape memory effects (SMEs) 348–349, 353
single-monolayer Langmuir-Blodgett (LB) film 402
single-walled carbon nanotubes (SWCNTs) 78–79, 94–95
singly charged polarons 368
singular value decomposition (SVD) 271, 275
sol-gel method 77
solar cells 367, 369, 386–388, 415, 421
solid interface 238–239, 250
solid polymer electrolytes (SPEs) 302, 309
soluble polymers 261, 367
solvent-induced crystallization 51–52
Sommer's criteria 385
spatial resolution 9, 29–30, 33, 34, 38, 40, 91, 94, 96, 117, 148, 150, 337, 347, 349, 369, 385, 417, 431
steered molecular dynamics (SMD) methodology 307
Stokes Raman scattering 189–190
stratified dipole-arrays (SDA) theory 400–402, 406, 408–409
structural dynamics 413, 418–420
styrene-butadiene rubber (SBR) 34–39, 81, 86, 95, 238, 244–250
sum frequency generation (SFG) vibrational spectroscopy 423
depth resolution 227
non-solvent interface 230–237
principle 228–229
solid interface 238–250
surface-enhanced Raman spectroscopy (SERS) spectroscopy 5–7, 9–10, 28–34, 227, 420–421
surface free energy 233
surface IR spectroscopy 39
surface modes of particles (SMP) 407–408
surface modes of phonon and polariton 405–408
surface/interface molecular orientation 413, 423–425
symmetry-adapted cluster-configuration interaction (SAC–CI) 166
synchrotron radiation (SR) 195
syndiotactic polystyrene (sPS) 51, 62, 64

t

Teflon® 307, 393
terahertz (THz) spectroscopy 4, 8, 107–120, 261, 298, 303, 425, 436
tetraethoxysilane (TEOS) 77
thermally-induced phase transitions 417–418
thermo-responsive polymers 259, 261–267, 284
thermoplastic polyurethanes (TPUs) 349–350, 353–356

thin-film transistors 367, 429–430
THz-time domain spectroscopy (THz-TDS) 8, 117, 120
time-dependent density functional theory (TD–DFT) 166–167, 169–170, 174, 178–179
time-dependent IR ATR spectra 406
time-resolved FT-IR spectroscopy 369, 388
time-resolved infrared spectroscopy 387, 418
time-resolved resonant Raman spectroscopy 420
time-resolved terahertz spectroscopy (TRTS) 425–426
tip-enhanced Raman scattering spectroscopy (TERS) 5–7, 9, 34–40, 88, 94–95
traditional composites 75
trans-polyacetylene 368
transmittance (Tr) absorption 165, 167
tunable infrared (IR) beams 227
two-dimensional correlation spectroscopy (2D-COS) 259–284
  application 259–284
  biodegradable polymers 261–262
  chemometric techniques 270–272
  2D gradient mapping method 269–270
  2D hetero-spectral correlation analysis 267–269
  hyperspectral imaging 278–283
  smooth factor analysis 272–275
  theory 260–261
  thermo-responsive polymers 262–267

u

ultraviolet (UV) light 165
unannealed PBTTT-C16 ILGT 380–383
UV resonance Raman (UVRR) spectroscopy
  characterization 197–203
  instrumentation 193–194
  polymeric hydrogels 203–215
  synchrotron-based setup 194–197

v

vacuum UV region 165
van der Waals forces 400
van der Waals radius of fluorine 393
vibrational circular dichroism (VCD) 298, 404
vibrational spectroscopy 6, 88–95, 413–431
  infrared absorption and Raman scattering 88
  infrared spectroscopy 3, 10, 89–91
  low-frequency 8, 16–20, 107, 108, 113, 117, 120
  Raman spectroscopy 91–95
  structural dynamics 418–420
  surface/interface molecular orientation 423–425
  thermally-induced phase transitions 417–418
viscoelastic length transitions (VLTs) 353–354
visible spectroscopy 4, 165

w

water-and-oil repellency 393, 400
weak hydrogen bonding
  CH-O 443–447

weak hydrogen bonding (*contd.*)
  PHB and PHV 439–441
  polyhydroxyalkanoate copolymers 442–443
  PHB 436–438
  vs. strong hydrogen bonds 438–439
wide-angle X-ray diffraction (WAXD) 108, 117–118, 303, 436, 443–445, 447

X
X-ray diffraction (XRD) 64, 113, 115, 153, 303, 322, 334, 393, 402, 437, 443

Z
zero damping 406